Proceedings of Symposia in
PURE MATHEMATICS

Volume 78

From Hodge Theory to Integrability and TQFT

tt*-geometry

International Workshop
From TQFT to tt* and integrability
May 25–29, 2007
University of Augsburg, Augsburg, Germany

Ron Y. Donagi
Katrin Wendland
Editors

American Mathematical Society
Providence, Rhode Island

2000 *Mathematics Subject Classification.* Primary 14C30, 14D07, 14D21, 14N35, 53C43, 53D45, 81R12, 81T30, 81T45.

Library of Congress Cataloging-in-Publication Data

International Workshop from TQFT to tt* and Integrability (2007 : Augsburg, Germany)
From Hodge theory to integrability and TQFT : tt*-geometry ; International Workshop from TQFT to tt* and Integrability, May 25–29, 2007, Augsburg, Germany / Ron Y. Donagi, Katrin Wendland, editors.
p. cm. — (Proceedings of symposia in pure mathematics ; v. 78)
Includes bibliographical references.
ISBN 978-0-8218-4430-4 (alk. paper)
1. Geometry, Algebraic—Congresses. 2. Hodge theory—Congresses. 3. Quantum field theory—Congresses. I. Donagi, Ron Y., 1956– II. Wendland, Katrin, 1970– III. Title.

QA564.I618 2008
516.3′5—dc22 2008027339

10 9 8 7 6 5 4 3 2 1 13 12 11 10 09 08

Contents

Introduction

Ideas from quantum field theory and string theory have had an enormous impact on geometry over the last two decades. One extremely fruitful source of new mathematical ideas can be traced back to works of Cecotti, Vafa, and their coauthors around 1991 on the Geometry of Topological Field Theory. The motivation for their "tt*-geometry" came from physics, but the work turned out to unify ideas that had been developing in such separate branches of mathematics as singularity theory, Hodge theory, integrable systems, matrix models, and Hurwitz spaces.

From one point of view, the geometry developed by Cecotti, Vafa, et al involved a refinement of the notion of Higgs bundle. This geometric structure had been previously introduced and studied by Hitchin. The refinement was closely related to what have since been codified by Dubrovin as Frobenius manifolds. Essentially the same objects had been previously introduced, on moduli spaces asscociated to quasihomogeneous singularities, by Kyoji Saito. They appeared also in Hodge theoretic work of Morihiko Saito and in numerous other approaches. In the papers by Cecotti, Vafa et al, these purely holomorphic structures were combined with a generalization of the Gauss -Manin connection, a generalization which nowadays is described as variations of harmonic Higgs bundles. Following ideas of Hitchin, Deligne and Simpson, the entire structure is most conveniently described by introducing a twistor line and a new, flat connection on the resulting space. The flatness equations for this connection are the tt*-equations of Cecotti and Vafa. The twistor structures capture the essence of the Corlette-Simpson non-Abelian Hodge theory, which establishes the equivalence of flat connections and Higgs bundles. The tt*-geometry combines algebro-geometric notions also with symplectic ones. Such a combination turns out to be tailor made for investigations of mirror symmetry and Gromov-Witten invariants.

These structures and equations have since appeared in many other branches of geometry and physics: Integrable systems have been closely related to the topic ever since Hitchin's works on moduli spaces of stable bundles, and Dubrovin has proved integrability of the tt*-equations. Almost simultaneously with the original work by Cecotti and Vafa, Simpson developed his notion of harmonic bundles. These have since evolved into mixed twistor structures, an essential contritubion to the field. Another key ingredient is Mochizuki's ground breaking work on tame –and more recently also wild– harmonic bundles. Hertling and collaborators have clarified and further developed the precise mathematical counterparts of the fundamental structures used in the physics papers of Cecotti and Vafa, introducing "TERP structures". On the physics side, the tt*-equations have been understood as governing tree level amplitudes in topological quantum field theories. Their generalization to higher-genus amplitudes by Bershadsky-Cecotti-Ooguri-Vafa is known as the holomorphic anomaly equation. The latter provides a recursive tool to calculate some of the quantities that are of interest in topological field theories – like the Gromov-Witten invariants and some of their cousins. The impact on enumerative geometry has been tremendous, and the full scope of ideas is yet to be understood.

The present volume is loosely based on the workshop "From tQFT to tt* and integrability", which we organized at the University of Augsburg in May 2007. We have tried to combine some of the above-mentioned, so far only partially understood aspects into a cohesive picture, and this proceedings volume aims to give an overview of these topics, leading to and including up-to-date research.

The first four papers of this volume give an overview of the underlying geometric structures along with recent developments of the latter. Claude Sabbah explains the notion of tt*-geometry in terms of a common extension of Saito structures and harmonic Higgs bundles. The existence of a canonical harmonic structure on the canonical Frobenius manifold attached to a convenient and nondegenerate Laurent polynomial is explained – a result that appears implicity already in the work of Cecotti and Vafa. The following paper by Kyoji Saito and Atsushi Takahashi addresses the holomorphic part of tt*-geometry in the context of singularity theory. The article gives a historical review and introduction to the ideas which led to the discovery of the structure of a Frobenius manifold on the base of the universal unfoldings of isolated singularities by means of primitive forms. Claus Hertling and Christian Sevenheck explain the full tt*-geometry for isolated hypersurface singularities and tame functions. They give the technical construction of all relevant geometric ingredients, introducing TERP structures, and explaining how the latter give rise to variations of twistor structures, which in turn they relate to harmonic bundles. Moreover, nilpotent orbits are discussed, and appropriate classifying spaces and limiting structures are introduced. The contribution by Vicente Cortés and Lars Schäfer emphasizes differential geometric aspects of tt*-geometry. A classical result of Dubrovin's, which describes solutions to the tt*-equations in terms of pluriharmonic maps, is explained and generalized. In particular, the relation of tt*-geometry to various other special geometries is worked out in detail.

All these diverse notions come together in the paper by Ludmil Katzarkov, Maxim Kontsevich and Tony Pantev. They generalize Hodge structures and their variation by developing an abstract theory of non-commutative Hodge structures, investigating existence and variations, and proposing explicit construction and classification techniques. Their work generalizes the algebraic constructions of Saito and Takahashi (as explained in the appendix to [Saito-Takahashi]) as well as the results from Hodge and singularity theories in the [Hertling-Sevenheck] and [Cortés-Schäfer] works. This combines tt*-geometry with the notions of weight systems, harmonic and wild bundles, and pure twistor structures, semi-infinite Hodge structures, and TERP structures. Moreover, connections to mirror symmetry and its applications are explicitly explained.

The Hodge theoretic approach is central also to the next contribution. Carlos Simpson studies the simplest cases (rank 1) of twistor spaces on open varieties. He points out that the space of local monodromy transformations around a divisor component at infinty has weight two – a new phenomenon in the non-compact case, suggesting one of the essential difficulties involved in constructing a complete mixed twistor theory.

The remaining papers of the volume focus on applications or generalizations of the ideas originally described by Cecotti, Vafa, and collaborators to integrability and to enumerative geometry.

The relation to integrable systems is the topic of the contributions by Luuk Hoevenaars and by Anton Gerasimov and Samson Shatashvili. While the former explains how the base of a special integrable system - the Neumann system - carries the structure of an almost Frobenius manifold, the latter gives evidence for the proposal that a certain quantum integrable system captures the geometry of the moduli space of Higgs bundles in an appropriate setting.

The paper by Vincent Bouchard and Marcos Mariño makes use of recent progress in the matrix model approach to topological field theories due to Eynard and Orantin. From it, the authors derive a conjectured recursive solution for Hurwitz numbers at all genera. Andrew Neitzke and Johannes Walcher study a further generalization of the tt*-equations beyond the Bershadsky-Cecotti-Ooguri-Vafa anomaly equations, namely to a setting on Calabi-Yau threefolds with background D-branes. The corresponding open topological partition functions are obtained by a shift of variables from the solutions to the ordinary holomorphic anomaly equations. This yields a proposal which interprets an auxiliary Hilbert space introduced by Witten as the space of states generated by all open topological partition functions.

Acknowledgements. We are grateful to the authors for the great effort they made to balance presentation of their latest results with careful introductions to the many facets of this deeply interconnected and fast evolving subject. We are also grateful to all the speakers and participants of the Augsburg conference "From tQFT to tt* and integrability", May 25-29, 2007. The conference was funded by NSF Focused Research Grant DMS 0139799 - The Geometry of Superstrings - of the Penn Math/Physics group (RD, Principal Investigator), and by DFG grant GZ 4851/142/07 of KW at the University of Augsburg.

We are grateful to the editors at the AMS for their help in putting this volume together: We cordially thank Sergei Gelfand and Christine Thivierge for their continuous guidance and help. Arthur Greenspoon has helped us with the editorial fine-tuning of every contribution and deserves our special gratitude.

Ron Donagi and Katrin Wendland

Proceedings of Symposia in Pure Mathematics
Volume **78**, 2008

Universal unfoldings of Laurent polynomials and tt* structures

Claude Sabbah

ABSTRACT. This article surveys the relations between harmonic Higgs bundles and Saito structures which lead to tt* geometry on Frobenius manifolds. We give the main lines of the proof of the existence of a canonical tt* structure on the base space of the universal unfolding of convenient and nondegenerate Laurent polynomials.

CONTENTS

2000 *Mathematics Subject Classification.* 53D45, 32S40, 14C30, 32L25, 34M55.
Key words and phrases. Higgs bundle, Saito structure, tt*-structure, Fourier-Laplace transform, Frobenius manifold, Laurent polynomial.

Introduction

The notion of a tt* structure on a holomorphic vector bundle is now understood, after the work of C. Hertling [**7**], as an enrichment of that of harmonic Higgs bundle previously introduced by N. Hitchin and C. Simpson. Given the Higgs field Φ and the harmonic metric h on the holomorphic bundle E on a complex manifold M, the new ingredients needed for a tt* structure are a real structure on the associated C^∞ bundle H, a holomorphic endomorphism $\mathscr{U}$ of E and a C^∞ endomorphism $\mathscr{Q}$ of H subject to some compatibility relations. In the following, we relax the condition for h to be positive definite, and only ask that it be Hermitian and nondegenerate. When needed, we will emphasize the positive definite case.

It has been highly enlightening in two ways to interpret (cf. [**31**]) harmonic Higgs bundles as variations of polarized twistor structures of weight 0: firstly, it makes the analogy with variations of Hodge structures more transparent and, secondly, it enables one to do geometry with the external parameter z added for this purpose. From this point of view, the relations satisfied by the endomorphisms $\mathscr{U}$ and $\mathscr{Q}$ in a tt* structure appear as expressing the complete integrability of a connection on the twistor bundle.

A nearby (purely holomorphic) notion, that of a Saito structure, has emerged from the work of K. Saito [**23**] and M. Saito [**25**] as a basic tool to produce Frobenius manifolds from singularities of holomorphic functions. While it is already present in [**7**], the bridge between these two notions is made more transparent in §1 by the introduction of a potential for the Higgs field.

When the tt* structure exists on the tangent bundle of M, we speak of tt* geometry, which is a generalization of special geometries on M. Of particular interest for us is the case where M is a Frobenius manifold. In such a case, a Saito structure exists on the tangent bundle together with supplementary symmetry properties, giving rise to a commutative and associative product with unit on TM. Adding a tt* structure in a compatible way (i.e., with the help of a potential for the Higgs field) leads to the structure of *harmonic Frobenius manifold.*

The main result we report here (cf. Theorem 4.7) is the existence of a canonical harmonic structure on the canonical Frobenius manifold attached to a convenient and nondegenerate Laurent polynomial. Moreover, the corresponding Hermitian form is *positive definite.*

In this survey article, which contains no original results, we first give (§§1 and 2) a quick overview of tt* structures, Saito structures and variations of twistor structures (a more detailed exposition can be found in [**7**] and [**9**]). In §3, we explain the Fourier-Laplace method for constructing polarized pure twistor structures starting from a variation of polarized Hodge structure. In §4, we show how to apply this technique to the Gauss-Manin connection of a Laurent polynomial, with the help of M. Saito's mixed Hodge theory [**26**]. One can find details for the results of §§3 and 4 in [**19, 22**], and many other results and applications in [**9**].

Acknowledgements. The author thanks Ron Donagi and Katrin Wendland, organizers of the conference "From tQFT to tt* and integrability", for having given him the opportunity to talk about the contents of this article.

1. Harmonic Frobenius manifolds

In this section we first recall the notion of a harmonic Higgs bundle, following C. Simpson [**30**], and we introduce supplementary structures which will be useful for defining harmonic Frobenius manifolds.

1.a. Harmonic Higgs bundles. Let M be a complex manifold and let E be a holomorphic bundle on M, equipped with a Hermitian nondegenerate sesquilinear form h (we do not impose at this stage that h is positive definite). We will say that (E, h) is a Hermitian holomorphic bundle. For any operator P acting linearly on E, we will denote by $P^\dagger$ its adjoint with respect to h.

By a holomorphic *Higgs field* Φ on E, we mean an $\mathscr{O}_M$-linear morphism $E \to \Omega^1_M \otimes_{\mathscr{O}_M} E$ satisfying the "integrability relation" $\Phi \wedge \Phi = 0$. We then say that (E, Φ) is a Higgs bundle.

Let (E, h) be a Hermitian holomorphic bundle with Higgs field Φ. Let H be the associated C^∞ bundle, so that $E = \operatorname{Ker} d''$, let $D = D' + D''$, with $D'' = d''$, be the Chern connection of h and let $\Phi^\dagger$ be the h-adjoint of Φ. We say that (E, h, Φ) is a *harmonic Higgs bundle* (or that h is *Hermite-Einstein* with respect to (E, Φ)) if ${}^V\!D := D + \Phi + \Phi^\dagger$ is an integrable connection on H. This is equivalent to a set of relations:

$$\begin{cases} (d'')^2 = 0, & d''(\Phi) = 0, & \Phi \wedge \Phi = 0, \\ (D')^2 = 0, & D'(\Phi^\dagger) = 0, & \Phi^\dagger \wedge \Phi^\dagger = 0, \\ D'(\Phi) = 0, & d''(\Phi^\dagger) = 0, & D'd'' + d''D' = -(\Phi \wedge \Phi^\dagger + \Phi^\dagger \wedge \Phi), \end{cases} \tag{1.1}$$

where the first line is by definition, the second one by h-adjointness from the first one, and the third line contains the remaining relations in the integrability condition of ${}^V\!D$. Let us notice that the holomorphic bundle $V = \operatorname{Ker}(d'' + \Phi^\dagger)$ is equipped with a flat holomorphic connection ${}^V\nabla$, which is the restriction of ${}^V\!D' := D' + \Phi$ to V.

1.b. Saito structures on holomorphic bundles. The notion of a Saito structure leads, together with a primitive homogeneous section, to the construction of Frobenius manifolds (sometimes one includes the data of the primitive homogeneous section in the Saito structure; we will not do this here). Let us recall it. By a Saito structure (of weight 0) on a holomorphic bundle E on M, we mean the data of $(E, \nabla, g, \Phi, \mathscr{U}, \mathscr{V})$ such that ∇ is a holomorphic connection on E, Φ is a holomorphic Higgs field, g is a nondegenerate symmetric holomorphic bilinear form on E, $\mathscr{U}$ and $\mathscr{V}$ are endomorphisms of E, satisfying (see e.g. [**17**, §VI.2.17–2.18]):

$$\begin{aligned} \nabla^2 = 0, \quad \nabla(\mathscr{V}) = 0, \quad \Phi \wedge \Phi = 0, \quad [\mathscr{U}, \Phi] = 0 \\ \nabla(\Phi) = 0, \quad \nabla(\mathscr{U}) - [\Phi, \mathscr{V}] + \Phi = 0, \\ \nabla(g) = 0, \quad \mathscr{V}^* + \mathscr{V} = 0 \\ \Phi^* = \Phi, \quad \mathscr{U}^* = \mathscr{U}. \end{aligned} \tag{1.2}$$

1.c. Harmonic Higgs bundles with supplementary structures. In this section, we consider supplementary structures on a harmonic Higgs bundle. These supplementary structures produce the data needed to construct a Saito structure (cf. [**7**] for more details).

Real structure on a Hermitian holomorphic bundle. Let (E, h) be a Hermitian holomorphic bundle, and let H be the associated C^∞ bundle. By a real structure we mean an *antilinear* isomorphism $\kappa : H \xrightarrow{\sim} H$ such that

$$\kappa^2 = \mathrm{Id}_H, \tag{1.3}$$

$$h(\kappa\bullet, \kappa\bullet) = \overline{h(\bullet, \bullet)} \quad \text{(equivalently, } h \text{ is real on } H_{\mathbb{R}} := \mathrm{Ker}(\kappa - \mathrm{Id}_H)), \tag{1.4}$$

$$D(\kappa) = 0. \tag{1.5}$$

(1.5) can also be read $D'\kappa = \kappa d''$, or equivalently according to (1.3), $\kappa D' = d''\kappa$. Let us consider the nondegenerate complex bilinear form $g(\bullet, \bullet) := h(\bullet, \kappa\bullet)$. Then, because h is Hermitian, (1.4) and (1.5) are respectively equivalent to

(1.4′) g is symmetric,

(1.5′) g is holomorphic on E and D' is compatible with g, i.e., $D(g) = 0$.

For any linear operator P on H, we denote by P^* the g-adjoint of P (which is holomorphic if P is so). Then

$$P^* = \kappa P^\dagger \kappa. \tag{1.6}$$

Potential harmonic Higgs bundles. Let (E, h, Φ) be a harmonic Higgs bundle. From (1.1), we have $d''\Phi^\dagger = 0$. Hence there exists, at least locally, a C^∞ endomorphism B of E such that $d''B = \Phi^\dagger$ or, equivalently by setting $A = B^\dagger$, $D'A = \Phi$. Let then (E, h) be a Hermitian holomorphic bundle and let A be a C^∞ endomorphism of the associated C^∞ bundle H. We say that (E, h, A) is a *potential harmonic Higgs bundle* if

$$(E, h, D'A) \text{ is a harmonic Higgs bundle.} \tag{1.7}$$

In other words, we assume here the global existence of A and a choice of it. (Except in simple examples such as Example 1.35, it is not expected that A is holomorphic.)

For a potential harmonic Higgs bundle (E, h, A), we set $\Phi = D'A$, which is a Higgs field by definition. We can associate to these data a new connection of type $(1, 0)$:

$$\nabla := D' + [A^\dagger, \Phi]. \tag{1.8}$$

From (1.1) we get that ∇ *is holomorphic*, i.e., $d''\nabla + \nabla d'' = 0$.

Real (potential) harmonic Higgs bundles. Let us consider (E, h, Φ, κ) (resp. (E, h, A, κ)), where (E, h, κ) is a real Hermitian holomorphic bundle and (E, h, Φ) (resp. (E, h, A)) is a (resp. potential) harmonic Higgs bundle. We say that (E, h, Φ, κ) is a real harmonic Higgs bundle if the following compatibility relation is satisfied:

$$\Phi^* = \Phi \quad \text{(resp. } A^* = A). \tag{1.9}$$

According to (1.6), this is equivalent to $\Phi^\dagger = \kappa\Phi\kappa$ (resp. $A^\dagger = \kappa A\kappa$) or, equivalently, to ${}^VD(\kappa) = 0$, i.e., κ is a real structure on the local system $\mathrm{Ker}\, {}^V\nabla$. In other words, if we denote, for any $z_o \in \mathbb{C}^*$, by $(V_{(z_o)}, {}^{V_{(z_o)}}\nabla)$ the flat bundle $\big(\mathrm{Ker}(D'' + z_o\Phi^\dagger), D' + \Phi/z_o\big)$, then the conditions (1.3), (1.5) and (1.9) on κ can be rephrased by saying that κ is a real structure on each of the locally constant sheaves $\mathrm{Ker}\, {}^{V_{(z_o)}}\nabla$, $z_o = \pm 1$ (or equivalently, $|z_o| = 1$).

Let us notice that, when $\Phi = D'A$, the condition $A^* = A$ implies $\Phi^* = \Phi$, as we have $D'(A^*) = (D'A)^*$ (because $D'(g) = 0$). Conversely, If $\Phi = D'A$ and $\Phi = \Phi^*$, we also have $\Phi = D'(A + A^*)/2$, so we can assume that $A = A^*$. Moreover, with this assumption, $[A^\dagger, \Phi]^* + [A^\dagger, \Phi] = 0$, so, according to (1.5′),

$$\nabla(g) = 0. \tag{1.10}$$

Lastly, from $D(\Phi) = 0$ given by (1.1) we also deduce

$$\nabla(\Phi) = 0. \tag{1.11}$$

Integrable (potential) harmonic Higgs bundles. Let (E, h, Φ) be a harmonic Higgs bundle. We say that it is *integrable*[1] if there exist two endomorphisms $\mathscr{U}$ and $\mathscr{Q}$ of H such that

$$\mathscr{U} \text{ is holomorphic, i.e., } d''(\mathscr{U}) = 0, \tag{1.12}$$

$$\mathscr{Q}^\dagger = \mathscr{Q}, \tag{1.13}$$

$$[\Phi, \mathscr{U}] = 0, \tag{1.14}$$

$$D'(\mathscr{U}) - [\Phi, \mathscr{Q}] + \Phi = 0, \tag{1.15}$$

$$D'(\mathscr{Q}) + [\Phi, \mathscr{U}^\dagger] = 0. \tag{1.16}$$

Let us note that $(\mathscr{U}, \mathscr{Q} + \lambda\,\mathrm{Id})$ satisfy the same equations for any $\lambda \in \mathbb{R}$.

If (E, h, Φ) is moreover potential (i.e., $\Phi = D'A$), then, in analogy with (1.8), we set

$$\mathscr{V} = \mathscr{Q} - [A^\dagger, \mathscr{U}]. \tag{1.17}$$

The equation $(1.16)^\dagger$ obtained from (1.16) by adjointness can be written $d''(\mathscr{V}) = 0$, showing that $\mathscr{V}$ is then a holomorphic endomorphism of E.

Let us also notice that (1.15) can as well be written as

$$\nabla(\mathscr{U}) - [\Phi, \mathscr{V}] + \Phi = 0. \tag{1.18}$$

tt bundles.* We now end with all the structures put together, that is, we consider tuples $(E, h, \Phi, \kappa, \mathscr{U}, \mathscr{Q})$ such that (E, h, κ) is a real Hermitian holomorphic bundle (so (1.3)–(1.5) are satisfied), $(E, h, \Phi, \mathscr{U}, \mathscr{Q})$ is an integrable harmonic Higgs bundle (i.e., (1.7) and (1.12)–(1.16) are satisfied), and moreover

$$\mathscr{U}^* = \mathscr{U}, \tag{1.19}$$

$$\mathscr{Q}^* + \mathscr{Q} = 0. \tag{1.20}$$

Let us note that (1.13) and (1.20) imply

$$\mathscr{Q} \text{ is purely imaginary, i.e., } \mathscr{Q}\kappa = -\kappa\mathscr{Q}. \tag{1.21}$$

One defines potential tt* bundles similarly. For such an object,

$$\mathscr{V}^* + \mathscr{V} = 0. \tag{1.22}$$

Data for a Saito structure. Given potential tt* bundle $(E, h, A, \kappa, \mathscr{U}, \mathscr{Q})$, we have associated to it a tuple $(E, \nabla, g, \Phi, \mathscr{U}, \mathscr{V})$ satisfying the relations (1.2) except possibly

$$\nabla^2 = 0, \quad \nabla(\mathscr{V}) = 0. \tag{1.23}$$

(Indeed, these relations were derived in (1.5′), (1.9), (1.10), (1.11), (1.18), (1.19) and (1.22).)

[1] This terminology will be justified in §2.c.

1.d. Saito structures with supplementary structures. Starting now from a Saito structure $(E, \nabla, g, \Phi, \mathscr{U}, \mathscr{V})$ on E as in §1.b, we will add supplementary structures κ and A and try to reconstruct a tt* bundle.

Real structure. We still denote by H the C^∞ bundle associated to E. We say that an antilinear isomorphism $\kappa : H \to H$ is a real structure if $\kappa^2 = \mathrm{Id}_H$. We define the nondegenerate sesquilinear form h by $h(\bullet, \kappa\bullet) = g(\bullet, \bullet)$. We impose the following compatibility relations:

- g is real on $H_{\mathbb{R}} := \mathrm{Ker}(\kappa - \mathrm{Id}_H)$,
- $D'(g) = 0$, where $D = D' + d''$ is the Chern connection of h.

We then speak of a *real Saito structure*. We say that the real Saito structure $(E, \nabla, g, \Phi, \mathscr{U}, \mathscr{V}, \kappa)$ is *harmonic* if (E, h, Φ) is a harmonic Higgs bundle.

Potential for a Saito structure. Let $(E, \nabla, g, \Phi, \mathscr{U}, \mathscr{V})$ be a Saito structure and let $A^\dagger$ be a C^∞ endomorphism of E such that $A^{\dagger *} = A^\dagger$ (here, $\dagger$ makes no reference to any adjointness). We then set $\Phi^\dagger = d''A^\dagger$, which thus satisfies $\Phi^{\dagger *} = \Phi^\dagger$. Let us then define

$$\begin{aligned} D' &= \nabla - [A^\dagger, \Phi], \\ \mathscr{Q} &= \mathscr{V} + [A^\dagger, \mathscr{U}]. \end{aligned}$$

We notice that, with this definition of D' and $\Phi^\dagger$, the third line of (1.1) is satisfied and that $D'(g) = 0$ (hence $D(g) = 0$, if we set $D = D' + d''$). We also have $\mathscr{Q}^* + \mathscr{Q} = 0$, (1.15) is fulfilled as well as $(1.16)^\dagger$.

We say that $A^\dagger$ is a harmonic potential for the Saito structure $(E, \nabla, g, \Phi, \mathscr{U}, \mathscr{V})$ if the second line of (1.1) is satisfied.

Harmonic potential real Saito structure. Let $(E, \nabla, g, \Phi, \mathscr{U}, \mathscr{V})$ be a Saito structure equipped with a harmonic real structure κ (and associated harmonic Hermitian form h) and a harmonic potential $A^\dagger$. We say that the two structures are compatible if the following properties are fulfilled:

(1) h and $A^\dagger$ define the same connection D',
(2) the h-adjoint of Φ is $\Phi^\dagger := d''A^\dagger$,
(3) $\mathscr{Q}^\dagger = \mathscr{Q}$.

We then speak of a harmonic potential real Saito structure $(E, \nabla, g, \Phi, \mathscr{U}, \mathscr{V}, \kappa, A^\dagger)$.

PROPOSITION 1.24. *There is a one-to-one correspondence between harmonic potential real Saito structures $(E, \nabla, g, \Phi, \mathscr{U}, \mathscr{V}, \kappa, A^\dagger)$ and potential tt* bundles $(E, h, A, \kappa, \mathscr{U}, \mathscr{Q})$ satisfying* (1.23). □

1.e. Harmonic Frobenius manifolds. A structure of Frobenius manifold (or Saito manifold) on a complex manifold M consists in giving a Saito structure on the tangent bundle M with the supplementary conditions that

(1.25) ∇ is torsion-free,

(1.26) Φ is symmetric,

(1.27) $\exists\, e \in \Gamma(M, TM)$, $\nabla(e) = 0$, e is an eigenvector of $\mathscr{V}$ and $\Phi_e = -\mathrm{Id}_{TM}$.

Let us recall that a Higgs field on TM defines a product $\star$ on TM by setting $\xi \star \eta = -\Phi_\xi \eta$ and that saying Φ is symmetric means that $\star$ is commutative (and also associative because of the Higgs condition). The vector field e is then a unit for $\star$. The *Euler field* $\mathfrak{E}$ is defined as $\mathscr{U}(e)$, and $\mathscr{U}$ is identified with the multiplication by $\mathfrak{E}$. If $c \in \mathbb{C}$ is defined by $\mathscr{V}e = ce$, we then have $\nabla\mathfrak{E} = \mathscr{V} + (1-c)\,\mathrm{Id}$, so that $\nabla_e \mathfrak{E} = e$. Moreover, we have $\mathrm{Lie}_{\mathfrak{E}}(g) = 2(1-c)$ (cf. e.g. [**17**, §VII.1.21]).

DEFINITION 1.28 (Harmonic Frobenius manifold). A structure of *harmonic Frobenius manifold* (or *harmonic Saito manifold*) on a complex manifold M consists in giving a *harmonic potential real Saito structure* on the tangent bundle M with the supplementary conditions (1.25)–(1.27) and such that $c \in \mathbb{R}$.

The new data added to a Frobenius manifold in order to make it harmonic consist of a real structure κ on TM and a potential $A^\dagger$ (or A), satisfying the relations described in §1.d.

PROPOSITION 1.29. *A harmonic Frobenius manifold is a manifold with a CDV-structure, in the sense of* [**7**, Def. 1.2].

PROOF. Firstly, (1.5) in loc. cit. means $\kappa\Phi\kappa = \Phi^\dagger$, which is equivalent to $\Phi^* = \Phi$ (and is true by assumption, cf. (1.2)).

As $\nabla e = 0$ and ∇ is torsion-free, we have $\nabla_e = \mathrm{Lie}_e$. As $\Phi_e = -\mathrm{Id}$, (1.8) thus implies $\mathrm{Lie}_e = \nabla_e = D'_e$. Therefore $\mathrm{Lie}_e(h) = 0$, so [**7**, (1.6)] is satisfied.

We will now show that, if we denote by c the eigenvalue of $\mathscr{V}$ with respect to e as above, we have

$$\mathscr{Q} = D_{\mathfrak{E}} - \mathrm{Lie}_{\mathfrak{E}} - (1-c)\,\mathrm{Id}\,. \tag{1.30}$$

Let us first remark that, as $\mathfrak{E}$ is holomorphic, $d''_{\mathfrak{E}}$ acting on TM is zero, so that we can replace $D_{\mathfrak{E}}$ with $D'_{\mathfrak{E}}$ in (1.30). Then, by (1.17), $\mathscr{Q} = \mathscr{V} + [A^\dagger, \mathscr{U}] = \nabla\mathfrak{E} + [A^\dagger, \mathscr{U}] - (1-c)\,\mathrm{Id}$. But on the other hand, using the torsion-freeness of ∇,

$$\nabla\mathfrak{E} = \nabla_{\mathfrak{E}} - \mathrm{Lie}_{\mathfrak{E}} = D'_{\mathfrak{E}} + [A^\dagger, \Phi_{\mathfrak{E}}] - \mathrm{Lie}_{\mathfrak{E}} = D'_{\mathfrak{E}} - [A^\dagger, \mathscr{U}] - \mathrm{Lie}_{\mathfrak{E}},$$

hence (1.30). In other words, [**7**, (1.8)] is satisfied.

We use [**33**, Rem. 3.6] to remark that $\mathscr{Q}^\dagger = \mathscr{Q}$ and $c \in \mathbb{R}$ imply $\mathrm{Lie}_{\mathfrak{E}-\overline{\mathfrak{E}}}(h) = 0$, so [**7**, (1.7)] is satisfied.

Lastly, the integrability property [**7**, (1.9)] is a consequence of (1.12)–(1.16) and the adjoint relations. □

Following the method of K. Saito, the infinitesimal period mapping with respect to a primitive section (cf. e.g. [**7**, Th. 5.12], [**20**, Chap. VII]) gives (cf. [**7**, Th. 5.15]):

COROLLARY 1.31. *Let $(E, \nabla, g, \Phi, \mathscr{U}, \mathscr{V}, \kappa, A^\dagger)$ be a harmonic potential real Saito structure on E and let ω be a ∇-horizontal section of E which is primitive and real homogeneous, i.e., such that*

(1) *$\varphi_\omega(\bullet) := -\Phi_\bullet(\omega) : TM \to E$ is an isomorphism,*
(2) *ω is an eigenvector of $\mathscr{V}$ with real eigenvalue.*

Then φ_ω equips M with the structure of a harmonic Frobenius manifold. □

1.f. Examples.

EXAMPLE 1.32 (Compact Kähler manifolds, after K. Corlette and C. Simpson). The main source of examples, when M is compact Kähler, comes from the fundamental result of K. Corlette [**2**] and C. Simpson [**28**] saying that a flat bundle $(V, {}^V\nabla)$ admits a harmonic *metric* (i.e., positive definite) if and only if the local system $\mathrm{Ker}\,{}^V\nabla$ is *semi-simple*. The metric is then uniquely determined (up to a positive constant) on any simple summand. Similarly, a Hermite-Einstein metric h exists for (E, Φ) iff (E, Φ) is *polystable with vanishing Chern classes*.

In general, such a metric is not determined explicitly, but is given by a nontrivial global existence result.

EXAMPLE 1.33 (Quasi-projective varieties, after T. Mochizuki). The results of Corlette and Simpson have been generalized by C. Simpson in dimension one [**29**] and then by T. Mochizuki [**15**] in arbitrary dimension. If M is a smooth quasi-projective variety and $\widetilde{M}$ is a smooth projective variety containing M as a Zariski dense subset and such that $D := \widetilde{M} \smallsetminus M$ is a divisor with normal crossings in $\widetilde{M}$, then one can introduce (cf. [**29**] if $\dim M = 1$) the notion of tameness for a Higgs field on a holomorphic vector bundle E on M: the eigenvalues of the Higgs field should have at most logarithmic growth along D. Similarly, given a flat bundle $(V, {}^V\nabla)$ with metric h on M, tameness means that the h-norm of horizontal sections has at most a moderate growth along D. One of the main results of [**15**] is a one-to-one correspondence between semi-simple local systems on M and harmonic flat bundles $(V, {}^V\nabla, h)$ for which the metric is tame and the eigenvalues of the residue of the Higgs field along D are purely imaginary. Here also, the metric is not determined explicitly, but is given by a nontrivial global existence result.

EXAMPLE 1.34 (Variations of complex Hodge structures of weight 0). Let H be a C^∞ vector bundle on M, equipped with a flat connection ${}^V\!D$ and a decomposition $H = \oplus_{p\in\mathbb{Z}} H^p$ into C^∞ subbundles. We assume that the Griffiths transversality relations hold:

$$ {}^V\!D' H^p \subset (H^p \oplus H^{p-1}) \otimes_{\mathscr{O}_M} \Omega^1_M, \quad {}^V\!D'' H^p \subset (H^p \oplus H^{p+1}) \otimes_{\mathscr{O}_{\overline{M}}} \Omega^1_{\overline{M}}. $$

We denote by $D_{|H^p}$ the composition of ${}^V\!D_{|H^p}$ with the projection to H^p, by $\Phi_{|H^p}$ the composition of ${}^V\!D'_{|H^p}$ with the projection to H^{p-1} and by $\Phi^\dagger$ that of ${}^V\!D''_{|H^p}$ with the projection to H^{p+1}. We then set $D = \oplus_p D_{|H^p}$, $\Phi = \oplus_p \Phi_{|H^p}$ and $\Phi^\dagger = \oplus_p \Phi^\dagger_{|H^p}$. Considering the holomorphic type together with the degree with respect to the decomposition, we find that the relations (1.1) are satisfied because of the flatness of ${}^V\!D$.

Assume that we are given a nondegenerate Hermitian form k such that ${}^V\!D(k) = 0$ and the decomposition $H = \oplus_{p\in\mathbb{Z}} H^p$ is k-orthogonal. Consider the nondegenerate Hermitian form $h = \oplus_p (-1)^p k_{|H^p}$. Then $D(h) = 0$ and $\Phi^\dagger$ is the h-adjoint of Φ. In such a case, we say that $(H = \oplus_p H^p, {}^V\!D, k)$ is a variation of complex Hodge structure of weight 0. In particular, (H, D'', h, Φ) is a harmonic Higgs bundle. Let us show that it is integrable. We set $\mathscr{Q} = \oplus_p p \operatorname{Id}_{H^p}$ and $\mathscr{U} = 0$. By definition, D is compatible with the decomposition, hence $D(\mathscr{Q}) = 0$ and, as p is real, we have $\mathscr{Q}^\dagger = \mathscr{Q}$. Lastly, we have $[\Phi, \mathscr{Q}] = \Phi$.

By a real structure κ, we mean an anti-linear involution $\kappa : H \to H$ which is ${}^V\!D$-horizontal (thus defining a real structure on the local system $\operatorname{Ker} {}^V\!D$) such that $\kappa(H^p) = H^{-p}$ for any p. Then $D(\kappa) = 0$ and $\Phi^\dagger = \kappa\Phi\kappa$. The previous data thus define a tt* bundle.

EXAMPLE 1.35 (The rank-one case on a punctured disc). In order to have a concrete example at hand, we will give details on the rank-one case on the punctured unit disc Δ^* (having coordinate t), with a positive definite Hermitian form h (i.e., a Hermitian metric). Although very simple, the rank-one case gives indications on the relations between the eigenvalues of the Higgs field and that of the monodromy of the integrable connection ${}^V\nabla$. Let us first introduce some notation.

In the following, we denote by $\mathfrak{B}$ the set of equivalence classes of affine forms $\mathfrak{b} : z \mapsto uz + v$ modulo the addition of a purely imaginary complex number: we have $\mathfrak{b} \equiv \mathfrak{b}' \Leftrightarrow (u = u')$ and $(v - v' \in i\mathbb{R})$. Given an affine form $\mathfrak{b}$, we consider the two functions[2]

$$z \longmapsto b_z := \operatorname{Re} \mathfrak{b}(z),$$
$$z \longmapsto \beta_z := z \text{ “Re” } \mathfrak{b}(z),$$

where, if $\mathfrak{b}(z) = uz + v$, we set[3]

$$z \text{ “Re” } \mathfrak{b}(z) = \tfrac{1}{2} z\big[uz + v - \overline{u}z^{-1} + \overline{v}\big] = \tfrac{1}{2}uz^2 + (\operatorname{Re} v)z - \tfrac{1}{2}\overline{u}.$$

Therefore, the two functions $b_\bullet$ and $\beta_\bullet$ only depend on the class of $\mathfrak{b}$ in $\mathfrak{B}$. Moreover, the function $\beta_\bullet$ uniquely determines the class of $\mathfrak{b} \in \mathfrak{B}$: indeed, the coefficients of $\beta_\bullet$ give u and $\operatorname{Re} v$.

(1.36) *The set of isomorphism classes of rank-one harmonic Higgs bundles (or harmonic flat bundles) on Δ^* is in one-to-one correspondence with the set of pairs $(\psi, \mathfrak{b})$, with $\psi \in \mathscr{O}(\Delta^*)$ and $\psi dt/t$ having no residue and $\mathfrak{b} \in \mathfrak{B} \bmod \mathbb{Z}$. Moreover, meromorphic Higgs bundles correspond to pairs $(\psi, \mathfrak{b})$ with ψ meromorphic at 0.*

PROOF. We will do the proof for harmonic Higgs bundles. The case of harmonic flat bundles is similar and the proof will show the 1-1 correspondence between the two. Let us start with a rank-one harmonic Higgs bundle (H, D'', Φ, h). Let e' be some holomorphic frame of $E = \operatorname{Ker} D''$. The Higgs field Φ can be written as

$$\Phi e' = \varphi(t)\, e'\, dt, \quad \varphi(t) \text{ holomorphic on } \Delta^*.$$

Let us set $\varphi(t) = \partial_t \psi(t) + \beta_0/t$ with $\psi \in \mathscr{O}(\Delta^*)$ and $\psi dt/t$ having no residue at 0 and $\beta_0 \in \mathbb{C}$. Let us note here that φ is meromorphic iff ψ is so. Let us also set $\|e'\|_h = \exp(\lambda(t))$ for some C^∞ real function λ on Δ^*. Then we have $D'e' = 2\partial_t\lambda \cdot e'\, dt$ and the harmonicity condition on the Higgs bundle is that λ is *harmonic* on Δ^*. Indeed, in the orthonormal frame $\varepsilon = \exp(-\lambda(t)) \cdot e'$, we have

$$D'\varepsilon = \partial_t\lambda\, \varepsilon\, dt, \quad D''\varepsilon = -\partial_{\overline{t}}\lambda\, \varepsilon\, d\overline{t}, \quad \Phi\varepsilon = \varphi(t)\, \varepsilon\, dt, \quad \Phi^\dagger\varepsilon = \overline{\varphi}(t)\, \varepsilon\, d\overline{t}. \tag{1.37}$$

Therefore,

$$^V\!D'\varepsilon = (\varphi(t) + \partial_t\lambda)\, \varepsilon\, dt, \quad {}^V\!D''\varepsilon = (\overline{\varphi}(t) - \partial_{\overline{t}}\lambda)\, \varepsilon\, d\overline{t}. \tag{1.38}$$

The flatness of $^V\!D$ is then equivalent to $2\partial_t\overline{\partial}_t\lambda = 0$.

If λ is harmonic on Δ^*, there exists (by working on the universal cover of Δ^*) a holomorphic function μ on Δ^* and a real number c, such that $\lambda(t) = 2(\operatorname{Re}\mu(t) + c \log|t|)$. We have $\partial_t\lambda = \partial_t\mu + c/t$ and $\partial_{\overline{t}}\lambda = \overline{\partial_t\lambda}$. Let us then replace e' with $e = \exp(-2\mu(t)) \cdot e'$, which has h-norm $|t|^{2c}$. So we can assume from the beginning that $\lambda = b_0 \log|t|$, $b_0 \in \mathbb{R}$. Therefore, $\varepsilon = |t|^{-b_0} e$ and we have

$$D'e = \frac{b_0}{t}\, e\, dt, \quad \Phi e = \Big(\partial_t\psi(t) + \frac{\beta_0}{t}\Big) e\, dt.$$

---

[2]In [**15**], Mochizuki uses the notation $-\mathfrak{p}, \mathfrak{e}$.

[3]Anticipating the notation introduced in §2, let us notice that “Re” $\mathfrak{b}(z) = \frac{1}{2}(\mathfrak{b}(z) + \overline{\mathfrak{b}(z)})$.

Now, (1.37) and (1.38) can be rewritten as

$$(1.39)\qquad \begin{aligned} D'\varepsilon &= \frac{b_0}{2t}\,\varepsilon\,dt, & D''\varepsilon &= -\frac{b_0}{2\overline{t}}\,\varepsilon\,d\overline{t},\\ \Phi\varepsilon &= \Big(\partial_t\psi(t)+\frac{\beta_0}{t}\Big)\varepsilon\,dt, & \Phi^\dagger\varepsilon &= \overline{\Big(\partial_t\psi(t)+\frac{\beta_0}{t}\Big)}\varepsilon\,d\overline{t}\end{aligned}$$

$$(1.40)\qquad {}^V\!D'\varepsilon = \Big(\partial_t\psi(t)+\frac{b_0+2\overline{\beta}_0}{2t}\Big)\varepsilon\,dt,\quad {}^V\!D''\varepsilon = \overline{\Big(\partial_t\psi(t)+\frac{2\beta_0-b_0}{2t}\Big)}\varepsilon\,d\overline{t}.$$

We will consider the affine form

$$\mathfrak{b}(z) = -2\overline{\beta}_0 z + b_0.$$

We will use the notation $b_1 = (\operatorname{Re}\mathfrak{b})(1) = b_0 - 2\operatorname{Re}\beta_0$, $\beta_1 = (z\text{ “Re” }\mathfrak{b})(1) = b_0+\beta_0-\overline{\beta}_0$. Let us consider the frame

$$v = e^{\psi-\overline{\psi}}|t|^{\mathfrak{b}(1)}\varepsilon.$$

It is ${}^V\!D''$-holomorphic and has h-norm $|t|^{b_1}$. We have

$${}^V\nabla v = \Big(2\partial_t\psi+\frac{\beta_1}{t}\Big)v\,dt.$$

Let us end the proof of (1.36). Let e' be another holomorphic frame of E such that $|t|^{-b_0'}e'$ has norm one. Then $e'=\nu(t)e$ for some holomorphic function $\nu(t)$ on Δ^* having moderate growth, which is thus meromorphic. We hence have $b_0'-b_0=k\in\mathbb{Z}$ and $e'=\nu t^k e$ with $|\nu|=1$. The Higgs field has the same expression in the bases e and e', thus fixing β_0 and ψ. □

REMARK 1.41. The reason for defining $\mathfrak{B}$ as equivalence classes modulo $i\mathbb{R}$ is to be able to conclude that the two categories of rank-one harmonic Higgs bundles and rank-one harmonic flat bundles are equivalent, even if the representative $\mathfrak{b}=uz+v$ chosen in each case is not the same. In the Higgs case, we choose $\operatorname{Im}v=0$, while in the flat case we choose $\operatorname{Im}v=-\operatorname{Im}u$.

In this example, tameness (as mentioned in Example 1.33) means ψ holomorphic on Δ, while the condition of being purely imaginary means $\beta_0\in i\mathbb{R}$, so $b_1=b_0$.

A real structure exists if and only if the residue of ${}^{V(z_o)}\nabla$ with $z_o=\pm1$ belongs to $\frac12\mathbb{Z}$. This reduces to β_1 and $-\beta_{-1}\in\frac12\mathbb{Z}$, that is, $b_0\in\frac12\mathbb{Z}$ and $\beta_0\in\mathbb{R}$.

A potential A is now a C^∞ function such that $\partial_tA=\partial_t\psi+\beta_0/t$. One must have $A=\psi+\beta_0\log|t|^2+\text{anti-holomorphic}$. However, as $A^\dagger$ only appears through a commutator, it is not needed in rank one, and one has $\nabla=D'$.

If $\beta_0\neq0$, integrability only holds on simply connected open sets of the punctured disc: $\mathscr{Q}$ must be a real constant and $\mathscr{U}$ has to be a holomorphic function which satisfies $\mathscr{U}'(t)=-(\partial_t\psi+\beta_0/t)$. In order to have integrability on the whole punctured disc while keeping a singularity at the origin, it is necessary to admit a nontame metric, i.e., ψ meromorphic and nonholomorphic.

2. Integrable variations of twistor structures

In this section, we explain a criterion, due to C. Hertling, for proving the harmonicity of a Frobenius manifold. We first recall the introduction of an extra parameter to explain either the relations defining a harmonic Higgs bundle or that defining a Saito structure.

All through this section, M will denote a complex manifold, $\overline{M}$ the conjugate manifold (with structure sheaf $\overline{\mathscr{O}_M}$), and $M_{\mathbb{R}}$ will denote the underlying real-analytic or C^∞-manifold. We will denote by $\mathbb{P}^1$ the Riemann sphere, covered by the two affine charts $\simeq \mathbb{A}^1$ with coordinates z and $1/z$, and by $p: M\times\mathbb{P}^1 \to M$ the projection.

The coordinate z being fixed, we denote by $\mathbf{S}$ the circle $|z|=1$, by Ω_0 an open neighbourhood of the closed disc $\Delta_0 := \{|z| \leqslant 1\}$ and by Ω_∞ an open neighbourhood of the closed disc $\Delta_\infty := \{|z| \geqslant 1\}$.

We will denote by $\sigma: \mathbb{P}^1 \to \overline{\mathbb{P}^1}$ the anti-holomorphic involution $z \mapsto -1/\overline{z}$. We assume that $\Omega_\infty = \sigma(\Omega_0)$. We denote by $\iota: \mathbb{P}^1 \to \mathbb{P}^1$ the holomorphic involution $z \mapsto -z$.

2.a. Families of vector bundles on the Riemann sphere. Let $\mathscr{H}''$ be a holomorphic vector bundle on $M \times \Omega_0$. Then $\overline{\mathscr{H}''}$ is a holomorphic bundle on the conjugate manifold $\overline{M} \times \overline{\Omega_0}$ and $\sigma^*\overline{\mathscr{H}''}$ is a holomorphic bundle on $\overline{M} \times \Omega_\infty$ (i.e., is an anti-holomorphic family of holomorphic bundles on Ω_∞). We will set $\overline{\overline{\mathscr{H}''}} := \sigma^*\overline{\mathscr{H}''}$.

By a *real-analytic* (resp. C^∞) *family* of holomorphic vector bundles on $\mathbb{P}^1$ parametrized by $M_{\mathbb{R}}$ we will mean the data of a triple $(\mathscr{H}', \mathscr{H}'', \mathscr{C}_{\mathbf{S}})$ consisting of holomorphic vector bundle $\mathscr{H}', \mathscr{H}''$ on $M \times \Omega_0$ and a nondegenerate morphism which is $\mathscr{O}_{M\times\mathbf{S}} \otimes_{\mathscr{O}_{\mathbf{S}}} \mathscr{O}_{\overline{M}\times\mathbf{S}}$-linear,

$$\mathscr{C}_{\mathbf{S}} : \mathscr{H}'_{|\mathbf{S}} \otimes_{\mathscr{O}_{\mathbf{S}}} \overline{\mathscr{H}''_{|\mathbf{S}}} \longrightarrow \mathscr{C}^{*,\mathrm{an}}_{M_{\mathbb{R}}\times\mathbf{S}},$$

where $\mathscr{C}^{*,\mathrm{an}}_{M_{\mathbb{R}}\times\mathbf{S}}$ is either the sheaf of real-analytic functions on $M_{\mathbb{R}} \times \mathbf{S}$ if $* = \mathbb{R}$-an, or the sheaf of C^∞ functions on $M_{\mathbb{R}} \times \mathbf{S}$ which are real analytic with respect to $z \in \mathbf{S}$ if $* = \infty$. The nondegeneracy condition means that $\mathscr{C}_{\mathbf{S}}$ defines $C^{*,\mathrm{an}}$-gluing between the dual $\mathscr{H}'^\vee$ of $\mathscr{H}'$ and $\overline{\overline{\mathscr{H}''}}$, giving rise to a $\mathscr{C}^{*,\mathrm{an}}_{M_{\mathbb{R}}\times\mathbb{P}^1}$-locally free sheaf of finite rank that we denote by $\mathscr{H}$, where $\mathscr{C}^{*,\mathrm{an}}_{M_{\mathbb{R}}\times\mathbb{P}^1}$ denotes the sheaf of C^∞ or real-analytic functions on $M_{\mathbb{R}} \times \mathbb{P}^1$ *which are holomorphic with respect to* z.

REMARK 2.1. We could have started from $\mathscr{C}^{*,\mathrm{an}}_{M_{\mathbb{R}}\times\Omega_0}$-locally free sheaves (cf. (2.3) below), but we insist here that these sheaves come from holomorphic bundles, even if the result $\mathscr{H}$ of the gluing is only a $\mathscr{C}^{*,\mathrm{an}}_{M_{\mathbb{R}}\times\mathbb{P}^1}$-locally free sheaf.

On the other hand, using a pairing instead of a gluing is not essential here, as we can replace $\mathscr{H}'$ with its dual bundle, and is mainly motivated by (2.5), as well as a better behaviour near singularities, when one develops the theory with singularities (twistor $\mathscr{D}$-modules, cf. [**20**]).

In order to give a uniform result on holomorphic, real-analytic or C^∞ families of holomorphic bundles on $\mathbb{P}^1$, we will give a similar presentation for holomorphic families. Let us denote by $\widetilde{\sigma}: \mathbb{P}^1 \to \mathbb{P}^1$ the involution $z \mapsto -1/z$. Given a holomorphic bundle $\mathscr{H}''$ on $M \times \Omega_0$, we now denote by $\overline{\overline{\mathscr{H}''}}$ the holomorphic bundle $\widetilde{\sigma}^*\mathscr{H}''$.

By a *holomorphic family* of vector bundles on $\mathbb{P}^1$ parametrized by M we will mean the data of a triple $(\mathscr{H}', \mathscr{H}'', \mathscr{C}_{\mathbf{S}})$, where $\mathscr{H}'$ (resp. $\mathscr{H}''$) is a holomorphic vector bundle on $M \times \Omega_0$ and $\mathscr{C}_{\mathbf{S}} : \mathscr{H}'_{|\mathbf{S}} \otimes_{\mathscr{O}_{M\times\mathbf{S}}} \overline{\overline{\mathscr{H}''_{|\mathbf{S}}}} \to \mathscr{O}_{M\times\mathbf{S}}$ is a nondegenerate $\mathscr{O}_{M\times\mathbf{S}}$-bilinear pairing between the sheaf-theoretic restrictions to $M \times \mathbf{S}$, defining a gluing between $\mathscr{H}'^\vee$ and $\overline{\overline{\mathscr{H}''}}$, hence a holomorphic vector bundle $\mathscr{H}$ on $M \times \mathbb{P}^1$.

Let us notice that a real analytic family as above, parametrized by the real analytic variety $M_{\mathbb{R}}$, can be extended to the germ of complexification $(M_{\mathbb{R}})_{\mathbb{C}}$ of $M_{\mathbb{R}}$ as a holomorphic family.

PROPOSITION 2.2. *Let $(\mathscr{H}', \mathscr{H}'', \mathscr{C}_{\mathbf{S}})$ be a holomorphic (resp. $C^{\mathbb{R}\text{-an,an}}$) family of holomorphic vector bundles on $\mathbb{P}^1$ parametrized by M (resp. $M_{\mathbb{R}}$).*

(1) *If $\mathscr{H}_{|\{x\}\times\mathbb{P}^1}$ is the trivial bundle on $\mathbb{P}^1$ for any $x \in M$, then $H := p_*\mathscr{H}$ is a holomorphic (resp. $C^{\mathbb{R}\text{-an}}$) vector bundle on M (resp. $M_{\mathbb{R}}$) and the natural morphism $p^*H = p^*p_*\mathscr{H} \to \mathscr{H}$ is an isomorphism.*

(2) *If M is connected, the subset Θ of M defined by the property that $\mathscr{H}_{|\{x\}\times\mathbb{P}^1}$ is the trivial bundle on $\mathbb{P}^1$ iff $x \in M \smallsetminus \Theta$ is either empty, or equal to M, or locally defined by a single holomorphic (resp. real-analytic) equation.*

Moreover, the first point also holds for C^∞ families.

PROOF. In the case of holomorphic families, one can use an argument on direct images of coherent sheaves, characterizing Θ as the support of the coherent sheaf $R^1p_*\mathscr{H}(-1)$ (cf. e.g. [**17**, Th. I.5.3]). The case of real analytic families is obtained by applying the case of holomorphic families to the complexified family.

For the first point in the case of C^∞ families, one can adapt the proof of [**13**, §4], which also applies to the holomorphic or real analytic case. One first notices that this is a local result on M, so that we can assume that M is a polydisc in $\mathbb{C}^n$. Then, applying Grauert's theorem (or a simple variant of it, as [**12**, Cor. 2.17]), we can assume that $\mathscr{H}'$ and $\mathscr{H}''$ are trivial holomorphic vector bundles. Let us denote by d the rank of $\mathscr{H}', \mathscr{H}''$. Fixing the trivializations, the pairing $\mathscr{C}_{\mathbf{S}}$ is now regarded as an invertible $d \times d$-matrix $C_x(z)$ defined on $M \times \{1/r < z < r\}$ with $r > 1$ (up to shrinking M once more), which is holomorphic with respect to z and C^∞ with respect to x.

The bundle $\mathscr{H}_{|\{x\}\times\mathbb{P}^1}$ is trivial if, up to reducing $r > 1$, there exist holomorphic invertible matrices $z \mapsto \Sigma'_x(z), \Sigma''_x(z)$ on $D_r := \{|z| < r\}$ such that, on $\mathbb{A}_r$, $C_x(z) = {}^t\Sigma'_x(z)\overline{\Sigma''_x(z)}$. Therefore, the proof consists in finding Σ', Σ'' depending in a C^∞ way on x. The conclusion follows by checking that the formula given in loc. cit. for Σ' depends in a C^∞ way on x. □

2.b. Variation of twistor structures. The notion of variation of twistor structure was introduced by C. Simpson [**31**]. He emphasized that it is convenient to express the relation between D and ${}^VD$ in the definition of a harmonic bundle by adding a parameter z and to express it with the notion of a z-connection, already suggested by P. Deligne.

By a real-analytic (resp. C^∞) *variation of twistor structure* on M we mean the data of a triple $(\mathscr{H}', \mathscr{H}'', \mathscr{C}_{\mathbf{S}})$ defining a real-analytic (resp. C^∞) family of holomorphic bundles on $\mathbb{P}^1$ as above, such that each of the holomorphic bundles $\mathscr{H}', \mathscr{H}''$ is equipped with a *relative* holomorphic connection

$$\nabla : \mathscr{H}'('') \longrightarrow \frac{1}{z}\Omega^1_{M\times\Omega_0/\Omega_0} \otimes_{\mathscr{O}_{M\times\Omega_0}} \mathscr{H}'('')$$

which has a pole along $z = 0$ and is integrable. Moreover, the pairing $\mathscr{C}_{\mathbf{S}}$ has to be compatible (in the usual sense) with the connections, i.e.,

$$d'\mathscr{C}_{\mathbf{S}}(m', \overline{m''}) = \mathscr{C}_{\mathbf{S}}(\nabla m', \overline{m''}) \quad \text{and} \quad d''\mathscr{C}_{\mathbf{S}}(m', \overline{m''}) = \mathscr{C}_{\mathbf{S}}(m', \overline{\nabla m''}).$$

Let us note that we can define $\overline{\nabla}$ as

$$\overline{\nabla}:\overline{\mathscr{H}''}\longrightarrow z\Omega^1_{\overline{M}\times\Omega_\infty/\Omega_\infty}\otimes_{\mathscr{O}_{\overline{M}\times\Omega_\infty}}\overline{\mathscr{H}''}.$$

If we regard $\mathscr{C}_{\mathbf{S}}$ as a $C^{*,\mathrm{an}}$-linear isomorphism

$$\mathscr{C}_{\mathbf{S}}:\mathscr{C}^{*,\mathrm{an}}_{M_{\mathbb{R}}\times\mathbf{S}}\otimes_{\mathscr{O}_{\overline{M}\times\mathbf{S}}}\overline{\mathscr{H}''_{|\mathbf{S}}}\xrightarrow{\ \sim\ }\mathscr{C}^{*,\mathrm{an}}_{M_{\mathbb{R}}\times\mathbf{S}}\otimes_{\mathscr{O}_{M\times\mathbf{S}}}\mathscr{H}'^{\vee}_{|\mathbf{S}},\tag{2.3}$$

the compatibility with ∇ means that $\mathscr{C}_{\mathbf{S}}$ is compatible with the connection $d'+\overline{\nabla}$ on the left-hand term and $\nabla^\vee+d''$ on the right-hand term, where d',d'' are the standard differentials with respect to M alone. The adjoint $(\mathscr{H}',\mathscr{H}'',\mathscr{C}_{\mathbf{S}})^\dagger$ is defined as $(\mathscr{H}'',\mathscr{H}',\mathscr{C}^\dagger_{\mathbf{S}})$, with

$$\mathscr{C}^\dagger_{\mathbf{S}}(m'',\overline{m'}):=\overline{\mathscr{C}_{\mathbf{S}}(m',\overline{m''})}.$$

With respect to (2.3), we can write $\mathscr{C}^\dagger_{\mathbf{S}}=\overline{\mathscr{C}_{\mathbf{S}}}^\vee$.

We say that the variation is

- *Hermitian* if $\mathscr{H}''=\mathscr{H}'$ and $\mathscr{C}_{\mathbf{S}}$ is "Hermitian", i.e., $\mathscr{C}^\dagger_{\mathbf{S}}=\mathscr{C}_{\mathbf{S}}$;
- *pure of weight* 0 if the restriction to each $x\in M$ defines a *trivial* holomorphic bundle on $\mathbb{P}^1$.

LEMMA 2.4 (C. Simpson [**31**]). *We have an equivalence between variations of Hermitian pure twistor structures of weight* 0 *and harmonic Higgs bundles, by taking* $\mathbb{P}^1$*-global sections.*

SKETCH OF PROOF (*see* [**31**] *for details, cf. also* [**20**, Lemma 2.2.2]). Let us start with a variation of Hermitian pure twistor structure of weight 0, that we denote by $(\mathscr{H}',\nabla,\mathscr{C}_{\mathbf{S}})$. According to Proposition 2.2, the $C^{*,\mathrm{an}}$ bundle $\mathscr{H}$ it defines is the pull-back by p of some C^* bundle, whose *conjugate* bundle we now denote by H. By definition, $\overline{H}$ is naturally included in the $C^{*,\mathrm{an}}$ bundles associated to $\mathscr{H}'^\vee$ and $\overline{\mathscr{H}'}$, and the inclusions are compatible with $\mathscr{C}_{\mathbf{S}}$. Taking conjugates, H is naturally included in the $C^{*,\mathrm{an}}$ bundles associated to $\overline{\mathscr{H}'^\vee}$ and $\mathscr{H}'$. The natural duality pairing $\mathscr{H}'\otimes\mathscr{H}'^\vee\to\mathscr{O}_{M\times\Omega_0}$ induces a pairing $h:H\otimes_{\mathscr{C}^*_{M_{\mathbb{R}}}}\overline{H}\to\mathscr{C}^*_{M_{\mathbb{R}}}$. That $\mathscr{C}_{\mathbf{S}}$ is Hermitian implies that h is Hermitian in the usual sense.

One defines E as the restriction of $\mathscr{H}'$ at $z=0$, Φ as the residue of ∇ along $z=0$, $D+\Phi+\Phi^\dagger$ as the restriction to $z=1$ of $\nabla+d''$.

Conversely, let us be given a harmonic Higgs bundle (E,h,Φ). Let H be the C^∞-bundle associated with E and let $\mathscr{H}_0=\mathscr{C}^{*,\mathrm{an}}_{M_{\mathbb{R}}\times\Omega_0}\otimes_{p^{-1}\mathscr{C}^*_{M_{\mathbb{R}}}}H$ be the pull-back of H by $p:M\times\Omega_0\to M$ (where, as above, $*$ is for $\mathbb{R}$-an or ∞, depending on whether h is real analytic or C^∞). This bundle is equipped with the d''-operator $d''+z\Phi^\dagger$, which defines a holomorphic bundle $\mathscr{H}'=\operatorname{Ker}(d''+z\Phi^\dagger)$, equipped with the meromorphic connection $\nabla=(D'+\Phi/z)_{|\mathscr{H}'}$. If we regard h as a $\mathscr{C}^*_{M_{\mathbb{R}}}$-linear morphism $H\otimes_{\mathscr{C}^*_{M_{\mathbb{R}}}}\overline{H}\to\mathscr{C}^*_{M_{\mathbb{R}}}$, then we can extend h as

$$h_{\mathbf{S}}:\mathscr{H}_{0|M_{\mathbb{R}}\times\mathbf{S}}\otimes_{\mathscr{C}^{*,\mathrm{an}}_{M_{\mathbb{R}}\times\mathbf{S}}}\overline{\mathscr{H}_{0|M_{\mathbb{R}}\times\mathbf{S}}}\longrightarrow\mathscr{C}^{*,\mathrm{an}}_{M_{\mathbb{R}}\times\mathbf{S}}\tag{2.5}$$

by $\mathscr{C}^{*,\mathrm{an}}_{M_{\mathbb{R}}\times\mathbf{S}}$-linearity and define $\mathscr{C}_{\mathbf{S}}$ to be its restriction to $\mathscr{H}'_{|\mathbf{S}}\otimes_{\mathscr{O}_{\mathbf{S}}}\overline{\mathscr{H}'_{|\mathbf{S}}}$. □

DEFINITION 2.6. Let $(\mathscr{H}',\nabla,\mathscr{C}_{\mathbf{S}})$ be a variation of Hermitian pure twistor structure of weight 0. We say that it is a *variation of polarized pure twistor structure of weight* 0 if h is positive definite.

2.c. Supplementary structures on a variation of twistor structure. We introduce new structures in a way analogous to that of §1.c. We assume that $(\mathscr{H}', \nabla, \mathscr{C}_{\mathbf{S}})$ is a variation of Hermitian pure twistor structure of weight 0.

Real structure. Let $\mathscr{P}$ be a nondegenerate $\mathscr{O}_{M\times\Omega_0}$-linear morphism $\mathscr{H}'\otimes_{\mathscr{O}_{M\times\Omega_0}} \iota^*\mathscr{H}' \to \mathscr{O}_{M\times\Omega_0}$, or in other words an isomorphism $\mathscr{H}'^\vee \xrightarrow{\sim} \iota^*\mathscr{H}'$. Let us notice that, on $\mathbf{S}$, σ and ι coincide. Therefore, restricting $\mathscr{P}$ to $M\times\mathbf{S}$ and composing with $\mathscr{C}_{\mathbf{S}}$ (cf. (2.3)) we get an isomorphism

$$\overline{\mathscr{K}_{\mathbf{S}}} : \mathscr{C}^{*,\mathbb{R}\text{-an}}_{M_{\mathbb{R}}\times\mathbf{S}} \otimes_{\mathscr{O}_{\overline{M}\times\mathbf{S}}} \overline{\mathscr{H}'_{|\mathbf{S}}} \xrightarrow{\sim} \mathscr{C}^{*,\mathbb{R}\text{-an}}_{M_{\mathbb{R}}\times\mathbf{S}} \otimes_{\mathscr{O}_{M\times\mathbf{S}}} \mathscr{H}'_{|\mathbf{S}}. \tag{2.7}$$

(Let us note that we use the usual conjugation functor here.)

We say that $\mathscr{P}$ is a *real structure* if the following holds:

(1) $\mathscr{K}_{\mathbf{S}}\overline{\mathscr{K}_{\mathbf{S}}} = \mathrm{Id}$,
(2) $\mathscr{P}$ is ι-Hermitian, i.e., $\mathscr{P} = \iota^*\mathscr{P}^\vee$,
(3) $\mathscr{P}$ is compatible with ∇ and $\iota^*\nabla^\vee$.

REMARK 2.8. We could have defined the real structure directly from $\mathscr{K}_{\mathbf{S}}$, but we emphasize here that the corresponding $\mathscr{P}_{|\mathbf{S}}$ should be the restriction of a ι-sesquilinear form $\mathscr{P}$ on $\mathscr{H}'$. In the integrable case below, $\mathscr{P}_{|\mathbf{S}}$ defines $\mathscr{P}$ on $M\times\Omega_0^*$, so our assumption here means that we are given an extension of $\mathscr{P}_{|\Omega_0^*}$ at $z=0$.

LEMMA 2.9. *The equivalence of Lemma* 2.4 *specializes to objects with real structures.* □

Integrability. A variation of twistor structure $(\mathscr{H}', \mathscr{H}'', \mathscr{C}_{\mathbf{S}})$ is *integrable* if the relative connection ∇ on $\mathscr{H}'$ comes from an absolute connection, also denoted by ∇, which has Poincaré rank one (cf. [**20**, Chap. 7]). In other words, $z\nabla$ should be an integrable meromorphic connection on $\mathscr{H}', \mathscr{H}''$ with a logarithmic pole along $z=0$. We also ask for a supplementary compatibility property of the absolute connection with the pairing in the following way:

$$z\nabla_{\partial_z}\mathscr{C}_{\mathbf{S}}(m', \overline{m''}) = \mathscr{C}_{\mathbf{S}}(z\nabla_{\partial_z}m', \overline{m''}) - \mathscr{C}_{\mathbf{S}}(m', \overline{z\nabla_{\partial_z}m''}).$$

EXAMPLE 2.10. Assume M is a point. A Hermitian pure twistor structure of weight 0 corresponds, according to Lemma 2.4, to a vector space H with a nondegenerate Hermitian form h. The integrability condition consists in giving a meromorphic connection on the trivial vector bundle $\mathscr{O}_{\mathbb{P}^1}\otimes H$ which has a pole of order at most 2 at 0 and ∞ and which satisfies a compatibility condition with h. The connection can be written as $d + (U/z - Q - zU')dz/z$, where U, U' and Q are endomorphisms of H. Compatibility with the pairing $\mathscr{C}_{\mathbf{S}}$ now translates as $U' = U^\dagger$ and $Q = Q^\dagger$, where † means adjoint with respect to h. Assume that h is positive definite. Then Q is semi-simple and its eigenvalues are real, so that H decomposes with respect to the eigenvalues of Q as $\bigoplus_{\alpha\in[0,1[}\bigoplus_{p\in\mathbb{Z}} H_{\alpha+p}$. If we set $H^{p,-p} = \bigoplus_{\alpha\in[0,1[} H_{\alpha+p}$, we get a polarized complex Hodge structure of weight 0 on H. The endomorphism Q induces a semi-simple endomorphism on each $H^{p,-p}$ with eigenvalues in $[p, p+1[$. We are also left with an endomorphism U of H with no particular relation with the polarized Hodge structure.

In case we start with a polarized complex Hodge structure of weight 0 on H, we choose $U=0$ and $Q = \bigoplus_{p\in\mathbb{Z}} p\,\mathrm{Id}_{H^{p,-p}}$ to recover an integrable polarized pure twistor structure of weight 0 (cf. Example 1.34).

Lemma 2.4 is now made precise in the case of integrable twistor structures.

LEMMA 2.11 (C. Hertling [7], cf. also [**20**, Cor. 7.2.6]). *The equivalence of Lemma* 2.4 *specializes to an equivalence between integrable variations of Hermitian pure twistor structures of weight* 0 *and integrable harmonic Higgs bundles, i.e., harmonic Higgs bundles equipped with endomorphisms* $\mathscr{U}, \mathscr{Q}$ *satisfying* (1.12)–(1.16). □

REMARK 2.12. The restriction to each point of M of an integrable variation of polarized pure twistor structures of weight 0 is an integrable polarized pure twistor structure of weight 0. Therefore, it is in particular a polarized complex Hodge structure plus supplementary data (Example 2.10). Nevertheless, it may not correspond, in general, to a variation of polarized Hodge structure of weight 0.

A sufficient condition to get such a variation of Hodge structure is that $\mathscr{U} \equiv 0$: in such a case, according to (1.13) and (1.16), $\mathscr{Q}$ is annihilated by D and its eigenvalues are constant (use an orthonormal frame for h in which the matrix of $\mathscr{Q}$ is diagonal); the eigenspace decomposition of H is left stable by D and is shifted by one by Φ, according to (1.15).

Such a condition is satisfied if we assume that M is compact Kähler, so that the notion of *integrable* variation of polarized twistor structure of weight 0 is a special case of that of variation of polarized complex Hodge structure (cf. [**20**, Cor. 7.2.8]).

The same holds if M is a punctured compact Riemann surface and the behaviour at the punctures is *tame* (cf. [**29**] for this notion): this has been proved by C. Hertling and Ch. Sevenheck.[4]

One can expect that such a result holds in the higher dimensional case, if we keep the tameness assumption.

Therefore, when M is quasi-projective, the notion of *integrable* variation of polarized twistor structure of weight 0 should be a new object only when a nontame (i.e., wild) behaviour occurs at infinity on M. In other words, this notion can be regarded as the right replacement of that of a polarized complex variation of Hodge structure of weight 0 when one wants to take into account a wild behaviour at infinity.

Let us now come back to the general notion of an integrable Hermitian variation of twistor structure. One can remark that, as ∇ is integrable, the restriction of $(\mathscr{H}', \nabla)$ to $M \times \Omega_0^*$ (where $\Omega_0^* := \Omega_0 \smallsetminus \{0\}$) is determined up to isomorphism by giving the corresponding locally constant sheaf $\operatorname{Ker} \nabla$. Similarly, on $M_{\mathbb{R}} \times \mathbf{S}$, the pairing $\mathscr{C}_{\mathbf{S}}$ is determined by its restriction to the corresponding local systems. Therefore, giving an integrable variation of twistor structure with a nondegenerate Hermitian pairing is equivalent to giving

(1) a holomorphic vector bundle $\mathscr{H}'$ on $M \times \Omega_0$ with a meromorphic connection ∇ having Poincaré rank one along $M \times \{0\}$ (i.e., $z\nabla$ has at most a logarithmic pole along $M \times \{0\}$),
(2) a nondegenerate pairing $\mathscr{C}_{\mathbf{S}} : \operatorname{Ker} \nabla_{|M\times\mathbf{S}} \otimes_{\mathbb{C}} \sigma^{-1}\overline{\operatorname{Ker} \nabla_{|M\times\mathbf{S}}} \to \mathbb{C}_{M\times\mathbf{S}}$.

The second part of the data is of a purely topological nature, as $\operatorname{Ker} \nabla_{|M\times\Omega_0^*}$ is a locally constant sheaf. Moreover, recall that ι coincides with σ on $\mathbf{S}$. As a direct application of Proposition 2.2 we get:

[4]This was obtained in [**20**, Cor. 7.2.10] as a consequence of the conjecture [**20**, Conj. 7.2.9]. Ch. Sevenheck sent a proof of this conjecture due to C. Hertling and himself to the author.

COROLLARY 2.13 (C. Hertling [7]). *Let us assume that M is 1-connected and let $x^o \in M$. Let $(\mathscr{H}', \nabla)$ be a holomorphic bundle on $M \times \Omega_0$ with a meromorphic connection ∇ having Poincaré rank one along $M \times \{0\}$, set $\mathscr{L}_{|\mathbf{S}} = \operatorname{Ker} \nabla_{|M\times\mathbf{S}}$, $\mathscr{L}^o_{|\mathbf{S}} = \mathscr{L}_{|\{x^o\}\times\mathbf{S}}$ and let $\mathscr{C}^o_{\mathbf{S}} : \mathscr{L}^o_{|\mathbf{S}} \otimes_{\mathbb{C}} \iota^{-1}\overline{\mathscr{L}^o_{|\mathbf{S}}} \to \mathbb{C}_{\mathbf{S}}$ be a nondegenerate Hermitian pairing. Let $\mathscr{C}_{\mathbf{S}} : \mathscr{L}_{|\mathbf{S}} \otimes_{\mathbb{C}} \iota^{-1}\overline{\mathscr{L}_{|\mathbf{S}}} \to \mathbb{C}_{M\times\mathbf{S}}$ be the unique nondegenerate Hermitian pairing extending $\mathscr{C}^o_{\mathbf{S}}$. It defines a real-analytic family of holomorphic bundles parametrized by M. Let us moreover assume that the twistor structure at x^o corresponding to these data is Hermitian and pure of weight 0.*

Then there exists a (possibly empty) real analytic subvariety $\Theta \not\ni x^o$ of M such that, on the connected component of $M \smallsetminus \Theta$ containing x^o, the Hermitian variation of twistor structure $(\mathscr{H}', \nabla, \mathscr{C}_{\mathbf{S}})$ is pure of weight 0. If moreover it is polarized at x^o, it is polarized on this connected component. □

REMARKS 2.14.

(1) This corollary shows that, given $(\mathscr{H}', \nabla)$ having Poincaré rank one along $z = 0$, it underlies an integrable variation of polarized pure twistor structure in some neighbourhood of x^o if and only if the fibre at x^o can be endowed with the structure of an integrable polarized twistor structure of weight 0.

(2) Let us start with an integrable pure twistor structure $(\mathscr{H}'^o, \nabla^o, \mathscr{C}^o_{\mathbf{S}})$ of weight 0 and let us assume that the residue of $z\nabla^o$ at $z = 0$ is *regular semi-simple* (i.e., has pairwise distinct eigenvalues). Then (cf. [13], see also [17, Th. III.2.10]) there exists a universal deformation $(\mathscr{H}', \nabla)$ of $(\mathscr{H}'^o, \nabla^o)$ parametrized by the universal covering M of $\mathbb{C}^d \smallsetminus$ diagonals, if d is the rank of $\mathscr{H}'^o$, and $\mathscr{C}^o_{\mathbf{S}}$ extends in a unique way to a Hermitian pairing $\mathscr{C}_{\mathbf{S}}$. We can apply the conclusion of Corollary 2.13 to the object $(\mathscr{H}', \nabla, \mathscr{C}_{\mathbf{S}})$.

(3) Let us denote by L^o the space of global multivalued sections of $\mathscr{L}^o$ (i.e., of global multivalued ∇^o-horizontal sections of $\mathscr{H}'^o_{|\{x^o\}\times\mathbb{C}^*}$). Giving $\mathscr{C}^o_{\mathbf{S}}$ is equivalent (after choosing the lifting $\zeta \mapsto \zeta + \pi i$ of $\iota : z = e^\zeta \mapsto -z$) to giving a nondegenerate Hermitian pairing $C^o : L^o \otimes_{\mathbb{C}} \overline{L^o} \to \mathbb{C}$. However, the purity condition and the positivity condition are not conditions on C^o (i.e., on ∇^o-horizontal sections). They depend strongly on the holomorphic bundle $\mathscr{H}'^o$. More precisely, the purity and positivity conditions are equivalent to the existence of a global holomorphic frame $\boldsymbol{\varepsilon}^o$ of $\mathscr{H}'^o$ which is orthonormal with respect to $\mathscr{C}^o_{\mathbf{S}} : \mathscr{H}'^o_{|\mathbf{S}} \otimes_{\mathscr{O}_{\mathbf{S}}} \overline{\mathscr{H}'^o_{|\mathbf{S}}} \to \mathscr{O}_{\mathbf{S}}$. In other words, for any $z_o \in \mathbf{S}$, the matrix $\big(\mathscr{C}^o_{\mathbf{S}}(\varepsilon_i^{o,(z_o)}, \varepsilon_j^{o,(-z_o)})\big)$ should be equal to the identity matrix (cf. [20, Rem. 2.1.4 and 2.2.3]).

Integrability with real structure. Let $(\mathscr{H}', \nabla, \mathscr{C}_{\mathbf{S}}, \mathscr{P})$ be a variation of Hermitian pure twistor structure of weight 0 with real structure $\mathscr{P}$. We say that it is *integrable* if $(\mathscr{H}', \nabla, \mathscr{C}_{\mathbf{S}})$ is integrable, as defined above, and $\mathscr{P}$ is compatible with the absolute connections (not only the relative ones).

The conjunction of Lemmas 2.9 and 2.11 holds. Similarly, starting with an *integrable* $(\mathscr{H}', \nabla, \mathscr{P})$, we get as in corollary 2.13 that, if M is 1-connected, an integrable variation of real Hermitian twistor structure on M with underlying $(\mathscr{H}', \nabla, \mathscr{P})$ is determined by a real structure $\mathscr{K}^o_{\mathbf{S}} : \mathscr{L}^o_{|\mathbf{S}} \xrightarrow{\sim} \overline{\mathscr{L}^o_{|\mathbf{S}}}$ at x^o, and if it is pure of weight 0 (resp. pure of weight 0 and polarized) at x^o, it is so on $M \smallsetminus \Theta$.

Potentiality. We now consider the analogue of the potentiality property. Let $(\mathscr{H}', \nabla, \mathscr{C}_{\mathbf{S}})$ be a variation of Hermitian pure twistor structure of weight 0. By a *potential* we will mean an isomorphism of holomorphic bundles $p^*E \xrightarrow{\sim} \mathscr{H}'$ inducing the identity when restricted to $z = 0$. Recall that $\mathscr{H}'$ is the holomorphic bundle $(p^*H, d'' + z\Phi^\dagger)$. On the other hand, $p^*E = (p^*H, d'')$. Therefore, such an isomorphism can be expanded as $\mathrm{Id}_H + zB_1 + \cdots + z^k B_k + \cdots$, where the B_k are C^∞ endomorphisms of H and the condition is

$$(\mathrm{Id}_H + zB_1 + \cdots)^{-1} \circ d'' \circ (\mathrm{Id}_H + zB_1 + \cdots) = d'' + z\Phi^\dagger.$$

In particular, the order-one condition is $d''B_1 = \Phi^\dagger$, that is, the potentiality condition for the harmonic Higgs bundle corresponding to $(\mathscr{H}', \nabla, \mathscr{C}_{\mathbf{S}})$. But the potentiality condition for the variation of twistor structure is stronger than that for harmonic Higgs bundles.

2.d. A criterion for a Saito structure to be harmonic. Let us consider a Saito structure $(E, \nabla, g, \Phi, \mathscr{U}, \mathscr{V})$ of weight 0. We consider on $M \times \Omega_0$ the pull-back bundle p^*E via $p : M \times \Omega_0 \to M$, equipped with the meromorphic connection

$$\nabla = p^*\nabla + \frac{\Phi}{z} + \Big(\frac{\mathscr{U}}{z} - \mathscr{V}\Big)\frac{dz}{z}. \tag{2.15}$$

The bilinear form g then defines then a nondegenerate bilinear pairing

$$\mathscr{P} : (p^*E, \nabla) \otimes_{\mathscr{O}_{M\times\Omega_0}} \iota^*(p^*E, \nabla) \longrightarrow (\mathscr{O}_{M\times\Omega_0}, d), \tag{2.16}$$

which is Hermitian with respect to the involution ι. If we denote by $\mathscr{L}$ the local system $\operatorname{Ker}\nabla_{|M\times\mathbb{C}^*}$ and by $\mathscr{L}_{|\mathbf{S}}$ its restriction to $M \times \mathbf{S}$, where $\mathbf{S} = \{z \mid |z| = 1\}$, we get a nondegenerate pairing $\mathscr{P}_{|\mathbf{S}} : \mathscr{L}_{|\mathbf{S}} \otimes_{\mathbb{C}} \iota^{-1}\mathscr{L}_{|\mathbf{S}} \to \mathbb{C}_{M\times\mathbf{S}}$.

COROLLARY 2.17 (cf. [**7**, Th. 5.15]). *In this situation, assume moreover that there exists a real structure* $\mathscr{K}_{\mathbf{S}}$ *on* $\mathscr{L}_{|\mathbf{S}}$*, that is, an isomorphism* $\mathscr{K}_{\mathbf{S}} : \mathscr{L}_{|\mathbf{S}} \to \overline{\mathscr{L}_{|\mathbf{S}}}$ *such that* $\mathscr{K}_{\mathbf{S}}\overline{\mathscr{K}_{\mathbf{S}}} = \mathrm{Id}$*, such that, setting* $\mathscr{C}_{\mathbf{S}}(\bullet, \mathscr{K}_{\mathbf{S}}\bullet) = \mathscr{P}(\bullet, \bullet)$*,* $(p^*E, \nabla, \mathscr{C}_{\mathbf{S}})$ *defines an integrable variation of Hermitian pure twistor structure of weight* 0*. Then the Saito structure* $(E, \nabla, g, \Phi, \mathscr{U}, \mathscr{V})$ *is harmonic.*

PROOF. As the variation of twistor structure is defined on p^*E, we have a given identification $\mathscr{H}' = p^*E$ and the potentiality property is fulfilled. The corollary is then a consequence of the extensions of Lemma 2.4 given above. □

REMARKS 2.18.

(1) By the corollaries 2.13 and 2.17, if there exists $x^o \in M$ such that the data $(\mathscr{O}_{\Omega_0} \otimes_{\mathbb{C}} E^o, d + (\mathscr{U}^o/z - \mathscr{V})dz/z, \mathscr{P}^o, \mathscr{K}^o_{\mathbf{S}})$ defines a pure twistor structure of weight 0, then $(E, \nabla, g, \Phi, \mathscr{U}, \mathscr{V})$ is harmonic on some neighbourhood of x^o. The same conclusion holds with the polarization property.

(2) This proposition can be applied to the Saito structure attached to a Frobenius manifold, and gives a criterion for a Frobenius manifold to be harmonic.

3. Twistor structures through Fourier-Laplace transform

According to Corollary 2.13, an important step in constructing an integrable variation of pure twistor structure of weight 0 is to construct it at one point, i.e., to construct an integrable pure twistor structure of weight 0. Checking that $(\mathscr{H}'^o, \nabla^o, \mathscr{C}^o_{\mathbf{S}})$ is such an object is not straightforward. We indicate in this section how to produce such an integrable pure twistor structure of weight 0 starting from a variation of polarized Hodge structure of weight 0 on a punctured Riemann sphere, by using the Fourier-Laplace transform. This will produce an integrable polarized pure twistor structure of weight 0: the polarization property is essential here, to get the purity property.

3.a. Laplace transform. Let V^{an} be a holomorphic vector bundle on the punctured affine line $\mathbb{A}^1 \smallsetminus P$, where $P = \{p_1, \dots, p_n\}$ is a finite set of points. We assume that V^{an} is equipped with a holomorphic connection ∇^{an}. By a classical result of Deligne, there exists a unique meromorphic extension (V, ∇) of $(V^{\mathrm{an}}, \nabla^{\mathrm{an}})$ on $\mathbb{P}^1$ such that ∇ has only regular singularities at $\{p_1, \dots, p_n, \infty\}$. Taking global sections of V on $\mathbb{P}^1$ gives a $\mathbb{C}[t]$-module M, where t denotes a chosen coordinate on $\mathbb{A}^1$. Moreover, the connection ∇ enables one to extend the $\mathbb{C}[t]$ action to an action of the Weyl algebra $\mathbb{C}[t]\langle\partial_t\rangle$ by setting $\partial_t m := \nabla_{\partial_t} m$.

The Laplace transform ${}^F\!M$ is the $\mathbb{C}$-vector space M equipped with the action of the Weyl algebra $\mathbb{C}[\tau]\langle\partial_\tau\rangle$ defined by the formulas

$$\tau m := \partial_t m, \quad \partial_\tau m = -tm.$$

We will be mainly interested in the behaviour of ${}^F\!M$ near $\tau = \infty$ and we will set $z = 1/\tau$, so that $z^2\partial_z = t$. It is known that $G := \mathbb{C}[\tau, \tau^{-1}] \otimes_{\mathbb{C}[\tau]} {}^F\!M$ is a free $\mathbb{C}[\tau, \tau^{-1}]$-module, or equivalently a free $\mathbb{C}[z, z^{-1}]$-module, i.e., an algebraic vector bundle on the torus $\mathbb{C}^*$. Moreover, the action of $t = -\partial_\tau = z^2\partial_z$ defines a connection ${}^F\nabla$ on it, which is known to have a regular singularity at $\tau = 0$ and, in general, an irregular one at $z = 0$ (see e.g. [**14**] or [**17**, Chap. V]).

Assume now that we are given a $\mathbb{C}[t]$-submodule of finite type $L \subset M$ such that $M = L + \partial_t L + \cdots + \partial_t^k L + \cdots$. Then we can regard L as a subspace of ${}^F\!M$ and, taking localization, we can consider its image L' in G. Let us then set $G_0^{(L)} = L' + zL' + \cdots + z^k L' + \cdots \subset G$. By construction, $G_0^{(L)}$ is a $\mathbb{C}[z]$-submodule of G and, L being a $\mathbb{C}[t]$-module, it is naturally equipped with a compatible action of $z^2\partial_z$. One can show (see e.g. [**17**, §V.2.c]) that, because ∇ has a regular singularity at infinity, $G_0^{(L)}$ is a $\mathbb{C}[z]$-module of finite rank. As it is equipped with an action of $z^2\partial_z$ (because L' is so), and as $G = \mathbb{C}[z, z^{-1}] \otimes_{\mathbb{C}[z]} G_0^{(L)}$, we find that the connection ${}^F\nabla$ on G has at most Poincaré rank one at $z = 0$ (in any $\mathbb{C}[z]$-basis of $G_0^{(L)}$, the matrix of ${}^F\nabla$ has a pole of order $\leqslant 1$ because $G_0^{(L)}$ is stable under $z^2\partial_z$).

We will define $(\mathscr{H}'^o, \nabla^o)$ to be the analytization of $(G_0^{(L)}, {}^F\nabla)$.

3.b. Fourier transform of a sesquilinear pairing. Let us assume that the holomorphic bundle V^{an} is equipped with a ∇-*horizontal* nondegenerate Hermitian pairing $\widetilde{k}$ (but not necessarily positive definite), that is, a $\mathscr{O}_{\mathbb{A}^1 \smallsetminus P} \otimes_{\mathbb{C}} \mathscr{O}_{\overline{\mathbb{A}^1} \smallsetminus P}$-linear morphism

$$\widetilde{k} : V^{\mathrm{an}} \otimes_{\mathbb{C}} \overline{V^{\mathrm{an}}} \longrightarrow \mathscr{C}^\infty_{\mathbb{A}^1 \smallsetminus P}$$

such that $d'\widetilde{k}(v_1, \overline{v_2}) = \widetilde{k}(\nabla v_1, \overline{v_2})$ and $d''\widetilde{k}(v_1, \overline{v_2}) = \widetilde{k}(v_1, \overline{\nabla v_2})$. Let us moreover assume that $\widetilde{k}$ extends to $M \otimes_{\mathbb{C}} \overline{M}$ as a Hermitian sesquilinear pairing k with values in the Schwartz space $\mathscr{S}'(\mathbb{A}^1)$ of tempered distributions on $\mathbb{A}^1$ (i.e., distributions on $\mathbb{A}^1$ having moderate growth at infinity), in a way compatible with the natural action of $\mathbb{C}[t]\langle\partial_t\rangle \otimes_{\mathbb{C}} \overline{\mathbb{C}[t]\langle\partial_t\rangle}$ on $\mathscr{S}'(\mathbb{A}^1)$.

REMARK 3.1 (A criterion for k to be nondegenerate). While the nondegeneracy property of $\widetilde{k}$ is easy to define (by restricting to points of $\mathbb{A}^1 \smallsetminus P$), that of k needs some care. The first step consists in considering the $\mathbb{C}[t]\langle\partial_t\rangle$-module $M^\dagger := \mathrm{Hom}_{\mathbb{C}[\overline{t}]\langle\overline{\partial}_t\rangle}\big(\overline{M}, \mathscr{S}'(\mathbb{A}^1)\big)$, where the $\mathbb{C}[t]\langle\partial_t\rangle$-structure comes from that on $\mathscr{S}'(\mathbb{A}^1)$. It is called the *Hermitian dual* of M and was introduced by M. Kashiwara in [**10**]. Then giving k is equivalent to giving a $\mathbb{C}[t]\langle\partial_t\rangle$-linear morphism $M \to M^\dagger$. We say that k is *nondegenerate* if this morphism is an isomorphism.

In general, if M is holonomic, then $M^\dagger$ is so: this is proved in [**10**] if M has only regular singularities and in [**18**] in general (but on a manifold of dimension one, $\mathbb{A}^1$ here).

Moreover, the previous construction can be localized in the analytic topology of $\mathbb{P}^1$; in particular, we replace $\mathscr{S}'(\mathbb{A}^1)$ with the sheaf of distributions on $\mathbb{A}^1$ having moderate growth at infinity. Then, on restriction to $(\mathbb{A}^1 \smallsetminus P)^{\mathrm{an}}$, the nondegeneracy of k in the sense of $\mathscr{D}$-modules is equivalent to that of $\widetilde{k}$ in the usual sense. In other words, saying that $\widetilde{k}$ is nondegenerate is equivalent to saying that the kernel and the cokernel of $k : M \to M^\dagger$ are supported on P.

At this point, let us recall a definition: we say that M is a *minimal extension* (one also says an intermediate extension, or a middle extension, with reference to perverse sheaves) if M has neither a nonzero submodule nor a nonzero quotient module supported on P.

As the construction † is functorial (in a contravariant way) for $\mathbb{C}[t]\langle\partial_t\rangle$-modules (see loc. cit.), M is a minimal extension iff $M^\dagger$ is so. In this case, the kernel and cokernel of k, if supported at P, must be zero. In other words, *if $\widetilde{k}$ is nondegenerate and is extended to a sesquilinear pairing k on the minimal extension M of (V, ∇), then k is also nondegenerate.*

This motivates the change of the definition of M in §3.a, and the replacement of the $\mathbb{C}[t]\langle\partial_t\rangle$-module naturally associated to M with its minimal extension.

Fourier transform of k. We then denote by ${}^F\!k$ the composition of k with the Fourier transform of tempered distributions with kernel $\exp(\overline{t\tau} - t\tau)\frac{i}{2\pi} dt \wedge d\overline{t}$. By standard properties of the Fourier transform, ${}^F\!k$ is a $\mathbb{C}[\tau]\langle\partial_\tau\rangle \otimes_{\mathbb{C}} \overline{\mathbb{C}[\tau]\langle\partial_\tau\rangle}$-linear morphism on ${}^F\!M \otimes_{\mathbb{C}} \iota^* \overline{{}^F\!M}$ (note the ι^* here, due to $\overline{\exp(\overline{t\tau} - t\tau)} = \exp -(\overline{t\tau} - t\tau)$).

Considering the Hermitian duality functor, we see that, composing with the Fourier transform $\mathscr{S}'(\mathbb{A}^1_t) \to \mathscr{S}'(\mathbb{A}^1_\tau)$ gives an isomorphism of $\mathbb{C}$-vector spaces

$$\mathrm{Hom}_{\mathbb{C}[\overline{t}]\langle\overline{\partial}_t\rangle}\big(\overline{M}, \mathscr{S}'(\mathbb{A}^1_t)\big) \longrightarrow \mathrm{Hom}_{\mathbb{C}[\overline{\tau}]\langle\overline{\partial}_\tau\rangle}\big(\overline{{}^F\!M}, \mathscr{S}'(\mathbb{A}^1_\tau)\big)$$

which is in fact a canonical identification $\iota^*({}^F\!M)^\dagger = {}^F(M^\dagger)$.

If we regard k as a linear morphism $M \to M^\dagger$, taking its Laplace transform gives a linear morphism ${}^F\!M \to {}^F(M^\dagger)$, and using Fourier identification above gives ${}^F\!k : {}^F\!M \to \iota^*({}^F\!M)^\dagger$. In particular, if k is nondegenerate, then so is ${}^F\!k$.

Twistorization. In particular, restricting to $\tau \neq 0$ and sheafifying in the analytic topology gives a ∇^o-flat nondegenerate Hermitian pairing (with respect to the composition of the involution ι and the conjugation) $\mathscr{H}'^o_{|\mathbb{C}^*} \otimes_{\mathbb{C}} \iota^* \overline{\mathscr{H}'^o_{|\mathbb{C}^*}} \to \mathscr{C}^\infty_{\mathbb{C}^*}$. By flatness, giving such a pairing is equivalent to giving the pairing induced on ∇^o-horizontal sections

$${}^F\!k : \mathscr{L}^o \otimes_{\mathbb{C}} \iota^{-1}\overline{\mathscr{L}^o} \longrightarrow \mathbb{C}_{\mathbb{C}^*}.$$

Finally, as $\mathbf{S} \hookrightarrow \mathbb{C}^*$ is a homotopy equivalence, giving this pairing is equivalent to giving the restricted pairing

$$\mathscr{C}^o_{\mathbf{S}} := {}^F k_{|\mathbf{S}} : \mathscr{L}^o_{|\mathbf{S}} \otimes_{\mathbb{C}} \iota^{-1}\overline{\mathscr{L}^o_{|\mathbf{S}}} \longrightarrow \mathbb{C}_{\mathbf{S}}.$$

In conclusion (and extending slightly the previous construction), *given* (M, L, k), *where* M *is a holonomic* $\mathbb{C}[t]\langle\partial_t\rangle$*-module having a regular singularity at infinity and which is a minimal extension at* P, L *is a* $\mathbb{C}[t]$*-submodule of finite type of* M *which generates it over* $\mathbb{C}[t]\langle\partial_t\rangle$, *and* k *is a nondegenerate sesquilinear Hermitian pairing on* M *with values in* $\mathscr{S}'(\mathbb{A}^1)$, *we obtain, by Fourier-Laplace transform, a twistor structure* $(\mathscr{H}'^o, \nabla^o, \mathscr{C}^o_{\mathbf{S}})$.

Remark 3.2 (Conjugate to Remark 2.8). In the previous construction, $\mathscr{C}^o_{\mathbf{S}}$ is the restriction to $\mathbf{S}$ of a sesquilinear pairing ${}^F k$ which is defined at $z = 0$ and taking value there in the space of distributions having moderate growth $z = 0$. More generally, given a free $\mathbb{C}\{z\}$-module $\mathscr{G}_0$ equipped with a connection having Poincaré rank one, and with a Hermitian nondegenerate sesquilinear pairing c on $\mathscr{G}_0 \otimes_{\mathbb{C}} \iota^*\overline{\mathscr{G}_0}$ taking values in the space of germs of distributions with moderate growth at $z = 0$, c being compatible with the ∂_z- and $\overline{\partial}_z$-actions, one can ask for a criterion ensuring that, if we extend flatly $\mathscr{G}_0$ and c to Ω_0, the triple $(\mathscr{G}_0, \partial_z, c_{\mathbf{S}})$ is pure of weight 0 (and polarized).

Let us also recall that $\mathscr{G}_0[1/z]$ is completely determined in terms of the formal meromorphic connection $\widehat{\mathscr{G}_0}[1/z] := \mathbb{C}[\![z]\!] \otimes_{\mathbb{C}\{z\}} \mathscr{G}_0[1/z]$ and the Stokes structure, and that $\widehat{\mathscr{G}_0}[1/z]$ has a decomposition into simple objects (Turrittin-Levelt decomposition). Moreover, according to a result of Malgrange, $\mathscr{G}_0$ is completely determined by $\widehat{\mathscr{G}_0} := \mathbb{C}[\![z]\!] \otimes_{\mathbb{C}\{z\}} \mathscr{G}_0$.

One can ask whether c can be determined by the data of sesquilinear pairings on the objects of the Turrittin-Levelt decomposition of $\widehat{\mathscr{G}_0}[1/z]$ (that we could denote by $\widehat{c}$) and a correspondence between Stokes matrices and their transpose-conjugate (to go from c to a sesquilinear pairing on the objects of the Turrittin-Levelt decomposition, one can use [**18**] after twisting by suitable exponential factors; on the other hand, describing the correspondence between the Stokes cocycle for $\mathscr{G}_0[1/z]$ and $\mathscr{G}_0[1/z]^\dagger$ is the content of the proof of [**18**, Th. II.3.1.2]). If this were the case, it would be interesting to give a criterion for the purity or polarization of $(\mathscr{G}_0, c_{\mathbf{S}})$ in terms of $(\widehat{\mathscr{G}_0}, \widehat{c})$ and the Stokes matrices (cf. [**9**, Conjecture 10.2] for a precise conjecture in this direction and §8 of loc. cit. for a more precise analysis).

3.c. Application to variations of Hodge structures. Let us assume that $(V^{\mathrm{an}}, \nabla)$ underlies a variation of polarized complex Hodge structure of weight 0 with flat Hermitian pairing $\widetilde{k}$ and associated metric h (cf. Example 1.34). Let $\widetilde{V}$ be the subsheaf of j_*V^{an} ($j : \mathbb{A}^1 \smallsetminus P \hookrightarrow \mathbb{P}^1$) consisting of sections whose h-norm has moderate growth near $P \cup \{\infty\}$, and let $\widetilde{L} \subset \widetilde{V}$ be the subsheaf consisting of sections of $\widetilde{V}$ whose h-norm is L^1_{loc} near any $p_i \in P$ (with respect to the local Euclidean metric at p_i). Classical results of Schmid's theory [**27**] on variations of Hodge structures, suitably extended to variations of complex Hodge structures, imply that $\widetilde{V}$ is an algebraic vector bundle on $\mathbb{A}^1 \smallsetminus P$ with a connection having only regular singularities, and $\widetilde{L}$ is an algebraic vector bundle on $\mathbb{A}^1$.

Last, let us denote by M the $\mathbb{C}[t]\langle\partial_t\rangle$-submodule of $V := \Gamma(\mathbb{P}^1, \widetilde{V})$ generated by $L = \Gamma(\mathbb{P}^1, \widetilde{L})$. Then it can be shown that M is a minimal extension at each point of P.

Moreover, the sesquilinear pairing $\widetilde{k}$ extends to M with values in $\mathscr{S}'(\mathbb{A}^1)$ (see the details in [**22**, §3g]), in a nondegenerate way as we have seen in Remark 3.1.

THEOREM 3.3 ([**22**]). *Under these conditions, the associated twistor structure* $(\mathscr{H}'^o, \nabla^o, \mathscr{C}^o_{\mathrm{S}})$ *obtained by Fourier-Laplace transform and twistorization is pure of weight* 0 *and polarized.*

SKETCH OF PROOF. Let us first explain in other terms the construction above, in particular the twistorization step, which looks mysterious. To the variation of Hodge structure one associates an integrable variation of twistor structure, by adding the external parameter z (cf. Lemma 2.11).

There is a well-defined notion of Fourier-Laplace transform of a variation of twistor structure (integrable or not) and we are mainly interested in the fibre at $\tau = 1$ of this Fourier-Laplace transform (where τ is the parameter introduced in §3.a). The fundamental result, independent of the integrability condition, is

THEOREM 3.4. *The Fourier-Laplace transform of a tame variation of polarized pure twistor structure of weight* 0 *is a variation of polarized pure twistor structure of weight* 0 *on* $\mathbb{C}^* = \{\tau \neq 0\}$ *with a tame behaviour at* $\tau = 0$. *In particular, the fibre at* $\tau = 1$ *of this variation is a polarized pure twistor structure of weight* 0.

On the other hand, the behaviour at $\tau = \infty$ is usually not tame, and is understood in particular cases by S. Szabo [**32**].

Let us now use the integrability condition. Starting from an integrable variation of twistor structure, its Fourier-Laplace transform remains integrable. Therefore, the fibre at $\tau = 1$ is an integrable twistor structure. When we more specifically start from a variation of Hodge structure, where we have a Hodge filtration, this integrable twistor structure is canonically identified with the twistor structure obtained from $(G_0^{(L)}, {}^F\nabla, {}^Fk)$. In other words, although the variation of twistor structure uses two a priori distinct variables τ^{-1} and z, this variation is produced from the fibre at $\tau = 1$ by rescaling the variable z, the rescaling parameter being τ. This is analogous (without considering any real structure however) to the orbits of TERP structures considered in [**9**] and explains the rather mysterious twistorization construction above (cf. the proof of the main theorem 1.32 in [**22**]).

Let us now come back to the proof of Theorem 3.4, which is the main point for getting purity. Let us emphasize that we get purity together with polarization, and that, without the polarization assumption on the original variation of twistor structure, the L^2 techniques are not enough.

According to Lemma 2.4, we consider a holomorphic bundle E on $\mathbb{A}^1 \smallsetminus P$, equipped with a Hermitian metric h and a holomorphic Higgs field Φ. We assume that (E, h, Φ) is a *tame harmonic bundle* in the sense of Simpson [**29**] (i.e., (E, h, Φ) is a harmonic Higgs bundle on $\mathbb{A}^1 \smallsetminus P$ and we impose a growth condition on the eigenvalues of the Higgs field at the points of $P \cup \{\infty\}$).

Let us consider the exponentially twisted harmonic bundle $(E, h, \Phi - dt)$. This remains a harmonic bundle which is tame at P, but *wild* at ∞. The main point in the proof of Theorem 3.4 is to find (using the notation of Theorem 3.3) a finite dimensional $\mathbb{C}$-vector space H^o in $\mathscr{H}'^o$ which contains an orthonormal basis for $\mathscr{C}^o_{\mathrm{S}}$. This is done using L^2-theory:

THEOREM 3.5 ([**19**]). *The space of* L^2 *harmonic sections of* $E \otimes \mathscr{A}^1_{\mathbb{A}^1 \smallsetminus P}$ *(1-forms with values in* E*), with respect to the metric* h *and a metric on* $\mathbb{A}^1 \smallsetminus P$ *equivalent* *to the Poincaré metric near $P \cup \{\infty\}$, and with respect to the Laplace operator of $d'' + \Phi - dt + z(D' + \Phi^\dagger - d\bar{t})$ is finite dimensional and independent of z.*

REMARK 3.6. The proof of this theorem that I first gave by using a degeneration argument $\tau \to 0$ was only valid for $(E, h, \Phi - \tau dt)$ for $|\tau|$ small (depending on (E, h, Φ)). The proof I gave can be made correct (cf. the erratum in [**19**]) by using in particular an argument whose proof is due to the the referee of [**22**] (note also that S. Szabo proves a similar result in [**32**, Lemma 2.32], with different methods however):

The harmonic sections have an exponential decay when $t \to \infty$.

Regarding the parameter z in Theorem 3.5 as the external parameter of a twistor structure (the metric being given by the L^2 metric), the result of this theorem is the L^2-construction of a polarized pure twistor structure of weight 0.

In order to realize our L^2-twistor structure as the fibre at $\tau = 1$ of the Fourier-Laplace transform of the original variation of twistor structure, one uses a technique which goes back to S. Zucker [**34**] which provides an identification with L^2 meromorphic sections (see the details in [**19, 22**]). In fact, this identification is already needed in the proof of Theorem 3.5 to ensure the finite dimensionality of the spaces of L^2 sections which are involved. □

4. The harmonic Frobenius manifold attached to a Laurent polynomial

Let U be the torus $\operatorname{Spec} \mathbb{C}[u_1, \ldots, u_n, u_1^{-1}, \ldots, u_n^{-1}] \simeq (\mathbb{C}^*)^n$ and let $f \in \mathbb{C}[u_1, \ldots, u_n, u_1^{-1}, \ldots, u_n^{-1}]$ be a Laurent polynomial. Throughout this section, we will assume that f is convenient and nondegenerate in the sense of Kouchnirenko [**11**]. Let us recall that the convenience assumption means that the origin belongs to the interior of the Newton polyhedron of f. A nice example is $f_w(u_1, \ldots, u_n) = u_1 + \cdots + u_n + 1/u_1^{w_1} \cdots u_n^{w_n}$ with $w_i \in \mathbb{N}^*$ (and, for simplicity, $\gcd(w_1, \ldots, w_n) = 1$).

We will also use a stronger convenience assumption.

DEFINITION 4.1 (A. Douai). We say that f is strongly convenient if, denoting by $0 = \alpha_0, \alpha_1, \ldots, \alpha_r$ the integral points contained in the interior of the Newton polyhedron of f, the polynomial ring on $u^{\alpha_1}, \ldots, u^{\alpha_r}$ surjects onto $\mathbb{C}[u, u^{-1}]/(\partial f)$.

The assumption holds for instance if there exists a $\mathbb{Z}$-basis $e_1, \ldots, e_n$ of $\mathbb{Z}^n$ such that $\{\pm e_1, \ldots, \pm e_n\}$ belong to the interior of the Newton polyhedron of f.

To any convenient and nondegenerate Laurent polynomial is attached in a canonical way a Frobenius manifold, under one of the following assumptions:

(4.2) f has only nondegenerate critical points and the critical values are distinct,

(4.3) f is strongly convenient.

Let us note that the Laurent polynomial f_w satisfies the first assumption but not the second one.

We recall in §4.a the construction of this canonical Frobenius manifold, which is due to A. Douai ([**3, 4**]) in the strongly convenient case (4.3). In §4.b, we enrich this structure to a structure of harmonic Frobenius manifold. The main argument consists in endowing the Brieskorn lattice of f with a polarized pure twistor structure. This point is not specific to convenient and nondegenerate Laurent polynomials, but is a general result for cohomologically tame functions on affine manifolds. We thus take the general point of view in §4.b.

4.a. The Frobenius manifold attached to a Laurent polynomial. Let f be a convenient and nondegenerate Laurent polynomial. The $\mathbb{C}$-vector space $\mathbb{C}[u, u^{-1}]/(\partial f)$ is finite dimensional and its dimension $\mu(f)$ is the sum of the Milnor numbers of f at each of its critical points. We will construct a canonical Frobenius manifold structure on the analytic germ M along a submanifold N of the $\mathbb{C}$-vector space $\mathbb{C}[u, u^{-1}]/(\partial f)$.

The method developed by A. Douai to construct such a structure consists in

(1) choosing a smooth affine subvariety $N \subset \mathbb{C}[u, u^{-1}]/(\partial f)$ canonically attached to f,
(2) constructing a Saito structure $(E, \nabla, g, \Phi, \mathscr{U}, \mathscr{V})$ on a holomorphic vector bundle E of rank $\mu(f)$ on N and canonically attached to f,
(3) giving a ∇-horizontal section ω of E, which is homogeneous with respect to $\mathscr{V}$, such that the infinitesimal period map $\varphi_\omega : TN \to E$ is injective,
(4) applying the result of [**13**] in Case (4.2) or [**8**] in Case (4.3) concerning the existence of a universal unfolding of a differential system to show that φ_ω extends in a unique way as a primitive homogenous section on some analytic neighbourhood M of N in $\mathbb{C}[u, u^{-1}]/(\partial f)$,
(5) obtaining the Frobenius manifold structure on M by carrying through φ_ω the Saito structure on E to a Saito structure on TM.

The important point is that, on N, the construction is completely algebraic, while the transcendental part of the construction is hidden in the application of the results of [**13**] or [**8**]. This much improves the construction given in [**5**], where a universal unfolding of the function itself was used, making the construction less canonical.

The choice of N. By the results of Kouchnirenko [**11**], the subspace $\mathbb{A}^r$ of $\mathbb{C}[u, u^{-1}]$ generated by the monomials u^{α_i}, where $(\alpha_i)_{i=1,\dots,r}$ denote the integral points of $\mathbb{Z}^n$ contained in the interior of the Newton polyhedron of f, injects into the Jacobi space $\mathbb{C}[u, u^{-1}]/(\partial f)$.

We consider the deformation

$$F(u, x) = f(u) + \sum_{i=1}^{r} x_i u^{\alpha_i}$$

parametrized by the affine space $\mathbb{A}^r$. Clearly, for any $x \in \mathbb{A}^r$, the function $F_x(u) := F(u, x)$ is convenient and nondegenerate. If $f = F_0$ satisfies (4.2), there exists a Zariski dense open subset N of $\mathbb{A}^r$ such that F_x satisfies (4.2) for any $x \in N$. If F_0 satisfies (4.3), then so does F_x for any $x \in N := \mathbb{A}^r$.

The Saito structure on N. We will first describe the canonical Saito structure parametrized by the complex manifold $\mathbb{A}^r$ attached to F.

We denote by $\Omega^k(U)[x]$ the space of algebraic differential k-forms on $U \times \mathbb{A}^r$ relative to $\mathbb{A}^r$ (i.e., with no dx) and by d_u the relative differential. The Brieskorn lattice $G_{F,0}$ of F is the quotient

$$\Omega^n(U)[x, z]/(zd_u - d_u F\wedge)\Omega^{n-1}(U)[x, z].$$

Under our assumption on f, this is a free $\mathbb{C}[z, x_1, \dots, x_n]$-module of rank $\mu(f)$, equipped with a connection ∇ having Poincaré rank one along $z = 0$ (cf. [**3**]). The connection is determined by its action on the class $[\eta]$ of differential forms $\eta \in \Omega^n(U)$:

$$z^2\nabla_{\partial_z}[\eta] = [F\eta], \quad z\nabla_{\partial_{x_i}}[\eta] = -[F'_{x_i}\eta] \quad (i = 1, \dots, r).$$

For any $x^o \in \mathbb{A}^r$ (for instance $x^o = 0$), let us consider the corresponding object attached to the function $F^o(u) = F(u, x^o)$. By using a construction of M. Saito together with Hodge theory for the function F^o (cf. [**25, 21, 6, 5**]) one can define in a canonical way a trivialization $G_{F^o,0} \simeq \mathbb{C}[z]^{\mu(f)}$ so that the connection ∇^o takes the form $d+(\mathscr{U}^o/z - \mathscr{V}^o)dz/z$, $\mathscr{V}^o$ is semi-simple and its eigenvalues form the *spectrum at infinity* of F^o translated by $-n/2$ (that is, centered at 0), together with a canonical pairing $\mathscr{P}^o$ obtained by microlocalization from the Poincaré duality in the fibres of F^o.

A. Douai moreover shows (cf. [**3**] after the proof of Prop. 3.1.2 and [**4**, Th. 3.3.1]) that this construction can be made for F over $\mathbb{A}^r$, is compatible with base change, and gives a matrix for the connection ∇ as in (2.15) and a pairing $\mathscr{P}$ as in (2.16).

The pre-primitive section. Now, the (constant w.r.t. x) complex volume form $\omega = \dfrac{du_1}{u_1} \wedge \cdots \wedge \dfrac{du_n}{u_n}$ defines all along $\mathbb{A}^r$ a real homogeneous pre-primitive form (it corresponds to the eigenvalue $-n/2$ of $\mathscr{V}$).

The choice of N (possibly distinct from $\mathbb{A}^r$) is made to apply, in Case (4.2), the theorem of Malgrange at each point of N. In Case (4.3), we can apply the result of [**8**] all along $\mathbb{A}^r$.

4.b. Twistor structures associated to tame functions. We now consider more generally an n-dimensional smooth affine variety U and a regular function $f : U \to \mathbb{C}$, that we assume to be "cohomologically tame": this property means that the critical points of f in U are isolated and that no critical value of f comes from a critical point at infinity (see e.g. [**21**]). For instance, if U is a torus and f is a Laurent polynomial, tameness is implied by the property of being convenient and nondegenerate. We will attach below to f a canonical twistor structure $(\mathscr{H}'^o, \nabla^o, \mathscr{C}^o_{\mathrm{S}})$ of weight 0.

Let us first define $(\mathscr{H}'^o, \nabla^o)$. The bundle $\mathscr{H}'^o$ is the analytization of the Brieskorn lattice

$$\Omega^n(U)[z]/(zd - df\wedge)\Omega^{n-1}(U)[z],$$

which is known to be a free $\mathbb{C}[z]$-module of finite rank $\mu(f)$ (cf. e.g. [**21**]) equipped with the connection ∇^o defined by

$$z^2\nabla^o_{\partial_z}\big[\textstyle\sum_k \omega_k z^k\big] = \big[\textstyle\sum_k k\omega_k z^{k+1} + \sum_k f\omega_k z^k\big].$$

On the other hand, it is classical (after the work of F. Pham [**16**]) that the local system $\mathscr{L}^o$ is identified with the locally constant sheaf of Lefschetz co-thimbles

$$H^n_{\Phi_z}(U, \mathbb{C}),$$

where $H^*_{\Phi_z}$ denotes the cohomology with support in the family Φ_z of closed sets in U on which $\mathrm{Re}(f(u_1, \ldots, u_n)/z) \leqslant c < 0$.

There is a natural intersection pairing (Poincaré duality pairing made sesquilinear)

$$\widehat{P}_z : H^n_{\Phi_z}(U, \mathbb{C}) \otimes \overline{H^n_{\Phi_{-z}}(U, \mathbb{C})} \longrightarrow \mathbb{C}. \tag{4.4}$$

DEFINITION 4.5 (C. Hertling). The twistor structure canonically attached to f is $(\mathscr{H}'^o, \nabla^o, \mathscr{C}^o_{\mathrm{S}})$, where $(\mathscr{H}'^o, \nabla^o)$ is the analytization of the Brieskorn lattice of f and $\mathscr{C}^o_{\mathrm{S}} := \dfrac{(-1)^{n(n-1)/2}}{(2\pi i)^n}\widehat{P}$.

THEOREM 4.6 (cf. [**22**, Th. 4.10]). *The twistor structure* $(\mathscr{H}'^{o}, \nabla^{o}, \mathscr{C}^{o}_{\mathrm{S}})$ *attached to* f *is pure of weight* 0 *and polarized.*

INDICATION FOR THE PROOF.

Reduction of the problem to dimension one. How can one prove such a positivity statement? One should start with a variation of polarized twistor structure of weight 0 and get our twistor structure by a natural operation from the previous one. Here is an example of such a result:

HODGE-SIMPSON THEOREM. *Given a variation of polarized twistor structure of weight* 0 *on a compact Kähler manifold* X*, its de Rham cohomology carries a polarized twistor structure (of some weight).*

The main ingredient in the proof of the previous theorem is the fact that, for any $z \in \mathbb{C}$, the Laplace operator Δ_z relative to the operator $D_z = d''+\Phi+z(D'+\Phi^\dagger)$ and the Kähler metric is essentially constant: $\Delta_z = (1+|z|^2)\Delta_0$ (this is the analogue of the classical Kähler identity $\Delta_d = 2\Delta_{d'} = 2\Delta_{d''}$). Hence, the space of harmonic sections does not depend on z. This will give the pure weight 0 property. The positivity is obtained by a standard argument of Hodge theory, on primitive sections first.

In Theorem 4.6, one can use a similar argument (taking direct image by the constant map, that is, taking de Rham cohomology): one can obtain $(\mathscr{H}'^{o}, \nabla^{o})$ by

(1) considering the trivial variation of twistor structure $(\mathscr{O}_U[z], zd)$ (the Higgs field is equal to 0),
(2) twisting it by $e^{-f/z}$, that is, adding a new Higgs field $\Phi = -df$,
(3) and taking the de Rham cohomology of this new variation.

The corresponding operator D_z is $d'' - df + z(d' - d\overline{f})$. We are now faced with two problems: U is noncompact and f is not bounded on U (so that e^{-f} can have exponential growth). The Hodge theory for the corresponding Laplacian can be difficult to develop (although it has been developed in some special cases).

Instead, we use *Horatio's method*: if we face numerous enemies, we fake escaping by running fast, then kill the enemy running faster when he reaches us, then kill the next one, etc. Here, we escape by falling down along the fibres of $f : U \to \mathbb{A}^1$.

Let t be the coordinate on $\mathbb{A}^1$. The Gauss-Manin connection of f gives bundles with connection on $\mathbb{A}^1 \smallsetminus \{\text{critical values of } f\}$. The interesting bundle V^{an} has fibre $H^{n-1}(f^{-1}(t), \mathbb{C})$. It underlies a variation of mixed Hodge structure (M. Saito). The assumption made on f (cohomological tameness) implies that this mixed Hodge structure is an extension of pure Hodge structures for which one subquotient is a variation of polarized Hodge structure (whose generic fibre is identified with the intersection cohomology of a suitable compactification of $f^{-1}(t)$) and any other quotient is a trivial variation of Hodge structure on $\mathbb{A}^1$.

The variation of polarized Hodge structure induces a variation of polarized twistor structure of the same weight.

End of the proof of the theorem. We will give the main steps (cf. [**22**] for details).

Starting from $f : U \to \mathbb{A}^1$, we consider the Gauss-Manin system M of f: $M = \Omega^n(U)[\partial_t]/(d - \partial_t df\wedge)\Omega^{n-1}(U)[\partial_t]$ with its natural structure of $\mathbb{C}[t]\langle\partial_t\rangle$-module. According to a general result of M. Saito [**26**], it underlies a mixed Hodge module on $\mathbb{A}^1$ (hence it is equipped with a Hodge filtration $F^{\mathrm{H}}_{\bullet}M$). This mixed Hodge module has a subquotient $M_{!*}$ (in the category of mixed Hodge modules) which is a polarized pure Hodge module of weight n, and M is an extension of $M_{!*}$ by *constant* Hodge modules (i.e., isomorphic to powers of $\mathbb{C}[t]$ with its natural $\mathbb{C}[t]\langle\partial_t\rangle$-action and trivial Hodge filtration up to a shift): this is a consequence of the tameness assumption.

Let us set $F_\ell M_{!*} := F^{\rm H}_{\ell-n} M_{!*}$ and let us choose an index ℓ_0 such that $M_{!*}/F_{\ell_0}M_{!*}$ is supported on a finite set of points (this is always possible by definition of a good filtration). We use $F_{\ell_0}M_{!*}$ as a submodule L considered in §3.a.

We note that, considering the localized Fourier-Laplace transforms of M and $M_{!*}$, we have $G = G_{!*}$, as the localized Fourier-Laplace transform of $\mathbb{C}[t]$ is zero. But in a less trivial way, if we denote by $G_{!*,0}$ the $\mathbb{C}[z]$-module $z^{\ell_0}G^{(L)}_{!*,0}$ (as defined in §3.a) and keep the notation G_0 for the Brieskorn lattice of f, we have (cf. [**22**, Lemma 4.7]) $G_{!*,0} = G_0$. In other words, this equality gives a definition of the Brieskorn lattice purely in terms of Hodge theory.

We now apply the construction of §3 to $M_{!*}$ and $G_{!*,0}$. Theorem 3.3 shows that it is a polarized pure twistor structure of weight 0.

It remains to identify the $\mathscr{C}_{!*,\mathbf{S}}$ with $\mathscr{C}^o_{\mathbf{S}}$ given in Definition 4.5. This is done by analyzing the definition of the polarization given by M. Saito [**24**]. □

4.c. The canonical harmonic Frobenius manifold structure. We are now in position to state the main theorem. Let f be a convenient and nondegenerate Laurent polynomial. Assume that f satisfies one of the conditions (4.2) or (4.3) and let M be the germ along N of the $\mathbb{C}$-vector space $\mathbb{C}[u,u^{-1}]/(\partial f)$, equipped with its canonical Frobenius manifold structure, as explained in §4.a. For any $x^o \in N$, the twistor structure given by Theorem 4.6 at x^o extends locally, by using the flat extension of the pairing $\mathscr{C}^o_{\mathbf{S}}$ of Definition 4.5 and, according to Corollary 2.13, it defines a positive definite Hermitian form on the Saito structure attached to F in some neighbourhood of x^o. Transporting this Hermitian form by the infinitesimal period mapping associated to the primitive real homogeneous section $\dfrac{du_1}{u_1}\wedge\cdots\wedge\dfrac{du_n}{u_n}$ endows the germ (M,x^o) with a canonical harmonic Frobenius manifold structure (cf. Corollary 1.31) with a positive definite harmonic metric.

THEOREM 4.7. *Under these conditions, the canonical harmonic Frobenius manifold structures on (M,x^o) glue together when x^o varies in N and define a canonical harmonic Frobenius manifold structure with a positive definite harmonic metric on the germ (M,N).*

PROOF. The point is to show that the flat extension $\mathscr{C}_{\mathbf{S}}$ of $\mathscr{C}^o_{\mathbf{S}}$ restricts, up to the factor $(-1)^{n(n-1)/2}/(2\pi i)^n$, to the pairing (4.4) defined from F_x by Poincaré duality. In other words, one should give a definition of the pairing (4.4) with parameter and show the compatibility with base change. Let us set $\widetilde{F}(u,x) = (F(u,x),x) : U\times\mathbb{A}^r \to \mathbb{A}^1\times\mathbb{A}^r$. Then $R\widetilde{F}_!\mathbb{C}_{U\times\mathbb{A}^r}$ is a constructible complex on $\mathbb{A}^1\times\mathbb{A}^r$ which is compatible with base change with respect to $x\in\mathbb{A}^r$. It is also Poincaré-Verdier dual to $R\widetilde{F}_*\mathbb{C}_{U\times\mathbb{A}^r}$ up to a shift.

ASSERTION. *Given any $x^o\in\mathbb{A}^r$, there exists a disc D^o in $\mathbb{A}^1$ such that the cohomology sheaves of $R\widetilde{F}_!\mathbb{C}_{U\times\mathbb{A}^r}$ (or $R\widetilde{F}_*\mathbb{C}_{U\times\mathbb{A}^r}$) are local systems on $(\mathbb{A}^1\smallsetminus D^o)\times \mathrm{nb}(x^o)$.*

Let us postpone the proof of the assertion. The partial Laplace transform of $R\widetilde{F}_!\mathbb{C}_{U\times\mathbb{A}^r}$ (or $R\widetilde{F}_*\mathbb{C}_{U\times\mathbb{A}^r}$) with respect to $\mathbb{A}^1$ then gives, by the same computations as in [**22**, §1b], a constructible complex on $(\mathbb{A}^1_\tau\setminus\{0\})\times\mathbb{A}^r$ whose cohomologies are local systems compatible with base change with respect to $x\in\mathbb{A}^r$, hence there is only one nonzero cohomology local system.

Note that the pairing (4.4) at any $x\in\mathbb{A}^r$ comes by Laplace transform from the Poincaré duality pairing between $RF_{x,!}\mathbb{C}_U$ and $RF_{x,*}\mathbb{C}_U$ made sesquilinear, and also from its restriction between $RF_{x,!}\mathbb{C}_U$ and $RF_{x,!}\mathbb{C}_U$ because $RF_{x,*}\mathbb{C}_U$ and $RF_{x,!}\mathbb{C}_U$ have the same Laplace transform. It is therefore obtained by base change from the partial Laplace transform of the restriction of the Poincaré duality pairing (made sesquilinear) between $R\widetilde{F}_!\mathbb{C}_{U\times\mathbb{A}^r}$ and itself. By definition, the latter object is the flat extension of $\widehat{P}_x$. □

Sketch of a proof for the assertion. Assume first that there exists a basis of $\mathbb{Z}^n$ contained in the interior of the Newton polygon of f. We will work with the corresponding coordinates u_i and argue as in [**1**]. The Milnor number of any F_x is constant with respect to $x\in\mathbb{A}^r$ and so is the Milnor number of $F_x^w := F_x+\sum_i w_iu_i$ (note that this deformation of f, taken with parameters x and w, is somewhat redundant). As a consequence, given a compact neighbourhood $\overline{B}$ of the critical points of f, there exists a compact neighbourhood $\mathrm{nb}(x^o)$ of x^o such that the pull-back by $(\partial F/\partial u_1,\dots,\partial F/\partial u_n):(\mathbb{C}^*)^n\times\mathrm{nb}(x^o)\to\mathbb{C}^n$ of some compact neighbourhood of the origin is contained in $\overline{B}\times\mathrm{nb}(x^o)$. In other words, $|\partial F/\partial u_i|$ are uniformly bounded from below on $[(\mathbb{C}^*)^n\setminus B]\times\mathrm{nb}(x^o)$. Moreover, as $\overline{B}$ is compact, the previous set is contained in $\widetilde{F}^{-1}\big((\mathbb{A}^1\setminus D^o)\times\mathrm{nb}(x^o)\big)$ for a suitable disc $D^o\subset\mathbb{A}^1$. Now the argument used in [**1**, p. 230] to show the fibration property extends to the present situation and shows that $\widetilde{F}$ is a locally trivial topological fibration above $(\mathbb{A}^1\setminus D^o)\times\mathrm{nb}(x^o)$, hence the result.

If such a basis does not exist, we first perform a covering $\rho:V\to U$ of the kind $u_i=v_i^m$, for m large enough to reduce to the previous case. The assertion then holds for $R\widetilde{F}_!\rho_*\mathbb{C}_{V\times\mathbb{A}^r}$ and we take invariants with respect to the covering group action to obtain the result. □

References

1. S.A. Broughton, *Milnor number and the topology of polynomial hypersurfaces*, Inventiones Math. **92** (1988), 217–241.
2. K. Corlette, *Flat G-bundles with canonical metrics*, J. Differential Geom. **28** (1988), 361–382.
3. A. Douai, *Construction de variétés de Frobenius via les polynômes de Laurent: une autre approche*, Singularités, Rev. Inst. Élie Cartan, vol. 18, Univ. Nancy, Nancy, 2005, pp. 105–123.
4. ———, *A canonical Frobenius structure*, arXiv: `0709.0186v1`, 2007.
5. A. Douai and C. Sabbah, *Gauss-Manin systems, Brieskorn lattices and Frobenius structures (I)*, Ann. Inst. Fourier (Grenoble) **53** (2003), no. 4, 1055–1116.
6. C. Hertling, *Frobenius manifolds and moduli spaces for singularities*, Cambridge Tracts in Mathematics, vol. 151, Cambridge University Press, 2002.
7. ———, *tt* geometry, Frobenius manifolds, their connections, and the construction for singularities*, J. reine angew. Math. **555** (2003), 77–161.
8. C. Hertling and Yu.I. Manin, *Unfoldings of meromorphic connections and a construction of Frobenius manifolds*, Frobenius manifolds (Quantum cohomology and singularities) (C. Hertling and M. Marcolli, eds.), Aspects of Mathematics, vol. E 36, Vieweg, 2004, pp. 113–144.
9. C. Hertling and Ch. Sevenheck, *Nilpotent orbits of a generalization of Hodge structures*, J. reine angew. Math. **609** (2007), 23–80, arXiv: `math.AG/0603564`.

10. M. Kashiwara, *Regular holonomic $\mathscr{D}$-modules and distributions on complex manifolds*, Complex analytic singularities, Adv. Stud. Pure Math., vol. 8, North-Holland, Amsterdam, 1987, pp. 199–206.
11. A.G. Kouchnirenko, *Polyèdres de Newton et nombres de Milnor*, Invent. Math. **32** (1976), 1–31.
12. J. Leiterer, *Holomorphic vector bundles and the Oka-Grauert principle*, Several complex variables. IV. Algebraic aspects of complex analysis, Encycl. Math. Sci., vol. 10, Springer-Verlag, 1990, pp. 63–103.
13. B. Malgrange, *Sur les déformations isomonodromiques, I, II*, Séminaire E.N.S. Mathématique et Physique (L. Boutet de Monvel, A. Douady, and J.-L. Verdier, eds.), Progress in Math., vol. 37, Birkhäuser, Basel, Boston, 1983, pp. 401–438.
14. ———, *Équations différentielles à coefficients polynomiaux*, Progress in Math., vol. 96, Birkhäuser, Basel, Boston, 1991.
15. T. Mochizuki, *Asymptotic behaviour of tame harmonic bundles and an application to pure twistor D-modules*, vol. 185, Mem. Amer. Math. Soc., no. 869-870, American Mathematical Society, Providence, RI, 2007.
16. F. Pham, *La descente des cols par les onglets de Lefschetz avec vues sur Gauss-Manin*, Systèmes différentiels et singularités (Luminy, 1983) (A. Galligo, J.-M. Granger, and Ph. Maisonobe, eds.), Astérisque, vol. 130, Société Mathématique de France, 1985, pp. 11–47.
17. C. Sabbah, *Déformations isomonodromiques et variétés de Frobenius*, Savoirs Actuels, CNRS Éditions & EDP Sciences, Paris, 2002, English Transl.: Universitext, Springer & EDP Sciences, 2007.
18. ———, *Vanishing cycles and Hermitian duality*, Proc. Steklov Inst. Math. **238** (2002), 194–214.
19. ———, *Fourier-Laplace transform of irreducible regular differential systems on the Riemann sphere*, Russian Math. Surveys **59** (2004), no. 6, 1165–1180, erratum: `http://math.polytechnique.fr/~sabbah/sabbah_Fourier-irred_err.pdf`.
20. ———, *Polarizable twistor $\mathscr{D}$-modules*, Astérisque, vol. 300, Société Mathématique de France, Paris, 2005.
21. ———, *Hypergeometric periods for a tame polynomial*, Portugal. Math. **63** (2006), no. 2, 173–226, arXiv: `math.AG/9805077`.
22. ———, *Fourier-Laplace transform of a variation of polarized complex Hodge structure*, J. reine angew. Math. (to appear), arXiv: `math.AG/0508551`, 36 pages.
23. K. Saito, *Period mapping associated to a primitive form*, Publ. RIMS, Kyoto Univ. **19** (1983), 1231–1264.
24. M. Saito, *Modules de Hodge polarisables*, Publ. RIMS, Kyoto Univ. **24** (1988), 849–995.
25. ———, *On the structure of Brieskorn lattices*, Ann. Inst. Fourier (Grenoble) **39** (1989), 27–72.
26. ———, *Mixed Hodge Modules*, Publ. RIMS, Kyoto Univ. **26** (1990), 221–333.
27. W. Schmid, *Variation of Hodge structure: the singularities of the period mapping*, Invent. Math. **22** (1973), 211–319.
28. C. Simpson, *Constructing variations of Hodge structure using Yang-Mills theory and applications to uniformization*, J. Amer. Math. Soc. **1** (1988), 867–918.
29. ———, *Harmonic bundles on noncompact curves*, J. Amer. Math. Soc. **3** (1990), 713–770.
30. ———, *Higgs bundles and local systems*, Publ. Math. Inst. Hautes Études Sci. **75** (1992), 5–95.
31. ———, *Mixed twistor structures*, Prépublication Université de Toulouse & arXiv: `math.AG/9705006`, 1997.
32. S. Szabo, *Nahm transform of meromorphic integrable connections on the Riemann sphere*, Ph.D. thesis, Université Louis Pasteur, Strasbourg, juillet 2005, arXiv: `0511471v1`.
33. A. Takahashi, *tt* geometry of rank two*, Internat. Math. Res. Notices (2004), no. 22, 1099–1114.
34. S. Zucker, *Hodge theory with degenerating coefficients: L_2-cohomology in the Poincaré metric*, Ann. of Math. **109** (1979), 415–476.

UMR 7640 du CNRS, Centre de Mathématiques Laurent Schwartz, École polytechnique, F–91128 Palaiseau cedex, France
E-mail address: sabbah@math.polytechnique.fr
URL: http://www.math.polytechnique.fr/~sabbah

Proceedings of Symposia in Pure Mathematics
Volume **78**, 2008

From Primitive Forms to Frobenius manifolds

Kyoji Saito and Atsushi Takahashi

CONTENTS

1. Introduction

In the early 1980s, the first named author, in the study of the period map associated with a primitive form [**S3, S4**], found a certain differential geometric structure on the base space of a universal unfolding of an isolated hypersurface singularity. He called it a *flat structure* because the structure contains a flat metric (bilinear form) on the tangent bundle of the base space as its basic ingredient. The introduction of the flat structure was motivated from the study of the period map [**S6**], where one needs a particular coordinate system on the deformation parameter space, called the flat coordinates (see [**S10**] and its references for the classical examples). The discovery of the flat structure was also inspired by the same structure already found by the same author on the quotient variety of a finite reflection group [**S1, S-Y-S**] where the invariant functions were called the flat invariants (cf. [**Du-Z, S7, S9, Sat**]).

The same structure was found independently by B. Dubrovin in his study of 2-dimensional topological field theories [**Du**] in order to study the WDVV equations and their relation with integrable systems. The structure is axiomatized under the

2000 *Mathematics Subject Classification.* Primary 32S20, 32S35; Secondary 32J25, 53D45.

This work is partly supported by Grant-in Aid for Scientific Research grant numbers 16340016, 17654015 and 17740036 from the Ministry of Education, Culture, Sports, Science and Technology, Japan.

name of a *Frobenius manifold structure* because of the Frobenius ring structure on the tangent bundle (see [**He1, Ma1, Sab**]). In the present paper, for simplicity, we shall call a manifold with a flat structure, or, equivalently, a manifold with a Frobenius structure, a *Frobenius manifold.*

As we observed, there are several constructions of Frobenius manifolds (cf. [**Ma1**]). There are roughly three classes of geometric origins: invariant theories of Weyl groups, quantum cohomologies, and deformation theories of complex varieties. Frobenius manifolds coming from different classes look, at a glance, different, however, they sometimes happen to coincide. Actually, mirror symmetry gives an isomorphism between the Frobenius manifold from the quantum cohomology of a variety and that from the deformation theory of another variety. It is interesting to study the isomorphism since it provides unexpected information such as numbers of curves in the variety [**CDGP**].

By Kontsevich's homological mirror symmetry conjecture [**Ko**], these isomorphisms of Frobenius manifolds are expected to be derived from equivalences of triangulated (or A_∞-) categories; in particular it is expected that we can construct Frobenius manifolds from deformations of these categories; in a unified way.

Although the original construction in [**S3, S4**] of the flat structure by the use of a primitive form is formulated only for the deformation space of germs of isolated hypersurface singularities, the construction itself, using the semi-infinite Hodge structure, is of a general nature. It can be applied as the prototype to many other settings including the one in homological mirror symmetry, and has inspired several works on semi-infinite Hodge theory ([**G, Ba, Ba-Ko, Do-Sab 1, Do-Sab 2, He2, I, Ka-Ko-P, Ko-So**]).

Since this way of construction of the Frobenius manifold was not stated explicitly in the literature, the present paper tries to fill this gap. Namely, in Section 2 of the present paper, we recall the axioms of a Frobenius manifold. In Sections 3-6, we recall the construction and the definition of primitive forms by use of semi-infinite Hodge structures, and then in Section 7 we show how a Frobenius manifold structure is deduced from a primitive form.

Acknowledgment. The authors express their gratitude to Ron Donagi for his invitation and support to contribute the present paper. This work is partly supported by Grant-in Aid for Scientific Research grant numbers 16340016, 17654015 and 17740036 from the Ministry of Education, Culture, Sports, Science and Technology, Japan.

2. Flat structure and Frobenius Manifold Structure

In the present section, we recall the axioms of a Frobenius manifold structure, i.e. that of a flat structure, on a manifold M. For simplicity, we shall call such M a Frobenius manifold. Then, we state some of the immediate consequences of the axioms.

2.1. Axioms. ([**Du**])

DEFINITION 2.1. Let $M = (M, \mathcal{O}_M)$ be a connected complex manifold of dimension μ, whose holomorphic tangent sheaf is denoted by $\mathcal{T}_M$. A *Frobenius manifold structure* of rank μ on M is a tuple $(\eta, \circ, e, E)$, where η is a non-degenerate $\mathcal{O}_M$-symmetric bilinear form on $\mathcal{T}_M$, $\circ$ is $\mathcal{O}_M$-bilinear product on $\mathcal{T}_M$, defining an associative and commutative $\mathcal{O}_M$-algebra structure with the unit e, and E is a holomorphic vector field on M, called the *Euler vector field*, which are subject to the following axioms:

- The product $\circ$ is self-adjoint with respect to η: that is,

$$\eta(\delta \circ \delta', \delta'') = \eta(\delta, \delta' \circ \delta''), \quad {}^{\forall}\delta, \delta', \delta'' \in \mathcal{T}_M. \tag{2.1}$$

- The Levi-Civita connection $\nabla : \mathcal{T}_M \to \mathcal{T}_M \otimes \Omega^1_M$ with respect to η is flat: that is,

$$[\nabla_\delta, \nabla_{\delta'}] = \nabla_{[\delta,\delta']} \quad \forall \delta, \delta' \in \mathcal{T}_M. \tag{2.2}$$

- The tensor $C : \mathcal{T}_M \to \mathrm{End}_{\mathcal{O}_M} \mathcal{T}_M$ defined by $C_\delta \delta' := \delta \circ \delta'$ $(\delta, \delta' \in \mathcal{T}_M)$ is flat: that is,

$$\nabla C = 0. \tag{2.3}$$

- The unit element e of the $\circ$-algebra is a ∇-flat holomorphic vector field: that is,

$$\nabla e = 0. \tag{2.4}$$

- The product $\circ$ and the metric η are homogeneous of degree 1 and $2-d$ $(d \in \mathbb{C})$, respectively, with respect to the Lie derivative Lie_E of the Euler vector field E: that is,

$$\mathrm{Lie}_E(\circ) = \circ, \quad \mathrm{Lie}_E(\eta) = (2-d)\eta. \tag{2.5}$$

2.2. Flat Coordinates and Flat Potentials.

Let us consider the space of horizontal sections of the connection ∇:

$$\mathcal{T}_M^f := \{\delta \in \mathcal{T}_M \mid \nabla_{\delta'}\delta = 0 \ \forall \delta' \in \mathcal{T}_M\}$$

which is, due to (2.2), a local system of rank μ on M such that the metric η takes constant values $\eta(\delta, \delta')$ for all $\delta, \delta' \in \mathcal{T}_M^f$. Since $\mathcal{T}_M^f$ is abelian under the bracket product, we have:

DEFINITION 2.2 (**Flat coordinates**). At each point of M, there exists a local coordinate system $(t_0, \ldots, t_{\mu-1})$, called *flat coordinates*, such that
i) $e = \partial_0$ and ii) the $\partial_i := \frac{\partial}{\partial t_i}$ $(i = 0, \cdots, \mu-1)$ span $\mathcal{T}_M^f$ over $\mathbb{C}$.

NOTE 2.3. The vector field $e = \partial_0$ is sometimes called a *primitive vector field*, or a *primitive derivation*, in connection with the theory of primitive forms.

Let us state a consequence of the axiom $\nabla C = 0$ (2.3), implying the existence of a "potential" for the flat metric η in the following sense:

PROPOSITION 2.4 (**Potential**). *At each point of M, there exists a local holomorphic function $\mathcal{F}$, called the flat potential, satisfying*

$$\eta(\partial_i \circ \partial_j, \partial_k) = \eta(\partial_i, \partial_j \circ \partial_k) = \partial_i\partial_j\partial_k\mathcal{F}, \quad {}^{\forall}i, j, k = 0, \ldots, \mu-1, \tag{2.6}$$

for any system of flat coordinates. In particular, one has

$$\eta_{ij} := \eta(\partial_i, \partial_j) = \partial_0\partial_i\partial_j F.$$

□

2.3. Spectrum, Exponents and Weight.

Let M be a Frobenius manifold of rank μ. The gradient ∇E of the Euler vector field E is a tensor in $\mathrm{End}_{\mathcal{O}_M}\mathcal{T}_M$. From the Axioms, one sees easily that it is flat: $\nabla(\nabla E) = 0$, and, hence, it induces a $\mathbb{C}$-endomorphism of the flat subspace $\mathcal{T}_M^f$. In the remainder of the present paper, we shall assume that *the endomorphism is semi-simple.* Let $1-q_0, \cdots, 1-q_{\mu-1}$ be the eigenvalues of the endomorphism at any point (independent of the point due to the connectedness of M), where we normalize $q_0 = 0$ since $\nabla_e E = e$. By using suitable flat coordinates $(t_0, \ldots, t_{\mu-1})$ such that $\nabla_{\partial_i} E = (1-q_i)\partial_i$, the Euler vector field is expressed as $E = \sum_{i=0}^{\mu-1}\{(1-q_i)t^i + r_i\}\partial_i$, for suitable constants r_i depending on the choice of the center of the flat coordinates.

DEFINITION 2.5 (**Spectrum, Weight and Exponents**). Suppose ∇E is semi-simple.

(i) The set $\{q_0, q_1, \ldots, q_{\mu-1}\}$ is called the *spectrum*.
(ii) Let us fix a non-negative integer $w \in \mathbb{Z}_{\geq 0}$, called a *weight* of the manifold M. Depending on a choice of a weight w, we define the *minimal exponent* by $r := (w - d)/2 \in \mathbb{C}$ and the set of *exponents* by the eigenvalues $r + q_0, \ldots, r + q_{\mu-1}$ of the endomorphism $N := -\nabla E + (r+1) \cdot \mathrm{id}_{\mathcal{T}_M}$ (see Section 7 for the meaning of r and N).

3. Universal Unfolding of Isolated Singularities

In the following Sections 3-6, we introduce the concept of a primitive form.

In the present section, we define a universal unfolding of an isolated singularity and give its basic properties. As we shall see, the concept of a universal unfolding already induces an algebra structure on the tangent bundle of its base space and the concept of the Euler vector field on it. The fiber dimension of the unfolding gives the weight in the definition of the exponents in Definition 2.5. However, the minimal exponent and the flat metric will be introduced later in Section 6 depending on a choice of a primitive form.

3.1. Universal Unfolding.

Let $f(\underline{x}) = f(x_0, \ldots, x_n)$ be a holomorphic function defined on a neighborhood of the origin in $\mathbb{C}^{n+1}$ with $f(0) = 0$. Assume that

$$\mathcal{O}_{\mathbb{C}^{n+1},0} \Big/ \Big(\frac{\partial f}{\partial x_0}, \ldots, \frac{\partial f}{\partial x_n}\Big)$$

is of finite rank, say μ, over $\mathbb{C}$. This is equivalent to the assumption that the function f is defined on $(\mathbb{C}^{n+1}, 0)$ and has an isolated critical point at the origin 0, or equivalently, the germ of the hypersurface $X_0 := \{f = 0\} \subset (\mathbb{C}^{n+1}, 0)$ has at most an isolated singular point at the origin.

The universal unfolding of f is an extension F of f defined on $\mathbb{C}^{n+1} \times \mathbb{C}^\mu$ extended by a μ dimensional parameter $\underline{t} \in \mathbb{C}^\mu$ defined as follows, where we denote by

$$p : \mathbb{C}^{n+1} \times \mathbb{C}^\mu \longrightarrow \mathbb{C}^\mu, \quad (\underline{x}, \underline{t}) \mapsto \underline{t}$$

the projection map from the total space to the parameter space.

DEFINITION 3.1 (R. Thom [**Th**]). A holomorphic function F defined on a neighborhood $\mathcal{X}$ of the origin $(0, 0)$ of $\mathbb{C}^{n+1} \times \mathbb{C}^\mu$ is called a *universal unfolding* of f, if it satisfies:

(i) $F|_{p^{-1}(0)} = f$ in a neighborhood of the origin in $p^{-1}(0) \simeq \mathbb{C}^{n+1}$.
(ii) There exists an $\mathcal{O}_{\mathbb{C}^\mu,0}$-isomorphism

$$\mathcal{T}_{\mathbb{C}^\mu,0} \to p_* \mathcal{O}_{\mathbb{C}^{n+1}\times\mathbb{C}^\mu,0} \Big/ \left(\frac{\partial F}{\partial x_0}, \ldots, \frac{\partial F}{\partial x_n}\right), \quad \delta \mapsto \widehat{\delta} F, \tag{3.1}$$

where $\widehat{\delta}$ is a lifting to $\mathbb{C}^{n+1} \times \mathbb{C}^\mu$ of the vector field δ with respect to the projection p.

FACT 3.2. For any germ of a holomorphic function f defining an isolated singularity at the origin, there exists a universal unfolding.

PROOF. Choose a system of holomorphic functions $1 =: \phi_0(\underline{x}), \phi_1(\underline{x}), \ldots, \phi_{\mu-1}(\underline{x})$ such that it forms a $\mathbb{C}$-linear basis of $\mathcal{O}_{\mathbb{C}^{n+1},0}/(\frac{\partial f}{\partial x_0}, \ldots, \frac{\partial f}{\partial x_n})$ and let

$$F(\underline{x}, \underline{t}) := f(\underline{x}) + \sum_{i=0}^{\mu-1} t_i \cdot \phi_i(\underline{x}),$$

defined in a neighborhood of $(0,0)$, where $(t_0, \ldots, t_{\mu-1})$ is a coordinate system of $\mathbb{C}^\mu$. □

Let $F(\underline{x}, \underline{t})$ be a universal unfolding of f. By fixing a domain $\mathcal{X}$ and a range $\mathbb{C}_\delta$ of F and a (small) open set $S \subset \mathbb{C}^\mu$ of the unfolding parameters $\underline{t} \in \mathbb{C}^\mu$, we consider the map $\varphi = (F, p) : \mathcal{X} \to \mathbb{C}_\delta \times S$, which defines a flat family of hypersurfaces where the fiber over 0 is isomorphic to the singularity $X_0 := \{f = 0\}$ and generic fibers are smoothings of X_0 containing vanishing cycles for the singularity. This is achieved as follows. Let us fix Euclidean norms $\|\cdot\|$ on $\mathbb{C}^{n+1} \times \mathbb{C}^\mu$, $\mathbb{C}$ and $\mathbb{C}^\mu$. For positive real numbers r, δ and ϵ, put

$$S_\epsilon := \{\underline{t} \in \mathbb{C}^\mu \mid \|\underline{t}\| < \epsilon\} \tag{3.2}$$

$$\mathbb{C}_\delta := \{w \in \mathbb{C} \mid \|w\| < \delta\} \tag{3.3}$$

$$X_{r,\delta,\epsilon} := \{(\underline{x}, \underline{t}) \in \mathbb{C}^{n+1} \times \mathbb{C}^\mu \mid \|(\underline{x}, \underline{t})\| < r, \|F(\underline{x}, \underline{t})\| < \delta\} \cap p^{-1}(S_\epsilon) \tag{3.4}$$

For the choice $1 \gg r \gg \delta \gg \epsilon > 0$ of the radius, we have

(i) $p : X_{r,\delta,\epsilon} \to S_\epsilon$ is a smooth Stein map, which is topologically acyclic.
(ii) The holomorphic map

$$\varphi : X_{r,\delta,\epsilon} \longrightarrow \mathbb{C}_\delta \times S_\epsilon, \quad (\underline{x}, \underline{t}) \mapsto (w, \underline{t}) := (F(\underline{x}, \underline{t}), \underline{t}) \tag{3.5}$$

is a flat Stein map, whose fibers are smooth and transverse to $\partial X_{r,\delta,\epsilon}$ at each point of $\partial X_{r,\delta,\epsilon}$. The fiber $\varphi^{-1}(0) \cap X_{r,\delta,\epsilon}$ over 0 has only an isolated singular point isomorphic to X_0 and is contractible to a point.

We study some basic properties of the family φ (3.5) in the following Sections 3.2, 3.3 and 3.4. In each step and also for the study of de Rham cohomology groups in Section 4, we need to choose the constants r, δ and ϵ suitably. Since it is cumbersome to state the choice each time, we fix the constants once and for all and put $\mathcal{X} := X_{r,\delta,\varepsilon}$ and $S := S_\varepsilon$.

3.2. Critical Set $\mathcal{C}$ and Product Structure on $\mathcal{T}_S$.

Let the setting be as in Section 3.1. Let us denote by $\mathcal{C}$ the critical set of the family (3.5), or, equivalently, the relative critical set of F with respect to the projection p. That is, $\mathcal{C}$ is the support of the sheaf:

$$\mathcal{O}_\mathcal{C} := \mathcal{O}_\mathcal{X} \Big/ \left(\frac{\partial F}{\partial x_0}, \ldots, \frac{\partial F}{\partial x_n}\right). \tag{3.6}$$

The assumption that f has an isolated singularity implies $\mathcal{C} \cap p^{-1}(0) = \{(0,0)\}$, and, hence, $p|_{\mathcal{C}}$ is a proper finite map. Actually, $\frac{\partial F}{\partial x_0}, \ldots, \frac{\partial F}{\partial x_n}$ form a regular sequence at the origin and $\mathcal{C}$ is Cohen-Macaulay of dimension equal to that of S so that $\mathcal{C} \to S$ is a flat finite cover. Therefore, $p_*\mathcal{O}_{\mathcal{C}}$ is a locally free sheaf on S of rank μ.

The universality (ii) of the unfolding F implies the following $\mathcal{O}_S$-isomorphism:

$$\mathcal{T}_S \;\simeq\; p_*\mathcal{O}_{\mathcal{C}}, \qquad \delta \;\mapsto\; \widehat{\delta}F|_{\mathcal{C}}. \tag{3.7}$$

By this isomorphism, the tangent bundle $\mathcal{T}_S$ naturally obtains an $\mathcal{O}_S$-algebra structure.

DEFINITION 3.3 (**Product on $\mathcal{T}_S$**). We shall denote by $\circ$ the induced product structure on $\mathcal{T}_S$. Namely, for $\delta, \delta' \in \mathcal{T}_S$, we have

$$\widehat{(\delta \circ \delta')}F|_{\mathcal{C}} \;:=\; \widehat{\delta}F|_{\mathcal{C}} \cdot \widehat{\delta'}F|_{\mathcal{C}}, \tag{3.8}$$

where $\circ$ is called the *residual product.*

NOTE 3.4. Originally [**S2, S3, S4, S5**], the product $\circ$ was denoted by $*$ and called the $*$-product.

NOTE 3.5. Consider the complex $(\wedge^{\bullet}\mathcal{T}_{\mathcal{X}/S}, \, dF \vdash)$ of $\mathcal{O}_{\mathcal{X}}$-modules with $\mathcal{O}_{\mathcal{X}}$ in degree 0:

$$0 \longrightarrow \wedge^{n+1}\mathcal{T}_{\mathcal{X}/S} \stackrel{dF\vdash}{\longrightarrow} \wedge^{n}\mathcal{T}_{\mathcal{X}/S} \stackrel{dF\vdash}{\longrightarrow} \cdots \stackrel{dF\vdash}{\longrightarrow} \mathcal{T}_{\mathcal{X}/S} \stackrel{dF\vdash}{\longrightarrow} \mathcal{O}_{\mathcal{X}} \longrightarrow 0, \tag{3.9}$$

where $df \vdash$ denotes the interior product. Since $\frac{\partial F}{\partial x_0}, \ldots, \frac{\partial F}{\partial x_n}$ form a regular sequence, the complex (3.9) is quasi-isomorphic to

$$p_*\mathcal{O}_{\mathcal{X}} \,/ p_* \left(dF \wedge \mathcal{T}_{\mathcal{X}/S}\right) \;\simeq\; p_*\mathcal{O}_{\mathcal{C}}. \tag{3.10}$$

3.3. Primitive Vector Field and Euler Vector Field.

Through the isomorphism (3.7), we introduce two particular vector fields on S.

DEFINITION 3.6. The vector field δ_0 on S corresponding to the unit $1 \in \mathcal{O}_{\mathcal{C}}$ is called the *primitive vector field.* That is, $\widehat{\delta_0}F|_{\mathcal{C}} = 1$. The vector field E on S corresponding to $F|_{\mathcal{C}} \in \mathcal{O}_{\mathcal{C}}$ is called the *Euler vector field.* That is, $(\widehat{E}F)|_{\mathcal{C}} = F|_{\mathcal{C}}$.

3.4. Discriminant Locus and Milnor's Fibration Theorem.

The image of the finite map $\varphi|_{\mathcal{C}}$ is a one-codimensional analytic subvariety (i.e. a divisor) in $\mathbb{C}_\delta \times S$, which we shall denote by $\mathcal{D}$ and call the discriminant locus of the unfolding F. The following is due to J. Milnor and H. Hamm (cf. [**Mi**]):

THEOREM 3.7. *The restriction $\varphi|_{\mathcal{X}\setminus\varphi^{-1}(\mathcal{D})}$ is a locally trivial fibration, whose general fiber $\varphi^{-1}(w, \underline{t})$ for $(w, \underline{t}) \in \mathbb{C}_\delta \times S \setminus \mathcal{D}$ is homotopic to a bouquet of μ copies of n-spheres S^n.* □

COROLLARY 3.8. *Fix a point $(w_0, \underline{t}) \in \partial\mathbb{C}_\delta \times S$ and set $\mathcal{X}_{\underline{t}} := p^{-1}(\underline{t})$ and $B_{\underline{t}} := \varphi^{-1}(w_0, \underline{t})$. Then one has*

$$H_{n+1}(\mathcal{X}_{\underline{t}}, B_{\underline{t}}; \mathbb{Z}) \simeq H_n(B_{\underline{t}}; \mathbb{Z}) \simeq \mathbb{Z}^{\mu}. \tag{3.11}$$

□

This leads us to study the relative de Rham cohmology groups of the family $(F, p : \mathcal{X} \to S)$ as the residues of the forms on $\mathcal{X}$, and, hence, to a primitive form.

4. Filtered de Rham Cohomology Groups

We study the relative de Rham cohomology group of the family $p : \mathcal{X} \to S$ in a non-classical manner. That is, compared with the classical de Rham theory, we introduce an auxiliary formal variable δ_w. This variable was necessarily introduced in [**S5**] in order to define the higher residue pairings (see Section 5) on the filtered de Rham cohomology groups. Because of this new variable, the study of the classical Hodge theory switches to the study of the semi-infinite Hodge theory (cf. Givental [**G**], Barannikov [**Ba**], Barannikov-Kontsevich [**Ba-Ko**], Hertling [**He2**], Iritani [**I**], Katzarkov-Kontsevich-Pantev [**Ka-Ko-P**], Kontsevich-Soibelman [**Ko-So**], and others).

4.1. Filtered de Rham Cohomology Group $\mathcal{H}_F$.

Let the setting $(F, p : \mathcal{X} \to S)$ be as in Section 3. Let us consider the complex:

$$(\Omega^\bullet_{\mathcal{X}/S}[[\delta_w^{-1}]][\delta_w],\ \delta_w^{-1} d_{\mathcal{X}/S} + dF\wedge) \tag{4.1}$$

of sheaves on $\mathcal{X}$, where $d_{\mathcal{X}/S}$ is the classical de Rham differential relative to the morphism $p : \mathcal{X} \to S$, δ_w is a formal variable which commutes with the wedge product and with the differentiation $d_{\mathcal{X}/S}$, and we set $\Omega^i_{\mathcal{X}/S}[[\delta_w^{-1}]][\delta_w] := \Omega^i_{\mathcal{X}/S} \otimes_{\mathcal{O}_S} \mathcal{O}_S[[\delta_w^{-1}]][\delta_w]$.

Since p is a Stein map and the singularities of the family φ are isolated, the direct image $\mathbb{R}p_*$ of (4.1) is pure $(n+1)$-dimensional, that is, its only non-vanishing cohomology group is $\mathbb{R}^{n+1}p_*(\Omega^\bullet_{\mathcal{X}/S}[[\delta_w^{-1}]][\delta_w], \delta_w^{-1}d + dF\wedge)$ (see also Note 3.5).

DEFINITION 4.1. We set

$$\mathcal{H}_F := \mathbb{R}^{n+1}p_*(\Omega^\bullet_{\mathcal{X}/S}[[\delta_w^{-1}]][\delta_w], \delta_w^{-1} d_{\mathcal{X}/S} + dF\wedge) \tag{4.2}$$

and call it the *filtered de Rham cohomology group* of the universal unfolding $(F, p : \mathcal{X} \to S)$.

The direct image $\mathbb{R}p_*$ of the truncation $(\Omega^\bullet_{\mathcal{X}/S}[[\delta_w^{-1}]]\delta_w^{-k}, \delta_w^{-1}d + dF\wedge)$ of the complex (4.1) for $k \in \mathbb{Z}$ is again pure $(n+1)$-dimensional. Put

$$\mathcal{H}_F^{(-i)} := \mathbb{R}^{n+1}p_*(\Omega^\bullet_{\mathcal{X}/S}[[\delta_w^{-1}]]\delta_w^{-i}, \delta_w^{-1}d + dF\wedge). \tag{4.3}$$

4.2. Degeneration of Hodge to de Rham.

It will be very important to consider $\mathcal{H}_F^{(-i)}$ as a deformation of a sheaf on the formal line $\mathbb{A}^1_{\delta_w^{-1}}$ over the parameter space S when we shall construct a primitive form. By Corollary 3.8 of Milnor's fibration theorem, we have the following:

PROPOSITION 4.2. *$\mathcal{H}_F^{(-i)}$ is an $\mathcal{O}_S[[\delta_w^{-1}]]$-free module of rank μ.* □

NOTE 4.3. This seems to be one of the motivations to formulate the *Hodge to de Rham degeneration conjecture* in the semi-infinite Hodge theory for Calabi–Yau A_∞-categories (cf. [**Ko-So**]).

One can show further that $\mathcal{H}_F^{(-i)}$, $i \in \mathbb{Z}$, are naturally submodules of $\mathcal{H}_F$ so that they form an increasing and exhaustive filtration of $\mathcal{H}_F$

$$\cdots \subset \mathcal{H}_F^{(-i-1)} \subset \mathcal{H}_F^{(-i)} \subset \cdots \subset \mathcal{H}_F^{(-1)} \subset \mathcal{H}_F^{(0)} \subset \cdots \subset \mathcal{H}_F, \tag{4.4}$$

$$\bigcup_{i\in\mathbb{Z}} \mathcal{H}_F^{(-i)} = \mathcal{H}_F \quad \text{and} \quad \bigcap_{i\in\mathbb{Z}} \mathcal{H}_F^{(-i)} = \{0\} \tag{4.5}$$

such that $\mathcal{H}_F$ is complete with respect to the filtration.

The multiplication by δ_w induces an $\mathcal{O}_S$-isomorphism

$$\delta_w : \mathcal{H}_F^{(-i)} \simeq \mathcal{H}_F^{(-i+1)}. \tag{4.6}$$

The k-th graded piece $\mathrm{gr}^{(k)}(\mathcal{H}_F) := \mathcal{H}_F^{(-k)}/\mathcal{H}_F^{(-k-1)}$ of the filtration (4.4) is isomorphic to

$$\Omega_F := p_*\Omega^{n+1}_{\mathcal{X}/S} \Big/ p_* \left(dF \wedge \Omega^n_{\mathcal{X}/S}\right) \tag{4.7}$$

by a homomorphism $r^{(k)} := r^{(0)} \circ \delta_w^k$, where $r^{(0)}$ is given by the natural short exact sequence:

$$0 \to \mathcal{H}_F^{(-1)} \subset \mathcal{H}_F^{(0)} \xrightarrow{r^{(0)}} \Omega_F \to 0. \tag{4.8}$$

NOTE 4.4. Let π be the natural projection: $\mathbb{C} \times S \to S$. The $\mathcal{O}_S\{\delta_w^{-1}\}$-free module $\mathcal{H}_0 := \pi_*(\mathbb{R}^n\varphi_*(\Omega^\bullet_{\mathcal{X}/\mathbb{C}_\delta \times S}))$ was first considered by Brieskorn [**B**], where he further introduced two $\mathcal{O}_S\{\delta_w^{-1}\}$-free modules $\mathcal{H}_0' := \pi_*(\varphi_*(\Omega^n_{\mathcal{X}/\mathbb{C}_\delta\times S})/ d(\varphi_*(\Omega^{n-1}_{\mathcal{X}/\mathbb{C}_\delta\times S})))$ and $\mathcal{H}_0'' := \pi_*(\varphi_*(\Omega^{n+1}_{\mathcal{X}/\mathbb{C}_\delta\times S})/dF \wedge d(\varphi_*(\Omega^{n-1}_{\mathcal{X}/\mathbb{C}_\delta\times S})))$ such that

$$\mathcal{H}_0 \subset \mathcal{H}_0' \subset \mathcal{H}_0'',$$

where $\mathcal{H}_0''$ is known as the *Brieskorn lattice*. Note that δ_w here is not just a "symbol" but can be considered as the derivative in w, i.e., $\delta_w = d/dw$.

We have the isomorphisms of the Brieskorn lattices with the filters of the filtered de Rham cohomology groups as follows (see [**S5**] 3.)

$$\begin{aligned} \mathcal{H}_0 \otimes_{\mathcal{O}_S\{\delta_w^{-1}\}} \mathcal{O}_S[[\delta_w^{-1}]] &\simeq \mathcal{H}_F^{(-2)}, \\ \mathcal{H}_0' \otimes_{\mathcal{O}_S\{\delta_w^{-1}\}} \mathcal{O}_S[[\delta_w^{-1}]] &\simeq \mathcal{H}_F^{(-1)}, \\ \mathcal{H}_0'' \otimes_{\mathcal{O}_S\{\delta_w^{-1}\}} \mathcal{O}_S[[\delta_w^{-1}]] &\simeq \mathcal{H}_F^{(0)}, \end{aligned}$$

where we set $\mathcal{O}_S\{\delta_w^{-1}\} := \mathcal{O}_S \otimes_{\mathbb{C}} \mathbb{C}\{\delta_w^{-1}\}$, and $\mathbb{C}\{\delta_w^{-1}\}$ is the ring of convergent power series in δ_w^{-1}. This gives the geometric significance to the filtered cohomology group $\mathcal{H}_F$ (as the de Rham cohmology group of the Milnor fibration (3.5)).

Like $\mathcal{H}_F^{(0)}, \mathcal{H}_F^{(-1)}$ and $\mathcal{H}_F^{(-2)}$, all $\mathcal{O}_S[[\delta_w^{-1}]]$-modules treated in this paper are completions of some $\mathcal{O}_S\{\delta_w^{-1}\}$-modules. This holomorphicity plays an important role when we construct primitive forms in Section 6.

NOTE 4.5. Comparing (3.10) with (4.7), we observe that Ω_F is a $p_*\mathcal{O}_C$ module of rank 1. That is, by choosing a form $\zeta \in \Gamma(\mathcal{X}, \Omega^{n+1}_{\mathcal{X}/S})$, we have $\mathcal{O}_S$-isomorphisms of three modules:

$$\mathcal{T}_S \simeq p_*\mathcal{O}_C \simeq \Omega_F, \qquad \delta \mapsto \widehat{\delta}F|_C \mapsto \widehat{\delta}F|_C\zeta, \tag{4.9}$$

where the last isomorphism in (4.9) depends on the class of ζ in Ω_F. This means that the $\mathcal{O}_S$-module $\mathcal{T}_S$ is equipped with the three structures:

(i) Lie algebra structure on $\mathcal{T}_S$ with the unit vector field and Euler vector field,
(ii) associative commutative algebra structure coming from that on $p_*\mathcal{O}_C$,
(iii) non-degenerate $\mathcal{O}_S$-symmetric bilinear pairing coming from the residue pairing on Ω_F (cf. [**Ha**]):

$$\eta_F(\omega_1, \omega_2) := \mathrm{Res}_{\mathcal{X}/S} \begin{bmatrix} \phi_1\phi_2 dx_0 \wedge \cdots \wedge dx_n \\ \dfrac{\partial F}{\partial x_0} \cdots \dfrac{\partial F}{\partial x_n} \end{bmatrix} \tag{4.10}$$

for the classes of $\omega_1 = \phi_1 dx_0 \wedge \cdots \wedge dx_n$ and $\omega_2 = \phi_2 dx_0 \wedge \cdots \wedge dx_n$ in Ω_F.

This is the first approximation to the flat structure on $\mathcal{T}_S$ (i.e., we have a Frobenius "algebra" structure). In order to assert the integrability of the metric η, we need to choose and lift the class of ζ in Ω_F to a good object in $\mathcal{H}_F^{(0)}$, which is the primitive form, introduced in Section 6.

4.3. Gauß–Manin Connection.

As in the case of classical de Rham cohomology groups, we introduce the Gauß–Manin connection on the filtered de Rham cohomology group $\mathcal{H}_F$, where the covariant differentiation is defined not only for the vector fields on S but also for the derivation of the formal variable δ_w[1]. Namely, let us consider the following $\mathcal{O}_S[[\delta_w^{-1}]]$ free module of rank $\mu+1$:

$$\mathcal{T}_{\mathbb{A}^1_{\delta_w^{-1}}\times S} := \mathcal{O}_S[[\delta_w^{-1}]]\frac{d}{d\delta_w} \oplus \mathcal{O}_S[[\delta_w^{-1}]] \otimes_{\mathcal{O}_S} \mathcal{T}_S, \tag{4.11}$$

which naturally carries a Lie algebra structure by extending that on $\mathcal{T}_S$ and by letting $\frac{d}{d\delta_w}$ commute with $\mathcal{T}_S$, where we note that $\frac{d}{d\delta_w}\delta_w^{-1} = -\delta_w^{-2}$.

DEFINITION 4.6. We define a connection, which we again call the *Gauß–Manin connection*,

$$\nabla : \mathcal{T}_{\mathbb{A}^1_{\delta_w^{-1}}\times S} \times \mathcal{H}_F \to \mathcal{H}_F \tag{4.12}$$

by letting, for $\delta \in \mathcal{T}_S$ and $\zeta = [\phi dx_0 \wedge \cdots \wedge dx_n] \in \mathcal{H}_F$,

$$\nabla_{\frac{d}{d\delta_w}}\zeta := \left[\left(F(\underline{x},\underline{t})\phi + \frac{d\phi}{d\delta_w}\right) dx_0 \wedge \cdots \wedge dx_n\right], \tag{4.13a}$$

$$\nabla_\delta \zeta := \delta_w \left[(\delta_w^{-1}\delta\phi + \delta F \phi) dx_0 \wedge \cdots \wedge dx_n\right]. \tag{4.13b}$$

¿From the definition, we immediately have the following:

PROPOSITION 4.7. (i) *The Gauß–Manin connection is integrable:*

$$[\nabla_\delta, \nabla_{\delta'}] = \nabla_{[\delta,\delta']}, \quad {}^\forall\delta, \delta' \in \mathcal{T}_{\mathbb{A}^1_{\delta_w^{-1}}\times S}. \tag{4.14}$$

(ii) *The Gauß–Manin connection satisfies the following trasversality condition:*

$$\nabla : \mathcal{T}_S \otimes \mathcal{H}_F^{(-k-1)} \longrightarrow \mathcal{H}_F^{(-k)}. \tag{4.15}$$

(iii) *The covariant differentiation $\nabla_{\frac{d}{d\delta_w}}$ preserves the filtration*

$$\nabla_{\frac{d}{d\delta_w}}\left(\mathcal{H}_F^{(-k)}\right) \subset \mathcal{H}_F^{(-k)}. \tag{4.16}$$

This implies that $\nabla_{\frac{d}{d\delta_w^{-1}}} = -\delta_w^2 \nabla_{\frac{d}{d\delta_w}}$ has a pole of order 2 at $\delta_w^{-1} = 0$.

□

[1] In the original formulation [**S3, S4**], the covariant differentiation with respect to the variable δ_w was not explicitly used, but the use was implicit in the formula (4.13a) and the higher residue pairings (see Section 5, [**S5**]). In the present paper, we make explicit the use of that covariant differentiation.

5. Higher Residue Pairings

In this section, we claim the existence of a non-degenerate $\mathcal{O}_S[[\delta_w^{-1}]][\delta_w]$ pairing on the filtered de Rham cohomology group $\mathcal{H}_F$, which we call the higher residue pairing, since it extends the classical residue pairing (4.10) to infinite sequences.

5.1. Definition and Existence of Higher Residue Pairings.

Let $(F, p : \mathcal{X} \to S)$ be as in Section 3 the data of a universal unfolding of f, and let $\mathcal{H}_F$ be its filtered de Rham cohomology group given in Section 4. To state the result, let us fix notation: for $P = \sum_{k\in\mathbb{Z}} p_k \delta_w^k \in \mathcal{O}_S[[\delta_w^{-1}]][\delta_w]$, define $P^* := \sum_{k\in\mathbb{Z}} p_k(-\delta_w)^k$ such that $P = (P^*)^*$.

THEOREM 5.1 (K. Saito [**S5**]). *There exists a unique $\mathcal{O}_S$-bilinear form:*

$$K_F : \mathcal{H}_F \times \mathcal{H}_F \to \mathcal{O}_S[[\delta_w^{-1}]][\delta_w] \tag{5.1}$$

satisfying the following properties:

(i) *For all $\omega_1, \omega_2 \in \mathcal{H}_F$,*

$$K_F(\omega_1, \omega_2) = K_F(\omega_2, \omega_1)^*. \tag{K1}$$

(ii) *For any $P \in \mathcal{O}_S[[\delta_w^{-1}]][\delta_w]$ and for all $\omega_1, \omega_2 \in \mathcal{H}_F$,*

$$PK_F(\omega_1, \omega_2) = K_F(P\omega_1, \omega_2) = K_F(\omega_1, P^*\omega_2). \tag{K2}$$

(iii) *For all $\omega_1, \omega_2 \in \mathcal{H}_F^{(0)}$,*

$$K_F(\omega_1, \omega_2) \in \delta_w^{-n-1}\mathcal{O}_S[[\delta_w^{-1}]], \tag{5.2}$$

and the following diagram is commutative:

$$\begin{array}{ccc}
K_F : \mathcal{H}_F^{(0)} \times \mathcal{H}_F^{(0)} & \longrightarrow & \delta_w^{-n-1}\mathcal{O}_S[[\delta_w^{-1}]] \\
\downarrow {\scriptstyle r^{(0)} \times r^{(0)}} & & \downarrow {\scriptstyle \bmod\ \delta_w^{-n-2}\mathcal{O}_S[[\delta_w^{-1}]]} \\
\eta_F : \Omega_F \times \Omega_F & \longrightarrow & \mathcal{O}_S,
\end{array}$$

(iv) *For all $\omega_1, \omega_2 \in \mathcal{H}_F$,*

$$\frac{d}{d\delta_w} K_F(\omega_1, \omega_2) = K_F(\nabla_{\frac{d}{d\delta_w}}\omega_1, \omega_2) + K_F(\omega_1, \nabla_{\frac{d}{d\delta_w}}\omega_2). \tag{K3}$$

(v) *For all $\omega_1, \omega_2 \in \mathcal{H}_F$ and for any $\delta \in \mathcal{T}_S$,*

$$\delta K(\omega_1, \omega_2) = K(\nabla_\delta\omega_1, \omega_2) + K(\omega_1, \nabla_\delta\omega_2). \tag{K4}$$

We define $K_F^{(k)}$ for $k \in \mathbb{Z}$ by the k-th coefficient of the expansion in δ_w:

$$K_F(\omega_1, \omega_2) := \sum_k K_F^{(k)}(\omega_1, \omega_2)\delta_w^{-n-1-k}, \tag{5.3}$$

and call it the k-th *higher residue pairing.*

REMARK 5.2. The conditions (K3) and (K4) together are equivalent to stating that K_F is flat with respect to the Gauß–Manin connection ∇.

6. Primitive Forms

In the present section, we introduce the concept of a primitive form and show its local existence. Let $(F, p : \mathcal{X} \to S)$ be a universal unfolding of f.

6.1. Definition and Existence of Primitive Forms.

For an element $\zeta^{(0)} \in \Gamma(S, \mathcal{H}_F^{(0)})$ and $k \in \mathbb{Z}$, we put $\zeta^{(-k)} := (\delta_w)^{-k}\zeta^{(0)} \in \Gamma(S, \mathcal{H}_F^{(-k)})$.

DEFINITION 6.1 (K. Saito [**S3, S4**]). An element $\zeta^{(0)} \in \Gamma(S, \mathcal{H}_F^{(0)})$ is called a *primitive form* if it satisfies the following five conditions:

(i) $\zeta^{(0)}$ induces an $\mathcal{O}_S$-isomorphism

$$\mathcal{T}_S \otimes_{\mathcal{O}_S} \mathcal{O}_S[[\delta_w^{-1}]] \simeq \mathcal{H}_F^{(0)}, \quad \sum_{k=0}^{\infty} \delta_k \cdot \delta_w^{-k} \mapsto \sum_{k=0}^{\infty} \nabla_{\delta_k} \zeta^{(-1)} \cdot \delta_w^{-k}. \tag{P1}$$

(ii) For $k \geq 1$ and for all $\delta, \delta' \in \mathcal{T}_S$,

$$K^{(k)}(\nabla_\delta \zeta^{(-1)}, \nabla_{\delta'} \zeta^{(-1)}) = 0. \tag{P2}$$

(iii) There exists a constant $r \in \mathbb{C}$ such that

$$\nabla_{\frac{d}{d\delta_w}} \zeta^{(0)} = \nabla_E \zeta^{(-1)} - r\zeta^{(-1)}. \tag{P3}$$

(iv) For $k \geq 2$ and for all $\delta, \delta', \delta'' \in \mathcal{T}_S$,

$$K^{(k)}(\nabla_\delta \nabla_{\delta'} \zeta^{(-2)}, \nabla_{\delta''} \zeta^{(-1)}) = 0. \tag{P4}$$

(v) For $k \geq 2$ and for all $\delta, \delta' \in \mathcal{T}_S$,

$$K^{(k)}\left(\nabla_{\frac{d}{d\delta_w}} \nabla_\delta \zeta^{(-1)}, \nabla_{\delta'} \zeta^{(-1)}\right) = 0. \tag{P5}$$

THEOREM 6.2 (M. Saito [**SM**]). *There exists a primitive form $\zeta^{(0)}$.*

The construction of a primitive form $\zeta^{(0)}$ consists of two steps. The first step is to show the existence of "a primitive form at the origin". Let us denote the restrictions of $\mathcal{H}_F^{(0)}$, Ω_F, K_F, etc. to $0 \in S$ by $\mathcal{H}_f^{(0)}$, Ω_f,K_f, etc..

LEMMA 6.3 (M. Saito [**SM**]). *One has a basis $\{v_0, \ldots, v_{\mu-1}\}$ of $\mathcal{H}_f^{(0)}$ over $\mathbb{C}[[\delta_w^{-1}]]$ satisfying the following conditions:*

(i) *The class $r^{(0)}(v_0) \in \Omega_f$ of v_0 generates Ω_f over $p_*\mathcal{O}_\mathbb{C}|_{\underline{t}=0}$.*

(ii) *The values $K_f(v_i, v_j)$ for $i, j = 0, \ldots, \mu-1$ belong to $\mathbb{C}\delta_w^{-n-1}$.*

(iii) *There exist matrices $C_f \in M(\mu, \mathbb{C})$ and $N_0 \in M(\mu, \mathbb{Q})$ with N_0 diagonal, whose eigenvalues $\{\alpha_0, \ldots, \alpha_{\mu-1}\}$ of N_0 coincide with the exponents of the singularity f, such that*

$$\nabla_{\frac{d}{d\delta_w}} \vec{v} = C_f \vec{v} - \delta_w^{-1} N_0 \vec{v}, \tag{6.1}$$

where $\vec{v} := {}^t(v_0, \ldots, v_{\mu-1})$. □

REMARK 6.4. In this first step, one should choose and fix exponents of f, which is one of the most delicate issues in the construction of a primitive form.

NOTE 6.5. In order to prove the above Lemma, we have to solve the Riemann-Hilbert-Birkhoff problem: we have to show the existence of a flat vector bundle H_f of rank μ over $\mathbb{P}^1_{\delta_w} := \mathbb{C}^1_{\delta_w^{-1}} \cup \{\delta_w = 0\} = \mathbb{C}^1_{\delta_w^{-1}} \cup \mathbb{C}^1_{\delta_w}$ such that $\mathcal{O}(H_f|_{\mathbb{C}^1_{\delta_w^{-1}}}) \simeq \mathcal{H}_f^{(0)}$ and $\mathcal{O}(H_f|_{\mathbb{C}^1_{\delta_w}})$ has logarithmic poles along $\delta_w = 0$, i.e., $\nabla_{d/d\delta_w}\mathcal{O}(H_f|_{\mathbb{C}^1_{\delta_w}}) \subset \delta_w^{-1}\mathcal{O}(H_f|_{\mathbb{C}^1_{\delta_w}})$. This is done by choosing a good *opposite filtration* S to $\mathcal{H}_f^{(0)}$, that is, a subspace $S \subset \mathcal{H}_f$ satisfying $\mathcal{H}_f = \mathcal{H}_f^{(0)} \oplus S$, $\delta_w S \subset S$, $K_f(S, S) \subset \delta_w^{-n+1}\mathbb{C}[\delta_w]$ and

$\nabla_{d/d\delta_w} S \subset \delta_w^{-1} S$. Then, the vector space spanned by $\{v_0, \dots, v_{\mu-1}\}$ is obtained as the intersection $\mathcal{H}_f^{(0)} \cap \delta_w^{-1} S$. The notion of the above opposite filtration is implicitly used in [**SM**], and is now the basis of the construction of primitive forms for general settings (cf. [**Ba, He1, Do-Sab 1, Do-Sab 2, Sab**]).

The second step is to extend the above basis $\{v_0, \dots, v_{\mu-1}\}$ of $\mathcal{H}_f^{(0)}$ over $\mathbb{C}[[\delta_w^{-1}]]$ to a basis of $\mathcal{H}_F^{(0)}$ over $\mathcal{O}_S[[\delta_w^{-1}]]$. One needs to show the following Lemma:

LEMMA 6.6 (M. Saito [**SM**]). *One has a basis* $\{\zeta_0, \dots, \zeta_{\mu-1}\}$ *of* $\mathcal{H}_F^{(0)}$ *over* $\mathcal{O}_S[[\delta_w^{-1}]]$ *such that*

(i) $\zeta_i|_{t=0} = v_i$ *for all* $i = 0, \dots, \mu - 1$.

(ii) *There exists a matrix* $C_F \in M(\mu, \Gamma(S, \mathcal{O}_S))$ *with* $C_F|_{t=0} = C_f$ *such that*

$$\nabla_{\frac{d}{d\delta_w}} \vec{\zeta} = C_F \vec{\zeta} - \delta_w^{-1} N_0 \vec{\zeta}, \tag{6.2}$$

where $\vec{\zeta} := {}^t(\zeta_0, \dots, \zeta_{\mu-1})$.

(iii) *For any* $\delta \in \mathcal{T}_S$, *there exists a matrix* $C_\delta \in M(\mu, \Gamma(S, \mathcal{O}_S))$ *such that*

$$\delta_w^{-1} \nabla_\delta \vec{\zeta} = C_\delta \vec{\zeta}. \tag{6.3}$$

□

Once we have the above two Lemmas, we obtain a primitive form $\zeta^{(0)}$ as ζ_0.

REMARK 6.7. For a given universal unfolding of an isolated singularity, the primitive form may not be unique (up to a constant factor). For instance, a simpl elliptic singularity admits a one-parameter family of primitive forms ([**S3**]). It is still an open question to determine the space of all primitive forms for a given universal unfolding. The above construction depends on a choice of (the phases) of exponents, where some "wild" choice of exponents is still allowed. One expects that a certain "tamed" class of primitive forms is to be constructed through (infinite dimensional) Lie algebras, and some works are in progress ([**K-S-T 1, K-S-T 2, K-S-T 3, S8, T1, T2**]).

7. Construction of Frobenius Manifold Structures via Primitive Forms

Let $(F, p : \mathcal{X} \to S)$ be a universal unfolding of f. In the present section, we construct a Frobenius manifold structure on S depending on a choice of a primitive form $\zeta^{(0)}$.

First, we give some direct consequences of the notion of primitive forms.

7.1. Important Properties of Primitive Forms.

Let us fix a primitive form $\zeta^{(0)}$. Note that, by (P1),

$$\mathcal{T}_S \simeq \Omega_F \simeq p_* \mathcal{O}_\mathcal{C}, \quad \delta \mapsto \widehat{\delta} F|_\mathcal{C} \cdot r^{(0)}(\zeta^{(0)}) \mapsto \widehat{\delta} F|_\mathcal{C}, \tag{S1}$$

is an $\mathcal{O}_S$-isomorphism.

PROPOSITION 7.1. *The bilinear form (again denoted by* η_F*) on* $\mathcal{T}_S$ *defined by*

$$\eta_F : \mathcal{T}_S \times \mathcal{T}_S \to \mathcal{O}_S, \quad (\delta, \delta') \mapsto \eta_F \left(r^{(0)}(\nabla_\delta \zeta^{(-1)}), r^{(0)}(\nabla_{\delta'} \zeta^{(-1)}) \right) \tag{7.1}$$

is symmetric and nondegenerate. □

The property (P2) implies the following:

PROPOSITION 7.2. *For $\delta, \delta' \in \mathcal{T}_S$, we have*

$$\boxed{K_F(\nabla_\delta \zeta^{(-1)}, \nabla_{\delta'} \zeta^{(-1)}) \in \delta_w^{-n-1} \mathcal{O}_S.} \tag{7.2}$$

□

Moreover, the conditions (P1), (P2) and (P4) imply the following:

PROPOSITION 7.3. *There exists a connection $\nabla\!\!\!/ : \mathcal{T}_S \times \mathcal{T}_S \to \mathcal{T}_S$ such that*

$$\boxed{\nabla_\delta \nabla_{\delta'} \zeta^{(-2)} = \nabla_{\delta \circ \delta'} \zeta^{(-1)} + \nabla_{\nabla\!\!\!/_\delta \delta'} \zeta^{(-2)}.} \tag{S2}$$

□

Similarly, the conditions (P1), (P2) and (P5) imply the following:

PROPOSITION 7.4. *There exists an $\mathcal{O}_S$-endomorphism $N : \mathcal{T}_S \to \mathcal{T}_S$ such that*

$$\boxed{\nabla_{\frac{d}{d\delta_w}} \nabla_\delta \zeta^{(-1)} = \nabla_{E \circ \delta} \zeta^{(-1)} - \nabla_{N\delta} \zeta^{(-2)}, \quad \delta \in \mathcal{T}_S.} \tag{S3}$$

□

Therefore, we obtain the following main result of the present paper:

THEOREM 7.5. *Let $(F, p : \mathcal{X} \to S)$ be a universal unfolding of f. Choose a primitive form $\zeta^{(0)}$. Then $(\eta_F, \circ, \delta_0, E)$ is a Frobenius manifold structure of rank μ of dimension $d = n + 1 - 2r$.*

Readers can check the axioms of a Frobenius manifold structure by direct calculation using Propositions 7.1-7.4 together with the flatness of higher residue pairings with respect to ∇. It is a consequence of the abstract notion of higher residue parings and primitive forms. We give a down to earth proof of the theorem.

7.2. Identity Vector Field.

By setting $\delta' = \delta_0$ in (S2) and compare it to (S2), we obtain the following properties:

PROPOSITION 7.6 (**Flatness of the Identity**).

$$\nabla\!\!\!/ \delta_0 = 0. \tag{7.3}$$

□

PROPOSITION 7.7 (**Homogeneity of the Identity**).

$$N(\delta_0) = r \cdot \delta_0. \tag{7.4}$$

□

7.3. Flatness of Gauß–Manin Connection.

Flatness of the Gauß–Manin connection ∇ together with equations (S2) imply the following Lemma:

LEMMA 7.8. *For $\delta, \delta', \delta'' \in \mathcal{T}_S$, we have*

$$[\nabla_\delta, \nabla_{\delta'}] \nabla_{\delta''} \zeta^{(0)} = \nabla_{[\delta,\delta']} \nabla_{\delta''} \zeta^{(0)}$$

$$\iff \begin{cases} \delta \circ (\delta' \circ \delta'') - \delta' \circ (\delta \circ \delta'') = 0 \\ \nabla\!\!\!/_\delta (\delta' \circ \delta'') + \delta \circ (\nabla\!\!\!/_{\delta'} \delta'') - \nabla\!\!\!/_{\delta'} (\delta \circ \delta'') - \delta' \circ (\nabla\!\!\!/_\delta \delta'') = [\delta, \delta'] \circ \delta'' \\ \nabla\!\!\!/_\delta (\nabla\!\!\!/_{\delta'} \delta'') - \nabla\!\!\!/_{\delta'} (\nabla\!\!\!/_\delta \delta'') = \nabla\!\!\!/_{[\delta,\delta']} \delta'' \end{cases} .$$

(7.5)

□

By this Lemma, we have the following three consequences:

PROPOSITION 7.9 (**Connection $\nabla\!\!\!/$ is Flat and Torsion-Free**).

$$[\nabla\!\!\!/_{\delta}, \nabla\!\!\!/_{\delta'}] = \nabla\!\!\!/_{[\delta,\delta']}, \quad \delta, \delta' \in \mathcal{T}_S, \tag{7.6}$$

$$\nabla\!\!\!/_{\delta}\delta' - \nabla\!\!\!/_{\delta'}\delta = [\delta, \delta'], \quad \delta, \delta' \in \mathcal{T}_S. \tag{7.7}$$

□

PROPOSITION 7.10 (**Product $\circ$ is Potential**).

$$\nabla\!\!\!/ C = 0, \tag{7.8}$$

where we denote by C the $\mathcal{O}_{\mathcal{T}_S}$-linear map $C : \mathcal{T}_S \to \mathcal{T}_S \otimes \Omega^1_S$ defined by $C_\delta \delta' := \delta \circ \delta'$, $\delta, \delta' \in \mathcal{T}_S$. □

PROPOSITION 7.11 (**Product $\circ$ is of degree** 1).

$$\mathrm{Lie}_E(\circ) = \circ, \tag{7.9}$$

namely,

$$[E, \delta \circ \delta'] - [E, \delta] \circ \delta' - \delta \circ [E, \delta'] = \delta \circ \delta', \quad \delta, \delta' \in \mathcal{T}_S. \tag{7.10}$$

□

On the other hand, the flatness of the Gauß–Manin connection ∇ together with equations (S3) imply the following Lemma:

LEMMA 7.12. *For $\delta, \delta' \in \mathcal{T}_S$, we have*

$$\left[\nabla_{\delta}, \nabla_{\frac{d}{d\delta_w}}\right] \nabla_{\delta'} \zeta^{(0)} = 0$$
$$\Longleftrightarrow \quad \begin{cases} E \circ (\delta \circ \delta') = \delta \circ (E \circ \delta') \\ N(\delta \circ \delta') + \nabla\!\!\!/_{\delta}(E \circ \delta') = E \circ \nabla\!\!\!/_{\delta}\delta' + \delta \circ N(\delta') + \delta \circ \delta' \\ \nabla\!\!\!/_{\delta} N(\delta') = N(\nabla\!\!\!/_{\delta}\delta') \end{cases} . \tag{7.11}$$

□

By this Lemma, we have the following two consequences:

LEMMA 7.13. *For $\delta \in \mathcal{T}_S$, we have*

$$N(\delta) = -\nabla\!\!\!/_{\delta} E + (r + 1)\delta. \tag{7.12}$$

□

PROPOSITION 7.14 (**Gradient of the Euler Vector Field $\nabla\!\!\!/ E$ is Flat**).

$$\nabla\!\!\!/ N = 0. \tag{7.13}$$

□

7.4. Flatness of the Higher Residue Parings.
The property (S2) with the flatness $\nabla K_F = 0$ imply the following two propositions:

PROPOSITION 7.15 (**Product $\circ$ is η_F-invariant**).

$$\eta_F(\delta \circ \delta', \delta'') = \eta_F(\delta, \delta' \circ \delta''), \quad \delta, \delta', \delta'' \in \mathcal{T}_S. \tag{7.14}$$

□

PROPOSITION 7.16 (**Connection ∇ is metric**).

$$\delta\eta_F(\delta', \delta'') = \eta_F(\nabla_\delta \delta', \delta'') + \eta_F(\delta', \nabla_\delta \delta''), \quad \delta, \delta', \delta'' \in \mathcal{T}_S. \tag{7.15}$$

□

The property (S3) with the flatness $d/d\delta_w K_F = 0$ imply the following:

LEMMA 7.17 (**η_F is homogeneous**).

$$(n+1)\eta_F(\delta, \delta') = \eta_F(N(\delta), \delta') + \eta_F(\delta, N(\delta')), \quad \delta, \delta' \in \mathcal{T}_S. \tag{7.16}$$

□

Combining this with Lemma 7.13, we have

COROLLARY 7.18 ($d = n + 1 - 2r$).

$$\mathrm{Lie}_E(\eta_F) = (2 - (n + 1 - 2r))\eta_F. \tag{7.17}$$

□

We have verified all the axioms of a Frobenius manifold. □

Appendix

It is expected from the homological mirror symmetry conjecture that one can construct Frobenius manifolds from deformations of Calabi–Yau A_∞-categories in a unified way. As is already written in the Introduction, although the original construction, explained in this paper, of the Frobenius manifolds based on filtered de Rham cohomology groups, higher residue pairings and primitive forms is formulated only for universal unfoldings of isolated hypersurface singularities, several notions and the idea of this construction are of a general nature. It has inspired and motivated the semi-infinite Hodge theory for Calabi–Yau A_∞ categories (cf. [**Ka-Ko-P, Ko-So**]).

We give a list from which readers can find the way in which objects in the original construction will be generalized and rewritten in this categorical language.

objects in this paper	generalization	
$(f, \mathbb{C}^{n+1})$	$\mathcal{A}$	A_∞-category (compact, smooth, Calabi-Yau)
J_f	$HH^\bullet(\mathcal{A},\mathcal{A})$	Hochschild cohomology
$(\Omega^\bullet_{\mathbb{C}^{n+1}}, df\wedge)$	$(C_\bullet(\mathcal{A},\mathcal{A}), b)$	Hochschild chain complex
Ω_f	$HH_\bullet(\mathcal{A},\mathcal{A})$	Hochschild homology
d	B	Connes' differential
δ_w^{-1}	u	parameter of degree 2
$\mathcal{H}_f$	$HP_\bullet(\mathcal{A})$	periodic cyclic homology
$\mathcal{H}_f^{(0)}$	$HC_\bullet^-(\mathcal{A})$	negative cyclic homology
$\Omega_f[[\delta_w^{-1}]] \Rightarrow \mathcal{H}_f^{(0)}$	$HH_\bullet(\mathcal{A},\mathcal{A})[[u]] \Rightarrow HC_\bullet^-(\mathcal{A})$	Hodge to de Rham spectral sequence
$\frac{d}{d\delta_w}$	$-u^2\frac{d}{du}$	flat connection on $HC_\bullet^-(\mathcal{A})$ with poles of order 2 at $u=0$
K_f	?	higher residue pairings
$\mathcal{O}_C \overset{\zeta}{\simeq} \Omega_f$	$HH^\bullet(\mathcal{A},\mathcal{A}) \overset{\zeta}{\simeq} HH_\bullet(\mathcal{A},\mathcal{A})$	primitive form (at the origin $0\in S$)

If one takes $\mathcal{A}$ in the list as the dg enhancement of the category of the singularity, then one recovers all objects given in the left column. As a result, the way of constructing Frobenius manifolds for isolated hypersurface singularities and that of Barannikov–Kontsevich [**Ba, Ba-Ko**] for Calabi–Yau manifolds are unified.

References

[Ba] S. Barannikov, *Quantum periods. I. Semi-infinite variations of Hodge structures*, Internat. Math. Res. Notices 2001, **23**, 1243-1264.

[Ba-Ko] S. Barannikov and M. Kontsevich, *Frobenius manifolds and formality of Lie algebras of polyvector fields*, Internat. Math. Res. Notices 1998, no. 4, 201–215.

[Br] T. Bridgeland, *Stability conditions on triangulated categories*, Ann. Math. (2) **166** (2007), 317–345.

[B] E. Brieskorn, Die Monodromie der isolierten Singularitäten von Hyperflächen, Manuscripta Math. **2** (1970), 103–161.

[CDGP] P. Candelas, X. de la Ossa, P. Green and L. Parkes, *A pair of Calabi-Yau manifolds as an exactly soluble superconformal theory*, Nucl. Phys. **B 359** (1991) 21–74.

[Do-Sab 1] A. Douai and C. Sabbah, *Gauss-Manin systems, Brieskorn lattices and Frobenius structures. I*, Proceedings of the International Conference in Honor of Frederic Pham (Nice, 2002). Ann. Inst. Fourier (Grenoble) **53** (2003), no. 4, 1055–1116.

[Do-Sab 2] ———, *Gauss-Manin systems, Brieskorn lattices and Frobenius structures. II*, Frobenius manifolds, 1–18, Aspects Math., E36, Vieweg, Wiesbaden, 2004.

[Du] B. Dubrovin, *Geometry of 2D Topological Field Theories*, in Integrable system and Quantum Groups, Montecatini,Terme 1993. Lecture Note in Math. **1620**, Springer–Verlag 1996, 120–348.

[Du-Z] B. Dubrovin and Y. Zhang, *Extended affine Weyl groups and Frobenius manifolds*, Compositio Math. **111** (1998), no. 2, 167–219.

[G] A. Givental, *Gromov-Witten invariants and quantization of quadratic Hamiltonians*, Mosc. Math. J. **1** (2001), no. 4, 551–568, 645.

[Ha] R. Hartshorne, *Residues and Duality*, Lecture Notes in Mathematics, No. 20, Springer-Verlag, Berlin-New York 1966.

[He1] C. Hertling, *Frobenius manifolds and moduli spaces for singularities*, Cambridge University Press, 2002.

[He2] C. Hertling, *tt* geometry, Frobenius manifolds, their connections, and the construction for singularities*, J. Reine Angew. Math. **555** (2003), 77–161.

[I] H. Iritani, *Real and integral structures in quantum cohomology I: toric orbifolds*, arXiv:0712.2204.

[K-S-T 1] H. Kajiura, K. Saito and A. Takahashi, *Matrix Factorizations and Representations of Quivers II: type ADE case*, Adv. Math. **211** (2007), no. 1, 327–362.

[K-S-T 2] ———, *Triangulated categories of matrix factorizations for regular systems of weights with $\epsilon = -1$*, arXiv:0708.0210.

[K-S-T 3] ———, *Weighted projective lines associated to regular systems of weights*, in preparation.

[Ka-Ko-P] L. Katzarkov, M. Kontsevich, T. Pantev, *Calculating Gromov–Witten invariants from Fukaya category*, in preparation.

[Ko] M. Kontsevich, *Homological algebra of mirror symmetry*, Proceedings of ICM (Zurich,1994), 120-139, Basel: Birkhäuser (1995).

[Ko-So] M. Kontsevich and Soibelman, *Notes on $\mathcal{A}_\infty$-algebras, $\mathcal{A}_\infty$-categories and non-commutative geometry. I*, arXiv:math/0606241.

[Ma1] Y. Manin, *Frobenius manifolds, quantum cohomology, and moduli spaces*, American Math.Society Colloquium Publ. v.**47**, 1999.

[Mi] J. Milnor, *Singular points of complex hypersurfaces*, Ann. of Math. Studies, vol. **61**, Princeton Univ. Press, 1968.

[Sab] C. Sabbah, *Déformations isomonodromiques et variétés de Frobenius*, 2002, EDP Sciences.

[S1] K. Saito, *On a linear Structure of the Quotient variety by a Finite Reflection group*, RIMS-**288**, Kyoto Univ., June 1979.

[S2] ———, *On the periods of primitive integrals I*, preprint Harvard (1980).

[S3] ———, *Period Mapping Associated to a Primitive Form*, Publ. RIMS, Kyoto University **19** (1983), 1231–1264.

[S4] ———, *Primitive forms for a universal unfolding of a function with an isolated critical point*, J. Fac. Sci. Univ. Tokyo Sect. IA **28** (1982), .

[S5] ———, *The higher residue pairings $K_F^{(k)}$ for a family of hypersurface singular points*, Proceedings of Symposia in Pure Mathematics Vol. **40** (1983), part 2, pp. 441–463.

[S6] ———, *Primitive Automorphic Forms*, Mathematics Unlimited 2001 and Beyond, edited by Engquist & Schmidt, Springer Verlag, (2001) 1003–1018.

[S7] ———, *Uniformization of the Orbifold of a Finite Reflection Group*, in Frobenius Manifolds, C. Hertling, M. Marcolli (eds.), Aspect Math. **E36** (2004), 265–320.

[S8] ———, *Towards a categorical construction of Lie algebras*, Advanced Studies in Pure Mathematics **50**, (2008), Algebraic Geometry in east Asia-Hanoi 2005, 101-175.

[S9] ———, *Extended affine root systems, II. (Flat invariants)*, Publ. RIMS **26** (1990) no.2, 15-78.

[S10] ———, *On a linear Structure of the Quotient Variety by a Finite Reflection Group*, Publ. RIMS **29** (1993), 535-579.

[SM] M. Saito, *On the structure of Brieskorn lattice*, Ann. Inst. Fourier (Grenoble) **39** (1989), no.1, 27–72.

[S-Y-S] K. Saito, T. Yano and J. Sekiguchi, *On a certain generator system of the ring of invariants of a finite reflection group*, Commun. in Alg. **8(4)** (1980), 373-408.

[Sat] I. Satake, *Flat structure and the prepotential for the elliptic root system of type $D_4^{(1,1)}$*, Topological field theory, primitive forms and related topics (Kyoto, 1996), 427–452, Progr. Math., **160**, Birkhäuser Boston, Boston, MA, 1998.

[T1] A. Takahashi, *Matrix Factorizations and Representations of Quivers I*, math.AG/0506347.
[T2] ———, *Weighted Projective Lines Associated to Regular Systems of Weights of dual Type*, arXiv:0711.3907.
[Th] R. Thom, *Structural stability and morphogenesis.* Translated from the French by D. H. Fowler. With a foreword by C. H. Waddington. Second printing. W. A. Benjamin, Inc., Reading, Mass.-London-Amsterdam, 1976.

RIMS, Kyoto University, Kyoto 606-8502, Japan
E-mail address: saito@kurims.kyoto-u.ac.jp

Department of Mathematics, Graduate School of Science, Osaka University, Toyonaka Osaka, 560-0043, Japan
E-mail address: takahashi@math.sci.osaka-u.ac.jp

Proceedings of Symposia in Pure Mathematics
Volume **78**, 2008

Twistor structures, tt^*-geometry and singularity theory

Claus Hertling and Christian Sevenheck

ABSTRACT. We give an overview on the tt^*-geometry defined for isolated hypersurface singularities and tame functions via Brieskorn lattices. We discuss nilpotent orbits in this context, as well as classifying spaces of Brieskorn lattices and (limits of) period maps.

1. Introduction

tt^*-geometry is a generalization of variation of Hodge structures. It appeared first in papers of Cecotti and Vafa ([**CV91, CV93**]). Their work connects to singularity theory through Landau-Ginzburg models. Here the tt^*-geometry is defined by oscillating integrals of tame algebraic functions and a careful handling of the underlying real structure, which leads to antiholomorphic data. A formalization of this construction was given in [**Her03**], where the basic object is a twistor, i.e., a vector bundle on $\mathbb{P}^1$, with some additional structure, namely, a flat connection with poles of order 2 at 0 and ∞. Such a twistor is is a holomorphic object, however, a family (or variation) of it over some parameter space is not.

The tame functions in Landau-Ginzburg models have isolated singularities, and the local geometry of these singularities has been studied for almost 40 years. The transcendental data of these germs of functions, i.e., Gauß-Manin connection, Brieskorn lattice and the polarized mixed Hodge structure of Steenbrink and Varchenko fit very well into the new framework provided by tt^*-geometry. In the case of germs of functions, it is possible to imitate oscillating integrals by a Fourier-Laplace transformation. The holomorphic theory of oscillating integrals for global tame functions had been studied by Pham and, more recently, by Sabbah and Douai ([**Sab06, DS03**]). Sabbah also recovered a basic positivity result of Cecotti and Vafa for that case ([**Sab05a**], see also Theorem 4.11).

A second, even older starting point is the work of Simpson on harmonic bundles and twistors ([**Sim88, Sim97**]), although the notion of harmonic bundles is slightly weaker. His work has been continued and generalized by Sabbah ([**Sab05b**]) and Mochizuki ([**Moc07**]) in Sabbah's notion of polarized twistor $\mathcal{D}$-modules.

2000 *Mathematics Subject Classification.* 14D07, 32S30, 32S40, 53C07, 32G20.
Key words and phrases. Brieskorn lattices, twistor structures, tt^*-geometry, mixed Hodge structures, TERP-structures, classifying spaces, curvature, compactifications, hyperbolicity.

Mochizuki proved results which generalize to a large extent Schmid's work on variations of Hodge structures, nilpotent orbits and limit mixed Hodge structures.

The theory of singularities of local and global functions lies at the crossroad of all these developments. In this survey we concentrate on the local functions (this is referred to as the regular singular case later on). We show how the work of Schmid and Mochizuki applies, we discuss nilpotent orbits, classifying spaces and (limits of) period maps.

The main references for the material covered here are the articles [**Her03, HS07b, HS07a**] and [**HS08**]. We will give precise definitions and complete statements for all the results presented, but almost no proofs for which we refer to the above papers.

Here is a short outline of the contents of this survey. In the next two sections, we recall classical material: the basic definition and properties of Brieskorn lattices of isolated hypersurface singularities and tame functions on affine varieties, their Fourier-Laplace transformation, and how to construct (variation of) twistor structures from them (along with the definition of the latter). An axiomatic framework for the various types of Brieskorn lattices is given under the name TERP-structure (an abbreviation for "twistor", "extension", "real structure" and "pairing"). Section 4 discusses the relation between (polarized) twistor structures and (polarized mixed) Hodge structures defined by filtrations associated to a Brieskorn lattice. The next two sections (5 and 6) investigate the twistor geometry on classifying spaces of Brieskorn lattices (resp. TERP-structures). These spaces were constructed in order to study Torelli theorems for so-called μ-constant families of singularities. We give a result on the curvature of the Hermitian metric induced on them. Moreover, we construct partial compactifications of these spaces on which the above metrics are complete. This allows us to deduce some results known for variations of Hodge structures in the case of TERP/twistor-structures using standard techniques from complex hyperbolic analysis. Finally, the last section discusses period mappings for hypersurface singularities, mainly for families over a punctured disc, and their behaviour at the boundary point $0 \in \Delta$. We show some consequences of Mochizuki's construction of a limit polarized mixed twistor and give statements analogues to those obtained for variations of Hodge structures.

2. Brieskorn lattices

We will give here a very brief reminder on the theory of Brieskorn lattices arising from isolated hypersurface singularities or polynomial functions. The goal is to motivate the constructions in the later sections, and to provide the main class of examples.

Consider first the classical situation of a germ of a holomorphic function $f : (\mathbb{C}^{n+1}, 0) \to (\mathbb{C}, 0)$ with an isolated critical point. By choosing appropriate representatives X and S for $(\mathbb{C}^{n+1}, 0)$ and $(\mathbb{C}, 0)$, this function germ yields a mapping $f : X \to S$ with the property that its restriction to $X' := f^{-1}(S')$, where $S' := S \backslash \{0\}$, is a locally trivial fibration, called the Milnor fibration. Its fibres are homotopic to a bouquet of μ spheres of (real) dimension n, where $\mu := \dim_{\mathbb{C}}(\mathcal{O}_{\mathbb{C}^{n+1},0}/J_f)$ is the Milnor number; here $J_f := (\partial_{x_0} f, \ldots, \partial_{x_n} f)$. In particular, $R^n f_* \mathbb{C}_{X'}$ is a local system, in other words, the sheaf $\mathcal{F} := R^n f_* \mathbb{C}_{X'} \otimes_{\mathbb{C}_{S'}} \mathcal{O}_{S'}$ defines a holomorphic bundle with a flat connection ∇. It was Brieskorn's idea to consider extensions of $(\mathcal{F}, \nabla)$ over the whole of S.

THEOREM 2.1 ([**Bri70, Mal74, Gre75**]). *Consider the following three sheaves and mappings between them:*

$$\mathcal{H}^{(-2)} := \mathcal{H}^n(f_*\Omega^\bullet_{X/S}) \overset{\alpha}{\hookrightarrow} \mathcal{H}^{(-1)} := \frac{f_*\Omega^n_{X/S}}{df \wedge f_*\Omega^{n-1}_{X/S}} \overset{\beta}{\hookrightarrow} \mathcal{H}^{(0)} := \frac{f_*\Omega^{n+1}_X}{df \wedge df_*\Omega^{n-1}_X}$$

(Note that these are not Brieskorn's original notations). Here α is the natural inclusion and $\beta(\widetilde{\omega}) := df \wedge \widetilde{\omega}$. All of them are $\mathcal{O}_S$-locally free and restrict to $\mathcal{F}$ on S'. The quotients $Coker(\alpha)$ and $Coker(\beta)$ are isomorphic to $f_*\Omega^{n+1}_{X/S}$ ($\beta : Coker(\alpha) \overset{\cong}{\to} f_*\Omega^{n+1}_{X/S}$), in particular, supported on $0 \in S$ and of dimension μ. Define differential operators

$$\begin{array}{cccccccc} \nabla_t : \mathcal{H}_0^{(-2)} & \longrightarrow & \mathcal{H}_0^{(-1)} & \qquad & \nabla_t : \mathcal{H}_0^{(-1)} & \longrightarrow & \mathcal{H}_0^{(0)} \\ \omega & \longmapsto & \eta & & \widetilde{\omega} & \longmapsto & d\widetilde{\omega} \end{array}$$

where η is such that $d\omega = df \wedge \eta$, and where t is the coordinate on S. These give an extension of the topological connection ∇ on $\mathcal{F}$ and define meromorphic connections with a regular singularity at $0 \in S$ on $\mathcal{H}^{(-2)}$, $\mathcal{H}^{(-1)}$ and $\mathcal{H}^{(0)}$, because $t^{-k}\mathcal{H}^{(-2)} \supset \mathcal{H}^{(-1)}$ and $t^{-k}\mathcal{H}^{(-1)} \supset \mathcal{H}^{(0)}$ where k is minimal with $f^k \in J_f$. Moreover, both are isomorphisms of $\mathbb{C}$-vector spaces. In particular, the inverse $\partial_t^{-1} := \beta \circ \nabla_t^{-1}$ is well-defined on $\mathcal{H}_0^{(0)}$ and moreover, the germ $\mathcal{H}_0^{(0)}$ is free of rank μ over the ring $\mathbb{C}\{\{\partial_t^{-1}\}\}$ of microdifferential operators with constant coefficients.

The most interesting of the three extensions is $\mathcal{H}^{(0)}$, mainly because Ω^{n+1}_X is a line bundle, so for fixed coordinates, we may represent elements in $\mathcal{H}^{(0)}$ by (classes of) functions. $\mathcal{H}^{(0)}$ is called the **Brieskorn lattice** of the function germ $(f, 0)$.

In most of the applications, one is interested in the case of families of function germs.

THEOREM 2.2 ([**Gre75**]). *Let $F : (\mathbb{C}^{n+1} \times M, 0) \to (\mathbb{C}, 0)$ be a germ where $M \subset \mathbb{C}^m$ is open such that $f := F_{|\mathbb{C}^{n+1}\times\{0\}}$ has an isolated singularity as before, then the relative Brieskorn lattice*

$$\mathcal{H}^{(0)} := \frac{\varphi_*\Omega^{n+1}_{\mathcal{X}/M}}{dF \wedge d\varphi_*\Omega^{n+1}_{\mathcal{X}/M}}$$

(where $F : \mathcal{X} \to S$ is a good representative as before, and $\varphi := F \times pr_M : \mathcal{X} \to S \times M$) is $\mathcal{O}_{S\times M}$-locally free and carries an integrable connection, which is meromorphic along the discriminant

$$\mathcal{D} := \big\{(t, \underline{y}) \,|\, \varphi^{-1}(t, \underline{y}) \text{ singular } \big\} \subset S \times M$$

It is defined by $\nabla_X(\omega) := Lie_{\widetilde{X}}(\omega) - \partial_t(Lie_{\widetilde{X}}(F)\omega)$, where $(\widetilde{X}, X) \in \mathcal{T}_{\mathcal{X}} \times \mathcal{T}_{S\times M}$ such that $d\varphi_*(\widetilde{X}) = X$ (in particular, $X \in \Theta_{S\times M}(\log \mathcal{D})$).

For a given germ $(f, 0)$, there are two particularly interesting cases of such deformations, namely:

- Take M to be an open ball in $\mathbb{C}^\mu$, and put $F := f + \sum_{i=1}^{\mu} y_i g_i$, with $(g_1, \ldots, g_\mu) \in (\mathbb{C}\{x\})^\mu$ representing a basis of $\mathbb{C}\{x\}/J_f$. Then F is called the semi-universal unfolding of f. The pair $(\mathcal{H}^{(0)}, \nabla)$ has a logarithmic pole along $\mathcal{D}$ in this case.

- Take any family F as above with the property that for each $\underline{y} \in M$, the only critical value of $F_{\underline{y}}$ is zero. A result of Gabrielov, Lazzeri and Lê shows that then the only critical point of $F_{\underline{y}}$ is the origin of $\mathbb{C}^{n+1}$. These families are called "μ-constant deformations" of f.

The second class of objects we are interested in is closely related to these function germs, but in this case phenomena of a more global nature occur. We start with a smooth affine manifold X and we want to study regular functions $f : X \to \mathbb{A}^1_{\mathbb{C}}$. There is a condition one needs to impose in order to make sure that no change of topology is caused by singularities at infinity (of a partial compactification $\overline{f} : \overline{X} \to \mathbb{A}^1_{\mathbb{C}}$).

DEFINITION 2.3 ([**NZ92**]). Let $f : X \to \mathbb{A}^1_{\mathbb{C}}$ be a regular function; then f is called M-tame iff for some embedding $X \subset \mathbb{A}^N$ and some $a \in \mathbb{A}^N$, we have that for any $\eta > 0$ there is some $R(\eta) > 0$ such that for any $r \geq R(\eta)$, the spheres $\|x - a\|^2 = r$ are transversal to all fibres $f^{-1}(t)$ for $|t| < \eta$; here $\| \cdot \|$ denotes the Euclidean distance on $\mathbb{C}^N$.

For any regular function f as above, we define

$$M_0 := \frac{\Omega_X^{alg,n+1}}{df \wedge d\Omega_X^{alg,n-1}}$$

(where $\Omega^{alg,k}$ are the algebraic differential k-forms on X) to be the algebraic Brieskorn lattice of f. The basic structure result ([**NS99, Sab06**]) is the following.

THEOREM 2.4. *If f is M-tame, then M_0 is a free $\mathbb{C}[t]$-module.*

Note that the rank of M_0 is not necessarily equal to $\mu := \sum_{x \in X} \mu(X, x)$, where $\mu(X, x)$ is the local Milnor number of the germ $f : (X, x) \to (\mathbb{C}, f(x))$. It is so if the space X is contractible, e.g., $X = \mathbb{A}^{n+1}$.

The next step is to define the Fourier-Laplace transformation of the Brieskorn lattice. In fact, the proper framework to carry this out requires the study of the Gauß-Manin system, which is an (algebraic or analytic) $\mathcal{D}$-module, in which the Briekorn lattice is embedded. One first defines the Fourier-Laplace transformation of this object and has to analyze what happens with the Brieskorn lattice under this operation. Details can be found in [**Sab06**] or [**Her03**].

We will need here and in the sequel the notion of the Deligne extensions of a flat bundle over a divisor. More precisely, let H' be a flat bundle on $\mathbb{C}^*$ (let t be a coordinate on $\mathbb{C}$), write $H^\infty = \oplus_{\lambda \in \mathbb{C}} H^\infty_\lambda$ for the generalized eigen-decomposition of the space of flat sections with respect to the monodromy, then we denote by V^α (resp. $V^{>\alpha}$) the locally free extension of H' to $\mathbb{C}$ generated by (the so-called elementary) sections $t^{\beta - \frac{N}{2\pi i}} A$, where N is the logarithm of the unipotent part of the monodromy of H', $A \in H^\infty_{e^{-2\pi i \beta}}$ and $\beta \geq \alpha$ (resp. $\beta > \alpha$). We denote similarly by V_α (resp. $V_{<\alpha}$) the corresponding Deligne extensions of H' over infinity. Finally, let $V^{>-\infty} := \cup_\alpha V^\alpha$ (resp. $V_{<\infty} := \cup_\alpha V_\alpha$) be the meromorphic extensions consisting of sections of H' with moderate growth at zero, resp. infinity.

THEOREM 2.5 ([**Sab06, Her03**]). *For a local singularity $f : (\mathbb{C}^{n+1}, 0) \to (\mathbb{C}, 0)$, consider $M_0 := H^0(\mathbb{P}^1, l_* V_{<\infty} \cap \widetilde{i}_* \mathcal{H}^{(0)})$, where $\widetilde{i} : \mathbb{C} \hookrightarrow \mathbb{P}^1$ and $l : \mathbb{P}^1 \backslash \{0\} \hookrightarrow \mathbb{P}^1$. Then M_0 is a free $\mathbb{C}[z]$-module, as is the Brieskorn lattice M_0 defined for a tame function. In both cases, define the following operators:*

$$\tau := \nabla_t \qquad \textit{and} \qquad \nabla_\tau := -t$$

In the local case, denote by G_0 the vector space M_0, seen as a $\mathbb{C}[\tau^{-1}]$-module. If f is a tame function, define $G_0 := M_0[\partial_t^{-1}]$. Then in both cases G_0 is a free $\mathbb{C}[\tau^{-1}]$-module, invariant under $-\nabla_\tau$, i.e., a free $\mathbb{C}[z]$-module, invariant under $z^2\nabla_z$, where $z := \tau^{-1}$. In both cases it has rank μ, in contrast to M_0 in the global case.

We will mainly work with analytic objects, therefore we consider the holomorphic vector bundle over $\mathbb{C}$ corresponding to G_0, denoted by H in the sequel. By definition it is equipped with a connection with a pole of order at most two at zero. Its restriction to $\mathbb{C}^*$ is then necessarily flat; we denote this restriction by H' (note that in general, H' is not equal to the local system F from the beginning of this section, although they are related, as we will see).

We are going to describe briefly another approach to the bundle H which is of more topological and analytical nature. For simplicity, we restrict to the case of tame functions, the local case is treated in detail in [**Her03**, Ch. 8]. We start with a direct construction of the flat bundle H', then we outline how to obtain the extension H. Let as above $f : X \to \mathbb{A}^1_{\mathbb{C}}$ be tame, and suppose moreover that all critical points of f are non-degenerate (although this seems to be a severe restriction, it is relatively easy to show that the construction for the general case can always be reduced to the one described by a small deformation, called Morsification). Choose an embedding $X \subset \mathbb{A}^N$ and $\eta > 0$ such that all critical values of f are contained in $S := \{t \mid |t| < \eta\}$, and choose $R(\eta) > 0$ sufficiently large. Then $U := \{x \in X \mid \|x\| < R(\eta), f(x) \in S\}$ contains all critical points of f. For any $z \in \mathbb{C}^*$, choose non-intersecting paths γ_i inside $S \backslash \{f(Crit(f))\}$ from $\eta\frac{z}{|z|}$ to the critical values of f, and construct for each such path a continuous family Γ_i (called the *Lefschetz thimble*) of vanishing cycles in $f^{-1}(\gamma_i)$. Then we have that $\Lambda_z := H_n(U, f^{-1}(\eta\frac{z}{|z|}), \mathbb{Z}) \cong \oplus_i \mathbb{Z}[\Gamma_i]$ and we put $H'_z := \mathrm{Hom}_{\mathbb{Z}}(\Lambda_z, \mathbb{C})$, which defines a local system of rank μ on $\mathbb{C}^*$. Denote by $\mathcal{H}'$ its sheaf of holomorphic sections, then one can prove that the intersection of Lefschetz thimbles on opposite fibres is well-defined (as it takes place in a compact subset of U), which gives a perfect pairing $\Lambda_z \times \Lambda_{-z} \to \mathbb{Z}$, and induces a flat pairing

$$P : \mathcal{H}' \otimes j^*\mathcal{H}' \longrightarrow \mathcal{O}_{\mathbb{C}^*}$$

Define a map

$$\begin{array}{rcl} \mathrm{osc} : \Omega_X^{alg,n+1} & \longrightarrow & i_*\mathcal{H}' \\ \omega & \longmapsto & \Big[\underbrace{z}_{\in\mathbb{C}^*} \mapsto (\underbrace{[\Gamma_i]}_{\in\Lambda_z} \mapsto \underbrace{\int_{\widetilde{\Gamma}_i} e^{-\frac{t}{z}}\omega)}_{\in\mathbb{C}}\Big] \end{array}$$

Here $i : \mathbb{C}^* \hookrightarrow \mathbb{C}$ is the inclusion and $\widetilde{\Gamma}_i$ denotes a family of vanishing cycles in $f^{-1}(\widetilde{\gamma}_i)$ extending Γ_i, where $\widetilde{\gamma}_i$ is an extension of γ_i to a path from a critical value to infinity, with asymptotic direction $\frac{z}{|z|}$. It follows from work of Pham ([**Pha83, Pha85**]) that these integrals are well-defined. Put $\mathcal{H} := \mathrm{Im}(\mathrm{osc})$, this gives a vector bundle over $\mathbb{C}$, and one checks easily that the connection has a pole of order at most two.

The following picture summarizes and illustrates the geometry used above in the definition of $\mathcal{H}'$, P and $\mathcal{H}$.

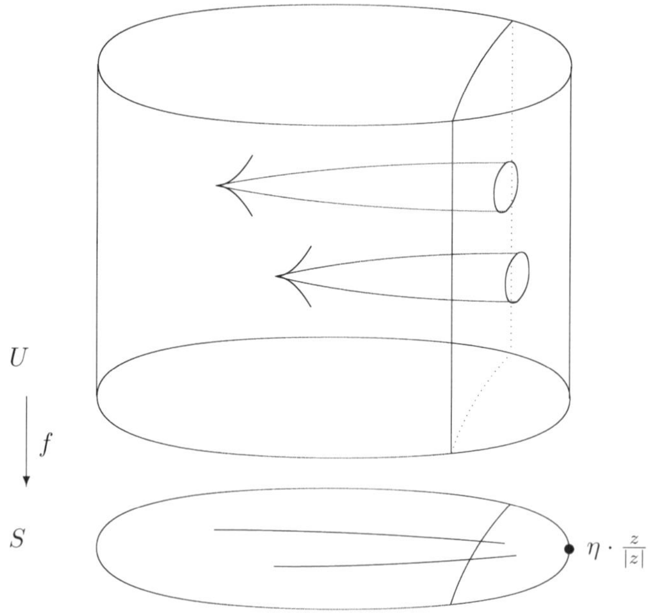

Proposition 2.6 ([**Her03**], Section 8.1). *The tuple $(\mathcal{H}, \nabla)$ constructed above using Lefschetz thimbles and oscillating integrals is exactly the Fourier-Laplace-transformation of the Brieskorn lattice M_0.*

In the case of a (deformation of a) local singularity or a non-tame deformation of a tame function, a good representative $F_{\underline{y}} : U \to S$ exists, but no extension to ∞. So one cannot work directly with oscillating integrals as above. But one can imitate them on a cohomological level [**Her03, DS03**]. In any case, if we are starting with an unfolding of a germ or of a tame function parameterized by M, then a variation of the above construction yields a holomorphic vector bundle H over $\mathbb{C} \times M$, with a connection with pole of type 1 (also called Poincaré rank one) along $\{0\} \times M$, i.e., the sheaf $\mathcal{H}$ is stable under $z^2\nabla_z$ and $z\nabla_X$ for any $X \in \mathcal{T}_M$.

3. Twistor structures

Here we define and study a framework which encompasses the objects described in the last section. The main objects are twistor and TERP-structures. The former have been introduced in [**Sim97**], for the latter, see [**Her03**] and [**HS07b**].

Definition 3.1. Let M be a complex manifold. A variation of TERP-structures on M of weight $w \in \mathbb{Z}$ is a tuple $(H, H'_\mathbb{R}, \nabla, P)$ consisting of:

(1) a holomorphic vector bundle H on $\mathbb{C} \times M$, equipped with a meromorphic connection $\nabla : \mathcal{H} \to \mathcal{H} \otimes z^{-1}\Omega^1_{\mathbb{C}\times M}(\log\{0\} \times M)$, satisfying $\nabla^2 = 0$. The restriction $H' := H_{|\mathbb{C}^* \times M}$ is then necessarily flat, corresponding to a local system given by a monodromy automorphism $M \in \mathrm{Aut}(H^\infty)$, where H^∞ is the space of flat multivalued sections. For simplicity, we will make the

assumption that the eigenvalues of M are in S^1, which is virtually always the case in applications.

(2) a flat real subbundle $H'_{\mathbb{R}}$ of maximal rank of the restriction $H' := H_{\mathbb{C}^* \times M}$. In particular, M is actually an element in $\mathrm{Aut}(H^\infty_{\mathbb{R}})$.

(3) a $(-1)^w$-symmetric, non-degenerate and flat pairing

$$P : \mathcal{H}' \otimes j^*\mathcal{H}' \to \mathcal{O}_{\mathbb{C}^*}$$

where $j(z,x) = (-z,x)$, and which takes values in $i^w\mathbb{R}$ on the real subbundle $H'_{\mathbb{R}}$. It extends to a pairing on $i_*\mathcal{H}'$, where $i : \mathbb{C}^* \times M \hookrightarrow \mathbb{C} \times M$, and has the following two properties on the subsheaf $\mathcal{H} \subset i_*\mathcal{H}'$:

(a) $P(\mathcal{H},\mathcal{H}) \subset z^w \mathcal{O}_{\mathbb{C}\times M}$.

(b) P induces a non-degenerate symmetric pairing

$$[z^{-w}P] : \mathcal{H}/z\mathcal{H} \otimes \mathcal{H}/z\mathcal{H} \to \mathcal{O}_M.$$

If no confusion is possible, we will denote a variation of TERP-structure by its underlying holomorphic bundle. A variation H over $M = \{pt\}$ is called a single TERP-structure.

A TERP-structure is called regular singular if ∇ has a regular singularity along $\{0\} \times M$.

THEOREM 3.2 ([**Her03, DS03**]). *The Fourier-Laplace transformation of the Brieskorn lattice of an isolated hypersurface singularity $f : (\mathbb{C}^{n+1}, 0) \to (\mathbb{C}, 0)$ or a tame function $f : X \to \mathbb{A}^1_{\mathbb{C}}$ underlies a TERP-structure, where P is induced by the intersection form on Lefschetz thimbles, and $w = n+1$. It is regular singular in the first case and in general irregular in the second case. Any unfolding F of a local singularity or a tame function yields a variation of TERP-structures, which is regular singular only if f is local and the unfolding is a μ-constant deformation. We will denote any such (variation of) TERP-structure(s) by TERP(f) (respectively TERP(F)).*

We will see that any TERP-structure gives rise to a twistor structure. This connects singularity theory to an priori completely different area, namely, the theory of harmonic bundles. We first recall the basic definitions, our main reference is [**Sim90, Sim97, Moc07**].

DEFINITION 3.3. A twistor is a holomorphic bundle on $\mathbb{P}^1$. A family of twistors over M is a complex vector bundle F on $\mathbb{P}^1 \times M$, equipped with a $\mathbb{P}^1$-holomorphic structure (i.e. the corresponding sheaf $\mathcal{F}$ of sections is a locally free $\mathcal{O}_{\mathbb{P}^1}\mathcal{C}^\infty_M$-module). F is called pure (of weight zero) if it is fibrewise trivial, i.e., if $\mathcal{F} = p^*p_*\mathcal{F}$ (here $p : \mathbb{P}^1 \times M \to M$ is the projection and the functor $\mathcal{O}_{\mathbb{P}^1}\mathcal{C}^\infty_M \otimes_{p^{-1}\mathcal{C}^\infty_M} p^{-1}(-)$ is denoted by p^*). To define a variation and a polarization of twistor structures, we need the map $\sigma : \mathbb{P}^1 \times M \to \mathbb{P}^1 \times M$, $\sigma(z,x) = (-\overline{z}^{-1}, x)$ and the two sheaves of meromorphic functions $\mathcal{O}_{\mathbb{P}^1}(1,0) = l_*\mathcal{O}_{\mathbb{P}^1\backslash\{0\}} \cap \widetilde{i}_* z^{-1}\mathcal{O}_{\mathbb{C}}$ and $\mathcal{O}_{\mathbb{P}^1}(0,1) = l_* z\mathcal{O}_{\mathbb{P}^1\backslash\{0\}} \cap \widetilde{i}_*\mathcal{O}_{\mathbb{C}}$ where $l : \mathbb{P}^1\backslash\{0\} \hookrightarrow \mathbb{P}^1$ and $\widetilde{i} : \mathbb{C} \hookrightarrow \mathbb{P}^1$. A variation of twistor structures is a family F of twistor structures, together with an operator

$$\mathbf{D} : \mathcal{F} \longrightarrow \mathcal{F} \otimes (\mathcal{O}_{\mathbb{P}^1}(1,0) \otimes \mathcal{A}^{1,0}_M \oplus \mathcal{O}_{\mathbb{P}^1}(0,1) \otimes \mathcal{A}^{0,1}_M)$$

such that $\mathbf{D}$ is $\mathcal{O}_{\mathbb{P}^1}$-linear and $\mathbf{D}_{|\{z\}\times M}$ is a flat connection for any $z \in \mathbb{C}^*$. Then $\mathbf{D}''_{|\mathbb{C}\times M}$ defines a holomorphic structure on $\mathcal{F}_{|\mathbb{C}\times M}$ and $\mathbf{D}'_{|\mathbb{C}\times M}$ gives a family of flat connections, with a pole of order 1 along $\{0\} \times M$, and $\mathbf{D}'_{|(\mathbb{P}^1-\{0\})\times M}$ defines an antiholomorphic structure on $\mathcal{F}_{|(\mathbb{P}^1-\{0\})\times M}$ and $\mathbf{D}''_{|(\mathbb{P}^1-\{0\})\times M}$ gives a family of flat connections, with antiholomorphic pole of order 1 along $\{\infty\}\times M$. In the case of a variation $\mathcal{F}$ of pure twistors (of weight zero) a polarization of $\mathcal{F}$ is a symmetric $\mathbf{D}$-flat non-degenerate pairing

$$S : \mathcal{F}\otimes\sigma^*\mathcal{F}\longrightarrow\mathcal{O}_{\mathbb{P}^1}$$

which is a morphism of twistors. It induces a Hermitian pairing h on $E := p_*\mathcal{F}$. Then F is called polarized if h is positive definite.

Given a pure variation, the connection $\mathbf{D}$ induces operators on E. If the variation is polarized these operators satisfy some natural compatibility conditions. More precisely, the metric h on E is harmonic, meaning that it corresponds to a pluriharmonic map from the universal cover of M to the space of positive definite Hermitian matrices. In terms more closely related to our situation, we can define (E,h) to be harmonic if $(D'+D''+\theta+\overline{\theta})^2=0$, where D'' defines the holomorphic structure on E which corresponds to $F_{|\{0\}\times M}$, θ is defined by the class of $z\mathbf{D}'_{\partial_z}\in\mathcal{E}nd_{\mathcal{O}_M}(\mathcal{F}/z\mathcal{F})$, $D'+D''$ is h-metric and $\overline{\theta}$ is the h-adjoint of θ (see also lemma 3.7 below).

Vice versa, given (E,h) with such operators D'' and θ, one might reconstruct the whole variation $(F:=p^*E,\mathbf{D},S)$ from them. The basic correspondence due to Simpson can be stated as follows.

THEOREM 3.4 ([**Sim97**], Lemma 3.1). *The category of variations of pure polarized twistor structures on M is equivalent to the category of harmonic bundles.*

We are going to construct a twistor for any given TERP-structure H. For this purpose, define $\gamma:\mathbb{P}^1\times M\to\mathbb{P}^1\times M$ to be $\gamma(z,x)=(\overline{z}^{-1},x)$. Consider the bundle $\overline{\gamma^*H}$ (here $\bar{\cdot}$ denotes the conjugate complex structure in the fibres of a bundle). It is a holomorphic bundle on the complex manifold $(\mathbb{P}^1\backslash\{0\})\times\overline{M}$ (there is no need for a conjugation in the $\mathbb{P}^1$-direction as this is already built-in in the definition of the bundle $\overline{\gamma^*H}$). We define an identification $\tau: H_{|\mathbb{C}^*\times M}\xrightarrow{\cong}\overline{\gamma^*H}_{|\mathbb{C}^*\times M}$ by

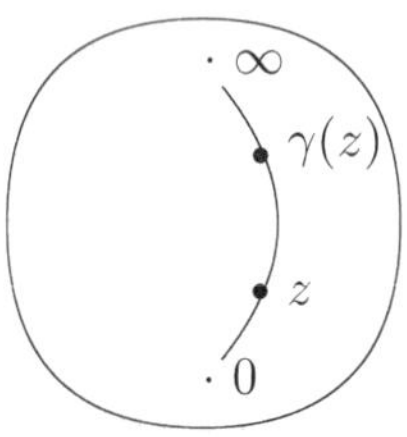

$$\begin{aligned}\tau: H_{z,x} &\longrightarrow \overline{H}_{\gamma(z),x}\\ s &\longmapsto \nabla\text{-parallel transport of } \overline{z^{-w}s}.\end{aligned}$$

PROPOSITION 3.5 ([**Her03**], Chapter 2). *τ is a linear involution, it identifies $H_{|\mathbb{C}^*\times M}$ with $\overline{\gamma^*H}_{|\mathbb{C}^*\times M}$ and defines a (variation of) twistor(s) $\widehat{H}:=H\cup_\tau\overline{\gamma^*H}$. Moreover, τ acts as an anti-linear involution on $E=p_*\widehat{H}$. We obtain a pairing S on the twistor $\widehat{H}$ by $S:=z^{-w}P(-,\tau-)$. Each bundle $\widehat{H}_{|\mathbb{P}^1\times\{x\}}$, $x\in M$, has degree zero.*

By analogy with the notion of twistors, we will make the following definition.

DEFINITION 3.6. H is called a pure TERP-structure iff $\widehat{H}$ is pure, i.e. fibrewise trivial. In that case, it is called pure polarized iff $\widehat{H}$ is so, i.e., iff $h:=S_{|p_*\widehat{H}}$ is positive definite.

The following lemma is a straightforward calculation; it gives a reformulation of the whole structure in terms of the bundle E, if the structure is pure. In particular, it allows one to define a very interesting object associated to a TERP-structure.

LEMMA 3.7 ([**Her03**], Theorem 2.19). *For any variation of pure TERP-structures, the connection operator ∇ takes the following form on fibrewise global sections*

$$\nabla = D' + D'' + z^{-1}\theta + z\overline{\theta} + \frac{dz}{z}\left(\frac{1}{z}\mathcal{U} - \mathcal{Q} + \frac{w}{2} - z\tau\mathcal{U}\tau\right)$$

The symbols used in this expressions are mappings

$$\begin{aligned} D', \theta : \mathcal{C}^\infty(E) &\longrightarrow \mathcal{C}^\infty(E) \otimes \mathcal{A}^{1,0}_M \\ D'', \overline{\theta} : \mathcal{C}^\infty(E) &\longrightarrow \mathcal{C}^\infty(E) \otimes \mathcal{A}^{0,1}_M \\ \mathcal{U}, \mathcal{Q} &\in \mathcal{E}nd_{\mathcal{C}^\infty_M}(E) \end{aligned}$$

where D' and D'' satisfy the Leibniz rule whereas θ and $\overline{\theta}$ are linear over $\mathcal{C}^\infty_M$. These objects satisfy the following relations, summarized under the name CV$\oplus$-structure in [**Her03**].

$$\begin{array}{lrcl}
(3.1) & h(\theta-,-) - h(-,\overline{\theta}) = 0, & & (D' + D'')(h) = 0 \\
(3.2) & (D'' + \theta)^2 = 0, & & (D' + \overline{\theta})^2 = 0 \\
(3.3) & D'(\theta) = 0, & & D''(\overline{\theta}) = 0 \\
(3.4) & D'D'' + D''D' & = & -(\theta\overline{\theta} + \overline{\theta}\theta) \\
(3.5) & [\theta, \mathcal{U}] = 0, & & D'(\mathcal{U}) - [\theta, \mathcal{Q}] + \theta = 0 \\
(3.6) & D''(\mathcal{U}) = 0, & & D'(\mathcal{Q}) + [\theta, \tau\mathcal{U}\tau] = 0 \\
 & \tau\theta\tau = \overline{\theta} & & (D' + D'')(\tau) = 0 \\
 & h(\mathcal{U}-,-) = h(-,\tau\mathcal{U}\tau-), & & h(\mathcal{Q}-,-) = h(-,\mathcal{Q}-) \\
 & \mathcal{Q} & = & -\tau\mathcal{Q}\tau
\end{array}$$

As already remarked, equations (3.1) to (3.4) say that the metric h on E is harmonic. The two identities (3.3) and (3.4) were called tt^*-equations in [**CV91**]. Finally, variation of twistor structures corresponding to harmonic bundles with operators $\mathcal{U}, \mathcal{Q}, \tau\mathcal{U}\tau$ are studied under the name "integrable" in [**Sab05b**, Chapter 7], and the identities (3.5) and (3.6) are called "integrability equations" in loc.cit.

A particularly interesting piece of this structure is the endomorphism $\mathcal{Q}$.

LEMMA 3.8. [**Her03**, Lemma 2.18] *If H is pure polarized, then $\mathcal{Q}$ is a Hermitian endomorphism of the bundle (E, h) and its real-analytically varying (real) eigenvalues are distributed symmetrically around zero.*

$\mathcal{Q}$ was already considered in [**CFIV92**] under the name "new supersymmetric index". We will describe some more results concerning its eigenvalues in section 7.

4. Nilpotent orbits

In this section we will consider a particular one-parameter variation of TERP-structures which gives rise to a very interesting correspondence with polarized mixed Hodge structures (PMHS). This relies on the construction of a filtration on the space H^∞. We will give two versions of this construction adapted to different situations. It can be easily checked that these families are in fact variations of TERP-structures on $\mathbb{C}^*$.

Let $(H, H'_{\mathbb{R}}, \nabla, P)$ be a given TERP-structure. We define two one-parameter variations K, resp. K', by taking the pull-backs $K := \pi^* H$, resp. $K' := (\pi')^* H$, where $\pi, \pi' : \mathbb{C} \times \mathbb{C}^* \to \mathbb{C}$ are defined by $\pi(z, r) = zr$ and $\pi'(z, r) = zr^{-1}$.

DEFINITION 4.1. K (resp. K') is called a nilpotent orbit (resp. Sabbah orbit) of TERP-structures iff its restriction K, resp. K', to $\Delta^*_r := \{z \in \mathbb{C}^* \mid |z| < r\}$ is pure polarized for $|r| \ll 1$.

As explained in section 2, the main sources of TERP-structures are Brieskorn lattices defined by holomorphic functions or germs. In the local case, a filtration on H^∞ has been defined in [**Var80**] and was later modified in [**SS85**]. In our language, these definitions are valid for regular singular TERP-structures. In the general case, we give a definition due to Sabbah [**Sab06**].

We use the Deligne extensions defined earlier. They induce a filtration, called a V-filtration, on the Brieskorn lattice so that one might consider the corresponding graded object. These graded parts determine the filtrations we are looking for.

DEFINITION 4.2. Let $(H, H'_{\mathbb{R}}, \nabla, P)$ be a given TERP-structure of weight w.

(1) Suppose that (H, ∇) is regular singular at zero (i.e. $\mathcal{H} \subset V^{>-\infty}$). Then put for $\alpha \in (0, 1]$

$$F^p H^\infty_{e^{-2\pi i\alpha}} := z^{p+1-w-\alpha+\frac{N}{2\pi i}} Gr_V^{\alpha+w-1-p} \mathcal{H} \tag{4.1}$$

(2) Let (H, ∇) be arbitrary and put $G_0 := H^0(\mathbb{P}^1, l_* V_{<\infty} \cap \widetilde{i}_* \mathcal{H})$ (where, as before, $\widetilde{i} : \mathbb{C} \hookrightarrow \mathbb{P}^1$ and $l : \mathbb{P}^1 \backslash \{0\} \hookrightarrow \mathbb{P}^1$). Then G_0 is a free $\mathbb{C}[z]$-module and the Deligne extensions at infinity induce a filtration on it so that we can define for $\alpha \in (0, 1]$

$$F^p_{Sab} H^\infty_{e^{-2\pi i\alpha}} := z^{p+1-w-\alpha+\frac{N}{2\pi i}} Gr^V_{\alpha+w-1-p} G_0 \tag{4.2}$$

We will actually not work with $F^\bullet$ and $F^\bullet_{Sab}$ directly, but with a twisted version defined by $\widetilde{F}^\bullet := G^{-1} F^\bullet$ and similarly for $F^\bullet_{Sab}$, where $G := \sum_{\alpha\in(0,1]} G^{(\alpha)} \in \operatorname{Aut}\left(H^\infty = \oplus_\alpha H^\infty_{e^{-2\pi i\alpha}}\right)$ is defined as follows (see [**Her03**, (7.47)]):

$$G^{(\alpha)} := \sum_{k\geq 0} \frac{1}{k!} \Gamma^{(k)}(\alpha) \left(\frac{-N}{2\pi i}\right)^k =: \Gamma\left(\alpha \cdot \mathrm{id} - \frac{N}{2\pi i}\right).$$

Here $\Gamma^{(k)}$ is the k-th derivative of the gamma function. In particular, G depends only on H' and induces the identity on $Gr^W_\bullet$, $W_\bullet$ being the weight filtration of N.

Note that for $H = TERP(f)$, where f is a local singularity, Steenbrink's Hodge filtration from [**SS85**] is exactly our $\widetilde{F}^\bullet$. It was defined in loc.cit. by a formula similar to (4.1), but using the Brieskorn lattice $\mathcal{H}^{(0)}$ and its V-filtration. On the other hand, the filtration $F^\bullet$ is defined using the Fourier-Laplace transform $\mathcal{H}$ of $\mathcal{H}^{(0)}$, so that G can be seen as the "topological part" of the Fourier-Laplace transformation.

The next ingredient we need is a polarizing form S induced on H^∞ by the pairing P. It can be defined as follows: P induces a pairing (denoted by the same symbol) on the local system $(H')^\nabla$, then given $A, B \in H^\infty$, we put $S(A, B) :=$

$(-1)(2\pi i)^w P(A, t(B))$ where

$$t(B) = \begin{cases} (M - \mathrm{Id})^{-1}(B) & \forall B \in H^\infty_{\neq 1} \\ -(\sum_{k\geq 1} \frac{1}{k!} N^{k-1})^{-1}(B) & \forall B \in H^\infty_1. \end{cases}$$

S is nondegenerate, monodromy invariant, $(-1)^w$-symmetric on H^∞_1, $(-1)^{w-1}$-symmetric on $H^\infty_{\neq 1}$, and it takes real values on $H^\infty_{\mathbb{R}}$ [**Her03**, Lemma 7.6]. A basic result concerning $\widetilde{F}^\bullet$ and S is the following.

THEOREM 4.3 ([**Var80, SS85, Her02**]). *For a local singularity f, the tuple $(H^\infty, H^\infty_{\mathbb{R}}, -N, S, \widetilde{F}^\bullet)$ is a (sum of) polarized mixed Hodge structure(s) (of weight w on H^∞_1 and of weight $w-1$ on $H^\infty_{\neq 1}$).*

We come back to our special situation of a family $K := \pi^* H$. Suppose first that it is regular singular (it suffices actually to check this at any value r, e.g., for $H = K_{|r=1}$).

THEOREM 4.4 ([**Her03, HS07b**]). *The family K is a nilpotent orbit of TERP-structures, i.e., it is pure polarized on a disc with sufficiently small radius iff $(H^\infty, H^\infty_{\mathbb{R}}, -N, S, \widetilde{F}^\bullet)$ defines a PMHS of weight $w-1$ on $H^\infty_{\neq 1}$ and a PMHS of weight w on H^∞_1.*

Let us give some comments on the proof: The direction "$\Leftarrow$" is proved in [**Her03**], and relies on the corresponding correspondence between PMHS and nilpotent orbits of Hodge structures ([**Sch73, CKS86**]). The opposite direction "$\Rightarrow$" is done in [**HS07b**], and uses a different strategy: It is built on a central result in [**Moc07**] which states that a tame variation of pure polarized twistors on, say Δ^*, degenerates to what is called a polarized mixed twistor structure (see section 7). In our situation, this limit object corresponds exactly to the (sum of) PMHS('s) on H^∞, which shows the desired result.

For arbitrary TERP-structures, we have a corresponding statement for Sabbah orbits, which is proven in exactly the same way.

THEOREM 4.5 ([**HS07b**]). *Let $(H, H'_{\mathbb{R}}, \nabla, P)$ be arbitrary. Then the family K' is a Sabbah orbit on Δ^*_r for sufficiently small r iff $(H^\infty, H^\infty_{\mathbb{R}}, N, S, \widetilde{F}^\bullet_{Sab})$ defines a PMHS of weight $w-1$ on $H^\infty_{\neq 1}$ and a PMHS of weight w on H^∞_1.*

REMARK 4.6. In [**HS07b**] a conjecture is stated which generalizes the correspondence in Theorem 4.4 to the case of general (i.e. irregular) TERP-structures. The notion of a nilpotent orbit is the same, but the other side, the PMHS, has to be replaced by more subtle conditions: First, it is assumed that the formal decomposition of the germ $(\mathcal{H}, \nabla)_0$ into model bundles exists without ramification. Second, Stokes structure and real structure shall be compatible. Third, the regular singular parts of the model bundles shall induce PMHS just as above. The implication from this generalization of a PMHS to a nilpotent orbit is proved in [**HS07b**], the other implication is still open. Notice that in the case of a TERP-structure of a deformation of a germ or a tame function the three conditions for the generalization of a PMHS are all satisfied: a ramification is not necessary, the regular singular parts of the model bundles are the (Fourier-Laplace transforms of the) Brieskorn lattices of the local singularities, Theorem 4.3 gives their PMHS, and the real and Stokes structure are compatible because both are defined by Lefschetz thimbles.

We will give some comments on applications of these results. First note the following simple lemma, which follows from the definition of the Gauß-Manin connection on the Brieskorn lattice given in section 2. It shows how nilpotent orbits arise naturally in singularity theory.

LEMMA 4.7 ([**HS07b**], lemma 11.3). *Let f be a germ of a holomorphic function with isolated singularities or a tame function on an affine manifold. Then for any $r \in \mathbb{C}$*

$$TERP(r \cdot f) = \pi'(TERP(f))_{|r}$$

Moreover, in the local case, the Hodge filtration $\widetilde{F}^\bullet(f)$ associated to $TERP(f)$ satisfies.

$$\widetilde{F}^\bullet(r \cdot f) = r^{\frac{N}{2\pi i}} \widetilde{F}^\bullet(f).$$

An easy consequence of the second part is the following result, which uses only classical facts about nilpotent orbits of Hodge structures.

COROLLARY 4.8. *For any function germ f and any sufficiently large $r \in \mathbb{R}$, we have that $\widetilde{F}^\bullet(r \cdot f)$ gives a pure polarized Hodge structure on H^∞ (of weight w on H^∞_1 and of weight $w-1$ on $H^\infty_{\neq 1}$).*

The following application uses the generalization to the irregular case discussed above and the geometry on the unfolding space of the germ f as discussed in detail in [**Her02**].

COROLLARY 4.9. *Let $F : (\mathbb{C}^{n+1} \times M, 0) \to (\mathbb{C}, 0)$ be a semi-universal unfolding of an isolated hypersurface singularity $f = F_{|\mathbb{C}^{n+1}\times\{0\}} : (\mathbb{C}^{n+1}, 0) \to (\mathbb{C}, 0)$. Then for any $\underline{y}_0 \in M$, $TERP(r \cdot F_{\underline{y}_0})$ is pure polarized for $r \gg 0$. Note that the families $\{r \cdot F_{\underline{y}} \mid r \in \mathbb{C}^*\}$ correspond to the orbits of the Euler field E on M, so this result says that going sufficiently far in M along $\mathrm{Re}(E)$, one always arrives at pure polarized TERP-structures.*

In the case of tame functions, we can use the other direction (of Theorem 4.5) to strengthen a result of Sabbah.

COROLLARY 4.10 ([**HS07b**], Corollary 11.4). *Let $f : X \to \mathbb{A}^1_{\mathbb{C}}$ be a given tame function, then the tuple $(H^\infty, H^\infty_{\mathbb{R}}, N, S, \widetilde{F}^\bullet_{Sab})$ associated to $TERP(f)$ is a polarized mixed Hodge structure.*

In [**Sab06**] it was shown that $(H^\infty, H^\infty_{\mathbb{R}}, N, F^\bullet_{Sab})$ is a mixed Hodge structure. The corollary follows from Theorem 4.5 using another, more recent result, which we believe to be of fundamental importance in the theory of twistor structures associated to tame functions. It already appears in a completely different language in [**CV91**] and [**CV93**].

THEOREM 4.11 ([**Sab05a**], see also [**Sab08b**]). *Let $f : X \to \mathbb{A}^1_{\mathbb{C}}$ be a tame function. Then $TERP(f)$ is pure polarized.*

5. Classifying spaces and curvature

In order to study period mappings for μ-constant deformations of hypersurface singularities, appropriate classifying spaces for Brieskorn lattices were discussed in [**Her99**]. We first give a short review of this construction, then we show how to endow a part (called pure polarized) of these classifying spaces with a Hermitian metric.

An important result in the theory of variation of Hodge structures is that the curvature of the Hermitian metric on the classifying spaces of Hodge structures is negative along horizontal directions. It turns out that a similar result holds for variation of TERP-structures; this will also be explained in the current section.

The first ingredient for constructing the classifying spaces is a classical cohomological invariant, the spectrum, attached to an isolated hypersurface singularity, which can in fact be defined for any regular singular TERP-structure. Note that our definition gives real numbers shifted by one compared to the original definition of Varchenko and Steenbrink. These shifted numbers were used by M. Saito under the name exponents.

DEFINITION 5.1. Let $(H, H'_{\mathbb{R}}, \nabla, P)$ be a regular singular TERP-structure of weight w.

(1) The spectrum of (H, ∇) at zero is defined as $\mathrm{Sp}(H,\nabla) = \sum_{\alpha\in\mathbb{Q}} d(\alpha)\cdot\alpha \in \mathbb{Z}[\mathbb{Q}]$ where

$$d(\alpha) := \dim_{\mathbb{C}}\left(\frac{Gr_V^{\alpha}\mathcal{H}}{Gr_V^{\alpha} z\mathcal{H}}\right) = \dim_{\mathbb{C}} Gr_F^{\lfloor w-\alpha\rfloor} H^{\infty}_{e^{-2\pi i\alpha}}.$$

We also write $\mathrm{Sp}(H,\nabla)$ as a tuple $\alpha_1,\ldots,\alpha_\mu$ of μ numbers (with $\mu = \mathrm{rank}(H)$), ordered by $\alpha_1 \leq \ldots \leq \alpha_\mu$. We have that $\alpha_i = w - \alpha_{\mu+1-i}$ and that α is a spectral number only if $e^{-2\pi i\alpha}$ is an eigenvalue of the monodromy M of H' (in particular, all α_i are real by assumption).

(2) The spectral pairs are a refinement of the spectrum taking into account the weight filtration $W_\bullet(N)$ (Here the restriction of $W_\bullet(N)$ to H^∞_1 is centered around w, and the restriction to $H^\infty_{\neq 1}$ is centered around $w-1$). They are given by $\mathrm{Spp}(H,\nabla) := \sum_{\alpha\in\mathbb{Q}} \widetilde{d}(\alpha,k)\cdot(\alpha,k) \in \mathbb{Z}[\mathbb{Q}\times\mathbb{Z}]$, where

$$\widetilde{d}(\alpha,k) := \begin{cases} \dim_{\mathbb{C}} Gr_F^{\lfloor w-\alpha\rfloor} Gr_k^W H^\infty_{e^{-2\pi i\alpha}} & \forall\alpha\notin\mathbb{Z} \\ \dim_{\mathbb{C}} Gr_F^{w-\alpha} Gr_{k+1}^W H^\infty_1 & \forall\alpha\in\mathbb{Z} \end{cases}$$

For the remainder of this section, we fix $H^\infty, H^\infty_{\mathbb{R}}, S, M, w, \mathrm{Spp}$ which we call initial data. Note that $(H^\infty, H^\infty_{\mathbb{R}}, S, M)$ are equivalent to the datum of a flat bundle H' over $\mathbb{C}^*$ with a flat real subbundle and a pairing P as in Definition 3.1, where S is induced from P as in section 4. For notational convenience, let $n := \lfloor\alpha_\mu - \alpha_1\rfloor$. We also fix a reference filtration $F_0^\bullet$ such that $(H^\infty, H^\infty_{\mathbb{R}}, N, S, F_0^\bullet)$ is a PMHS. Denote by $P_l \subset Gr_l^W$ the primitive subspace of the weight filtrations $W_\bullet$ of N. Then put

$$\begin{aligned}\check{D}_{PMHS} \;:=\; & \{F^\bullet H^\infty \mid \dim F^p P_l = \dim F_0^p P_l;\ S(F^p, F^{w+1-p}) = 0; \\ & N(F^p)\subset F^{p-1}, \text{ and all powers of } N \text{ are strict with respect to } F^\bullet\}\end{aligned}$$

There is a projection map $\check{\beta} : \check{D}_{PMHS} \to \check{D}_{PHS}$, sending $F^\bullet$ to $(F^\bullet P_l)_{l\in\mathbb{Z}}$, here $\check{D}_{PHS}$ is a product of classifying spaces of Hodge-like filtrations on the primitive subspaces P_l. $\check{D}_{PHS}$ is a complex homogeneous space, and it contains the open submanifold D_{PHS} which is a product of classifying spaces of polarized Hodge structures, and which has the structure of a real homogeneous space. $\check{\beta}$ is a locally trivial fibration with affine spaces as fibres. Define D_{PMHS} to be the restriction of this fibration to

D_{PHS}. Then also $\check{D}_{PMHS}$ (resp. D_{PMHS}) is a complex (resp. real) homogeneous space.

One can construct period maps for singularities by associating to a singularity f its Hodge filtration $\widetilde{F}^\bullet$. However, it is readily seen that in general the information encoded in the Hodge structure is too weak for, say, Torelli theorems. The basic philosophy is that the whole Brieskorn lattice should be sufficiently rich to determine the singularity. This calls for a classifying space for Brieskorn lattices, which is constructed as follows. Put

$$\begin{aligned} \check{D}_{BL} \quad := \quad & \{\mathcal{H} \subset V^{\alpha_1} \,|\, H \to \mathbb{C} \text{ holomorphic vector bundle}, H_{|\mathbb{C}^*} = H', \\ & (z^2\nabla_z)(\mathcal{H}) \subset \mathcal{H}, P(\mathcal{H}_0, \mathcal{H}_0) \subset z^w\mathbb{C}\{z\} \text{ non-degenerate}, \widetilde{F}^\bullet_H \in \check{D}_{PMHS}\,\} \end{aligned}$$

Remember that $F^\bullet_H$ is the filtration defined by H using formula (4.1) on H^∞, and $\widetilde{F}^\bullet_H = G^{-1}F^\bullet_H$. As before, we obtain a projection map $\check{\alpha} : \check{D}_{BL} \to \check{D}_{PMHS}$ by $H \mapsto \widetilde{F}^\bullet_H$. One of the main results of [**Her99**] is that this map is still a locally trivial fibration where the fibres are affine spaces. Again we put $D_{BL} := \check{\alpha}^{-1}(D_{PMHS})$. The situation can be visualized in the following diagram.

$$\begin{array}{ccc} D_{BL} & \hookrightarrow & \check{D}_{BL} \\ \downarrow \alpha & & \downarrow \check{\alpha} \\ D_{PMHS} & \hookrightarrow & \check{D}_{PMHS} \\ \downarrow \beta & & \downarrow \check{\beta} \\ D_{PHS} & \overset{\gamma}{\hookrightarrow} & \check{D}_{PHS} \end{array}$$

Note that neither $\check{D}_{BL}$ nor D_{BL} are homogeneous. However, there is a good $\mathbb{C}^*$-action on the fibers of $\check{D}_{BL}$; the corresponding zero section $\check{D}_{PMHS} \hookrightarrow \check{D}_{BL}$ consists of regular singular TERP-structures in $\check{D}_{BL}$ which are generated by elementary sections, see [**Her99**, Theorem 5.6].

Now suppose that we have in addition to the initial data a lattice $H^\infty_\mathbb{Z} \subset H^\infty_\mathbb{R}$ such that $M \in \mathrm{Aut}(H^\infty_\mathbb{Z})$. Then the discrete group $G_\mathbb{Z} := \mathrm{Aut}(H^\infty_\mathbb{Z}, S, M)$ acts properly discontinuously on D_{PMHS} and D_{BL} with normal complex spaces as quotients [**Her99**, Prop. 2.6 and Cor. 5.7]. For any μ-constant family of isolated hypersurface singularities parameterized by a manifold M, we obtain period maps $\phi_{PMHS} : M \to D_{PMHS}/G_\mathbb{Z}$ and $\phi_{BL} : M \to D_{BL}/G_\mathbb{Z}$ which are holomorphic and locally liftable (here the fact, due to Varchenko [**Var82**], that for such a family the spectral pairs are constant is used).

From the very construction of the classifying spaces, we have the following fact.

LEMMA 5.2 ([**HS07a**], Lemma 3.1). *The space $\check{D}_{BL}$ comes equipped with a universal bundle $H \to \mathbb{C} \times \check{D}_{BL}$ of Brieskorn lattices. H underlies a family of TERP-structures on $\check{D}_{BL}$, but not a variation; more precisely, we have a connection operator*

$$\nabla : \mathcal{H} \longrightarrow \mathcal{H} \otimes \left(z^{-2}\Omega^1_{\mathbb{C}\times\check{D}_{BL}/\check{D}_{BL}} \oplus z^{-n}\Omega^1_{\mathbb{C}\times\check{D}_{BL}/\mathbb{C}}\right)$$

We put

$$\check{D}^{pp}_{BL} := \{x \in \check{D}_{BL} \,|\, \widehat{H}_{|x} \text{ is pure polarized }\}$$

We obtain a Hermitian metric on $p_*\widehat{H}_{|\check{D}^{pp}_{BL}}$ by definition. A nice feature of the space $\check{D}^{pp}_{BL}$ is that this metric can be transferred canonically to its tangent bundle.

PROPOSITION 5.3. *The tangent sheaf* $\mathcal{T}_{\check{D}_{BL}}$ *is naturally a subbundle of*

$$\widetilde{\mathcal{H}} := \mathcal{H}om_{\mathcal{O}_{\mathbb{C}\times\check{D}_{BL}}}\left(\mathcal{H}, \frac{z^{-n}\mathcal{H}}{\mathcal{H}}\right).$$

Here $\mathcal{H}$ *and* $z^{-n}\mathcal{H}$ *are seen as submodules of* $\mathcal{H}\otimes\mathcal{O}_{\mathbb{C}\times\check{D}_{BL}}(*\{0\}\times\check{D}_{BL})$. *There is a coherent subsheaf* $\mathcal{T}^{hor}_{\check{D}_{BL}}$ *of* $\mathcal{T}_{\check{D}_{BL}}$ *such that for any variation of TERP-structures on a simply connected space* M, *we have that* $d\phi_{BL}(\mathcal{T}_M) \subset \phi^*_{BL}(\mathcal{T}^{hor}_{\check{D}_{BL}})$ *(e.g. this holds for the local lifts of the period map from above). On the subspace* $\check{D}^{pp}_{BL} \subset \check{D}_{BL}$, *the tangent bundle* $\mathcal{T}_{\check{D}^{pp}_{BL}}$ *is naturally equipped with a positive definite Hermitian metric* h.

PROOF. First note that even though $\mathcal{H}$ is an $\mathcal{O}_{\mathbb{C}\times\check{D}_{BL}}$-module, the quotient $z^{-n}\mathcal{H}/\mathcal{H}$ is z-torsion and so is $\widetilde{\mathcal{H}}$, which might thus be considered as an $\mathcal{O}_{\check{D}_{BL}}$-module. Then the inclusion $\mathcal{T}_{\check{D}_{BL}} \subset \widetilde{\mathcal{H}}$ is given by the Kodaira-Spencer map

$$\begin{array}{rcl} KS : \mathcal{T}_{\check{D}_{BL}} & \longrightarrow & \widetilde{\mathcal{H}} \\ X & \longmapsto & [s \mapsto \nabla_X(s)]\,. \end{array}$$

Injectivity follows from the universality of the classifying space $\check{D}_{BL}$. The subsheaf $\mathcal{T}^{hor}_{\check{D}_{BL}}$ is defined by

$$\mathcal{T}^{hor}_{\check{D}_{BL}} := \mathcal{T}_{\check{D}_{BL}} \cap \mathcal{H}om_{\mathcal{O}_{\mathbb{C}\times\check{D}_{BL}}}\left(\mathcal{H}, \frac{z^{-1}\mathcal{H}}{\mathcal{H}}\right).$$

from which its coherence is obvious. Computation of examples shows that it is not necessarily locally free, except if $n = 1$, then $\mathcal{T}^{hor}_{\check{D}_{BL}} = \mathcal{T}_{\check{D}_{BL}}$. Consider now the tangent bundle on the open subspace $\check{D}^{pp}_{BL}$. By definition, $H_{|\mathbb{C}\times\check{D}^{pp}_{BL}}$ is pure polarized, and $p_*\widehat{\mathcal{H}}_{|\check{D}^{pp}_{BL}}$ is a locally free $\mathcal{C}^\infty_{\check{D}^{pp}_{BL}}$-module of rank μ. Let us denote by $\mathcal{R}$ the sheaf of rings $\mathcal{O}_{\mathbb{C}}\mathcal{C}^\infty_{\check{D}^{pp}_{BL}}$ for a moment. We write $\mathcal{H}^{sp}$ for the sheaf $p_*\widehat{\mathcal{H}}_{|\check{D}^{pp}_{BL}}$. It gives a splitting of the projection $\mathcal{R}\otimes\mathcal{H}_{|\mathbb{C}\times\check{D}^{pp}_{BL}} \twoheadrightarrow \mathcal{R}\otimes(\mathcal{H}/z\mathcal{H})_{|\mathbb{C}\times\check{D}^{pp}_{BL}}$, i.e. $k^{-1}(\mathcal{R}\otimes\mathcal{H}_{|\mathbb{C}\times\check{D}^{pp}_{BL}}) = \mathcal{H}^{sp}\oplus k^{-1}(\mathcal{R}\otimes(z\mathcal{H})_{|\mathbb{C}\times\check{D}^{pp}_{BL}})$, where $k : \check{D}^{pp}_{BL} \hookrightarrow \mathbb{C}\times\check{D}^{pp}_{BL}$, $t \mapsto (0,t)$. This implies that

$$\begin{aligned} &\mathcal{C}^\infty_{\check{D}^{pp}_{BL}} \otimes \widetilde{\mathcal{H}}_{|\check{D}^{pp}_{BL}} \cong \mathcal{H}om_{\mathcal{R}}\left(\mathcal{R}\otimes\mathcal{H}_{|\mathbb{C}\times\check{D}^{pp}_{BL}}, \mathcal{R}\otimes(\tfrac{z^{-n}\mathcal{H}}{\mathcal{H}}{}_{|\mathbb{C}\times\check{D}^{pp}_{BL}})\right) \cong \\ (5.1)\qquad &\bigoplus_{k=1}^n \mathcal{H}om_{\mathcal{C}^\infty_{\check{D}^{pp}_{BL}}}\left(\mathcal{H}^{sp}, z^{-k}\mathcal{H}^{sp}\right) \cong \left[\mathcal{E}nd_{\mathcal{C}^\infty_{\check{D}^{pp}_{BL}}}\left(\mathcal{H}^{sp}\right)\right]^n. \end{aligned}$$

The Hermitian metric on $p_*\widehat{\mathcal{H}}_{|\check{D}^{pp}_{BL}}$ induces a metric on $\mathcal{E}nd_{\mathcal{C}^\infty_{\check{D}^{pp}_{BL}}}(p_*\widehat{\mathcal{H}}_{|\check{D}^{pp}_{BL}})$ (by putting $h(\phi,\psi) := \mathrm{Tr}(\phi\cdot\psi^*)$, where ψ^* is the Hermitian adjoint of ψ) and on its powers and this metric induces by restriction a positive definite metric on $\mathcal{T}_{\check{D}^{pp}_{BL}}$. □

The main result of [**HS07a**] is the following.

THEOREM 5.4 ([**HS07a**], Theorem 4.1). *Consider the holomorphic sectional curvature*

$$\begin{array}{rcl} \kappa : T_{\check{D}^{pp}_{BL}}\backslash\{\textit{zero-section}\} & \longrightarrow & \mathbb{R} \\ v & \longmapsto & h(R(v,\overline{v})v,v)/h^2(v,v) \end{array}$$

of the metric h. Here $R \in \mathcal{E}nd_{\mathcal{C}^\infty_{\check{D}^{pp}_{BL}}}(\mathcal{T}^{\mathbb{C}}_{\check{D}^{pp}_{BL}}) \otimes \mathcal{A}^{1,1}_{\check{D}^{pp}_{BL}}$ *is the curvature tensor of the Chern connection of h. Then the restriction of* κ *to* $T^{hor}_{\check{D}^{pp}_{BL}} \backslash \{zero\ section\}$ *(i.e., the complement of the zero section of the linear space associated to the coherent sheaf* $\mathcal{T}^{hor}_{\check{D}^{pp}_{BL}}$*) is bounded from above by a negative number.*

The following is a direct consequence and proved as in the Hodge case using Ahlfors' lemma.

COROLLARY 5.5 ([**HS07a**], Proposition 4.3). *Any horizontal mapping* $\phi : M \to \check{D}^{pp}_{BL}$ *from a complex manifold M is distance decreasing with respect to the Kobayashi pseudodistance on M and the distance induced by h on* $\check{D}^{pp}_{BL}$.

As an application, we obtain the following rigidity result.

COROLLARY 5.6 ([**HS07a**], Corollary 4.5). *Any variation of regular singular pure polarized TERP-structures with constant spectral pairs on* $\mathbb{C}^m$ *is trivial, i.e., flat in the parameter direction.*

This result follows directly from the last corollary as the Kobayashi pseudodistance of $\mathbb{C}^m$ is zero, so that the corresponding period map must be constant (see, e.g., [**Kob05**]).

6. Compactifications

A drawback of the construction of the last section is that the metric h on $\check{D}^{pp}_{BL}$ is not complete. The reason is that there exist families (even variations) of regular singular pure polarized TERP-structures, where the spectral numbers jump for special parameters. Period mappings of such families lead naturally to a partial compactification of $\check{D}^{pp}_{BL}$. The following example illustrates this phenomenon of jumping spectral numbers.

EXAMPLE 6.1 ([**HS08**], Section 9.2). We will describe a variation of TERP-structures over $\mathbb{C}$ which extends to a variation over $\mathbb{P}^1$ with jumping spectral numbers at ∞. Consider a three-dimensional real vector space $H^\infty_{\mathbb{R}}$, its complexification $H^\infty := H^\infty_{\mathbb{R}} \otimes \mathbb{C}$, choose a basis $H^\infty = \oplus_{i=1}^3 \mathbb{C}A_i$ such that $\overline{A_1} = A_3$, $A_2 \in H^\infty_{\mathbb{R}}$. Choose a real number $\alpha_1 \in (-3/2, -1)$, put $\alpha_2 := 0$, $\alpha_3 := -\alpha_1$ and let $M \in \mathrm{Aut}(H^\infty_{\mathbb{C}})$ be given by $M(\underline{A}) = \underline{A} \cdot \mathrm{diag}(\lambda_1, \lambda_2, \lambda_3)$ where $\underline{A} := (A_1, A_2, A_3)$ and $\lambda_i := e^{-2\pi i \alpha_i}$ (then M is actually an element in $\mathrm{Aut}(H^\infty_{\mathbb{R}})$). Let, as before, $(H', H'_{\mathbb{R}}, \nabla)$ be the flat holomorphic bundle on $\mathbb{C}^* \times \mathbb{C}$ with real flat subbundle corresponding to $(H^\infty, H^\infty_{\mathbb{R}}, M)$, and put $s_i := z^{\alpha_i} A_i \in \mathcal{H}'$. Denote by r a coordinate on $\mathbb{C}$, and define $\mathcal{H} := \oplus_{i=1}^3 \mathcal{O}_{\mathbb{C}^2} v_i$, where

$$\begin{aligned} v_1 &:= s_1 + rz^{-1}s_2 + \tfrac{r^2}{2} z^{-2} s_3 \\ v_2 &:= s_2 + rz^{-1} s_3 \\ v_3 &:= s_3 \end{aligned}$$

Moreover, define the pairing P with $P : \mathcal{H}' \otimes j^*\mathcal{H}' \to \mathcal{O}_{\mathbb{C}^* \times \mathbb{C}}$ by $P(\underline{s}^{tr}, \underline{s}) := (\delta_{i+j,4})_{i,j \in \{1,\ldots,3\}}$.

It can be checked by direct calculations that $(H, H'_{\mathbb{R}}, \nabla, P)$ is a variation of regular singular TERP-structures on $\mathbb{C}$. Moreover, the Hodge filtration induced on H^∞ is constant in r and gives a sum of pure polarized Hodge structures of weights 0 and -1 on H^∞_1 and $H^\infty_{\neq 1}$; namely, we have that

$$\{0\} = F^2 \subsetneq F^1 := \mathbb{C}A_1 \subsetneq F^0 := \mathbb{C}A_1 \oplus \mathbb{C}A_2 = F^{-1} \subsetneq F^{-2} := H^\infty$$

The polarizing form S is defined by P via [**Her03**, (7.51), (7.52)], namely, we have:

$$S(\underline{A}^{tr}, \underline{A}) := \begin{pmatrix} 0 & 0 & \gamma \\ 0 & 1 & 0 \\ -\gamma & 0 & 0 \end{pmatrix},$$

where $\gamma := \frac{-1}{2\pi i}\Gamma(\alpha_1+2)\Gamma(\alpha_3-1)$. In particular, we have for $p=1$

$$i^{p-(-1-p)}S(A_1, A_3) = (-1)iS(A_1, A_3) = \frac{\Gamma(\alpha_1+2)\Gamma(\alpha_3-1)}{2\pi} > 0$$

and for $p=0$

$$i^{p-(-p)}S(A_2, A_2) = S(A_2, A_2) > 0$$

so that $F^\bullet$ indeed induces a pure polarized Hodge structure of weight -1 on $H^\infty_{\neq 1} = \mathbb{C}A_1 \oplus \mathbb{C}A_2$ and a pure polarized Hodge structure of weight 0 on $H^\infty_1 = \mathbb{C}A_2$.

The situation is visualized in the following diagram, where each column represents a space generated by elementary sections (that is, a space isomorphic to $V^\alpha/V^{>\alpha}$).

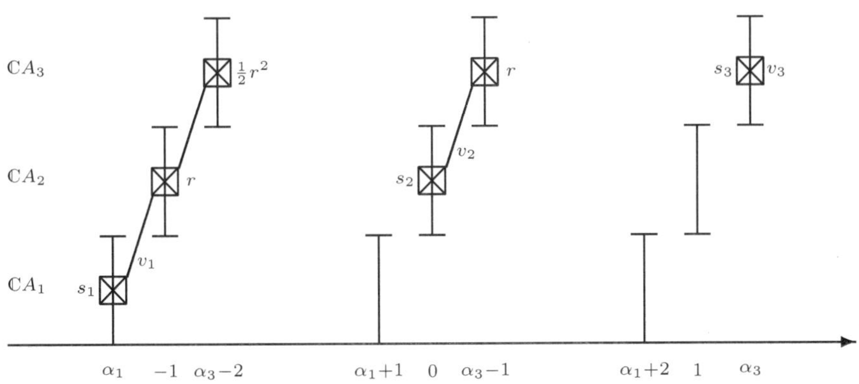

Let us calculate the variation of twistors associated to this example. We have

$$\begin{aligned} \tau v_1 &:= s_3 + \overline{r}zs_2 + \frac{\overline{r}^2}{2}z^2 s_1 \\ \tau v_2 &:= s_2 + \overline{r}zs_1 \\ \tau v_3 &:= s_1, \end{aligned}$$

where τ is the morphism defined before Proposition 3.5. A basis for the space $H^0(\mathbb{P}^1, \widehat{\mathcal{H}})$ is given for any r with $|r| \neq \sqrt{2}$ by $w_1 := v_1$, $w_2 := \overline{r}zs_1 + (1+\frac{1}{2}r\overline{r})s_2 + $ $+rz^{-1}s_3$, $w_3 := \tau v_1$. The metric h on $p_*\widehat{\mathcal{H}}$ is in this basis $H := h(\underline{w}^{tr}, \underline{w}) = \mathrm{diag}\left((1-\frac{1}{2}|r|^2)^2\right)$. This means that the space of points $r \in \mathbb{C}$ on which $\mathcal{H}$ is pure has two connected components, and that $\mathcal{H}$ is pure polarized on each of them.

We are going to compute a limit TERP-structure, when the parameter r tends to infinity. First note that the following sections are elements of $\mathcal{H}$: $zs_1 + \frac{r}{2}s_2 = zv_1 - \frac{r}{2}v_2$ and $z^2 s_1 = z^2 v_1 - rzv_2 + \frac{r^2}{2}v_3$. Taking the basis $(\widetilde{v}_1, \widetilde{v}_2, \widetilde{v}_3)$, where

$$\begin{aligned} \widetilde{v}_1 &= (r')^2 s_1 + r'z^{-1}s_2 + \tfrac{1}{2}z^{-2}s_3 \\ \widetilde{v}_2 &= r'zs_1 + \tfrac{1}{2}s_2 \\ \widetilde{v}_3 &= z^2 s_1 \end{aligned}$$

and $r' := r^{-1}$, of $\mathcal{H}$ outside of $r = 0$ shows that we obtain an extension of the family to $\mathbb{P}^1$. We see (by setting $r' := 0$) that the fibre $\mathcal{H}_{|r=\infty}$ is the TERP-structure generated by the elementary sections $z^{-2}s_3, s_2, z^2 s_1$. In particular, the spectral numbers, which are constant for finite r (namely, $\alpha_1, \alpha_2, \alpha_3$) have changed to $\alpha_3 - 2, \alpha_2, \alpha_1 + 2$.

One might analyze this example further by calculating the classifying space D_{BL} attached to the initial data used above, namely, for the spectrum $\alpha_1, \alpha_2, \alpha_3$. We give the result, without carrying out the details.

LEMMA 6.2. *For the given initial data from above, we have that* $D_{BL} \cong \mathbb{C}^2$, *with the universal family of Brieskorn lattices given by* $\mathcal{H} = \oplus_{i=1}^3 \mathcal{O}_{\mathbb{C}^3} v_i$, *where*

$$\begin{array}{rcl} v_1 & := & s_1 + rz^{-1}s_2 + \frac{r^2}{2}z^{-2}s_3 + tz^{-1}s_3 \\ v_2 & := & s_2 + rz^{-1}s_3 \\ v_3 & := & s_3. \end{array}$$

The family of twistors $\widehat{\mathcal{H}}$ *is pure outside of the real-analytic hypersurface* $(1-\rho)^4 - \theta = 0$, *where* $\rho = \frac{1}{2}r\bar{r}$ *and* $\theta = t\bar{t}$.

The complement has three components. The family of twistors is polarized on two of them, those which contain $\{(r,0) \mid |r| < \sqrt{2}\}$ *and respectively* $\{(r,0) \mid |r| > \sqrt{2}\}$. *On the third component the metric on* $p_*\widehat{\mathcal{H}}$ *has signature* $(+,-,-)$. *The restriction of the metric on* $\mathcal{T}_{\check{D}_{BL}^{pp}}$ *to the tangent space of* $\{(r,0) \mid |r| \neq \sqrt{2}\}$ *is given by*

$$h(\partial_r, \partial_r) = 2\frac{1+\rho^2}{(1-\rho)^4}.$$

So we see that the above family $\mathcal{H}(r)$ is in fact the restriction of the universal bundle on D_{BL} to the subspace $t = 0$. The fact that the family $\mathcal{H}(r)$ does not have a limit inside D_{BL} (and not even inside $\check{D}_{BL}$, which is equal to D_{BL} in this case) when $r \to \infty$ has the consequence that the metric on $\mathcal{T}_{\check{D}_{BL}^{pp}}$ is not complete. This can be seen directly, namely, we have

$$h(\partial_{r'}, \partial_{r'}) = h(-r^2\partial_r, -r^2\partial_r) = 8\rho^2\frac{1+\rho^2}{(1-\rho)^4} \overset{r\to\infty}{\longrightarrow} 8.$$

For applications, it is important that the target domain of the period map is a complete Hermitian space. For that reason, we are going to describe a partial compactification of $\check{D}_{BL}^{pp}$ on which we are able to construct such a complete Hermitian metric.

The main idea is to relax slightly the initial data to be fixed by requiring only that the spectral numbers be contained in an interval. More precisely, we fix as before $H^\infty, H^\infty_{\mathbb{R}}, S, M, w$ and a rational number α_1 (for notational convenience, we denote $\alpha_\mu := w - \alpha_1$).

THEOREM 6.3 ([**HS08**], Section 7). *There is a projective variety* $\mathcal{M}_{BL}$ *containing* $\check{D}_{BL}$ *as a locally closed subset and which comes equipped with a universal locally free sheaf* $\mathcal{H}$ *whose fibres are regular singular TERP-structures with spectral numbers in* $[\alpha_1, \alpha_\mu]$ *and initial data* $(H^\infty, H^\infty_{\mathbb{R}}, S, M)$. *On* $\check{D}_{BL}$ *it reduces to the universal bundle considered above.* $\mathcal{M}_{BL}$ *is stratified by locally closed subsets of regular singular TERP-structures with fixed spectral pairs.*

REMARK 6.4. We will give some remarks on the construction of $\mathcal{M}_{BL}$ and the proof of the theorem. The basic idea is that by fixing the range of spectral numbers to lie in the interval $[\alpha_1, \alpha_\mu]$, one is able to reduce the setting to a finite-dimensional situation and to control the possible regular singular TERP-structures with spectral numbers in that range. More precisely, the pairing P induces a symplectic structure on the quotient $V^{\alpha_1}/V^{>\alpha_\mu - 1}$ and the operators $z\cdot$ and $z^2\nabla_z$ give rise to nilpotent endomorphisms of $V^{\alpha_1}/V^{>\alpha_\mu-1}$. Then we consider the subvariety of the Lagrangian Grassmannian of $V^{\alpha_1}/V^{>\alpha_\mu-1}$ consisting of subspaces K which are invariant under these two nilpotent endomorphisms. For any regular singular TERP-structure H the space $K := H/V^{>\alpha_\mu-1}$ gives a point in this subvariety and we obtain a bijective correspondence between regular singular TERP-structures (with spectral numbers in $[\alpha_1, \alpha_\mu]$) and these subspaces K. It is clear that this subvariety is projective, but it might be very singular, and even non-reduced. We define $\mathcal{M}_{BL}$ to be this variety with its reduced scheme structure. By construction $\mathcal{M}_{BL}$ carries a universal sheaf of Brieskorn lattices, with an induced V-filtration, so that fixing the spectral pairs gives an intersection of open and closed conditions. This yields a stratification where each space $\check{D}^{pp}_{BL}$ (for fixed spectral pairs such that its spectral numbers are contained in $[\alpha_1, \alpha_\mu]$) is a locally closed stratum.

Let

$$\Theta := \left\{ x \in \mathcal{M}_{BL} \,|\, \widehat{\mathcal{H}} \text{ not pure} \right\}.$$

It is clear that Θ is a real analytic subvariety of $\mathcal{M}_{BL}$. The complement $\mathcal{M}_{BL}\backslash\Theta$ splits into connected components, on which the form $h : p_*\widehat{\mathcal{H}} \otimes \overline{p_*\widehat{\mathcal{H}}} \to \mathcal{C}^\infty_{\mathcal{M}_{BL}}$ is defined. Consider the union of those components on which it is positive definite, denote this space by $\mathcal{M}^{pp}_{BL}$, then we have the following, which is one of the main results in [**HS08**].

THEOREM 6.5 ([**HS08**], Theorem 8.6). *The positive definite Hermitian metric h on $(p_*\widehat{\mathcal{H}})_{|\mathcal{M}^{pp}_{BL}}$ induces a **complete** distance d_h on $\mathcal{M}^{pp}_{BL}$, which defines its topology. The same holds for the closed subspace $\overline{\check{D}}_{BL} \cap \mathcal{M}^{pp}_{BL} \subset \mathcal{M}^{pp}_{BL}$, which therefore is a partial compactification of $\check{D}^{pp}_{BL}$ with complete distance.*

If the monodromy respects a lattice in $H^\infty_\mathbb{R}$, we have the following.

PROPOSITION 6.6 ([**HS08**], Theorem 8.8). *Suppose that $H^\infty_\mathbb{Z} \subset H^\infty_\mathbb{R}$ is a lattice with $M \in \mathrm{Aut}(H^\infty_\mathbb{Z})$. Then the discrete group $G_\mathbb{Z} := \mathrm{Aut}(H^\infty_\mathbb{Z}, S, M)$ acts properly discontinuously on $\mathcal{M}^{pp}_{BL}$ and on $\overline{\check{D}}_{BL} \cap \mathcal{M}^{pp}_{BL}$ so that the quotients $\mathcal{M}^{pp}_{BL}/G_\mathbb{Z}$ and $(\overline{\check{D}}_{BL} \cap \mathcal{M}^{pp}_{BL})/G_\mathbb{Z}$ are complex spaces.*

In the following proposition, we describe the structure of the space $\mathcal{M}_{BL}$ for the example considered in the beginning of this section.

PROPOSITION 6.7. *The space $\mathcal{M}_{BL}$ for the initial data from example 6.1 is the weighted projective space $\mathbb{P}(1,1,4)$, where the embedding $\check{D}_{BL} \hookrightarrow \mathbb{P}(1,1,4) = \mathrm{Proj}\,\mathbb{C}[X,Y,Z]$ is given by $X := r, Y := 1, Z := t$ (here $\deg(X) = 1, \deg(Y) = 1, \deg(Z) = 4$). The only singular point of $\mathbb{P}(1,1,4)$ is $(0:0:1)$, in particular, $\mathcal{M}^{pp}_{BL}$ is smooth in this case.*

The following is a direct application of Theorem 6.5.

THEOREM 6.8 ([**HS08**], Theorem 9.5). *Let X be a complex variety, $Z \subset X$ a complex space of codimension at least two. Suppose that the complement $Y := X\backslash Z$ is simply connected. Let $(H, H'_\mathbb{R}, \nabla, P)$ be a variation of regular singular* *pure polarized TERP-structures on the complement Y which has constant spectral pairs. Then this variation extends to the whole of X, with possibly jumping spectral numbers over Z.*

This theorem can be proved exactly as in the case of Hodge structures, namely, the given variation defines a period map $Y \to \check{D}^{pp}_{BL}$. In particular, due to Corollary 5.5, we know that this map is distance decreasing with respect to the (complete) distance d_h on $\overline{\check{D}}_{BL} \cap \mathcal{M}^{pp}_{BL}$ and the Kobayashi pseudometric on Y. This implies that it extends continuously and then also holomorphically to X (the proof is the same as in [**Kob05**, Corollary 3.5], note however that $\overline{\check{D}}_{BL} \cap \mathcal{M}^{pp}_{BL}$ is not a hyperbolic space, but the period map is horizontal by assumption, which is sufficient).

To finish this section, we describe two further examples of classifying spaces $\mathcal{M}_{BL}$. More examples can be found in [**HS08**].

EXAMPLE 6.9 ([**HS08**], Section 9.3). Here is an example where $\mathcal{M}_{BL}$ is reducible. Fix any number $n \in \mathbb{N}_{>0}$ and consider the initial data: $H^\infty_\mathbb{R} := \mathbb{R}B_1 \oplus \mathbb{R}B_2$, $M := \mathrm{id} \in \mathrm{Aut}(H^\infty_\mathbb{R})$, and $S(\underline{B}^{tr}, \underline{B}) = (-1)^n \frac{1}{2}\,\mathrm{diag}(1,1)$. Let $A_1 := B_1 + iB_2$, $A_2 := \overline{A}_1 = B_1 - iB_2$ and consider the following reference filtration

$$\{0\} = F_0^{n+1} \subsetneq F_0^n := \mathbb{C}A_1 = F_0^{n-1} = \cdots = F_0^{-n+1} \subsetneq F_0^{-n} = H^\infty$$

Then $\alpha_1 = -n$ and the classifying space is $\check{D}_{PMHS} = D_{PMHS} = \check{D}_{PHS} = D_{PHS}$, it consists of two points, namely $F_0^\bullet$ and $\overline{F}_0^\bullet$, which are both pure polarized Hodge structures of weight zero with Hodge decomposition $H^{n,-n} \oplus H^{-n,n}$. Put $s_1 := z^{-n}A_1$ and $s_2 := z^n A_2$. The classifying space D_{BL} is a disjoint union of two affine lines, namely, the universal family over the component above $F_0^\bullet$ is given by $\mathcal{H}_{-n}(r) := \mathcal{O}_{\mathbb{C}^2} v_1 \oplus \mathcal{O}_{\mathbb{C}^2} v_2$, where $v_1 := z^{-n}A_1 + rz^{n-1}A_2 = s_1 + rz^{-1}s_2, v_2 := z^n A_2 = s_2$, and the universal family over the other component is given by $\mathcal{H}_n(r) := \mathcal{O}_{\mathbb{C}^2}(z^{-n}A_2 + rz^{n-1}A_1) \oplus \mathcal{O}_{\mathbb{C}^2} z^n A_1$. The following diagram visualizes this situation.

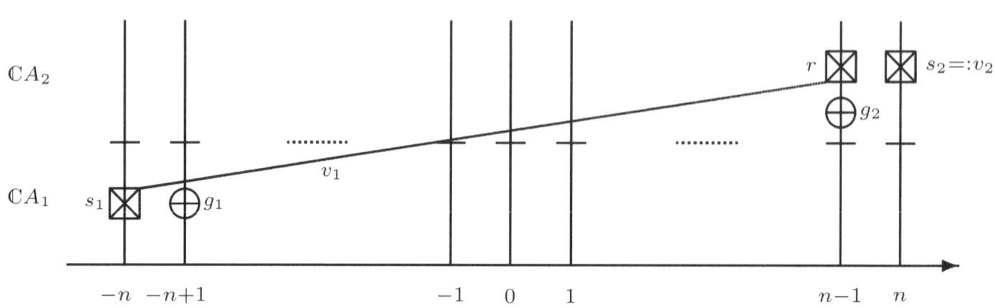

It is directly evident that the "limit TERP"-structure (when r approaches infinity), is given by $\mathcal{G}_{-n+1} := \mathcal{O}_\mathbb{C} g_1 \oplus \mathcal{O}_\mathbb{C} g_2$, where $g_1 := zs_1$ and $g_2 := z^{-1}s_2$. We see that $\mathcal{G}_{-n+1}$ is the origin in one of the two components of the classifying space $\check{D}_{BL}$ associated to the same data, but with $\alpha_1 := -n+1$. Note, however, that the two (conjugate) filtrations induced by $\mathcal{G}_{-n+1}$ and $\mathcal{G}_{n-1}$, respectively, are not pure polarized: the Hodge metric is negative definite. Taking the limit of the universal family for this classifying spaces yields TERP-structures $\mathcal{G}_{-n+2}$ and $\mathcal{G}_{n-2}$, respectively, and we can continue this procedure until we arrive at $\mathcal{G}_{-1}$ and $\mathcal{G}_1$. The limits $\lim_{r\to\infty} \mathcal{H}_{-1}(r)$ and $\lim_{r\to\infty} \mathcal{H}_1(r)$ are both equal to the lattice $\mathcal{G}_0 = V^0$. This shows that the space $\mathcal{M}_{BL}$ is a chain of $2n$ copies of $\mathbb{P}^1$, where the Hodge filtration gives pure polarized, resp. negative definite, pure Hodge structures on every other component of this chain.

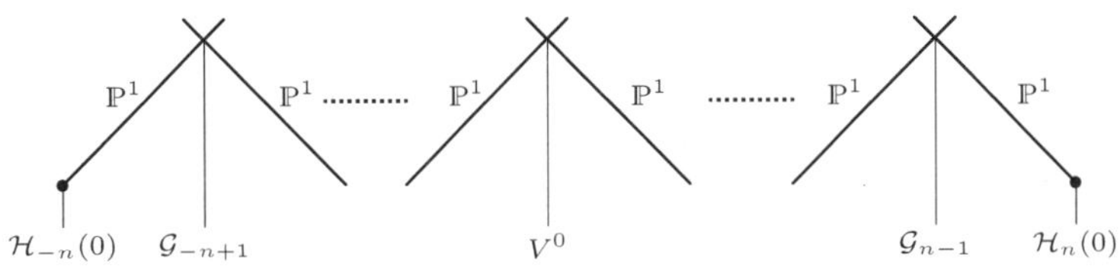

It is easy to calculate the associated twistors: For the original family $\mathcal{H}_{-n}$, we have

$$\widehat{\mathcal{H}}_{-n} := \mathcal{O}_{\mathbb{P}^1}\mathcal{C}^\infty_{\mathbb{C}}\underbrace{(s_1 + rz^{-1}s_2)}_{w_1} \oplus \mathcal{O}_{\mathbb{P}^1}\mathcal{C}^\infty_{\mathbb{C}}\underbrace{(s_2 + \overline{r}zs_1)}_{w_2}$$

and the metric is $h(w_1, w_2) = \operatorname{diag}(1 - |r|^2)$ (note that $P(s_1, s_2) = 1$ due to our choices), so that $\mathcal{H}_{-n}(r)$ is a variation of pure polarized TERP-structures on Δ^* (it is even a nilpotent orbit, which corresponds by Theorem 4.4 to the pure polarized Hodge structure $(H^\infty, H^\infty_{\mathbb{R}}, S, F^\bullet_0)$). The same holds for the family $\mathcal{H}_n(r)$.

Note however that due to $P(g_1, g_2) = -1$, the variation of twistors on the second left-(or right-)most $\mathbb{P}^1$ is pure polarized on $\mathbb{P}^1\backslash\overline{\Delta}$, where the origin is the TERP-structure $\mathcal{G}_{-n+1}$ (resp. $\mathcal{G}_{n-1}$). This means that in the above picture, the points of intersection on the lower level are pure polarized, but not those on the upper level. In particular, V^0 is pure polarized precisely if n is even (the above picture already supposes that n is odd), which can also be seen directly from the formula $S(A_1, A_2) = (-1)^n$.

EXAMPLE 6.10 (**[HS08]**, Section 9.1). This example shows that $\mathcal{M}_{BL}$ is not necessarily a union or a product of projective or weighted projective spaces. Consider the following initial data: $H^\infty_{\mathbb{R}} := \oplus_{i=1}^4 \mathbb{R}B_i$, $A_1 := B_1 + iB_4$, $A_2 := B_2 + iB_3$, $\overline{A}_1 = A_4$, $\overline{A}_2 = A_3$, $M(A_i) := e^{-2\pi i\alpha_i}A_i$, where we choose α_1 and α_2 with $-1 < \alpha_1 < \alpha_2 < -\frac{1}{2}$ and $\alpha_3 = -\alpha_2$, $\alpha_4 = -\alpha_1$. Moreover, we put

$$S(\underline{B}^{tr}, \underline{B}) := \begin{pmatrix} 0 & 0 & 0 & -\gamma_1 \\ 0 & 0 & -\gamma_2 & 0 \\ 0 & \gamma_2 & 0 & 0 \\ \gamma_1 & 0 & 0 & 0 \end{pmatrix}$$

where $\gamma_1 := \frac{-1}{4\pi}\Gamma(\alpha_1 + 1)\Gamma(\alpha_4)$ and $\gamma_2 := \frac{-1}{4\pi}\Gamma(\alpha_2 + 1)\Gamma(\alpha_3)$ (so that $S(A_1, A_4) = 2i\gamma_1$ and $S(A_2, A_3) = 2i\gamma_2$). Let $s_i := z^{\alpha_i}A_i$ and $\mathcal{H} := \oplus_{i=1}^4 \mathcal{O}_{\mathbb{C}^4}v_i$, where

$$\begin{aligned} v_1 &:= s_1 + rz^{-1}s_3 + qz^{-1}s_4 \\ v_2 &:= s_2 + pz^{-1}s_3 + rz^{-1}s_4 \\ v_3 &:= s_3 \\ v_4 &:= s_4 \end{aligned}$$

Then $(H, H'_{\mathbb{R}}, \nabla, P)$ is a variation of TERP-structures of weight zero on $\mathbb{C}^3$ with constant spectrum $\mathrm{Sp} = (\alpha_1, \dots, \alpha_4)$, it is in fact the universal family of the classifying space $\check{D}_{BL}$ associated to the initial data $(H^\infty, H^\infty_{\mathbb{R}}, S, M, \mathrm{Sp})$. The induced (constant) filtration is

$$\{0\} = F^1 \subsetneq F^0 := \mathbb{C}A_1 \oplus \mathbb{C}A_2 \subsetneq F^{-1} = H^\infty$$

from which one checks that $(H^\infty, H^\infty_{\mathbb{R}}, S, F^\bullet)$ is a pure polarized Hodge structure of weight -1. The associated diagram looks as follows.

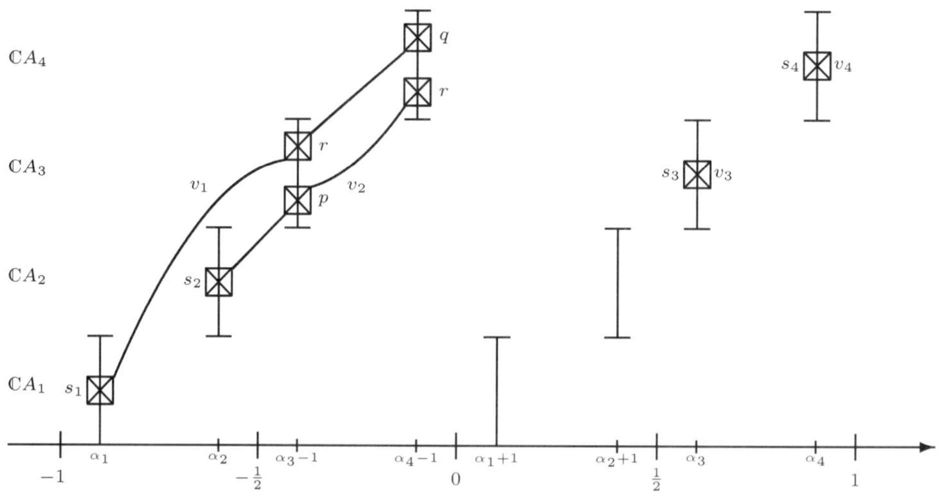

Let us calculate the space $\mathcal{M}_{BL}$ associated to the initial data $(H^\infty, H^\infty_{\mathbb{R}}, S, M, \alpha_1, w)$. It is very simple in this example, as we have $\lfloor \alpha_\mu - \alpha_1 \rfloor = 1$. This implies that $zv_i \in V^{>\alpha_4-1}$ and $(z^2\nabla_z)v_i \in V^{>\alpha_4-1}$ for $i = 1, \ldots, 4$, so that

$$\mathcal{M}_{BL} = \mathrm{LGr}(V^{\alpha_1}/V^{>\alpha_4-1}, [P^{(-1)}]),$$

where $[P^{(-1)}] = (z^{-1}s_4)^* \wedge (s_1)^* + (z^{-1}s_3)^* \wedge (s_2)^* \in \bigwedge^2 (V^{\alpha_1}/V^{>\alpha_4-1})^*$ is the symplectic form induced by P and $\mathrm{LGr}(V^{\alpha_1}/V^{>\alpha_4-1}, [P^{(-1)}])$ denotes the Lagrangian Grassmannian of subspaces of dimension two of $V^{\alpha_1}/V^{>\alpha_4-1}$ on which $[P^{(-1)}]$ vanishes. Using the Plücker embedding, one checks that this Lagrangian Grassmannian is a hyperplane section of the Plücker quadric in $\mathbb{P}^5$, i.e., a smooth quadric in $\mathbb{P}^4$ which is isomorphic neither to $\mathbb{P}^3$ nor to $\mathbb{P}^1 \times \mathbb{P}^2$. It is also clear that $\mathcal{M}_{BL}$ is indeed the closure of the three-dimensional affine classifying space $\check{D}_{BL}$ considered above, but it also contains other strata, e.g., the closure of the two-dimensional classifying space associated to (the same topological data and) the spectrum $\mathrm{Sp} = (\alpha_1, \alpha_3 - 1, \alpha_2 + 1, \alpha_4)$ is the weighted projective space $\mathbb{P}^1(1, 1, 2)$, which appears as a (further) hyperplane section of the Lagrangian Grassmannian $\mathcal{M}_{BL}$.

7. Period mappings and limit structures

In this section, we will give an interpretation of one of the main results of [**Moc07**] to the case of period mappings associated to isolated hypersurface singularities. We are considering variations of TERP-structures on non-compact varieties, and we are interested in studying their behaviour at the boundary. In order to keep the notation as simple as possible, we will restrict here to variations on a punctured disc. This is sufficient for many applications; the details for the higher dimensional case are worked out in [**HS08**]. We will denote throughout this section by r a coordinate on Δ.

LEMMA 7.1 ([**HS08**], Sections 3 and 5). *Let H be a variation of pure polarized regular singular TERP-structures on Δ^*. Then for any $a \in \mathbb{R}$, there are vector bundles ${}_a\mathcal{E}$ (resp. ${}_{<a}\mathcal{E}$) on $\mathbb{C} \times \Delta$ which extend H and such that the restriction $({}_a\mathcal{E})_{|\mathbb{C}^* \times \Delta}$ (resp. $({}_{<a}\mathcal{E})_{|\mathbb{C}^* \times \Delta}$) coincides with the Deligne extension $\mathcal{V}_\Delta^{-a}$ (resp. $\mathcal{V}_\Delta^{>-a}$) of order $-a$ (resp. bigger than $-a$) of H' over $\mathbb{C}^* \times \Delta$. Moreover, we have a connection*

$$\nabla : {}_a\mathcal{E} \longrightarrow {}_a\mathcal{E} \otimes z^{-1}\Omega^1_{\mathbb{C}\times\Delta}(\log(\{0\} \times \Delta \cup \mathbb{C} \times \{0\}))$$

extending the given connection operators on $\mathcal{H}$ and $\mathcal{V}_\Delta^{-a}$. There is a non-degenerate pairing

$$P : {}_a\mathcal{E} \otimes {}_{<1-a}\mathcal{E} \longrightarrow z^w \mathcal{O}_{\mathbb{C}\times\Delta}$$

which restricts to the given P on $\mathbb{C} \times \Delta^*$.

As a consequence, we can construct a limit object which turns out to be a TERP-structure.

THEOREM 7.2 ([**HS08**], Theorem 3.5). *Let H be as above. Then the $\mathbb{C}$-bundle*

$$\mathcal{G} := \bigoplus_{a\in(0,1]} {}_a\mathcal{E}/{}_{<a}\mathcal{E}$$

underlies a regular singular TERP-structure. It comes equipped with a nilpotent endomorphism

$$\mathcal{N} : \mathcal{G} \longrightarrow z^{-1}\mathcal{G}$$

given by the residue of the operator $r\nabla_r$ on $\mathcal{G}$ along $\mathbb{C} \times \{0\}$.

In [**Moc07**], a construction for a limit object starting from a general tame variation of pure polarized twistor structures is described. He proves that this is a polarized mixed twistor structure. It turns out that this limit can be identified with the twistor $\widehat{\mathcal{G}}$ constructed from the TERP-structure $\mathcal{G}$.

THEOREM 7.3 ([**HS08**], Theorem 3.7). *The $\mathbb{P}^1$-bundle $\widehat{\mathcal{G}}$ underlies a polarized mixed twistor structure.*

We refer to [**Moc07**] or [**HS07b**] for a precise definition of a polarized mixed twistor structure. It is essentially given by the additional datum of a weight filtration by subbundles $\widehat{\mathcal{W}}_k$ of $\widehat{\mathcal{G}}$ such that the quotients are pure twistors of weight k (i.e. isomorphic to a $\mathcal{O}^n_{\mathbb{P}^1}(k)$ for some $n \in \mathbb{N}$) and such that a positivity condition is satisfied on primitive subspaces. In our case $\widehat{\mathcal{W}}_\bullet$ is defined by the weight filtration $\mathcal{W}_\bullet$ on $\mathcal{G}$ associated to $\mathcal{N}$.

A very easy but important consequence is the following, which is the analogue of [**Sch73**, Corollary 4.11].

COROLLARY 7.4. *Let H be as above, and suppose that the monodromy around $\mathbb{C} \times \Delta^*$ of the local system H' is trivial. Then for any $a \in \mathbb{R}$, ${}_a\mathcal{E}$ underlies a variation of pure polarized regular singular TERP-structures on Δ, extending H. The limit $\mathcal{G}$ is naturally identified with the fibre ${}_a\mathcal{E}/r \cdot {}_a\mathcal{E}$. In particular, $\mathcal{G}$ is still pure polarized.*

We are mainly interested in TERP-structures defined by isolated hypersurface singularities. In that case, we obtain the following result.

THEOREM 7.5 ([**HS08**], Theorem 9.7). *Given a μ-constant family of isolated hypersurface singularities $F : (\mathbb{C}^{n+1} \times \Delta^*, 0) \to (\mathbb{C}, 0)$, suppose that the variation $TERP(F)$ over Δ^* is pure polarized.*

- *If the monodromy around $\mathbb{C} \times \{0\}$ is semi-simple, then the locally liftable period map $\phi_{BL} : \Delta^* \to \check{D}_{BL}^{pp}/G_{\mathbb{Z}} \subset \overline{\check{D}}_{BL} \cap \mathcal{M}_{BL}^{pp}/G_{\mathbb{Z}}$ extends holomorphically (not necessarily locally liftable) to $\phi_{BL} : \Delta \to \overline{\check{D}}_{BL} \cap \mathcal{M}_{BL}^{pp}/G_{\mathbb{Z}}$.*
- *Suppose that the monodromy around $\mathbb{C} \times \{0\}$ is reduced to the identity, in particular, there is a period map $\phi_{BL} : \Delta^* \to \check{D}_{BL}^{pp} \subset \overline{\check{D}}_{BL} \cap \mathcal{M}_{BL}^{pp}$. Then this map extends to*

$$\overline{\phi} : \Delta \to \overline{\check{D}}_{BL} \cap \mathcal{M}_{BL}^{pp}$$

and the variation $TERP(F)$ extends to a variation on Δ.

We finish this survey by a result concerning the "new supersymmetric index", i.e., the endomorphism $\mathcal{Q}$ defined at the end of chapter 3.

THEOREM 7.6 ([**Her03**], Theorem 7.20). *Consider a nilpotent orbit of regular singular TERP-structures K, i.e., a family over Δ^* (coordinate r), constructed as $K := \pi^* H$ from a single TERP-structure H, where $\pi : \mathbb{C} \times \mathbb{C}^* \to \mathbb{C}$, $(z, r) \mapsto zr$. By definition K is pure polarized for small $|r|$, and we have the Hermitian endomorphism $\mathcal{Q}$ of $p_* \widehat{\mathcal{K}}$. Then its eigenvalues tend for $r \to 0$ to $\frac{w}{2} - \mathrm{Sp}(H, \nabla)$.*

A generalization of this result to the irregular case can be found in the recent preprint [**Sab08a**].

References

[Bri70] Egbert Brieskorn, *Die Monodromie der isolierten Singularitäten von Hyperflächen*, Manuscripta Math. **2** (1970), 103–161.

[CFIV92] Sergio Cecotti, Paul Fendley, Ken Intriligator, and Cumrun Vafa, *A new supersymmetric index*, Nuclear Phys. B **386** (1992), no. 2, 405–452.

[CKS86] Eduardo Cattani, Aroldo Kaplan, and Wilfried Schmid, *Degeneration of Hodge structures*, Ann. of Math. (2) **123** (1986), no. 3, 457–535.

[CV91] Sergio Cecotti and Cumrun Vafa, *Topological–anti-topological fusion*, Nuclear Phys. B **367** (1991), no. 2, 359–461.

[CV93] ———, *On classification of $N = 2$ supersymmetric theories*, Comm. Math. Phys. **158** (1993), no. 3, 569–644.

[DS03] Antoine Douai and Claude Sabbah, *Gauss-Manin systems, Brieskorn lattices and Frobenius structures. I*, Ann. Inst. Fourier (Grenoble) **53** (2003), no. 4, 1055–1116.

[Gre75] G.-M. Greuel, *Der Gauß-Manin-Zusammenhang isolierter Singularitäten von vollständigen Durchschnitten*, Math. Ann. **214** (1975), 235–266.

[Her99] Claus Hertling, *Classifying spaces for polarized mixed Hodge structures and for Brieskorn lattices*, Compositio Math. **116** (1999), no. 1, 1–37.

[Her02] ———, *Frobenius manifolds and moduli spaces for singularities*, Cambridge Tracts in Mathematics, vol. 151, Cambridge University Press, Cambridge, 2002.

[Her03] ———, *tt* geometry, Frobenius manifolds, their connections, and the construction for singularities*, J. Reine Angew. Math. **555** (2003), 77–161.

[HS07a] Claus Hertling and Christian Sevenheck, *Curvature of classifying spaces for Brieskorn lattices*, preprint math.AG/0712.3691, 2007.

[HS07b] ———, *Nilpotent orbits of a generalization of Hodge structures.*, J. Reine Angew. Math. **609** (2007), 23–80.

[HS08] ———, *Limits of families of Brieskorn lattices and compactified classifying spaces*, work in progress, 2008.

[Kob05] Shoshichi Kobayashi, *Hyperbolic manifolds and holomorphic mappings*, second ed., World Scientific Publishing Co. Pte. Ltd., Hackensack, NJ, 2005.

[Mal74] Bernard Malgrange, *Intégrales asymptotiques et monodromie*, Ann. Sci. École Norm. Sup. (4) **7** (1974), 405–430 (1975).

[Moc07] Takuro Mochizuki, *Asymptotic behaviour of tame harmonic bundles and an application to pure twistor $\mathcal{D}$-modules, Part 1*, Mem. Amer. Math. Soc. **185** (2007), no. 869, xi+324.

[NS99] A. Némethi and C. Sabbah, *Semicontinuity of the spectrum at infinity*, Abh. Math. Sem. Univ. Hamburg **69** (1999), 25–35.

[NZ92] András Némethi and Alexandru Zaharia, *Milnor fibration at infinity*, Indag. Math. (N.S.) **3** (1992), no. 3, 323–335.

[Pha83] Frédéric Pham, *Vanishing homologies and the n variable saddlepoint method*, Singularities, Part 2 (Arcata, Calif., 1981), Proc. Sympos. Pure Math., vol. 40, Amer. Math. Soc., Providence, RI, 1983, pp. 319–333.

[Pha85] ———, *La descente des cols par les onglets de Lefschetz, avec vues sur Gauss-Manin*, Astérisque (1985), no. 130, 11–47, Differential systems and singularities (Luminy, 1983).

[Sab05a] Claude Sabbah, *Fourier-Laplace transform of a variation of polarized complex Hodge structure*, preprint math.AG/0508551, to appear in J. Reine Angew. Math., 2005.

[Sab05b] ———, *Polarizable twistor $\mathcal{D}$-modules*, Astérisque (2005), no. 300, vi+208.

[Sab06] ———, *Hypergeometric periods for a tame polynomial*, Port. Math. (N.S.) **63** (2006), no. 2, 173–226.

[Sab08a] ———, *Fourier-Laplace transform of a variation of polarized complex Hodge structure, II.*, preprint arXiv:0804.4328.

[Sab08b] ———, *Universal unfoldings of Laurent polynomials and tt^*-structures*, Proceedings to the conference "From tQFT to tt^* and integrability", Augsburg, May 2007, Proceedings of Symposia in Pure Mathematics, American Mathematical Society, 2008, this volume, also available as math.AG/0802.1259.

[Sch73] Wilfried Schmid, *Variation of Hodge structure: the singularities of the period mapping*, Invent. Math. **22** (1973), 211–319.

[Sim88] Carlos T. Simpson, *Constructing variations of Hodge structure using Yang-Mills theory and applications to uniformization*, J. Amer. Math. Soc. **1** (1988), no. 4, 867–918.

[Sim90] ———, *Harmonic bundles on noncompact curves*, J. Amer. Math. Soc. **3** (1990), no. 3, 713–770.

[Sim97] ———, *Mixed twistor structures*, preprint math.AG/9705006, 1997.

[SS85] J. Scherk and J. H. M. Steenbrink, *On the mixed Hodge structure on the cohomology of the Milnor fibre*, Math. Ann. **271** (1985), no. 4, 641–665.

[Var80] A. N. Varchenko, *Asymptotic behavior of holomorphic forms determines a mixed Hodge structure*, Dokl. Akad. Nauk SSSR **255** (1980), no. 5, 1035–1038.

[Var82] ———, *The complex singularity index does not change along the stratum $\mu = \text{const}$*, Funktsional. Anal. i Prilozhen. **16** (1982), no. 1, 1–12, 96.

Lehrstuhl VI für Mathematik, Universität Mannheim, A5 6, 68131 Mannheim, Germany

E-mail address: `Claus.Hertling@uni-mannheim.de`

Lehrstuhl VI für Mathematik, Universität Mannheim, A5 6, 68131 Mannheim, Germany

E-mail address: `Christian.Sevenheck@uni-mannheim.de`

Proceedings of Symposia in Pure Mathematics
Volume **78**, 2008

Differential geometric aspects of the tt*-equations

Vicente Cortés and Lars Schäfer

Abstract. We discuss the relation between pluriharmonic maps and solutions of the tt*-equations. The correspondence is obtained without assuming the integrability of the almost complex structure of the base manifold. We present examples which stem from special Kähler and nearly Kähler geometry. This paper is based on a lecture given by the first author on the Workshop 'From tQFT to tt* and integrability' (Augsburg, 2007).

Introduction

The tt*-equations were proposed by Cecotti and Vafa in 1991 [**CV**] as a description of the geometric structure on the moduli space of $2D$ $N = 2$ supersymmetric QFTs. It was discovered by Dubrovin in 1992 [**D**] that solutions of the tt*-equations correspond to pluriharmonic maps from a complex manifold M of dimension n into $\mathrm{GL}(n, \mathbb{R})/\mathrm{O}(n)$. This correspondence has been further developed by Simpson [**Si**] and the second author [**S1**].

In these notes we will present a general version of the tt*-equations purely in terms of differential geometry, explain the relation with pluriharmonic maps and discuss some classes of solutions. A proof of the above correspondence which does not assume the integrability of the almost complex structure of the base manifold is given.

1. tt*-bundles

Definition 1.1. A tt*-bundle (E, D, S) over an almost complex manifold (M, J) is a real vector bundle $E \to M$ endowed with a connection D and a section $S \in \Gamma(T^*M \otimes \mathrm{End}\, E)$, which satisfy the tt*-equation

$$R^\theta = 0,$$

for all $\theta \in \mathbb{R}$, where R^θ is the curvature of the connection

$$D^\theta_X := D_X + \cos\theta\, S_X + \sin\theta\, S_{JX}, \; X \in \Gamma(TM).$$

2000 *Mathematics Subject Classification.* Primary 53C43, Secondary 53C25.
Key words and phrases. pluriharmonic maps, tt*-geometry, pseudo-Riemannian symmetric spaces, nearly Kähler manifolds, special Kähler manifolds.

A metric/symplectic tt*-bundle (E, D, S, β) is a tt*-bundle (E, D, S) endowed with a parallel non-degenerate symmetric/skew-symmetric bilinear form β on the fibers of E such that S_X is β-symmetric for all $X \in \Gamma(TM)$. It is called unimodular if $\operatorname{tr} S_X = 0$ for all $X \in \Gamma(TM)$.

REMARKS 1.2. 1) Any oriented unimodular metric/symplectic tt*-bundle carries a canonical volume element[1] $\nu \in \Gamma(\Lambda^r E^*)$, $r = \operatorname{rk} E$, which is D^θ-parallel for all $\theta \in \mathbb{R}$.
2) The tt*-equation $R^\theta = 0$ for all $\theta \in \mathbb{R}$ is equivalent to the following system (1.1-1.3):

$$d^D S = d^D S_J = 0, \text{ where} \tag{1.1}$$

$$(d^D S)(X, Y) = D_X(S_Y) - D_Y(S_X) - S_{[X,Y]},$$

$$[S_X, S_Y] = [S_{JX}, S_{JY}], \quad \forall X, Y \in \Gamma(TM), \tag{1.2}$$

$$R^D(X, Y) + [S_X, S_Y] = 0, \quad \forall X, Y \in \Gamma(TM). \tag{1.3}$$

2. Pluriharmonic maps

DEFINITION 2.1. A smooth map $f : M \to N$ from a complex manifold M to a pseudo-Riemannian manifold N is called pluriharmonic if its restriction to any complex curve is harmonic.

This is equivalent to the differential equation

$$(\nabla df)^{1,1} = 0, \tag{2.1}$$

where ∇ is the connection on $T^*M \otimes f^*TN$ induced by a torsion-free complex connection D on M, i.e. a torsion-free connection D satisfying $DJ = 0$, and the Levi-Civita connection on N. More generally, any connection D on M such that D is complex and $(T^D)^{1,1} = 0$ leads to the same equation (2.1). Here, and in the following, T^D stands for the torsion of the connection D. Such connections D exist on any almost complex manifold (M, J) and we can define the notion of pluriharmonicity for maps with an almost complex source manifold by equation (2.1). We point out that a pluriharmonic map $f : M \to N$ is harmonic for any almost Hermitian metric g on M such that $\nabla^g - D$ is trace-free, where ∇^g is the Levi-Civita connection of g. This follows from $\operatorname{tr}_g \nabla^g df = \operatorname{tr}_g \nabla df = \operatorname{tr}_g (\nabla df)^{1,1} = 0$. In particular, any pluriharmonic map from a pseudo-Kähler manifold to a pseudo-Riemannian manifold is harmonic, since in that case one can choose $D = \nabla^g$. It is also clear that holomorphic maps between pseudo-Kähler manifolds are pluriharmonic. More generally, Rawnsley [**Ra**] has shown that a holomorphic map between almost Hermitian manifolds is pluriharmonic if the fundamental 2-forms of the domain and target manifolds satisfy $(d\omega)^{1,2} = 0$.

3. Correspondence between metric/symplectic tt*-bundles and pluriharmonic maps

In [**D**] Dubrovin developed a correspondence between certain metric tt*-bundles (E, D, S, g) with a positive-definite metric g over a complex manifold of complex dimension n and pluriharmonic maps $M \to \mathrm{GL}(n, \mathbb{R})/\mathrm{O}(n)$. In his case $E = (T^{1,0}M)^\rho$ consists of the real points of the holomorphic tangent bundle with respect to a real

[1] Notice that in the symplectic case the bundle is automatically oriented. The volume element is $\nu = \frac{\beta^r}{r!}$.

structure ρ. Independently, Simpson [**Si**] established a correspondence between Higgs bundles of rank r on M endowed with a harmonic metric and harmonic maps $M \to \mathrm{GL}(r,\mathbb{C})/\mathrm{U}(r)$. These results were generalised by the second author [**S1**], who proved the next theorem under the assumption that the almost complex structure on M is integrable. The result of Simpson, for instance, can be recovered from the next Theorem using the totally geodesic inclusion

$$\frac{\mathrm{GL}(r,\mathbb{C})}{\mathrm{U}(r)} \subset \frac{\mathrm{GL}(2r,\mathbb{R})}{\mathrm{O}(2r)}$$

as was shown in [**S2**]. The next theorem is proven in Section 5. In the following, $Sym_{p,q}(\mathbb{R}^r)$ stands for the cone of symmetric $r \times r$-matrices of signature (p,q), $p+q=r$. Similarly, for even r the symbol $Skew_{reg}(\mathbb{R}^r)$ stands for the cone of invertible skew-symmetric $r \times r$-matrices.

THEOREM 3.1. *Let (E, D, S, β) be a metric/symplectic tt*-bundle over a simply connected almost complex manifold (M, J). For any $\theta \in \mathbb{R}$ there exists a D^θ-parallel frame $(e_1^\theta, \ldots, e_r^\theta)$ of E and the correspondence*

$$x \mapsto f_\theta(x) := (\beta(e_i^\theta(x), e_j^\theta(x)))$$

defines a pluriharmonic map

$$f_\theta : M \to \begin{cases} Sym_{p,q}(\mathbb{R}^r) \cong \mathrm{GL}(r,\mathbb{R})/\mathrm{O}(p,q), \\ Skew_{reg}(\mathbb{R}^r) \cong \mathrm{GL}(r,\mathbb{R})/\mathrm{Sp}(r,\mathbb{R}), \end{cases}$$

$$\textit{if } \beta \textit{ is } \begin{cases} \textit{symmetric of signature } (p,q),\ p+q=r, \\ \textit{skew-symmetric, respectively.} \end{cases}$$

The target manifold is a pseudo-Riemannian symmetric space with the metric induced from the (bi-invariant) trace form, i.e. $\langle A, B\rangle = tr(AB)$, on $\mathfrak{gl}(r,\mathbb{R})$. If (E, D, S, β) is oriented and unimodular, f_θ takes values in the irreducible symmetric space

$$\begin{cases} \mathrm{SL}(r,\mathbb{R})/\mathrm{SO}(p,q), \\ \mathrm{SL}(r,\mathbb{R})/\mathrm{Sp}(r,\mathbb{R}), \textit{respectively.} \end{cases}$$

In all cases the pluriharmonic map $f = f_\theta$ has the following additional property:

$$R^N(df\, T^{1,0}M, df\, T^{1,0}M) = 0, \tag{3.1}$$

where R^N is the curvature tensor of the symmetric target manifold

$$N = \begin{cases} \mathrm{GL}(r,\mathbb{R})/\mathrm{O}(p,q), & \mathrm{SL}(r,\mathbb{R})/\mathrm{SO}(p,q), \\ \mathrm{GL}(r,\mathbb{R})/\mathrm{Sp}(r,\mathbb{R}), & \mathrm{SL}(r,\mathbb{R})/\mathrm{Sp}(r,\mathbb{R}), \ \textit{respectively.} \end{cases} \tag{3.2}$$

Conversely, any pluriharmonic map $f : M \to N$, into one of the symmetric target manifolds (3.2) *which has the property* (3.1) *and satisfies*

$$\bar\partial f(T(Z,W)) = df(T(Z,W)^{0,1}) = 0 \tag{3.3}$$

for all $Z, W \in T^{1,0}M$, is obtained from a metric/symplectic tt-bundle by the above construction.*

The property (3.1) is automatic if N is Riemannian, since there holds

$$0 = \langle R^N(Z,W)\bar W, \bar Z\rangle = -\|[Z,W]\|^2 \Rightarrow [Z,W] = 0,\ \forall Z, W \in T^{1,0}N.$$

We remark that, in general, $R^N(df\,T^{1,0}M, df\,T^{1,0}M) \subset TN$ is not necessarily zero but only isotropic.

4. Solutions

4.1. Special complex and special Kähler manifolds.

DEFINITION 4.1. A special complex manifold (M, J, ∇) is a complex manifold endowed with a flat torsion-free connection ∇, such that

$$(\nabla_X J)Y = (\nabla_Y J)X, \ \forall X, Y \in \Gamma(TM). \tag{4.1}$$

A special Kähler manifold (M, J, ∇, ω) is a special complex manifold endowed with a J-invariant ∇-parallel symplectic structure ω. The (pseudo-)Kähler metric $g = \omega(J\cdot, \cdot)$ is called the special Kähler metric of (M, J, ∇, ω).

We remark that the integrability of J follows from the properties of ∇. Hertling has observed in section 3.3 of [**He**] that a special complex manifold is the same as a *variation of Hodge structure* (VHS) on $TM \otimes \mathbb{C}$ and a special Kähler manifold is the same as a *polarized* such VHS.

Let Ω be the canonical symplectic form on $T^*\mathbb{C}^n = \mathbb{C}^{2n}$. We recall that a holomorphic immersion ϕ from a complex manifold of dimension n into $T^*\mathbb{C}^n$ is called Lagrangian if $\phi^*\Omega = 0$. Define the sesquilinear form $\gamma := i\Omega(\cdot, \bar{\cdot})$. Then ϕ is called non-degenerate if $\phi^*\gamma$ is non-degenerate.

THEOREM 4.2. [**ACD**] *Any simply connected special complex (resp. special Kähler) manifold M admits a holomorphic immersion (resp. a non-degenerate Lagrangian immersion) $\phi : M \to T^*\mathbb{C}^n$, $n = \dim_{\mathbb{C}} M$, inducing the special geometric structures on M. The immersion is unique up to an affine transformation of $T^*\mathbb{C}^n = \mathbb{C}^{2n}$ with linear part in* $\mathrm{GL}(2n, \mathbb{R})$ *(resp. in* $\mathrm{Sp}(2n, \mathbb{R})$*).*

Any special complex manifold (M, J, ∇) has a canonical torsion-free complex connection D. D is determined by $S := \nabla - D = \frac{1}{2} J \nabla J$. In fact, from $T^\nabla = 0$ and the symmetry of S it follows $T^D = 0$, and $DJ = 0$ is obtained from the following calculation: $DJ = \nabla J - [S, J] = \nabla J - 2SJ = \nabla J + 2JS = \nabla J - \nabla J = 0$. When (M, J, ∇, g) is a special Kähler manifold then D is the Levi-Civita connection of the metric g. In fact, from the skew-symmetry of J and ∇J it follows that S_X is ω-skew-symmetric. This implies $D\omega = 0$ and finally $Dg = 0$.

DEFINITION 4.3. A tt*-bundle (TM, D, S) over a complex manifold M is called special if for all θ the connection D^θ is torsion-free and special, i.e. $D^\theta J$ is symmetric.

The tt*-bundle (TM, D, S) is special if and only if S and S_J are symmetric, $T^D = 0$ and $DJ = 0$.

THEOREM 4.4. [**CS1**]

(i) *There exists a one-to-one correspondence*

$$\Phi : \left\{ \begin{array}{c} \textit{special complex manifolds} \\ (M, J, \nabla) \end{array} \right\} \to \left\{ \begin{array}{c} \textit{special tt*-bundles } (TM, D, S) \\ \textit{over complex manifolds } (M, J) \\ \textit{satisfying } a), b) \end{array} \right\}$$

a) $\{S_X, J\} = 0, \ \forall X \in TM$,
b) $DJ = 0$,
where Φ is given by

$$\Phi(M, J, \nabla) = (TM, D := \nabla - S, S = \frac{1}{2} J \nabla J)$$

and its inverse is given by

$$\Phi^{-1}(TM, D, S) = (M, J, \nabla := D + S).$$

(ii) *This correspondence* Φ *induces a one-to-one correspondence* Φ :

$$\left\{ \begin{array}{c} \textit{special Kähler} \\ \textit{manifolds} \\ (M, J, \nabla, g) \end{array} \right\} \to \left\{ \begin{array}{c} \textit{special metric tt*-bundles } (TM, D, S, g) \\ \textit{over pseudo-Hermitian manifolds } (M, J, g) \\ \textit{satisfying } a), b) \end{array} \right\}.$$

Notice that the metric tt*-bundle associated to a special Kähler manifold is automatically oriented by the complex structure J and unimodular, since $\nabla\omega = D\omega = 0$ implies $\operatorname{tr} S_X = 0$.

COROLLARY 4.5. *Let* (M, J, ∇, g) *be a special Kähler manifold. Any* ∇*-parallel frame* $s = (s_1, \ldots, s_n)$ *of volume* 1 *defines a pluriharmonic map*

$$G^{(s)} = (g(s_i, s_j)) : M \to Sym^1_{p,q}(\mathbb{R}^{2n}) \cong \mathrm{SL}(2n, \mathbb{R})/\mathrm{SO}(p, q),$$

where $(p, q) = (2k, 2l)$ *is the signature of the metric* g, $2n = p + q = \dim_{\mathbb{R}} M$, *and* $Sym^1_{p,q}(\mathbb{R}^{2n}) \subset Sym_{p,q}(\mathbb{R}^{2n})$ *stands for the subset consisting of matrices of determinant* 1.

Let us now describe the pluriharmonic map $G^{(s)}$ in terms of the holomorphic data of the special Kähler geometry.

Let (M, J, ∇, g) be a simply connected special Kähler manifold of complex dimension $n = k + l$, where g is of signature $(2k, 2l)$. According to Theorem 4.2 we have a non-degenerate holomorphic Lagrangian immersion

$$\Phi : M \to V = T^*\mathbb{C}^n = \mathbb{C}^{2n},$$

which is unique up to the action of the group $\mathrm{Aff}_{\mathrm{Sp}(\mathbb{R}^{2n})}(\mathbb{C}^{2n})$. This immersion induces a map

$$\begin{array}{rcl} L : M & \to & Gr_0^{k,l}(\mathbb{C}^{2n}), \\ p & \mapsto & d\phi_p \, T_pM \subset V, \end{array}$$

into the Grassmannian of complex Lagrangian subspaces $W \subset V$ of signature (k, l), i.e. such that $\gamma_{|W}$ is a Hermitian form of signature (k, l).

We call L the **dual Gauß map**. It is in fact dual to the **Gauß map**

$$\begin{array}{rcl} L^\perp : M & \to & Gr_0^{l,k}(\mathbb{C}^{2n}), \\ p & \mapsto & L(p)^\perp = \overline{L(p)} \underset{\Omega}{\cong} L(p)^*. \end{array}$$

The map L is holomorphic, whereas $L^\perp$ is anti-holomorphic.

The target of the dual Gauß map L is a pseudo-Hermitian symmetric space;

$$Gr_0^{k,l}(\mathbb{C}^{2n}) \cong \mathrm{Sp}(\mathbb{R}^{2n})/\mathrm{U}(k, l).$$

THEOREM 4.6. **[CS1]** *Let* (TM, D, S, g) *be the metric tt*-bundle associated to a simply connected special Kähler manifold* (M, J, ∇, g). *Then there exists a* ∇*-parallel frame* s *such that the pluriharmonic map*

$$G^{(s)} = (g(s_i, s_j)) : M \to \mathrm{SL}(2n, \mathbb{R})/\mathrm{SO}(2k, 2l),$$

takes values in the totally geodesic submanifold

$$\mathrm{Sp}(\mathbb{R}^{2n})/\mathrm{U}(k, l) \subset \mathrm{SL}(2n, \mathbb{R})/\mathrm{SO}(2k, 2l)$$

and coincides with the dual Gauß map L.

4.2. Nearly Kähler manifolds.

DEFINITION 4.7. A nearly Kähler manifold (M, J, g) is an almost complex manifold (M, J) endowed with a pseudo-Riemannian metric g such that

i) J is skew-symmetric with respect to g and
ii) $(D_X J)Y = -(D_Y J)X$, $\forall X, Y \in TM$, where D is the Levi-Civita connection of g.

By a result of Friedrich and Ivanov [**FI**] any nearly Kähler manifold has a unique connection ∇ with totally skew-symmetric torsion satisfying $\nabla g = 0$, $\nabla J = 0$ and $T^\nabla = -2\eta$ with $\eta = \frac{1}{2} JDJ$. The 3-form η satisfies $\{\eta_X, J\} = 0$ for all $X \in TM$.

It was observed by the second author [**S3**] that the bundle $(TM, D, D - \nabla = \eta, \omega)$ is a symplectic tt*-bundle provided that g (i.e. $D = \nabla + \eta$) is flat.

In joint work [**CS2**] the present authors have classified flat nearly Kähler manifolds: Any such manifold is locally the product of a flat pseudo-Kähler factor of maximal dimension and a strict nearly Kähler manifold of split signature $(2m, 2m)$ with $m \geq 3$. The geometry of the second factor is encoded in a three-form $\zeta \in \Lambda^3(\mathbb{C}^m)^*$; see Theorem 4.8 below. The first non-trivial example occurs in $\dim_\mathbb{R} M = 2n = 4m = 12$.

Any such strict nearly Kähler manifold comes with a non-trivial pluriharmonic map [**S3**] (in fact, J-holomorphic map):

$$M \to \mathrm{SO}(2m, 2m)/\mathrm{U}(m, m) \underset{\text{tot. geod.}}{\subset} \mathrm{SL}(4m, \mathbb{R})/\mathrm{Sp}(\mathbb{R}^{4m})$$

from the *almost complex manifold* (M, J) into the pseudo-Hermitian symmetric space $\mathrm{SO}(2m, 2m)/\mathrm{U}(m, m)$.

Let $\Lambda^3_{reg}(\mathbb{C}^m)^* \subset \Lambda^3(\mathbb{C}^m)^*$ denote the open subset of regular forms, that is, $\zeta \in \Lambda^3(\mathbb{C}^m)^*$ such that

$$\mathrm{span}\{\zeta(X, Y, \cdot) | X, Y \in \mathbb{C}^m\} = (\mathbb{C}^m)^*.$$

THEOREM 4.8. [**CS2**] *There exists a one-to-one correspondence between* $\mathrm{GL}(m, \mathbb{C})$*-orbits on* $\Lambda^3_{reg}(\mathbb{C}^m)^* \subset \Lambda^3(\mathbb{C}^m)^*$ *and isomorphism classes of complete flat simply connected nearly Kähler manifolds of real dimension* $4m \geq 12$ *and without pseudo-Kähler de Rham factor.*

In dimension 12, 16, 20 (i.e. $m = 3, 4, 5$) one has $\Lambda^3_{reg}(\mathbb{C}^m)^*/\mathrm{GL}(m, \mathbb{C}) = \{pt\}$. In dimension 24 (i.e. $m = 6$) one has the inclusion $\Lambda^3_{stab}(\mathbb{C}^6)^* \subset \Lambda^3_{reg}(\mathbb{C}^6)^*$, where $\Lambda^3_{stab}(\mathbb{C}^6)^* = \{\phi \in \Lambda^3(\mathbb{C}^6)^* \,|\, Q(\phi) \neq 0\}$ are the stable three-forms in the sense of Hitchin [**Hi**]. Here Q is the quartic $\mathrm{SL}(6, \mathbb{C})$-invariant. $\Lambda^3_{stab}(\mathbb{C}^6)^*$ is an open $\mathrm{GL}(6, \mathbb{C})$-orbit. The complement of $\Lambda^3_{stab}(\mathbb{C}^6)^* \subset \Lambda^3_{reg}(\mathbb{C}^6)^*$, is also an orbit. For the discussion of the orbit space we refer to [**Re**]. The real classification is given in appendix A of [**B**].

5. Proof of Theorem 3.1

5.1. Integrability conditions for maps into symmetric spaces. Let $N = G/K$ be a pseudo-Riemannian symmetric space and ∇^N be its Levi-Civita connection. Let M be a second smooth manifold and $f : M \to N$ be a smooth map. We put $F = df : TM \to B = f^*TN$. The vector bundle B is endowed with the pull-back connection $\hat{\nabla} = f^*\nabla^N$. The data $(B, \hat{\nabla}, F)$ satisfy the following structure equations:

$$\hat{\nabla}_V F(W) - \hat{\nabla}_W F(V) - F([V,W]) = 0, \tag{5.1}$$

$$\hat{R}(V,W)\zeta = \hat{\nabla}_V \hat{\nabla}_W \zeta - \hat{\nabla}_W \hat{\nabla}_V \zeta - \hat{\nabla}_{[V,W]}\zeta = R^N(FV, FW)\zeta, \tag{5.2}$$

for any vector fields V, W on M and any section ζ of B.

Conversely, let $B \to M$ be an abstract vector bundle which carries a connection $\hat{\nabla}$ and let N be a pseudo-Riemannian symmetric space. We say that B has the algebraic structure of N if it is endowed with a parallel bundle homomorphism $R^B : \Lambda^2 B \to \operatorname{End} B$ and a linear isomorphism $\Phi_o : B_{p_o} \to T_o N$ for some fixed $p_o \in M$ and $o \in N$, which maps $R^B_{p_o}$ to R^N_o. Then one has the following integration result for a bundle homomorphism $F : TM \to B$, which is a special case of a result proven by Eschenburg and Tribuzy.

THEOREM 5.1. [**ET1**] *Let M be a simply connected manifold and B be a vector bundle over M endowed with a connection $\hat{\nabla}$ having the algebraic structure of N and $F : TM \to B$ a vector bundle homomorphism satisfying the integrability conditions* (5.1) *and*

$$\hat{R}(V,W)\zeta = \hat{\nabla}_V \hat{\nabla}_W \zeta - \hat{\nabla}_W \hat{\nabla}_V \zeta - \hat{\nabla}_{[V,W]}\zeta = R^B(FV, FW)\zeta, \tag{5.3}$$

cf. (5.2). *Then there exists a smooth map $f : M \to N$ and a parallel bundle isomorphism $\Phi : B \to f^*TN$ with the initial value $\Phi|_{B_{p_o}} = \Phi_o$ mapping R^B_p to $R^N_{f(p)}$ for all $p \in M$ and such that $df = \Phi \circ F$.*

5.2. Relation between associated families and pluriharmonic maps. Now we consider an almost complex manifold (M, J). We put $\mathcal{R}_\theta := \exp \theta J \in \Gamma(\operatorname{End} TM)$. An associated family for f is a family of maps $f_\theta : M \to N, \theta \in \mathbb{R}$, such that

$$\Psi_\theta \circ df_\theta = df \circ \mathcal{R}_\theta, \quad \forall \theta \in \mathbb{R}, \tag{5.4}$$

for some bundle isomorphism $\Psi_\theta : f_\theta^* TN \to f^*TN, \theta \in \mathbb{R}$, which is parallel with respect to ∇^N in the sense that

$$\Psi_\theta \circ (f_\theta^* \nabla^N) = (f^*\nabla^N) \circ \Psi_\theta.$$

REMARKS 5.2. 1) Notice that we are not making any assumption on the dependence of the associated family on the parameter θ. We can always modify the family θ such that $f_0 = f$.
2) One observes that each member f_{θ_0} of an associated family f_θ itself admits an associated family $g_\theta = f_{\theta_0 + \theta}$.
3) Note that given an associated family f_θ one can always suppose that $f_\theta(p_o) = o$ by left-multiplication $L_{g(\theta)}$ with an element $g(\theta) \in G$, satisfying $L_{g(\theta)} f_\theta(p_o) = o$. The map Ψ_θ is then replaced by $\Psi_\theta \circ dL^{-1}_{g(\theta)}$. One can then choose the initial value $\Psi_\theta|_{p_0} = \mathrm{Id} : (f_\theta^* TN)_{p_0} = T_o N \to (f^*TN)_{p_0} = T_o N$ for all $\theta \in \mathbb{R}$.

The next theorem generalizes a result of Eschenburg and Tribuzy [**ET2**]:

THEOREM 5.3. *Let (M, J) be an almost complex manifold endowed with a complex connection ∇ such that the $(1,1)$-part of its torsion T vanishes and let N be a pseudo-Riemannian symmetric space. A smooth map $f : M \to N$ admits an associated family f_θ if and only if it is pluriharmonic and satisfies* (3.1) *and* (3.3).

PROOF. Let $f : M \to N$ be a pluriharmonic map which satisfies (3.1), (3.3). To show that f admits an associated family we put $B = f^*TN$, $\hat{\nabla} = f^*\nabla^N$, $F = df$, $F_\theta := F \circ \mathcal{R}_\theta : TM \to B$. We check that F_θ satisfies the integrability conditions of Theorem 5.1. First we observe that B has the algebraic structure of N. In fact, without loss of generality we can assume that $f(p_o) = o$ for some point $p_o \in M$. Then we define $\Phi_o : B_{p_o} = (f^*TN)_{p_o} \to T_oN$ as the canonical identification and $R^B : \Lambda^2 B \to \operatorname{End} B$ as the map which corresponds to $R^N : \Lambda^2 TN \to \operatorname{End} TN$ under the canonical identification $B_p \cong T_{f(p)}N$.

Next we analyse (5.1) for the family F_θ:

$$\begin{aligned} & \hat{\nabla}_V F_\theta(W) - \hat{\nabla}_W F_\theta(V) - F_\theta([V,W]) \\ = \; & \hat{\nabla}_V F(\mathcal{R}_\theta W) - \hat{\nabla}_W F(\mathcal{R}_\theta V) - F(\mathcal{R}_\theta [V,W]) \\ = \; & e^{+i\theta}\left[\hat{\nabla}_V F(W^{1,0}) - \hat{\nabla}_W F(V^{1,0}) - F([V,W]^{1,0})\right] \\ + \; & e^{-i\theta}\left[\hat{\nabla}_V F(W^{0,1}) - \hat{\nabla}_W F(V^{0,1}) - F([V,W]^{0,1})\right] \\ = \; & e^{+i\theta}\left[(\hat{\nabla}_V F)W^{1,0} - (\hat{\nabla}_W F)V^{1,0} + F\left(T(V,W)^{1,0}\right)\right] \\ + \; & e^{-i\theta}\left[(\hat{\nabla}_V F)W^{0,1} - (\hat{\nabla}_W F)V^{0,1} + F\left(T(V,W)^{0,1}\right)\right]. \end{aligned}$$

For the last equality we have extended the connection $\hat{\nabla}$ on B to a connection on $T^*M \otimes B$ using the connection ∇ on M.

If V, W have different type the pluriharmonic map equation, i.e. $(\hat{\nabla} F)^{1,1} = 0$, and the condition $T^{1,1} = 0$ yield that this expression vanishes identically.

If V, W have type $(1,0)$, one obtains the following expression:

$$\begin{aligned} & e^{+i\theta}\left[(\hat{\nabla}_V F)W^{1,0} - (\hat{\nabla}_W F)V^{1,0} + F\left(T(V,W)^{1,0}\right)\right] + e^{-i\theta} F\left(T(V,W)^{0,1}\right) \\ = \; & -e^{+i\theta} F\left(T(V,W)^{0,1}\right) + e^{-i\theta} F\left(T(V,W)^{0,1}\right) = -2i\sin(\theta)\, F\left(T(V,W)^{0,1}\right), \end{aligned}$$

where we used the integrability condition (5.1) for F. This vanishes for all θ if and only if $F\left(T(V,W)^{0,1}\right) = 0$.

If V, W have type $(0,1)$ it follows by the same line of arguments that condition (5.1) holds for all θ if and only if $F\left(T(V,W)^{1,0}\right) = 0$. The last equation is equivalent to (3.3) by complex conjugation.

Now we check equation (5.3) for the family F_θ. If V, W have different type then the two factors $e^{\pm i\theta}$ on the right-hand side cancel each other and the left-hand side does not depend on θ. Therefore, the equation (5.3) for the family F_θ follows from that for F if V, W have different type.
Let now V, W be of the same type. From (3.1) and (5.3) for F we have

$$\hat{R}(V,W) = R^N(FV, FW) = 0$$

and, hence,

$$R^N(F_\theta V, F_\theta W) = e^{\pm 2i\theta} R^N(FV, FW) = 0,$$

which implies $\hat{R}(V,W) = R^N(F_\theta V, F_\theta W) = 0$ for all V, W of same type. This shows the integrability condition (5.3) for F_θ.

Now Theorem 5.1 ensures the existence of a map $f_\theta : M \to N$ with $df_\theta = \Phi_\theta \circ F_\theta$ where Φ_θ is a parallel isomorphism which identifies $B = f^*TN$ and f_θ^*TN. This shows that f admits an associated family with $\Psi_\theta = \Phi_\theta^{-1}$.

Conversely, consider a smooth map $f : M \to N$ which admits an associated family f_θ. As above, we put $F_\theta = F \circ \mathcal{R}_\theta$. We use the integrability constraint (5.1) for $F = df$ and the vanishing of $T(V,W)$ for $V \in \Gamma(T^{1,0}M)$ and $W \in \Gamma(T^{0,1}M)$ to get that

$$0 = \hat{\nabla}_V F(W) - \hat{\nabla}_W F(V) - F([V,W]) = (\hat{\nabla}_W F)V - (\hat{\nabla}_W F)V. \tag{5.5}$$

Similarly, since Ψ_θ is parallel the integrability constraint (5.1) for df_θ yields

$$0 = \Psi_\theta[f_\theta^* \nabla_V^N df_\theta(W) - f_\theta^* \nabla_W^N df_\theta(V) - df_\theta([V,W])] = (\hat{\nabla}_W F_\theta)V - (\hat{\nabla}_W F_\theta)V. \tag{5.6}$$

Using the fact that $\mathcal{R}_\theta$ is parallel we obtain $(\hat{\nabla}_V F_\theta)W = e^{-i\theta}(\hat{\nabla}_V F)W$ and $(\hat{\nabla}_W F_\theta)V = e^{i\theta}(\hat{\nabla}_W F)V$ for all θ. In view of (5.5) and (5.6), this is possible only if $(\hat{\nabla}_V F)W = (\hat{\nabla}_W F)V = 0$. This means that the map f is pluriharmonic. □

In particular, we learn from Theorem 5.3 that, for maps from an almost complex manifold, the pluriharmonic map equation is weaker than the existence of an associated family. Therefore we introduce the following notion.

DEFINITION 5.4. A map f from an almost complex manifold M into a pseudo-Riemannian manifold N is called $\mathbb{S}^1$*-pluriharmonic* if it admits an associated family; see (5.4).

Consider a map

$$f : M \to N := \begin{cases} \mathrm{GL}(r,\mathbb{R})/\mathrm{O}(p,q), & p+q=r, \\ \mathrm{GL}(r,\mathbb{R})/\mathrm{Sp}(r,\mathbb{R}). \end{cases}$$

Denote the map which is induced by f by

$$\hat{f} : M \to \hat{N} := \begin{cases} Sym_{p,q}(\mathbb{R}^r), \\ Skew_{reg}(\mathbb{R}^r). \end{cases}$$

We shall identify the tangent space $T_{\hat{p}}\hat{N}$ at $\hat{p} \in \hat{N}$ with the (ambient) vector space of symmetric/skew-symmetric matrices in $\mathfrak{gl}(r,\mathbb{R})$. The tangent space T_pN of N at a point $p \in N$ which corresponds to $\hat{p} \in \hat{N}$ will be identified with the space of $\hat{p}$-symmetric matrices in $\mathfrak{gl}(r,\mathbb{R})$.

A careful analysis shows that there holds

$$-2df = \hat{f}^{-1}d\hat{f}. \tag{5.7}$$

The previous theorem reduces Theorem 3.1 to the following statement:

THEOREM 5.5. *Let (E,D,S,β) be a metric/symplectic tt*-bundle over a simply connected almost complex manifold (M,J). For any $\theta \in \mathbb{R}$ there exists a D^θ-parallel frame $(e_1^\theta,\ldots,e_r^\theta)$ of E and the correspondence*

$$x \mapsto \hat{f}_\theta(x) :- (\beta(e_i^\theta(x), e_j^\theta(x))) \in \hat{N}$$

defines an $\mathbb{S}^1$-pluriharmonic map

$$f_\theta : M \to N = \begin{cases} \mathrm{GL}(r,\mathbb{R})/\mathrm{O}(p,q), \\ \mathrm{GL}(r,\mathbb{R})/\mathrm{Sp}(r,\mathbb{R}), \end{cases}$$

$$\textit{if } \beta \textit{ is } \begin{cases} \textit{symmetric of signature } (p,q),\ p+q=r, \\ \textit{skew-symmetric, respectively.} \end{cases}$$

If (E,D,S,β) is oriented and unimodular f_θ takes values in the irreducible symmetric space

$$\begin{cases} \mathrm{SL}(r,\mathbb{R})/\mathrm{SO}(p,q), \\ \mathrm{SL}(r,\mathbb{R})/\mathrm{Sp}(r,\mathbb{R}), \textit{respectively.} \end{cases}$$

Conversely, any $\mathbb{S}^1$-pluriharmonic map $f : M \to N$, with

$$N = \begin{cases} \mathrm{GL}(r,\mathbb{R})/\mathrm{O}(p,q), & \mathrm{SL}(r,\mathbb{R})/\mathrm{SO}(p,q), \\ \mathrm{GL}(r,\mathbb{R})/\mathrm{Sp}(r,\mathbb{R}), & \mathrm{SL}(r,\mathbb{R})/\mathrm{Sp}(r,\mathbb{R}), \textit{ respectively,} \end{cases}$$

is obtained from a metric/symplectic tt-bundle by the above construction.*

Proof. Let us consider a metric/symplectic tt*-bundle (E,D,S,β). The connection $D^\theta = D + S^\theta$, $S^\theta_X := \cos(\theta)S_X + \sin(\theta)S_{JX}$, is flat and M is simply connected. Therefore, there exists a D^θ-parallel frame $e_\theta := (e^\theta_1,\ldots,e^\theta_r)$. Since S^θ_X is symmetric with respect to β and $D\beta = 0$, there follows

$$\begin{aligned} X\beta(e^\theta_i,e^\theta_j) &= \beta(D_X e^\theta_i, e^\theta_j) + \beta(e^\theta_i, D_X e^\theta_j) \\ &= -\beta(S^\theta_X e^\theta_i, e^\theta_j) - \beta(e^\theta_i, S^\theta_X e^\theta_j) \\ &= -2\beta(S^\theta_X e^\theta_i, e^\theta_j). \end{aligned}$$

Let $\hat{S}^\theta = e_\theta^{-1}\circ S^\theta \circ e_\theta$ and $\hat{f}_\theta : M \to \hat{N}$ be the matrix representation of S^θ and β in the frame $e_\theta : M\times\mathbb{R}^r \to E$. Denote by $f_\theta : M \to N$ the map induced by $\hat{f}_\theta$; then the above equation reads

$$-2df_\theta \overset{(5.7)}{=} (\hat{f}_\theta)^{-1}d\hat{f}_\theta = -2\hat{S}^\theta$$

or, equivalently,

$$df_\theta = e_\theta^{-1}\circ S^\theta\circ e_\theta. \tag{5.8}$$

This shows for $X\in\Gamma(TM)$

$$\begin{aligned} df_\theta(X) &\overset{(5.8)}{=} e_\theta^{-1}\circ S^\theta_X\circ e_\theta = e_\theta^{-1}\circ S_{\mathcal{R}_\theta X}\circ e_\theta \\ &\overset{(5.8)}{=} (e_\theta^{-1}e_0)\circ df_0(\mathcal{R}_\theta X)\circ(e_0^{-1}e_\theta) \\ &= \mathrm{Ad}^{-1}_{\alpha_\theta}\circ df_0(\mathcal{R}_\theta X) = \Psi_\theta^{-1}\circ df_0(\mathcal{R}_\theta X), \end{aligned}$$

where $\alpha_\theta = e_\theta^{-1}e_0$ is the frame change from e_θ to e_0 and $\Psi_\theta = \mathrm{Ad}_{\alpha_\theta}$, which is parallel with respect to the Levi-Civita connection on N. In other words f_θ is an associated family for f_0 and the maps f_θ are $\mathbb{S}^1$-pluriharmonic.

Conversely, let us consider an associated family $f_\theta : M \to N$ for a pluriharmonic map $f = f_0$ (see Remarks 5.2) and denote the induced maps by $\hat{f}_\theta : M \to \hat{N}$.

We define a metric/symplectic form β on the vector bundle $E = M \times \mathbb{R}^r$ by $\beta(\cdot,\cdot) = \langle \hat{f}_0 \cdot, \cdot \rangle$ and set $S^\theta = df \circ \mathcal{R}_\theta$ and $S = S^0$. Here $\langle \cdot,\cdot \rangle$ stands for the standard scalar product on $\mathbb{R}^r$. Denote by ∂ the canonical flat connection on E and define the connection D on E by

$$D_X s := \partial_X s - S_X s, \quad s \in \Gamma(E).$$

A direct calculation using equation (5.7) and the identification of T_pN with the vector space of $\hat{p}$-symmetric matrices yield that $D\beta = 0$ and that S is β-symmetric (see Lemma 1 [**S1**]).

Since f is $\mathbb{S}^1$-pluriharmonic we have

$$F_\theta := df_\theta = \Psi_\theta^{-1} \circ df \circ \mathcal{R}_\theta,$$

where Ψ_θ is parallel. Recall the first integrability condition (5.1) for the family f_θ :

$$\hat{\nabla}^\theta_V F_\theta(W) - \hat{\nabla}^\theta_W F_\theta(V) - F_\theta([V,W]) = 0, \quad \hat{\nabla}^\theta := f_\theta^* \nabla^N.$$

Evaluating this equation for $\theta = 0, \pi/2$ we obtain

$$\begin{aligned} 0 &= \hat{\nabla}_V F_0(W) - \hat{\nabla}_W F_0(V) - F_0([V,W]) \\ &= \Psi_0^{-1}\left(\hat{\nabla}_V df(W) - \hat{\nabla}_W df(V) - df([V,W])\right), \\ 0 &= \hat{\nabla}^{\frac{\pi}{2}}_V F_{\frac{\pi}{2}}(W) - \hat{\nabla}^{\frac{\pi}{2}}_W F_{\frac{\pi}{2}}(V) - F_{\frac{\pi}{2}}([V,W]) \\ &= \Psi_{\frac{\pi}{2}}^{-1}\left(\hat{\nabla}_V df(JW) - \hat{\nabla}_W df(JV) - df(J[V,W])\right). \end{aligned}$$

Next we use the embedding $(f^*TN, \hat{\nabla}) \subset (\mathrm{End}(E), D = \partial - S)$ as a parallel subbundle by $(f^*TN)_p = T_{f(p)}N = \mathrm{sym}(\hat{f}(p)) \subset \mathrm{End}(\mathbb{R}^r)$. Here $\mathrm{sym}(\hat{f}(p))$ stands for the vector space of $\hat{f}(p)$-symmetric matrices. (We refer to [**S1**] Proposition 2 and 5 for the calculation of the connection.)

Now we check that $(E, D, S = df, \beta = \langle \hat{f}\cdot,\cdot \rangle)$ is a metric/symplectic tt*-bundle. We already know that S is symmetric with respect to β and that $D\beta = 0$. It remains to check the tt*-equations for the data (D,S) on (M,J). In order to show that the tt*-equations are satisfied at the point $p_0 \in M$, we choose the associated family for f in such a way that $f_\theta(p_0) = o$ and $\Psi_\theta(p_0) = \mathrm{Id}$, see Remarks 5.2. Using $S = df$ we can rewrite the above integrability constraint as

$$\begin{aligned} 0 &= D_V(S_W) - D_W(S_V) - S_{[V,W]} = d^D S(X,Y), \\ 0 &= D_V(S_{JW}) - D_W(S_{JV}) - S_{J[V,W]} = d^D S_J(X,Y). \end{aligned}$$

Since $D + S = \partial$ is flat and $d^D S = 0$, we obtain

$$R^D + S \wedge S = R^D + d^D S + S \wedge S = R^{D+S} = 0.$$

In order to verify the tt*-equations (see Remarks 1.2) it only remains to show

$$[S_{JX}, S_{JY}] = [S_X, S_Y]. \tag{5.9}$$

The second integrability condition for the map f_θ can be brought to the form

$$\begin{aligned} \hat{R}(V,W) &= \Psi_\theta \circ R^N(F_\theta V, F_\theta W) \circ \Psi_\theta^{-1} \\ &= \Psi_\theta \circ R^N\left(\Psi_\theta^{-1}(df(\mathcal{R}_\theta V)), \Psi_\theta^{-1}(df(\mathcal{R}_\theta W))\right) \circ \Psi_\theta^{-1}, \end{aligned}$$

where $\hat{R}$ is the curvature of $\hat{\nabla}$ and V, W are vector fields on M. The left-hand side does not depend on the parameter θ. Setting $\theta = 0, \pi/2$ and evaluating the equation at the point p_0 we obtain $R^N(df(JV), df(JW)) = R^N(df(V), df(W))$, $V, W \in T_{p_0}M$. Since $S = df$ and $R^N(A, B)$, $A, B \in T_oN = \mathrm{sym}(\hat{f}(p_0))$, is proportional to $\mathrm{ad}_{[A,B]}$ we get (5.9). Notice that A, B are matrices which are symmetric with respect to $\hat{f}(p_0) = \hat{o}$, the standard scalar product of signature (p, q) on $\mathbb{R}^r$ ($r = p+q$), respectively, the standard symplectic form on $\mathbb{R}^r$, in which case r is necessarily even. $\square$

References

[ACD] D. V. Alekseevsky, V. Cortés and C. Devchand, *Special complex manifolds*, J. Geom. Phys. **42** (2002), 85-105.

[B] R. L. Bryant, *On the geometry of almost complex 6-manifolds,* Asian J. Math. **10** (2006), no. 3, 561-605.

[CS1] V. Cortés and L. Schäfer, *Topological-antitopological fusion equations, pluriharmonic maps and special Kähler manifolds,* in 'Complex, Contact and Symmetric Manifolds', eds. O. Kowalski, E. Musso and D. Perrone, Progress in Mathematics 234, Birkhäuser 20, 59–74.

[CS2] V. Cortés and L. Schäfer, *Flat nearly Kähler manifolds*, Ann. Global Anal. Geom. **32** (2007), no. 4, 379–389.

[CV] S. Cecotti , C. Vafa, *Topological-antitopological fusion,* Nuclear Physics **B 367** (1991), 351-461.

[He] C. Hertling, *tt* geometry, Frobenius manifolds, their connections, and the construction for singularities*, J. Reine Angew. Math. **555** (2003), 77–161.

[Hi] N. Hitchin, *The geometry of three-forms in six dimensions,* J. Differ. Geom. **55** (2000), no. 3, 547–576.

[D] B. Dubrovin, *Geometry and integrability of topological-antitopological fusion,* Commun. Math. Phys. **152** (1992), 539-564.

[ET1] J.-H Eschenburg and R. Tribuzy, *Existence and uniqueness of maps into affine homogeneous spaces,* Rend. Sem. Mat. Univ. Padova **89** (1993), 11–18.

[ET2] J.-H Eschenburg and R. Tribuzy, *Associated families of pluriharmonic maps and isotropy,* Manuscr. Math. **95** (1998), no. 3, 295-310.

[FI] Th. Friedrich and S. Ivanov, *Parallel spinors and connections with skew-symmetric torsion in string theory,* Asian J. Math. **6** (2002), no. 2, 303–335.

[Ra] J.H. Rawnsley, *f-structures, f-twistor spaces and harmonic maps*, Geometry seminar "Luigi Bianchi" II—1984, 85–159, Lecture Notes in Math., 1164, Springer, Berlin, 1985.

[Re] W. Reichel, *Über die Trilinearen Alternierenden Formen in 6 und 7 Veränderlichen,* Dissertation, Greifswald, (1907).

[S1] L. Schäfer, *tt*-bundles and pluriharmonic maps*, Ann. Global Anal. Geom. **28** (2005), no. 3, 285-300.

[S2] L. Schäfer, *Harmonic bundles, topological-antitopological fusion and the related puriharmonic maps*, J. Geom. Phys. **56** (2006), no. 5 , 830-842.

[S3] L. Schäfer, *tt*-geometry on the tangent bundle of an almost complex manifold,* J. Geom. Phys. **57** (2007), 999-1014.

[Si] C.T. Simpson, *Higgs-bundles and local systems*, Publ. Math. Inst. Hautes Études Sci. **75** (1992), 5-95.

Department Mathematik, Universität Hamburg, Bundesstrasse 55, D-20146 Hamburg, Germany

E-mail address: cortes@math.uni-hamburg.de, schaefer@math.uni-hamburg.de

Proceedings of Symposia in Pure Mathematics
Volume **78**, 2008

Hodge theoretic aspects of mirror symmetry

L. Katzarkov, M. Kontsevich, and T. Pantev

ABSTRACT. We discuss the Hodge theory of algebraic non-commutative spaces and analyze how this theory interacts with the Calabi-Yau condition and with mirror symmetry. We develop an abstract theory of non-commutative Hodge structures, investigate existence and variations, and propose explicit construction and classification techniques. We study the main examples of non-commutative Hodge structures coming from a symplectic or a complex geometry possibly twisted by a potential. We study the interactions of the new Hodge theoretic invariants with mirror symmetry and derive non-commutative analogues of some of the more interesting consequences of Hodge theory such as unobstructedness and the construction of canonical coordinates on moduli.

CONTENTS

2000 *Mathematics Subject Classification.* Primary 14C30 14J32; Secondary 14A22 19D55 34M40.

During the preparation of this work Ludmil Katzarkov was partially supported by the Focused Research Grant DMS-0652633 and a research grant DMS-0600800 from the National Science Foundation, and a FWF grant P20778. Tony Pantev was partially supported by NSF Focused Research Grant DMS-0139799, NSF Research Training Group Grant DMS-0636606, and NSF grant DMS-0700446.

1. Introduction

This paper is the first in a series aiming to develop a general procedure associating a 2-dimensional cohomological field theory in the sense of [**KM94**] (CohFT for short) to a certain structure in derived algebraic geometry. More precisely, for any Calabi-Yau A_∞-category satisfying appropriate finiteness conditions (smoothness and compactness), and such that a noncommutative analog of the Hodge $\Rightarrow$ de Rham spectral sequence collapses, we associate an infinite-dimensional family of CohFTs. The additional parameters needed to specify the CohFT are of a purely cohomological nature. Conjecturally, our procedure applied to the Fukaya category should give (higher genus) Gromov-Witten invariants of the underlying symplectic manifold.

This program was first outlined by the second author in several talks given in 2003-2004, and some aspects of it were later developed in depth by K.Costello [**Cos07b, Cos05, Cos07c, Cos07a**]. The whole picture turns out to be very intricate, and in the process of writing we realized that we have to address a large variety of problems. In this installment we do not discuss the general plan of our approach but rather focus on those features of A_∞ or dg categories that can be captured by Hodge theoretic constructions. We propose a formalism that starts with Homological Mirror Symmetry and extrapolates a geometric picture for the requisite categories that makes them amenable to study via old and new Hodge theory. Our hope is that this geometric treatment will provide new invariants and will expand the scope of possible applications in symplectic geometry and algebraic geometry.

Mirror symmetry was introduced in physics as a special duality between two $N = 2$ superconformal field theories. Traditionally a $N = 2$ superconformal field theory is constructed as a quantization of a non-linear σ-model with target a compact Calabi-Yau manifold equipped with a Ricci-flat Kähler metric and a closed 2-form - the so called B-field. Two Calabi-Yau manifolds X and Y form a *mirror pair* $X|Y$ if the associated $N = 2$ superconformal field theories are mirror dual to each other [**CK99**].

Homological Mirror Symmetry was introduced in 1994 by the second author for the case of Calabi-Yau manifolds, but today the realm of its applicability is much broader. In particular many of our considerations in the present work are governed by an analogue of Homological Mirror Symmetry for geometries with potentials. We study the effect of such mirror symmetry on the associated categories of D-branes and especially on the associated non-commutative Hodge structures on homological invariants, i.e. on the Hochschild and cyclic homology and cohomology of such categories. We study mirror pairs consisting of a compact manifold on one side, and of a Landau-Ginzburg model with a proper potential on a non-compact manifold having $c_1 = 0$ on the other. We formulate the mirror symmetry conjecture on the Hodge theoretic level in both directions. That is, we compare the non-commutative Hodge structures associated with a compact complex manifold and a mirror holomorphic Landau-Ginzburg model, and also the non-commutative Hodge structures associated with a compact complex manifold with a chosen smooth anticanonical divisor and with the mirror symplectic Landau-Ginzburg model. This picture is clearly non-symmetric and has to be generalized. In order to completely understand the Hodge theoretic aspect of mirror symmetry, one will have to allow for non-trivial potentials on both sides of the duality and include the novel toric dualities between formal Landau-Ginzburg models of Clarke [**Cla08**] and the new smooth variations of non-commutative Hodge structures of Calabi-Yau type that we attach to anticanonical $\mathbb{Q}$-divisors in Section 4.3. We plan to return to such a generalization in a future work.

Due to its foundational nature the paper comes out somewhat long-winded and technical, for which we apologize. It is organized in three major parts:

The first part introduces and develops the abstract theory of non-commutative (**nc**) Hodge structures. This theory is a variant of the formalism of semi-infinite Hodge structures that was introduced by Barannikov [**Bar01, Bar02a, Bar02b**]. We discuss the general theory of **nc**-Hodge structures in the abstract and analyze the various ways in which the Betti, de Rham and Hodge filtration data can be specified. In particular we compare **nc** and ordinary Hodge theory and explain how **nc**-Hodge theory fits within the setup of categorical non-commutative geometry. We also pay special attention to the **nc**-aspect of Hodge theory and its interaction with the classification of irregular connections on the line via topological data. One of the most useful technical results in this part is the gluing Theorem 2.35 which allows us to assemble **nc**-Hodge structures out of some simple geometric ingredients. This theorem is used later in the paper for constructing **nc**-Hodge structures attached to geometries with a potential.

The second part explains how symplectic and complex geometry give rise to **nc**-Hodge structures and how these structures can be viewed as interesting invariants of Gromov-Witten theory, projective geometry and the theory of algebraic cycles. In particular we analyze the Betti part of the **nc**-Hodge theory of a projective space (viewed as a symplectic manifold) and use this analysis to propose a general conjecture for the integral structure on the cohomology of the Fukaya category of a general compact symplectic manifold. The formula for the integral structure uses only genus zero Gromov-Witten invariants and characteristic classes of the tangent bundle. Our conjecture is in complete agreement with the recent work of Iritani [**Iri07**] who made a similar proposal based on mirror symmetry for toric Fano orbifolds. We also discuss in detail the origin of the Stokes data for holomorphic geometries with potentials and investigate the possible categorical incarnations of these data.

In the third part we study **nc**-Hodge structures and their variations under the Calabi-Yau condition. We extend and generalize the standard treatment of the deformation theory of Calabi-Yau spaces in order to get a theory which works equally well in the **nc**-context and to be able to properly define the canonical coordinates in Homological Mirror Symmetry. We approach the deformation-obstruction problem both algebraically and by Hodge theoretic means and we obtain unobstructedness results, generalized pre-Frobenius structures and some interesting geometric properties of period domains for **nc**-Hodge structures. We also study global and infinitesimal deformations and describe different constructions of Betti and de Rham **nc**-Hodge data for ordinary geometry, relative geometry, geometry with potentials and abstract **nc**-geometry.

Acknowledgments: Throughout the preparation of this work we have benefited from discussions with many people who generously shared their thoughts and insights with us. Special thanks are due to D. Auroux, M. Abouzaid, A. Bondal, R. Donagi, V. Golyshev, M. Gross, A. Losev, D. Orlov, C. Simpson, Y. Soibelman, Y. Tschinkel, A. Todorov, and B. Toën for expert help, encouragement and advice. We would also like to thank the University of Miami for providing the productive research environment in which most of this work was done. During various stages of this work we have enjoyed the hospitality of several outstanding research institutions. We thank the IAS, the IHES, the Centre Interfacultaire Bernoulli at the EPFL, and the ESI for the excellent working conditions they have provided. The first and third author would especially like to thank the organizers of the conference "From tQFT to tt^* and integrability" at the University of Augsburg for giving them an opportunity to speak and for the invitation to contribute to the proceedings volume of the conference.

2. Non-commutative Hodge structures

In this section we will discuss the notion of a pure non-commutative (**nc**) Hodge structure. The **nc**-Hodge structures are analogues of the classical notion of a pure Hodge structure on a complex vector space. Both the **nc**-Hodge structures discussed presently and Simpson's non-abelian Hodge structures [**Sim97a**] generalize classical Hodge theory. In Simpson's theory, one allows for non-linearity in the substrate of the Hodge structure: the non-abelian Hodge structures of [**Sim97a**] are given by imposing Hodge and weight filtrations on non-linear topological invariants of a Kähler space, e.g. on cohomology with non-abelian coefficients, or on the homotopy type. In contrast the **nc**-Hodge structures discussed in this paper consist of a novel filtration-type data (the twistor structure of [**Sim97b, Her03, Sab05b**]) which are still specified on a vector space, e.g. on the periodic cyclic homology of an algebra.

Similarly to ordinary Hodge theory **nc**-Hodge structures arise naturally on the de Rham cohomology of non-commutative spaces of categorical origin.

2.1. Hodge structures.

We will give several different descriptions of an **nc**-Hodge structure in terms of local data. We begin with the notion of rational and unpolarized **nc**-Hodge structures, ignoring for the time being the existence of polarizations and integral lattices.

2.1.1. ***Notation.*** The **nc**-Hodge structures will be described in terms of geometric data on the punctured complex line, so we fix once and for all a coordinate u on $\mathbb{C}$ and the compactification $\mathbb{C} \subset \mathbb{P}^1$. We will write $\mathbb{C}[[u]]$ for the algebra of formal power series in u, and $\mathbb{C}((u))$ for the field of formal Laurent series in u. Similarly, we will write $\mathbb{C}\{u\}$ for the algebra of power series in u having positive radius of convergence, and $\mathbb{C}\{u\}[u^{-1}]$ for the field of meromorphic Laurent series in u with a pole at most at $u = 0$.

2.1.2. ***Meromorphic connections on the complex line.*** We will need some standard notions and facts related to meromorphic connections on Riemann surfaces. We briefly recall those next. More details can be found in e.g. [**Sab02**, chapter II].

Let $\mathscr{M}$ be a finite dimensional vector space over $\mathbb{C}\{u\}[u^{-1}]$, and let ∇ be a ***meromorphic connection*** on $\mathscr{M}$. Explicitly, ∇ is given by a $\mathbb{C}$-linear map $\nabla_{\frac{d}{du}} : \mathscr{M} \to \mathscr{M}$ which satisfies the Leibniz rule for multiplication by elements in $\mathbb{C}\{u\}[u^{-1}]$. A ***holomorphic extension of*** $\mathscr{M}$ is a free finitely generated $\mathbb{C}\{u\}$-submodule $\mathscr{H} \subset \mathscr{M}$, such that

$$\mathscr{H} \otimes_{\mathbb{C}\{u\}} \mathbb{C}\{u\}[u^{-1}] = \mathscr{M}.$$

Traditionally (see e.g. [**Sab02**, section 0.8]) a holomorphic extension is called a *lattice*. We will avoid this classical terminology since it may lead to confusion with the integral lattice structures that we need.

As usual the data $(\mathscr{M}, \nabla)$ or $(\mathscr{H}, \nabla)$ should be viewed as local models for geometric data on a Riemann surface: $(\mathscr{M}, \nabla)$ is the local model of a germ of a meromorphic bundle with connection on a Riemann surface, and $(\mathscr{H}, \nabla)$ is the local model of a holomorphic bundle with meromorphic connection on a Riemann surface.

Suppose $(\mathscr{M}, \nabla)$ is a meromorphic bundle with connection over $\mathbb{C}\{u\}[u^{-1}]$ and let $\mathscr{H} \subset \mathscr{M}$ be a holomorphic extension. We say that $\mathscr{H}$ is **logarithmic** with respect to ∇ if $\nabla(\mathscr{H}) \subset \mathscr{H}\frac{du}{u}$. We say that $(\mathscr{M}, \nabla)$ has at most a **regular singularity at** 0 if we can find a holomorphic extension $\mathscr{H} \subset \mathscr{M}$ which is logarithmic with respect to ∇.

REMARK 2.1. The Riemann-Hilbert correspondence implies (see e.g. [**Sab02**, II.3.7]) that the functor of taking the germ at $0 \in \mathbb{P}^1$:

$$\left(\begin{array}{l}\text{finite rank algebraic vector bun-}\\ \text{dles with connections on } \mathbb{A}^1 - \{0\}\\ \text{with a regular singularity at } \infty\end{array}\right) \xrightarrow{\mathfrak{G}_0} \left(\begin{array}{l}\text{finite dimensional } \mathbb{C}\{u\}[u^{-1}]\text{-}\\ \text{vector spaces with meromor-}\\ \text{phic connections}\end{array}\right)$$

is an equivalence of categories. For future reference we choose once and for all an inverse $\mathfrak{B}_0$ of $\mathfrak{G}_0$. We will call $\mathfrak{B}_0$ the **algebraization functor** or the **Birkhoff extension functor**.

Suppose $\mathscr{H}$ is a *holomorphic bundle* over $\mathbb{C}\{u\}$ equipped with a meromorphic connection ∇. Let $\mathscr{M} = \mathscr{H} \otimes_{\mathbb{C}\{u\}} \mathbb{C}\{u\}[u^{-1}]$ and let $(M, \nabla) = \mathfrak{B}_0(\mathscr{M}, \nabla)$ be the corresponding Birkhoff extension. The algebraic bundle M on $\mathbb{A}^1 - \{0\}$ admits a natural extension to a holomorphic bundle H on $\mathbb{A}^1$: on a small punctured disc centered at $0 \in \mathbb{A}^1$, the bundle M is analytically isomorphic to $\mathscr{M}$, and so $\mathscr{H}$ gives us an extension to the full disc. In particular $\mathfrak{G}_0$ and $\mathfrak{B}_0$ can be promoted to a pair of inverse equivalences

$$\left(\begin{array}{l}\text{finite rank algebraic vector bun-}\\ \text{dles on } \mathbb{A}^1 \text{ equipped with a mero-}\\ \text{morphic connection with poles at}\\ \text{most at 0 and } \infty\text{, and a regular}\\ \text{singularity at } \infty\end{array}\right) \underset{\mathfrak{B}_0}{\overset{\mathfrak{G}_0}{\rightleftarrows}} \left(\begin{array}{l}\text{finite rank free } \mathbb{C}\{u\}\text{-modules}\\ \text{equipped with a meromorphic}\\ \text{connection}\end{array}\right)$$

which we will denote again by $\mathfrak{G}_0$ and $\mathfrak{B}_0$.

2.1.3. ***Stokes data.*** Let $(\mathscr{H}, \nabla)$ be a holomorphic bundle with meromorphic connection over $\mathbb{C}\{u\}$. We will need the Deligne-Malgrange description of the associated meromorphic connection $(\mathscr{M}, \nabla)$ via a filtered sheaf on the circle. We briefly recall this description next. More details can be found in [**Mal83**] and [**BV85**]. Let $(M, \nabla) := \mathfrak{B}_0((\mathscr{M}, \nabla))$ be the Birkhoff extension of $(\mathscr{M}, \nabla)$ to $\mathbb{P}^1$. Consider the circle $\boldsymbol{S}^1 := \mathbb{C}^\times/\mathbb{R}_+^\times$. The sheaf of local ∇-horizontal sections of M^{an} on $\mathbb{C}^\times$ is a locally constant sheaf on $\mathbb{C}^\times$, which by contractibility of $\mathbb{R}_+^\times$ induces a locally constant sheaf S of $\mathbb{C}$-vector spaces on $\boldsymbol{S}^1$.

The sheaf S is equipped with a natural local filtration by subsheaves $\{\mathsf{S}_{\leq\omega}\}_{\omega\in\mathsf{Del}}$, where

(i) Del is the complex local system on $\boldsymbol{S}^1$ for which for every open $U \subset \boldsymbol{S}^1$ the space of sections $\mathsf{Del}(U)$ is defined to be the space of all holomorphic one-forms ω on the sector

$$\mathrm{Sec}(U) := \left\{ re^{i\varphi} \middle| \, r > 0, \varphi \in U \right\}$$

which are of the form

$$\omega = \left(\sum_{\substack{a\in\mathbb{Q} \\ a<-1}} c_a u^a \right) du,$$

where at most finitely many $c_a \neq 0$ and the branches u^a are chosen arbitrarily.

Note that the germs of sections of Del are naturally ordered: if $\omega', \omega'' \in \mathsf{Del}(U)$, $\varphi \in U$, and if

$$\omega' - \omega'' = c_a u^a + \begin{pmatrix} \text{higher} \\ \text{order terms} \end{pmatrix},$$

then one sets

$$\omega' <_\varphi \omega'' \quad \Leftrightarrow \quad \mathrm{Re}\left(\frac{c_a e^{i\varphi(a+1)}}{a+1} \right) < 0.$$

(ii) For every $\varphi \in \boldsymbol{S}^1$ and every $\omega \in \mathsf{Del}_\varphi$ the stalk

$$\left(\mathsf{S}_{\leq\omega}\right)_\varphi \subset \mathsf{S}_\varphi$$

is defined to be the subspace

$$\left(\mathsf{S}_{\leq\omega}\right)_\varphi := \left\{ s \in \mathsf{S}_\varphi = \Gamma\left(\mathbb{R}_+^\times e^{i\varphi}, M^{\mathrm{an}}\right)^\nabla \;\middle|\; \begin{array}{l} e^{-\int \omega} s \text{ has moderate growth in} \\ \text{the direction } \varphi, \text{ i.e.} \\ \quad \left\| e^{-\int \omega} s \right\|_{|\mathbb{R}_+^\times e^{i\varphi}} = \boldsymbol{O}\left(r^{-N}\right), \\ \text{when } r \to 0,\ N \gg 0. \end{array} \right\}$$

Here $\| \bullet \|$ is the Hermitian norm of a section of M computed in some (any) meromorphic trivialization of M^{an} near $u = 0$.

DEFINITION 2.2. *The filtration we just defined is the* ***Deligne-Malgrange-Stokes filtration****, and the* Del*-filtered sheaf* S *is called the* ***Stokes structure*** *associated to* $(\mathscr{M}, \nabla)$.

REMARK 2.3. The Deligne-Malgrange-Stokes filtration satisfies the following property. First of all, there exists a covariantly local system of finite sets $\mathsf{Del}_{(\mathscr{M},\nabla)} \subset \mathsf{Del}$ canonically associated with $(\mathscr{M}, \nabla)$ such that the filtration by all of Del is determined by a filtration by all $\mathsf{Del}_{(\mathscr{M},\nabla)}(U)$ and all consecutive factors are non-trivial at all points of $\boldsymbol{S}^1$ except finitely many (called the directions of the Stokes rays). Outside the Stokes rays the set $\mathsf{Del}_{\mathscr{M},\nabla}(\phi)$ is totally ordered. It is easy to see that when we cross a Stokes ray then the order changes by flipping the order on several disjoint intervals (e.g. $\{1,2,3,4,5,6\} \to \{2,1,3,6,5,4\}$). Moreover, on the subquotients corresponding to these intervals, the two filtrations coming from the left and from the right of the anti-Stokes ray are opposed to each other. This implies that the graded pieces with respect to the Deligne-Malgrange-Stokes filtration are locally constant systems of vector spaces on the total space of stalks of the sheaf $\mathsf{Del}_{(\mathscr{M},\nabla)}$ (which is a disjoint union of finite coverings of $\boldsymbol{S}^1$).

REMARK 2.4. A fundamental theorem of Deligne and Malgrange [**Mal83**, Theorem 4.2], [**BV85**, Theorem 4.7.3] asserts that the functor $(\mathscr{M}, \nabla) \longmapsto (\mathsf{S}, \{\mathsf{S}_{\leq \omega}\}_{\omega \in \mathsf{Del}})$ is an equivalence between the category of meromorphic connections over $\mathbb{C}\{u\}[u^{-1}]$ and the category of Del-filtered local systems on $\boldsymbol{S}^1$ satisfying the conditions described in Remark 2.3. We will use this equivalence to define the Betti part of an **nc**-Hodge structure.

2.1.4. ***The definition of an* nc*-Hodge structure.*** After these preliminaries we are now ready to define **nc**-Hodge structures.

DEFINITION 2.5. *A* ***rational pure* nc*-Hodge structure*** *consists of the data* $(H, \mathscr{E}_B, \mathsf{iso})$, *where*

- H *is a* $\mathbb{Z}/2$*-graded algebraic vector bundle on* $\mathbb{A}^1$.
- $\mathscr{E}_B$ *is a local system of finite dimensional* $\mathbb{Z}/2$*-graded* $\mathbb{Q}$*-vector spaces on* $\mathbb{A}^1 - \{0\}$.
- iso *is an* analytic *isomorphism of holomorphic vector bundles on* $\mathbb{A}^1 - \{0\}$*:*

$$\mathsf{iso} : \mathscr{E}_B \otimes \mathcal{O}_{\mathbb{A}^1-\{0\}} \xrightarrow{\cong} H_{|\mathbb{A}^1-\{0\}}.$$

Note: *The isomorphism* iso *induces a natural flat holomorphic connection* ∇ *on* $H_{|\mathbb{A}^1-\{0\}}$.

These data have to satisfy the following axioms:

(nc-filtration axiom) ∇ *is a meromorphic connection on* H *with a pole of order* ≤ 2 *at* $u = 0$ *and a regular singularity at* ∞. *More precisely, there exist:*

- *a holomorphic frame of* H *near* $u = 0$ *in which*

$$\nabla = d + \left(\sum_{k \geq -2} A_k u^k \right) du$$

with $A_k \in \mathrm{Mat}_{r\times r}(\mathbb{C})$, $r = \mathrm{rank}\ H$.

- *a meromorphic frame of H near $u = \infty$ in which*

$$\nabla = d + \left(\sum_{k \geq -1} B_k u^{-k} \right) d(u^{-1})$$

and $B_k \in \mathrm{Mat}_{r\times r}(\mathbb{C})$.

($\mathbb{Q}$-structure axiom) *The $\mathbb{Q}$-structure $\mathscr{E}_B$ on (H, ∇) is compatible with the Stokes data. More precisely, let $\left(\mathsf{S}, \{\mathsf{S}_{\leq\omega}\}_{\omega\in\mathsf{Del}}\right)$ be the Stokes structure corresponding to the germ $(\mathscr{H}, \nabla) := \mathfrak{G}_0(H, \nabla)$, and let $\mathsf{S}_B \subset \mathsf{S}$ be the $\mathbb{Q}$-structure on S induced from $\mathscr{E}_B$ via the isomorphism* **iso**. *We require that the Deligne-Malgrange-Stokes filtration on S is defined over $\mathbb{Q}$, i.e.*

$$(\mathsf{S}_{\leq\omega} \cap \mathsf{S}_B) \otimes_{\mathbb{Q}} \mathbb{C} = \mathsf{S}_{\leq\omega}$$

for all local sections $\omega \in$ **Del**.

(opposedness axiom) *The $\mathbb{Q}$-structure S_B induces a real structure on S and hence a complex conjugation $\tau : \mathsf{S} \to \mathsf{S}$. Let $\widehat{H}$ be the holomorphic bundle on $\mathbb{P}^1$ obtained as the gluing of $H^{\mathrm{alg}}_{|\{|u|\leq 1\}}$ and $\left(\gamma^* \overline{H^{\mathrm{alg}}}\right)_{|\{|u|\geq 1\}}$ via τ, where $\gamma : \mathbb{P}^1 \to \mathbb{P}^1$ is the real structure on $\mathbb{P}^1$ which fixes the unit circle, i.e. $\gamma(u) := 1/\bar{u}$. Then we require that $\widehat{H}$ be holomorphically trivial, i.e. $\widehat{H} \cong \mathcal{O}_{\mathbb{P}^1}^{\oplus r}$.*

A morphism $\boldsymbol{f} : (H_1, \mathscr{E}_{B,1}, \mathbf{iso}_1) \to (H_2, \mathscr{E}_{B,2}, \mathbf{iso}_2)$ of **nc**-*Hodge structures is a pair $\boldsymbol{f} = (f, f_B)$, where $f : H_1 \to H_2$ is an algebraic map of vector bundles which intertwines the connections, and $f_B : \mathscr{E}_{B,1} \to \mathscr{E}_{B,2}$ is a map of $\mathbb{Q}$-local systems, such that $f \circ \mathbf{iso}_1 = \mathbf{iso}_2 \circ (f_B \otimes \mathrm{id}_{\mathcal{O}})$. We will write ($\mathbb{Q}$-**ncHS**) for the category of pure* **nc**-*Hodge structures.*

REMARK 2.6. The meromorphic connection (M, ∇) where $M = H \otimes_{\mathbb{C}[u]} \mathbb{C}[u, u^{-1}]$ can be thought of as the de Rham data of the **nc**-Hodge structure, the local system S_B of rational vector spaces over S^1 endowed with the rational Stokes filtration (see $\mathbb{Q}$-structure axiom) can be thought of as the Betti data, and the holomorphic extension H of M can be thought of as the analogue of the Hodge filtration.

2.1.5. ***Variations of* nc-*Hodge structures.*** One can also define variations of **nc**-Hodge structures:

DEFINITION 2.7. *Let S be a complex manifold. A* ***variation of pure* nc-*Hodge structures*** *over S is a triple $(H, \mathscr{E}_B, \mathbf{iso})$, where*

- *H is a holomorphic $\mathbb{Z}/2$-graded vector bundle on $\mathbb{A}^1 \times S$ which is algebraic in the $\mathbb{A}^1$-direction.*
- *$\mathscr{E}_B$ is a local system of $\mathbb{Z}/2$-graded $\mathbb{Q}$-vector spaces on $(\mathbb{A}^1 - \{0\}) \times S$.*
- **iso** *is an analytic isomorphism of holomorphic vector bundles*

$$\mathbf{iso} : \mathscr{E}_B \otimes \mathcal{O}_{(\mathbb{A}^1-\{0\})\times S} \xrightarrow{\cong} H_{|(\mathbb{A}^1-\{0\})\times S}.$$

Let ∇ be the induced meromorphic connection on H. The data $(H, \mathscr{E}_B, \mathbf{iso})$ should satisfy:

(nc-filtration axiom) *The connection* ∇ *has a regular singularity along* $\{\infty\} \times S$ *and Poincaré rank* ≤ 1 *along* $\{0\} \times S$*, i.e.*

$$u^2 \cdot \nabla_{\frac{\partial}{\partial u}} : H \to H$$

is a holomorphic differential operator on H *of order* ≤ 1*.*

(Griffiths transversality axiom) *For every locally defined vector field* $\xi \in T_S$ *we have that*

$$u \cdot \nabla_\xi : H \to H,$$

is a holomorphic differential operator on H *of order* ≤ 1*.*

($\mathbb{Q}$-structure axiom) *The Stokes structure on the local system* S *on* $\boldsymbol{S}^1 \times S$ *is well defined, i.e. the steps in the Deligne-Malgrange-Stokes filtration on* S *are sheaves on* $\boldsymbol{S}^1 \times S$*. Furthermore the* $\mathbb{Q}$*-structure* $\mathscr{E}_B$ *is compatible with the Stokes data as in Definition 2.5.*

(opposedness axiom) *The relative version of the gluing construction for* **nc***-Hodge structures gives a globally defined complex vector bundle* $\widehat{H}$ *on* $\mathbb{P}^1 \times S$*, which is holomorphically trivial in the* $\mathbb{P}^1$ *direction. Moreover, with respect to the extension* $\widehat{H}$ *the connection* ∇ *is meromorphic with Poincaré rank one along* $(\{0\} \times S) \cup (\{\infty\} \times S)$*.*

2.1.6. ***Relation to other definitions.*** Various special cases and partial versions of our notion of an **nc**-Hodge structure have been studied before in slightly different but related setups. We list a few of the relevant notions and references without going into detailed comparisons:

- A version of ($\mathbb{Z}$-graded) **nc**-Hodge structures appears in the fundamental work of K. Saito (see [**Sai83, Sai98b, Sai98a**] and references therein) on the Hodge theoretic invariants of quasi-homogeneous hypersurface singularities under the name ***weight system***.
- A version of the notion of a variation of (complex) **nc**-Hodge structure appears in the work of Cecotti-Vafa in Conformal Field Theory [**CV91, CV93a, CV93b, BCOV94**] under the name tt^****-geometry***.
- Various versions of the notion of a (complex, polarized) **nc**-Hodge structure appear in algebraic geometry and non-abelian Hodge theory in the works of C. Simpson [**Sim97a, Sim97b**] and T. Mochizuki [**Moc06a, Moc06b, Moc07a, Moc07b**] under the names of ***(tame or wild) harmonic bundle*** or ***pure twistor structure***, and in the work of C. Sabbah [**Sab05b**] under the name ***integrable pure twistor*** D***-module***.
- The analytic germ of a (complex) variation of **nc**-Hodge structures appears in mirror symmetry in the work of Barannikov [**Bar01, Bar02a, Bar02b**] and Barannikov and the second author [**BK98**] under the name ***semi-infinite Hodge structure***. The integral structures on semi-infinite Hodge structure were recently introduced and studied in the work of Iritani [**Iri07**].
- A version of the notion of a (real) **nc**-Hodge structure appears in singularity theory in the work of Hertling [**Her03, Her06**] and Hertling-Sevenheck [**HS07**] under the name ***TER structure***. Hertling and Sevenheck also

consider polarized and mixed **nc**-Hodge structures. Those appear under the names ***TERP structure*** and ***mixed TERP structure***, respectively. In particular in [**HS07**] Hertling and Sevenheck study the degenerations of TERP structures and prove a version of Schmid's nilpotent orbit theorem which gives rise to the notion of a limiting mixed TERP structure. Degenerations of variants of **nc**-Hodge structures, as well as limiting mixed **nc**-Hodge structures appear also in the works of C. Sabbah [**Sab05a**] and S. Szabo [**Sza07**].

2.1.7. ***Relation to usual Hodge theory.*** Recall (see e.g. [**Del71**]) that a ***pure rational Hodge structure*** of weight w is a triple $(V, F^{\bullet}V, V_{\mathbb{Q}})$ where:

- V is a complex vector space,
- $V_{\mathbb{Q}} \subset V$ is a $\mathbb{Q}$-subspace such that $V = V_{\mathbb{Q}} \otimes_{\mathbb{Q}} \mathbb{C}$, and
- $F^{\bullet}V$ is a ***Hodge filtration*** of weight w on V, i.e. $F^{\bullet}V$ is a decreasing finite exhaustive filtration by complex subspaces which satisfies $F^pV \oplus \overline{F^{w+1-p}V} = V$, where the complex conjugation on V is the one given by the real structure $V_{\mathbb{R}} = V_{\mathbb{Q}} \otimes \mathbb{R} \subset V$.

A ***pure Hodge structure*** is a direct sum of pure Hodge structures of various weights, and a morphism of pure Hodge structures is a linear map of complex vector spaces which maps the rational structures into each other and is strictly compatible with the filtrations. We will write ($\mathbb{Q}$-HS) for the category of pure rational Hodge structures. It is well known [**Del71**] that ($\mathbb{Q}$-HS) is an abelian $\mathbb{Q}$-linear tensor category. For every $w \in \mathbb{Z}$ we have a $\otimes$-invertible object in ($\mathbb{Q}$-HS) of pure weight $2w$: the ***Tate Hodge structure*** $\mathbb{Q}(w)$ given by $\mathbb{Q}(w) := (\mathbb{C}, F^{\bullet}, \mathbb{Q})$, where $F^i = \mathbb{C}$ for $i \leq w$ and $F^i = \{0\}$ for $i > w$.

It turns out that pure Hodge structures can be viewed as **nc**-Hodge structures. This is achieved through a version of the Rees module construction (see e.g. [**Sim97a**]) which converts a filtered vector space into a bundle over the affine line $\mathbb{A}^1$. Specifically, given a pure Hodge structure $(V, F^{\bullet}V, V_{\mathbb{Q}})$ of weight w we consider the rank one meromorphic bundle with connection

$$\boldsymbol{\mathcal{T}}_{\frac{w}{2}} := \left(\mathbb{C}\{u\}[u^{-1}], d - \frac{w}{2} \cdot \frac{du}{u}\right)$$

and we set

- $\mathscr{H} := \mathscr{H}^{w \bmod 2} := \sum_i u^{-i} F^i V\{u\}$ viewed as a $\mathbb{C}\{u\}$-submodule in $\mathbb{C}\{u\}[u^{-1}] \otimes_{\mathbb{C}} V$. Clearly, this submodule is preserved by the operator $\nabla_{u\frac{d}{du}}$ for the connection $\nabla := \left(d - \frac{w}{2} \cdot \frac{du}{u}\right) \otimes \mathrm{id}_V$, i.e. $(\mathscr{H}, \nabla)$ is a logarithmic holomorphic extension of the meromorphic bundle with connection $\boldsymbol{\mathcal{T}}_{\frac{w}{2}} \otimes_{\mathbb{C}} V$.
  Note: Consider the algebraization $(H, \nabla) = \mathfrak{B}_0(\mathscr{H}, \nabla)$ of $(\mathscr{H}, \nabla)$. The fiber $H_1 := H/(u-1)H$ of H at $1 \in \mathbb{A}^1$ is canonically identified with V. By definition the connection ∇ on H has monodromy $(-1)^w \mathrm{id}_V$ and so preserves any rational subspace in V.
- $\mathscr{E}_B := \mathscr{E}_B^{w \bmod 2}$– the $\mathbb{Q}$-local system on $\mathbb{A}^1 - \{0\}$ defined as the subsheaf $\mathscr{E}_B \subset H$ consisting of sections whose value at 1 is in $V_{\mathbb{Q}} \subset V = H/(u-1)H$. In other words $\mathscr{E}_B$ is the locally constant sheaf on $\mathbb{A}^1 - \{0\}$ with fiber $V_{\mathbb{Q}}$ and monodromy $(-1)^w \mathrm{id}_{V_{\mathbb{Q}}}$.

- **iso** is the isomorphism of complex local systems, corresponding to the embedding $\mathscr{E}_B \subset H$.

REMARK 2.8. On every simply connected open (in the analytic topology) subset $U \subset \mathbb{A}^1 - \{0\}$ the bundle with connection $\boldsymbol{T}_{\frac{w}{2}}$ has a horizontal section $u^{w/2}$. In particular on such opens we have $H_{|U} = \sum_i u^{w/2} u^{-i} F^i[u]$.

The data $(H, \mathscr{E}_B, \mathbf{iso})$ satisfy tautologically the **($\mathbb{Q}$-structure axiom)** and the **(opposedness axiom)** from Definition 2.5. Indeed, the **($\mathbb{Q}$-structure axiom)** is satisfied since by definition ∇ has a regular singularity at 0 and so $\mathbf{S}_{\leq\omega} = \mathbf{S}$ or 0 for all ω. The **(opposedness axiom)** is satisfied as it is equivalent in the case of regular singularities to the oposedness property in the definition of the usual Hodge structures.

Thus, the assignment $(V, F^\bullet V, V_{\mathbb{Q}}) \to (H, \mathscr{E}_B, \mathbf{iso})$ gives a functor

$$\mathfrak{n} : (\mathbb{Q}\text{-HS}) \to (\mathbb{Q}\text{-}\mathbf{ncHS})$$

which by definition factors through the orbit category (see e.g. [**Kel05**] for the definition of an orbit category)

$$\pi : (\mathbb{Q}\text{-HS}) \to (\mathbb{Q}\text{-HS})/(\bullet \otimes \mathbb{Q}(1)),$$

i.e we have $\mathfrak{N} = \mathfrak{n} \circ \pi$ for a functor

$$\mathfrak{N} : (\mathbb{Q}\text{-HS})/(\bullet \otimes \mathbb{Q}(1)) \to (\mathbb{Q}\text{-}\mathbf{ncHS}).$$

The proof of the following statement is an immediate consequence of the definition.

LEMMA 2.9. *The functor $\mathfrak{N}$ is fully faithful and its essential image consists of all* **nc**-*Hodge structures that have regular singularities and monodromy* $=$ id *on* H^0 *and* $= -$ id *on* H^1.

REMARK 2.10. It is straightforward to check that the functor $\mathfrak{N}$ can also be defined in families and embeds the category of variations of Hodge structures (modulo the Tate twist) into the category of variations of **nc**-Hodge structures.

2.1.8. ***nc-Hodge structures of exponential type.*** As we saw in Section 2.1.7 the usual Hodge structures give rise to special **nc**-Hodge structures with regular singularities. The **nc**-Hodge structures with regular singularities are also important because they can serve as building blocks of general **nc**-Hodge structures. Let $(H, \mathscr{E}_B, \mathbf{iso})$ be an **nc**-Hodge structure, let $(\mathscr{H}, \nabla) = \mathfrak{G}_0((H, \nabla))$ be the germ of (H, ∇) at zero, and assume that $A_{-2} \neq 0$, i.e. ∇ has a second order pole. According to the Turrittin-Levelt formal decomposition theorem (see e.g. [**Mal79**], [**BV85**], [**Sab02**, II.5.7 and II.5.9]) we can find a finite base change $p_N : \mathbb{C} \to \mathbb{C}$, $p_N(t) := t^N = u$, so that $p_N^*(\mathscr{H}, \nabla)[t^{-1}]$ is formally isomorphic to a direct sum of regular singular connections on meromorphic bundles multiplied by exponents of Laurent polynomials. More precisely we can find polynomial tails $g_i(t) \in \mathbb{C}[t^{-1}]$, $\mathbb{C}\{t\}[t^{-1}]$-vector spaces $\mathscr{R}_i$ and meromorphic connections

$$(\nabla_i)_{\frac{d}{dt}} : \mathscr{R}_i \to \mathscr{R}_i,$$

each with at most a regular singularity at 0, so that we have an isomorphism of formal meromorphic connections over $\mathbb{C}((t))$:

$$\Psi : p_N^*(\mathscr{H},\nabla) \bigotimes_{\mathbb{C}\{t\}[t^{-1}]} \mathbb{C}((t)) \xrightarrow{\cong} \left(\bigoplus_{i=1}^{m} \mathcal{E}^{g_i} \bigotimes_{\mathbb{C}\{t\}[t^{-1}]} (\mathscr{R}_i, \nabla_i) \right) \bigotimes_{\mathbb{C}\{t\}[t^{-1}]} \mathbb{C}((t)).$$

Here $\mathcal{E}^f$ denotes the rank one holomorphic bundle with meromorphic connection $(\mathbb{C}\{t\}, d - df)$, and $(\mathscr{R}_i, \nabla_i)$ denote meromorphic bundles with connections having regular singularities.

REMARK 2.11. The bundle $\mathcal{E}^f$ has a non-vanishing horizontal section, namely e^f. In particular the multivalued flat sections of $\mathcal{E}^{g_i} \otimes (\mathscr{R}_i, \nabla_i)$ are given by multiplying multivalued flat sections of $(\mathscr{R}_i, \nabla_i)$ by e^{g_i}.

In the examples coming from Mirror Symmetry that we are interested in, the base change p_N is not needed for the decomposition to work. In this case we can take $g_i(u) = \boldsymbol{c}_i/u$ where $\boldsymbol{c}_1, \ldots, \boldsymbol{c}_m \in \mathbb{C}$ denote the distinct eigenvalues of A_{-2}. Because of this we introduce the following definition (see also [**HS07**, Definition 8.1]):

DEFINITION 2.12. *We say that an* **nc***-Hodge structure* $(H, \mathscr{E}_B, \mathsf{iso})$ *is of* ***exponential type*** *if there exists a formal isomorphism*

$$\Psi : (\mathscr{H} \otimes_{\mathbb{C}\{u\}} \mathbb{C}[[u]], \nabla) \xrightarrow{\cong} \bigoplus_{i=1}^{m} \left(\mathcal{E}^{\boldsymbol{c}_i/u} \otimes (\mathscr{R}_i, \nabla_i) \right) \otimes_{\mathbb{C}\{u\}} \mathbb{C}[[u]]$$

where $(\mathscr{R}_i, \nabla_i)$ *are meromorphic bundles with connections with regular singularities and* $\boldsymbol{c}_1, \ldots, \boldsymbol{c}_m \in \mathbb{C}$ *denote the distinct eigenvalues of* A_{-2}.

REMARK 2.13. • There are various sufficient conditions that will guarantee that a given **nc**-Hodge structure is decomposable without base change. For instance, this will be the case if A_{-2} has distinct eigenvalues, or if $A_{-1} = 0$. More generally, it suffices to require that we can find holomorphic functions $\ell_i(u) \in \mathbb{C}\{u\}$ so that $\ell_i(0) = \boldsymbol{c}_i$ for $i = 1, \ldots, m$ and the characteristic polynomial of $u^2 A(u)$ is $\det\left(\boldsymbol{c} \cdot \mathrm{id} - u^2 A(u)\right) = \prod_{i=1}^{m} (\boldsymbol{c} - \ell_i)^{\nu_i}$.

• Not every irregular connection with a pole of order two is of exponential type. Indeed, the rank two connection

$$\nabla = d - \begin{pmatrix} 0 & u^{-2} \\ u^{-1} & \frac{u^{-1}}{2} \end{pmatrix}$$

has a horizontal section

$$\begin{pmatrix} e^{-2u^{-\frac{1}{2}}} \\ u^{\frac{1}{2}} e^{-2u^{-\frac{1}{2}}} \end{pmatrix},$$

and so one needs a quadratic base change for the formal decomposition to work for this connection.

• If an **nc**-Hodge structure $(H, \mathscr{E}_B, \mathsf{iso})$ is of exponential type, then one can check (see [**HS07**, Lemma 8.2]) that for each $i = 1, \ldots, m$ we can find a unique holomorphic extension $\mathscr{H}_{\boldsymbol{c}_i} \subset \mathcal{R}_i$ in which the connection has a second order pole and so that Ψ induces a formal isomorphism of holomorphic bundles with meromorphic connections

$$\Psi : (\mathscr{H}, \nabla) \otimes \mathbb{C}[[u]] \xrightarrow{\cong} \left(\bigoplus_{i=1}^{m} \mathcal{E}^{\boldsymbol{c}_i/u} \bigotimes_{\mathbb{C}\{u\}} (\mathscr{H}_{\boldsymbol{c}_i}, \nabla_i) \right) \otimes \mathbb{C}[[u]],$$

over $\mathbb{C}[[u]]$.

The **nc**-Hodge structures with regular singularities or the **nc**-Hodge structures of exponential type comprise full subcategories

$$(\mathbb{Q}\text{-}\mathbf{nc}\mathsf{HS})^{\mathrm{reg}} \subset (\mathbb{Q}\text{-}\mathbf{nc}\mathsf{HS})^{\mathrm{exp}} \subset (\mathbb{Q}\text{-}\mathbf{nc}\mathsf{HS})$$

in $(\mathbb{Q}\text{-}\mathbf{nc}\mathsf{HS})$. In fact, in the exponential type case one can state the **nc**-Hodge structure axioms in an easier way. The simplification comes from the fact that in this case the Deligne-Malgrange-Stokes filtration is given by subsheaves $\mathsf{S}_{\leq\lambda}$ of S that are labeled by $\lambda \in \mathbb{R}$ and consisting of solutions decaying faster than $\boldsymbol{O}\left(\exp\left(\frac{\lambda+o(1)}{r}\right)\right)$, $r = |u|$. Indeed, tracing through the definition one sees that in the exponential case for a ray defined by φ the jumps of the steps of the Deligne-Malgrange-Stokes filtration occur exactly at the numbers $\mathrm{Re}(\boldsymbol{c}_i e^{-i\varphi})$. Furthermore, the associated graded pieces for the filtration are local systems on the circle and in fact coincide with the regular pieces $(\mathscr{R}_i, \nabla_i)$ that appear in the formal decomposition of the connection. Hence one arrives at the following

DEFINITION 2.14. *A* ***rational pure* nc*-Hodge structure of exponential type*** *consists of the data* $(H, \mathscr{E}_B, \mathbf{iso})$*, where*

- H *is a* $\mathbb{Z}/2$*-graded algebraic vector bundle on* $\mathbb{A}^1$*.*
- $\mathscr{E}_B$ *is a local system of finite dimensional* $\mathbb{Z}/2$*-graded* $\mathbb{Q}$*-vector spaces on* $\mathbb{A}^1 - \{0\}$*.*
- **iso** *is an* analytic *isomorphism of holomorphic vector bundles on* $\mathbb{A}^1 - \{0\}$*:*

$$\mathbf{iso} : \mathscr{E}_B \otimes \mathcal{O}_{\mathbb{A}^1-\{0\}} \xrightarrow{\cong} H_{|\mathbb{A}^1-\{0\}}.$$

These data have to satisfy the following axioms:

(nc-filtration axiom)$^{\mathbf{exp}}$ *The connection* ∇ *induced from* **iso** *is a meromorphic connection of exponential type on* H *with a pole of order* ≤ 2 *at* $u = 0$ *and a regular singularity at* ∞*. More precisely, there exist:*

- *a holomorphic frame of* $\mathscr{H}$ *near* $u = 0$ *in which*

$$\nabla = d + \left(\sum_{k \geq -2} A_k u^k \right) du$$

with $A_k \in \mathrm{Mat}_{r\times r}(\mathbb{C})$*,* $r = \mathrm{rank}_{\mathbb{C}\{u\}} \mathscr{H}$*.*

- *a holomorphic frame of* $\mathscr{H}$ *near* $u = \infty$ *in which*

$$\nabla = d + \left(\sum_{k \geq -1} B_k u^{-k} \right) d(u^{-1})$$

and $B_k \in \mathrm{Mat}_{r\times r}(\mathbb{C})$*.*

- *a formal isomorphism over* $\mathbb{C}((u))$*:*
$$(\mathscr{H}[u^{-1}], \nabla) \overset{\cong}{\to} \bigoplus_{i=1}^{m} \boldsymbol{\mathcal{E}}^{\boldsymbol{c}_i/u} \otimes (\mathscr{R}_i, \nabla_i)$$
where $(\mathscr{R}_i, \nabla_i)$ *are meromorphic bundles with connections with regular singularities and* $\boldsymbol{c}_1, \ldots, \boldsymbol{c}_m \in \mathbb{C}$ *denote the distinct eigenvalues of* A_{-2}*.*

($\mathbb{Q}$-structure axiom)^{exp} *The* $\mathbb{Q}$*-structure* $\mathscr{E}_B$ *on* (H, ∇) *is compatible with Stokes data in the following sense. The filtration* $\{\mathsf{S}_{\leq\lambda}\}_{\lambda\in\mathbb{R}}$ *of* S *by the subsheaves* $\mathsf{S}_{\leq\lambda}$*, whose stalk at* $\varphi \in \boldsymbol{S}^1$ *is given by*
$$(\mathsf{S}_{\leq\lambda})_{\varphi} := \left\{ s \in \mathsf{S}_{\varphi} = \Gamma\left(\mathbb{R}_+^{\times} e^{i\varphi}, H\right) \;\middle|\; \begin{array}{l} s \text{ is a } \nabla\text{-horizontal section of } H \text{ over the} \\ \text{ray } \mathbb{R}_+^{\times} e^{i\varphi}, \text{ for which} \\ \left\| s\left(re^{i\varphi}\right)\right\| = \boldsymbol{O}\left(\exp\left(\dfrac{\lambda + \boldsymbol{o}(1)}{r}\right)\right) \\ \text{when } r \to 0. \end{array} \right\}$$
is defined over $\mathbb{Q}$*, i.e.*
$$(\mathsf{S}_{\leq\lambda} \cap \mathsf{S}_B) \otimes_{\mathbb{Q}} \mathbb{C} = \mathsf{S}_{\leq\lambda}$$
for all $\lambda \in \mathbb{R}$*.*

(opposedness axiom)^{exp} = (opposedness axiom)

REMARK 2.15. It is instructive to understand more explicitly the behavior of the Deligne-Malgrange-Stokes filtration for **nc**-Hodge structures (or more generally irregular connections) of exponential type. As before we denote by S the complex local system on the circle $\boldsymbol{S}^1$ corresponding to an **nc**-Hodge structure for which A_{-2} has distinct eigenvalues $\boldsymbol{c}_1, \ldots, \boldsymbol{c}_m$.

By definition, for every φ, the steps in the Deligne-Malgrange-Stokes filtration $(\mathsf{S}_{\leq\lambda})_{\varphi}$ jump exactly when λ crosses one of the numbers $\mathrm{Re}(\boldsymbol{c}_k e^{-i\varphi})$. More invariantly, the assignment $\varphi \in \boldsymbol{S}^1 \mapsto \{\mathrm{Re}(\boldsymbol{c}_1 e^{-i\varphi}), \ldots, \mathrm{Re}(\boldsymbol{c}_k e^{-i\varphi})\} \subset \mathbb{R}$ is a sheaf Λ of finite sets of real numbers (possibly with repetitions) on $\boldsymbol{S}^1$. For a general value of φ, the real numbers $\{\mathrm{Re}(\boldsymbol{c}_1 e^{-i\varphi}), \ldots, \mathrm{Re}(\boldsymbol{c}_k e^{-i\varphi})\}$ are all distinct but for finitely many special values of φ some of $\mathrm{Re}(\boldsymbol{c}_1 e^{-i\varphi}), \ldots, \mathrm{Re}(\boldsymbol{c}_k e^{-i\varphi})$ will coalesce. More precisely we have the Stokes rays $\mathbb{R}_{>0} \cdot i(\boldsymbol{c}_b - \boldsymbol{c}_a)$ and the associated set $\mathsf{SD} \subset [0, 2\pi)$ of Stokes directions: i.e. $\varphi \in \mathsf{SD}$ if and only if there is some pair $a \neq b$ such that $\boldsymbol{c}_a - \boldsymbol{c}_b = re^{i\left(\frac{\pi}{2} + \varphi\right)}$ for some $r > 0$. Clearly for every open arc $U \subset \boldsymbol{S}^1$ which does not intersect SD, the restriction $\Lambda_{|U}$ is a local system of finite sets of cardinality m. Moreover the values $\varphi \in \mathsf{SD}$ are precisely the ones for which some of $\mathrm{Re}(\boldsymbol{c}_1 e^{-i\varphi}), \ldots, \mathrm{Re}(\boldsymbol{c}_k e^{-i\varphi})$ become equal to each other.

Now recall that for any given $\varphi \in \boldsymbol{S}^1$, the subspaces $(\mathsf{S}_{\leq\lambda})_{\varphi} \subset \mathsf{S}_{\varphi}$ do not change if we move $\lambda \in \mathbb{R}$ continuously without passing through some element of Λ_{φ}. In other words, we can label the steps of the Deligne-Malgrange-Stokes filtration by local sections of Λ, and so that at each $\varphi \in \boldsymbol{S}^1$ the steps are ordered according to the order on Λ_{φ} induced from the embedding $\Lambda_{\varphi} \subset \mathbb{R}$. The finite set $\mathsf{SD} \subset \boldsymbol{S}^1$ of Stokes directions breaks the circle into disjoint arcs. Over each such arc U we have that $\Lambda_{|U}$ is a local system of finite sets of real numbers with m linearly ordered flat sections, and the steps of the Deligne-Malgrange-Stokes filtration of $\mathsf{S}_{|U}$ are labeled naturally by these sections. If we move from U to an adjacent arc U' by passing across a Stokes direction $\phi \in \mathsf{SD}$, then some of the elements in the labelling set get identified at ϕ and get reordered when we cross over to U' (see Figure 1).

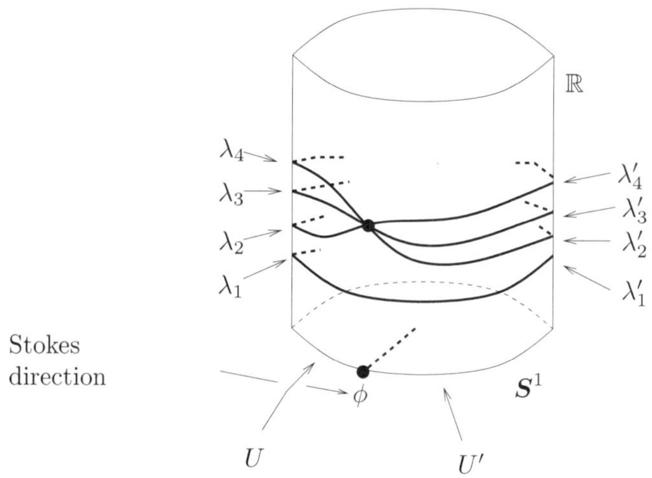

FIGURE 1. The system of labels for the Deligne-Malgrange-Stokes filtration.

In fact, if $\lambda_1 < \ldots < \lambda_m$ are the ordered flat sections of $\Lambda_{|U}$, and $\lambda_1' < \ldots < \lambda_m'$ are the ordered flat sections of $\Lambda_{|U'}$, then the transition from the λ's to the λ''s is always such that certain groups of consecutive λ's are totally reordered into groups of consecutive λ''s. For instance in Figure 1 the passage from $\{\lambda_1, \lambda_2, \lambda_3, \lambda_4\}$ to $\{\lambda_1', \lambda_2', \lambda_3', \lambda_4'\}$ across the Stokes point $\phi \in \mathsf{SD}$ has the effect of relabelling: $\lambda_1 \mapsto \lambda_1'$, $\lambda_2 \mapsto \lambda_4'$, $\lambda_3 \mapsto \lambda_3'$, and $\lambda_4 \mapsto \lambda_2'$.

This behavior of the labelling set and the behavior of the associated filtration can be systematized in the following:

DEFINITION 2.16. *Let S be a finite dimensional local system of $\mathbb{Z}/2$-graded complex vector spaces over $\boldsymbol{S}^1$. Let $\boldsymbol{c}_1, \ldots, \boldsymbol{c}_m$ be distinct complex numbers, let Λ be the sheaf of finite sets of real numbers on $\boldsymbol{S}^1$ given by $\varphi \mapsto \{\mathrm{Re}(\boldsymbol{c}_1 e^{-i\varphi}), \ldots, \mathrm{Re}(\boldsymbol{c}_1 e^{-i\varphi})\}$, and let $\mathsf{SD} \subset \boldsymbol{S}^1$ be the associated set of Stokes directions.*

*An **abstract Deligne-Malgrange-Stokes filtration of S of exponential type and exponents** $(\boldsymbol{c}_1, \ldots, \boldsymbol{c}_m)$ is a filtration by subsheaves $\mathsf{S}_{\leq\lambda}$ such that:*

- *$\mathsf{S}_{\leq\lambda}$ is labelled by local continuous sections λ of Λ and is locally constant on any arc which does not intersect SD.*
- *Suppose $\varphi \in \mathsf{SD}$, and let $U, U' \subset \boldsymbol{S}^1 - \mathsf{SD}$ be the two arcs adjacent at φ. Let $\lambda_1 < \cdots < \lambda_m$ and $\lambda_1' < \cdots < \lambda_m'$ be the ordered flat sections of $\Lambda_{|U}$ and $\Lambda_{|U'}$, respectively. Trivialize S on $U \cup U' \cup \{\varphi\}$ by identifying the flat sections with the elements of the fiber S_φ, and let*

$$0 \subset F_{\leq\lambda_1} \subset \cdots \subset F_{\leq\lambda_m} \subset \mathsf{S}_\varphi, \quad \text{and } 0 \subset F'_{\leq\lambda_1'} \subset \cdots \subset F'_{\leq\lambda_m'} \subset \mathsf{S}_\varphi$$

be the filtrations corresponding to this trivialization and the filtrations $\mathsf{S}_{\leq\lambda}$ *on* U *and* U', *respectively, i.e.*

$$F_{\leq\lambda_i} := \lim_{\substack{\psi\in U\\ \psi\to\varphi}} \left(\mathsf{S}_{\leq\lambda_i}\right)_\psi \qquad \textit{and} \qquad F'_{\leq\lambda'_i} := \lim_{\substack{\psi\in U'\\ \psi\to\varphi}} \left(\mathsf{S}_{\leq\lambda'_i}\right)_\psi$$

Let $1 \leq i_1 < j_1 \leq i_2 < j_2 \leq \cdots \leq i_k < j_k \leq m$ *be the sequence of integers such that* $\lambda_a = \lambda'_a$ *for* $a \notin [i_1, j_1] \cup [i_2, j_2] \cup \cdots \cup [i_k, j_k]$, *and for each interval* $[i_s, j_s]$ *we have that* $\lambda'_{j_s} = \lambda_{i_s}$, $\lambda_{j_s-1} = \lambda_{i_s+1}$, $\ldots \lambda'_{i_s} = \lambda_{j_s}$. *Then we require that:*

- *for each* $a \notin [i_1, j_1] \cup [i_2, j_2] \cup \cdots \cup [i_k, j_k]$ *we have* $F_{\leq\lambda_a} = F'_{\lambda'_a}$;
- *for each* $s = 1, \ldots, k$, $F_{\leq\lambda_{j_s}} = F_{\leq\lambda'_{j_s}}$ *and the filtrations*

$$\begin{array}{ccccccc} F_{\leq\lambda_{i_s}}/F_{\leq\lambda_{i_s-1}} & \subset & F_{\leq\lambda_{i_s+1}}/F_{\leq\lambda_{i_s-1}} & \subset & \cdots & \subset & F_{\leq\lambda_{j_s}}/F_{\leq\lambda_{i_s-1}} \\ & & & & & & \| \\ F'_{\leq\lambda'_{i_s}}/F'_{\leq\lambda'_{i_s-1}} & \subset & F'_{\leq\lambda'_{i_s+1}}/F'_{\leq\lambda'_{i_s-1}} & \subset & \cdots & \subset & F'_{\leq\lambda'_{j_s}}/F'_{\leq\lambda'_{i_s-1}} \end{array}$$

are $(j_s - i_s)$*-opposed.*

REMARK 2.17. The above discussion generalizes immediately from connections of exponential type to arbitrary meromorphic connections (see remark 2.3). One gets a collection of curves drawn on the boundary of the cylinder which can be interpreted as a projection to 0-jets of a Legendrian link in the contact manifold of 1-jets of functions on $\boldsymbol{S}^1$.

The categories of **nc**-Hodge structures, of **nc**-Hodge structures of exponential type, or of **nc**-Hodge structures with regular singularities all behave similarly to ordinary Hodge structures. For instance one can introduce the notion of polarization on **nc**-Hodge structures, which specializes to the usual notion in the case of ordinary Hodge structures. (This will not be needed for our discussion so we will not spell it out here. The interested reader may wish to consult [**Her06, HS07, Kon08**] for the details of the definition.) In fact we have the following

LEMMA 2.18. *The categories* $(\mathbb{Q}\text{-}\mathbf{nc}\mathsf{HS})^{\mathrm{reg}} \subset (\mathbb{Q}\text{-}\mathbf{nc}\mathsf{HS})^{\mathrm{exp}} \subset (\mathbb{Q}\text{-}\mathbf{nc}\mathsf{HS})$ *are* $\mathbb{Q}$*-linear abelian categories. The respective categories of polarizable* **nc**-*Hodge structures are semi-simple.*

Proof: The statement is a manifestation of Simpson's **Meta-Theorem** from [**Sim97b**]. The opposedness axiom implies that the respective categories are abelian and the existence of polarizations implies the semi-simplicity. The proofs follow verbatim the argument in usual Hodge theory or the argument in [**Sim97b**]. Alternatively one can use the comparison statement [**HS07**, Lemma 3.9] identifying the **nc**-Hodge structures with pure twistor structures and then invoke [**Sim97b**, Lemma 1.3 and Lemma 3.1]. □

The bundles with connections $(\mathscr{H}_{\mathbf{c}_i}, \nabla_i)$ can be thought of as the regular singular constituents of the **nc**-Hodge structure $(H, \mathscr{E}_B, \mathbf{iso})$. The $(\mathscr{H}_{\mathbf{c}_i}, \nabla_i)$'s are invariants of the **nc**-Hodge structure but of course they do not give a complete set of invariants (see the third point in 2.13). As usual we need additional Stokes data (see e.g. [**Sab02**]) in order to reconstruct the pair $(\mathscr{H}, \nabla)$ from its regular constituents. To understand how the rest of **nc**-Hodge structure arises from the constituents we need to understand how the rational structure $\mathscr{E}_B$ interacts with the Stokes data. This process is very similar to the interaction between Betti, de Rham and Dolbeault cohomology in ordinary Hodge theory and we will describe it in detail in section 2.3.

The **nc**-Hodge structures one finds in geometric examples are very often regular (e.g. in the case of ordinary Hodge structures) or at worst have exponential type. It is also expected that the **nc**-Hodge structures arising in mirror symmetry will always be of exponential type but at the moment this is only supported by experimental evidence.

We will discuss in detail some of this evidence in the subsequent sections. Before we get to the examples however, it will be instructive to comment on the reason for introducing the **nc**-Hodge structures at the first place. The geometric significance of these structures stems from the fact that they appear naturally on the cohomology of non-commutative spaces of categorical nature.

2.2. Hodge structures in nc geometry. The version of non-commutative geometry that is most relevant to **nc**-Hodge structures is the one in which a proxy for the notion of a non-commutative space (**nc**-space) is a category, thought of as the (unbounded) derived category of quasi-coherent sheaves on that space.

2.2.1. ***Categorical* nc*-geometry.*** The basic notion here is:

DEFINITION 2.19. *A **graded complex* nc*-space** (respectively a **complex* nc*-space**) is a $\mathbb{C}$-linear differential graded (respectively $\mathbb{Z}/2$-graded) category C which is homotopy complete and cocomplete.*

Notation: We will often write C_X for the category to signify that it describes the sheaf theory of some **nc**-space X, even when we do not have a geometric construction of X.

The categorical point of view on non-commutative geometry goes back to the works of Bondal [**Bon93**], Bondal-Orlov [**BO01, BO02**] with many non-trivial examples computed in the later works of Orlov [**Orl04, Orl05b, Orl05a**], Caldero-Keller [**CK05, CK06**], Auroux, Orlov, and the first author, [**AKO04, AKO06**], Kuznetsov [**Kuz05b, Kuz05a, Kuz06**], etc. More recently this approach to **nc**-geometry became the central part of a long term research program initiated by the second author and was studied systematically in the works of the second author and Soibelman [**KS06b, Kon08**], Toën [**Toë07a**], and Toën-Vaquie [**TV05**].

REMARK 2.20. **(i)** We do not spell out here the notions of homotopy completeness and cocompleteness in dg categories since on the one hand they are quite technical and on the other hand they will not be used later in the paper. It is worth mentioning though that some effort is required to define these notions. In the original approach of the second author described in his 2005 IAS lectures and in his 2007 course at the University of Miami the homotopy completeness and cocompletness in C was defined by a universal property for homotopy coherent diagrams of objects in the dg category labeled by simplicial sets. Alternatively [**Toë07c**], one may use the model category $(C^{\mathrm{op}} - \mathsf{mod})$ of C^{op}-dg modules, whose equivalences are the quasi-isomorphisms, and whose fibrations are the epimorphisms. In these

terms one says that C is homotopy complete if the full subcategory of $(C^{\mathrm{op}} - \mathsf{mod})$ consisting of quasi-representable objects is preserved by all small homotopy limits (defined via the given model structure). Similarly we say that C is homotopy cocomplete if C^{op} is homotopy complete.
(ii) Note that in the above definition the category C is automatically triangulated as follows already from the existence of *finite* homotopy limits, and Karoubi closed by the standard mapping telescope construction [**BN93**].

EXAMPLE 2.21. The two main types of **nc**-spaces are the following:

usual schemes: Usual complex schemes can be viewed as (graded) **nc**-spaces. Given a scheme X over $\mathbb{C}$, the corresponding category C_X is the derived category $D(\mathsf{Qcoh}(X)w)$ of quasi-coherent sheaves on X taken with an appropriate dg enhancement (see [**BK91**]). In particular, the closed point $\mathrm{pt} = \mathrm{Spec}(\mathbb{C})$ corresponds to the category C_{pt} of complexes of $\mathbb{C}$-vector spaces.

modules over an algebra: If A is a differential graded (or $\mathbb{Z}/2$-graded) unital associative algebra over $\mathbb{C}$, then we get an **nc**-space $\mathbf{ncSpec}(A)$ such that $C_{\mathbf{ncSpec}(A)} = (A - \mathsf{mod})$ is the category of dg modules over A which admit an exhaustive increasing filtration whose associated graded are sums of shifts of A.

To illustrate how the above notion of an **nc**-space fits with the **nc**Hodge structures we will concentrate on the case of **nc**-affine spaces, i.e. **nc**-spaces equivalent to $\mathbf{ncSpec}(A)$ for some differential $\mathbb{Z}/2$-graded algebra A over $\mathbb{C}$. Note that because of derived Morita equivalences an affine **nc**-space X does not determine an algebra A uniquely, i.e. different algebras can give rise to the same **nc**-space.

REMARK 2.22. The condition is not as restrictive as it appears at first glance. In fact almost all **nc**-spaces that one encounters in practice are affine. For instance usual quasi-compact quasi-separated schemes of finite type over $\mathbb{C}$ are affine when viewed as **nc**-spaces. This follows from a deep theorem of Bondal and van den Bergh [**BvdB03**] which asserts that for such a scheme X the category $C_X = D(\mathsf{Qcoh}(X))$ has a compact generator $\mathcal{E}$. That is, we can find an object $\mathcal{E} \in C_X$ so that

$$\mathrm{Hom}(\mathcal{E}, \bullet) : C_X \to C_{\mathrm{pt}}$$

commutes with homotopy colimits and has a zero kernel. In particular the dg algebra computing the category C_X is given in terms of the generator $\mathcal{E}$, i.e.

$$C_X \cong (\mathrm{Hom}(\mathcal{E}, \mathcal{E})^{\mathrm{op}} - \mathsf{mod}).$$

Suppose now that $X = \mathbf{ncSpec}(A)$. Recall that an object $\mathcal{E} \in C_X = (A - \mathsf{mod})$ is ***perfect*** if $\mathrm{Hom}(\mathcal{E}, \bullet)$ preserves small homotopy colimits. We will write Perf_X for the full subcategory of perfect objects in C_X. We now have the following definition (see e.g. [**KS06b, Kon08**] or [**TV05**]):

DEFINITION 2.23. *A complex differential $\mathbb{Z}/2$-graded algebra is called*

smooth: *if $A \in \mathsf{Perf}_{\mathbf{ncSpec}(A \otimes A^{\mathrm{op}})}$;*
compact: *if $\dim_{\mathbb{C}} H^\bullet(A, d_A) < +\infty$ or equivalently if $A \in \mathsf{Perf}_{\mathrm{pt}}$.*

Note: One can check (see e.g. [**KS06b**] or [**TV05**]) that the properties of X being smooth and compact do not depend on the choice of the algebra A which computes

C_X. Also, for a usual scheme X of finite type over $\mathbb{C}$, smoothness and compactness in the scheme-theoretic sense are equivalent to smoothness and compactness in the **nc**-sense.

2.2.2. ***The main conjecture.*** The analogy with commutative geometry suggests that one should look for pure **nc**-Hodge structures on the cohomology of smooth and proper **nc**-spaces. More precisely we have the following basic conjecture

CONJECTURE 2.24. *Let X be a smooth and compact* **nc***-space over $\mathbb{C}$. Then the periodic cyclic homology $HP_\bullet(C_X)$ of C_X carries a natural functorial pure $\mathbb{Q}$-***nc***-Hodge structure with regular singularities.*

Furthermore, if the $\mathbb{Z}/2$-grading on X can be refined to a $\mathbb{Z}$-grading, then the **nc***-Hodge structure on $HP_\bullet(C_X)$ is an ordinary pure Hodge structure, i.e. belongs to the essential image of the functor $\mathfrak{N}$.*

2.2.3. ***Cyclic homology.*** There are some natural candidates for the various ingredients of the conjectural **nc**-Hodge structure on $HP_\bullet(C_X)$. Assuming that $X \cong \mathbf{ncSpec}(A)$ is **nc**-affine, we can compute $HP_\bullet(C_X)$ in terms of A. Namely

$$HP_\bullet(C_X) = HP_\bullet(A) = HP_\bullet\left(C_\bullet^{\text{red}}(A,A)((u)), \partial + u\cdot B\right),$$

where

- u is an even formal variable (of degree 2 in the $\mathbb{Z}$-graded case);
- $C_{-k+1}^{\text{red}}(A,A)((u)) := A \otimes (A/\mathbb{C}\cdot 1_A)^{\otimes k} \otimes \mathbb{C}((u))$, for all $k \geq 0$;
- $\partial = b + \delta$, where

$$\begin{aligned} b(a_0 \otimes \cdots \otimes a_n) := & \sum_{i=0}^{n-1} (-1)^{\deg(a_0\otimes\cdots\otimes a_i)} a_0 \otimes \cdots \otimes a_i a_{i+1} \otimes \cdots \otimes a_n \\ & + (-1)^{\deg(a_0\otimes\cdots\otimes a_n)(\deg(a_n)+1)+1} a_n a_0 \otimes \cdots \otimes a_{n-1}, \end{aligned}$$

  is the Hochschild differential, and

$$\delta(a_0 \otimes \cdots \otimes a_n) := \sum_{i=0}^{n} (-1)^{\deg(a_0\otimes\cdots\otimes a_{i-1})} a_0 \otimes \cdots \otimes d_A a_i \otimes \cdots \otimes a_n$$

  is the differential induced by d_A via the Leibniz rule;
-

$$\begin{aligned} B(a_0 \otimes \cdots \otimes a_n) \quad := \quad & \sum_{i=0}^{n} (-1)^{(\deg(a_0\otimes\cdots\otimes a_i)-1)(\deg(a_{i+1}\otimes\cdots\otimes a_n)-1)} \\ & \cdot 1_A \otimes a_{i+1} \otimes \cdots \otimes a_n \otimes a_0 \otimes \cdots \otimes a_i \end{aligned}$$

  is Connes' cyclic differential.

2.2.4. ***The degeneration conjecture and the vector bundle part of the*** **nc*****-Hodge structure.*** Note that by construction $HP_\bullet(C_X)$ is a module over $\mathbb{C}((u))$. We can also look at the negative cyclic homology $HC_\bullet^-(C_X)$ of C_X. By definition $HC_\bullet^-(C_X)$ is the cohomology of the complex

$$\left(C_\bullet^{\text{red}}(A,A)[[u]], \partial + u\cdot B\right),$$

and so is a module over $\mathbb{C}[[u]]$. The specialization

$$HC_\bullet^-(C_X)/uHC_\bullet^-(C_X)$$

of this module at $u = 0$ maps to the cohomology of the complex

$$(C_\bullet^{\text{red}}(A, A), \partial)$$

of reduced Hochschild chains for A which by definition is the Hochschild homology $HH_\bullet(A)$ of A. The Hochschild-to-cyclic spectral sequence implies that

$$\dim_{\mathbb{C}((u))} HP_\bullet(A) \leq \dim_{\mathbb{C}} HH_\bullet(A). \tag{2.2.1}$$

If X is a smooth and compact **nc**-space, the Hochschild chain complex of C_X is the derived tensor product over $A \otimes A^{\text{op}}$ of a perfect complex with finite dimensional cohomology with itself. In particular $HH_\bullet(C_X) := HH_\bullet(A)$ is a finite dimensional $\mathbb{C}$-vector space, and so by (2.2.1) we have that $HP_\bullet(C_X)$ is finite dimensional over $\mathbb{C}((u))$. Thus the $\mathbb{C}[[u]]$-module $HC_\bullet^-(C_X)$ is finitely generated and so corresponds to the formal germ at $u = 0$ of an algebraic $\mathbb{Z}/2$-graded coherent sheaf on $\mathbb{A}^1_{\mathbb{C}}$. The fiber of this sheaf at $u = 0$ is $HH_\bullet(C_X)$ and the generic fiber is $HP_\bullet(C_X)$. In [**KS06b, Kon08**] the second author proposed the so called ***degeneration conjecture*** asserting that for a smooth and compact **nc**-space $X = \mathbf{ncSpec}(A)$ we have an equality of dimensions in (2.2.1). In other words the degeneration conjecture asserts that for a smooth and compact **nc**-space the $\mathbb{C}[[u]]$-module $HC_\bullet^-(C_X)$ is free of finite rank and thus corresponds to an algebraic vector bundle on the one-dimensional formal disc $\mathbb{D} := \text{Spf}(\mathbb{C}[[u]])$.

REMARK 2.25. There is a lot of evidence supporting the validity of this conjecture. The work of Weibel [**Wei96**] shows that if X is a usual quasi-compact and quasi-separated complex scheme the Hochschild and periodic cyclic homology of X viewed as a **nc**-spaces can be identified with the algebraic de Rham and Dolbeault cohomology of X, respectively. Combined with the degeneration of the Hodge-to-de Rham spectral sequence in the smooth proper case this shows that the degeneration conjecture holds true for usual schemes. Also recently in a very exciting sequence of papers [**Kal07a, Kal06**] Kaledin proved the degeneration conjecture for graded **nc**-spaces $X = \mathbf{ncSpec}(A)$ for which A is concentrated in non-negative degrees. The case of graded **nc**-spaces $X = \mathbf{ncSpec}(A)$ for which A is concentrated in non-positive degrees was also settled by Shklyarov [**Shk08**]. The general graded case and the $\mathbb{Z}/2$-graded case are still wide open.

2.2.5. ***The meromorphic connection in the u-direction.*** The next observation is that the $\mathbb{C}\{u\}[u^{-1}]$-module $HP_\bullet(C_X)$ comes equipped with a natural meromorphic connection. Indeed, recall that by the work of Getzler [**Get93**] there is a version of the Gauss-Manin connection which exists on the periodic cyclic homology of any flat family of differential graded algebras (see also [**Tsy07, Kal07b**]). An analogous statement holds in the $\mathbb{Z}/2$-graded case as explained e.g. in [**KS06b**, Section 11.5]. The Gauss-Manin connection for any family of dg algebras $\mathcal{A}_x$ over the formal disc $\text{Spf}\,\mathbb{C}[[x]]$ with a formal parameter x is an operator

$$\begin{aligned}\nabla^{\text{GM}}_{u\frac{\partial}{\partial x}} : H^\bullet(C^{\text{red}}(\mathcal{A}_x, \mathcal{A}_x)[[u, x]], \partial_{\mathcal{A}_x} + u \cdot B_{\mathcal{A}_x}) \\ \to H^\bullet(C^{\text{red}}(\mathcal{A}_x, \mathcal{A}_x)[[u, x]], \partial_{\mathcal{A}_x} + u \cdot B_{\mathcal{A}_x})\end{aligned}$$

satisfying the Leibniz rule with respect to the multiplications by u and x (compare this with the **(Griffiths transversality axiom)** in Definition 2.7 from Section 2.1.5).

Suppose now that A is a differential $\mathbb{Z}/2$-graded (d ($\mathbb{Z}/2$)g) algebra with product m_A, differential d_A, and a strict unit 1_A. Then we can form a flat family $\mathcal{A} \to \mathbb{A}^1 - \{0\}$ of differential $\mathbb{Z}/2$-graded algebras parameterized by the punctured affine line $\mathbb{A}^1 - \{0\}$. The fiber $\mathcal{A}_t$ of $\mathcal{A}$ over a point $t \in \mathbb{A}^1_{\mathbb{C}} - \{0\}$ is the d($\mathbb{Z}/2$)g algebra for which the underlying $\mathbb{Z}/2$-graded vector space is A and

$$\begin{aligned} m_{\mathcal{A}_t} &:= t \cdot m_A, \\ d_{\mathcal{A}_t} &:= t \cdot d_A, \\ 1_{\mathcal{A}_t} &:= t^{-1} \cdot 1_A. \end{aligned}$$

Looking at the scaling properties of ∂ and B we see that the identity morphism on the level of cochains induces a natural isomorphism

(2.2.2)

$$H^\bullet\left(C^{\text{red}}_\bullet(\mathcal{A}_t, \mathcal{A}_t)[[u]], \partial_{\mathcal{A}_t} + u \cdot B_{\mathcal{A}_t}\right) \xrightarrow{\cong} H^\bullet(C^{\text{red}}_\bullet(A, A)[[u]], \partial + ut^{-2} \cdot B).$$

This isomorphism does not come from a quasi-isomorphism of complexes, as the identity map is not a morphism of complexes: the differentials do not coincide but differ by the factor t. If A is smooth and compact, then the negative cyclic homology of the family of algebras $\mathcal{A}_t$ gives rise to an algebraic vector bundle $\widetilde{HC}^-$ on the product $(\mathbb{A}^1 - \{0\}) \times \mathbb{D}$. Here $\mathbb{D} := \operatorname{Spf} \mathbb{C}[[u]]$ denotes the one-dimensional formal disc. We will write (t, u) for the coordinates on $(\mathbb{A}^1 - \{0\}) \times \mathbb{D}$. We will be interested in fact only in the formal neighborhood of the point $t = 1$ where we can choose as a local coordinate $x := \log(t)$. The Getzler-Gauss-Manin connection can then be viewed as a relative holomorphic connection ∇^{GM} on $\widetilde{HC}^-$ which differentiates only along $\mathbb{A}^1_{\mathbb{C}} - \{0\}$. On the other hand the formal completion of the group $\mathbb{C}^\times$ at 1 acts on $(\mathbb{A}^1_{\mathbb{C}} - \{0\}) \times \mathbb{D}$ by $(t, u) \mapsto (\mu t, \mu^2 u)$ for $\mu \in \mathbb{C}^\times$. The isomorphism (2.2.2) gives rise to a $\mathbb{C}^\times$-equivariant structure on the vector bundle $\widetilde{HC}^-$ and the infinitesimal action of $d/d\mu$ associated with this equivariant structure gives a holomorphic differential operator $\mathbf{\Lambda} \in \operatorname{Diff}^{\leq 1}(\widetilde{HC}^-, \widetilde{HC}^-)$ with symbol equal to

$$\left(t\frac{\partial}{\partial t} + 2u\frac{\partial}{\partial u}\right) \cdot \operatorname{id}_{\widetilde{HC}^-}.$$

Hence

$$\nabla_{u^2 \frac{\partial}{\partial u}} := \frac{u}{2} \cdot \mathbf{\Lambda} - \nabla^{\text{GM}}_{\frac{ut}{2}\frac{\partial}{\partial t}}$$

is a first order differential operator on $\widetilde{HC}^-$ with symbol

$$u^2 \frac{\partial}{\partial u} \cdot \operatorname{id}_{\widetilde{HC}^-}$$

and so after restricting $\widetilde{HC}^-$ to $\{1\} \times \mathbb{D}$ this operator gives a meromorphic connection ∇ on the $\mathbb{C}[[u]]$-module $HC^-_\bullet(C_X)$ with at most a second order pole at $u = 0$. Note also that if the algebra A is $\mathbb{Z}$-graded, then the family $\mathcal{A}_t$ is easily seen to be trivial and the connection ∇ has a first order pole at $u = 0$ with monodromy equal to $(-1)^{\text{parity}}$.

2.2.6. ***The $\mathbb{Q}$-structure.*** The categorical origin of the rational (or integral) structure of the conjectural **nc**-Hodge structure is more mysterious. Conceptually, the correct rational structure should come from the Betti cohomology or, say, the topological K-theory of the **nc**-space. There are two natural approaches to constructing the rational structure $\mathscr{E}_B \subset HP_\bullet(C_X)$:

(a) The soft algebra approach ([Kon08]). Let again $X = \mathbf{ncSpec}(A)$ be an affine **nc**-space, and assume X is compact. By analogy with the classical geometric case one expects that there should exist a nuclear Fréchet d($\mathbb{Z}/2$)g algebra A_{C^∞} so that

- The K-theory of A_{C^∞} satisfies Bott periodicity, i.e. $K_i(A_{C^\infty}) = K_{i+2}(A_{C^\infty})$ for all $i \geq 0$.
- There is a homomorphism $\varphi : A \to A_{C^\infty}$ of d($\mathbb{Z}/2$)g algebras for which $\varphi_* : HP_\bullet(A) \to HP_\bullet(A_{C^\infty})$ is an isomorphism, and the image of the Chern character map
$$\mathrm{ch} : K_\bullet(A_{C_\infty}) \to HP_\bullet(A_{C^\infty})$$
is an integral lattice, and hence gives a rational structure $\mathscr{E}_B \subset HP_\bullet(A)$.

Note: If X is a smooth and compact complex variety and if $\mathcal{E} \in \mathsf{Perf}(X)$ is a vector bundle generating C_X, then one may take
$$A := A^{0,\bullet}(X, \mathcal{E}^\vee \otimes \mathcal{E}), \bar{\partial})$$
$$A_{C^\infty} := A^{0,0}(X, \mathcal{E}^\vee \otimes \mathcal{E}).$$
Note that the algebra A_{C^∞} is Morita equivalent to $C^\infty(X)$.

(b) The semi-topological K-theory approach (Bondal, Toën, [Toë07b]). Assume again that $X = \mathbf{ncSpec}(A)$ is a smooth and compact graded **nc**-affine **nc**-space. Consider the moduli stack $\mathscr{M}_X$ of all objects in Perf_X. This is an ∞-stack which by a theorem of Toën and Vaquie [**TV05**] is locally geometric and locally of finite presentation. Moreover, for any $a, b \in \mathbb{N}$ the substack $\mathscr{M}_X^{[a,b]} \subset \mathscr{M}_X$ consisting of objects of amplitude in the interval $[a, b]$ is a geometric $(b - a + 1)$-stack. The functor sending a complex scheme to the underlying topological space in the analytic topology gives rise by a left Kan extension to a topological realization functor
$$|\bullet| : \mathrm{Ho}(\mathsf{Stacks}/\mathbb{C}) \to \mathrm{Ho}(\mathsf{Top})$$
from the homotopy category of stacks to the homotopy category of complex spaces. Following Friedlander-Walker [**FW05**] we define the semi-topological K-group of the **nc**-space X to be
$$K_0^{st}(X) := \pi_0(|\mathscr{M}_X|).$$
The group structure here is induced by the direct sum $\oplus$ of A-modules: the monoid $(\pi_0(|\mathscr{M}_X|), \oplus)$ is actually a group. To see this note that for any A-module E we have that $[E \oplus E[1]] = 0$ in $\pi_0(|\mathscr{M}_X|)$. Indeed we have distinguished triangles
$$E \longrightarrow 0 \longrightarrow E[1] \longrightarrow E[1]$$
$$E \longrightarrow E \oplus E[1] \longrightarrow E[1] \longrightarrow E[1]$$

the first of which corresponds to $\mathrm{id} \in \mathrm{Ext}^1(E[1], E) = \mathrm{Hom}(E, E)$, and the second of which corresponds to $0 \in \mathrm{Ext}^1(E[1], E) = \mathrm{Hom}(E, E)$. Since $\mathrm{Ext}^1(E[1], E) = \mathrm{Hom}(E, E)$ is a vector space, it follows that id deforms to 0 continuously and so the second terms in the above triangles represent the same point in $\pi_0(|\mathscr{M}_X|)$.

More generally $\oplus$ makes $|\mathscr{M}_X|$ into an H-space $\mathbb{K}^{st}(X)$ which is the degree zero part of a natural spectrum. Using this one can define $K_i^{st}(X)$ for all $i \geq 0$.

Next note that since C_X is triangulated it is a module over the category $\mathsf{Perf}_{\mathrm{pt}}$ of complexes of $\mathbb{C}$-vector spaces with finite dimensional total cohomology. In particular $K_\bullet^{st}(X)$ is a graded module over $K_\bullet^{st}(\mathrm{pt})$. It can be checked that

$$\mathbb{K}^{st}(\mathrm{pt}) = BU = \mathbb{K}^{\mathrm{top}}(\mathrm{pt}),$$

and so $K_\bullet^{st}(X)$ is a graded $\mathbb{Z}[u]$-module ($\deg u = 2$).

Now we can define

$$K_\bullet^{\mathrm{top}}(X) := K_\bullet^{st}(X)[u^{-1}] = K_\bullet^{st}(X) \otimes_{\mathbb{Z}[u]} \mathbb{Z}[u, u^{-1}].$$

Again one expects that there is a Chern character map

$$\mathrm{ch} : K_\bullet^{\mathrm{top}}(X) \to HP_\bullet(C_X)$$

whose image gives a rational structure $\mathscr{E}_B$ on $HP_\bullet(C_X)$.

Note: If X is a smooth and compact complex variety, then the Friedlander-Walker comparison theorem [**FW01**] implies that $K^{\mathrm{top}}(D(\mathsf{QCoh}(X))) \cong K^{\mathrm{top}}(X^{\mathrm{top}})$, where X^{top} is the topological space underlying X.

2.2.7. ***Questions.*** Even though we have some good candidates for the ingredients H, ∇, $\mathscr{E}_B$ of the conjectural **nc**-Hodge structure associated with an **nc**-space, there are several important problems that need to be addressed before one can prove Conjecture 2.24:

- show that the connection ∇ has regular singularities (this is automatically true in the $\mathbb{Z}$-graded case);
- show that ∇ preserves the rational structure;
- show that the opposedness axiom holds.

One can show that for a smooth compact nc-space the coefficient A_{-2} in the u-connection is a nilpotent operator, which gives some evidence in favor of the regular singularity question.

In fact Conjecture 2.24 and the above questions are special cases of a general conjecture which predicts the existence of a general **nc**-Hodge structure on the periodic cyclic homology of a curved d($\mathbb{Z}/2$)g category which is formally smooth and compact. We will not discuss the general conjecture or the relevant constructions here but we will revisit these questions in some interesting geometric examples in Section 3.

2.3. Gluing data. In this section we discuss how general **nc**-Hodge structures of exponential type can be glued together out of **nc**-Hodge structures with regular singularities and additional gluing data.

2.3.1. **nc-*de Rham data.*** The de Rham part of an **nc**-Hodge structure of exponential type can be prescribed in three equivalent ways:

ncdR(i) A pair $(\mathscr{M}, \nabla)$, where $\mathscr{M}$ is a finite dimensional vector space over $\mathbb{C}\{u\}[u^{-1}]$ and ∇ is a meromorphic connection. These data should satisfy the following

Property ncdR(i): There exist:

- a frame $\underline{e} = (e_1, \ldots, e_r)$ of $\mathscr{M}$ over $\mathbb{C}\{u\}[u^{-1}]$ in which

$$\nabla = d + \left(\sum_{k \geq -2} A_k u^k \right) du$$

with $A_k \in \mathrm{Mat}_{r\times r}(\mathbb{C})$, $r = \mathrm{rank}_{\mathbb{C}\{u\}[u^{-1}]} \mathscr{M}$. In other words, there is a holomorphic extension $\mathscr{H} = \mathbb{C}\{u\}e_1 \oplus \cdots \oplus \mathbb{C}\{u\}e_r$ in which ∇ has at most a second order pole.

- a formal isomorphism over $\mathbb{C}((u))$:

$$(\mathscr{M}, \nabla) \otimes_{\mathbb{C}\{u\}[u^{-1}]} \mathbb{C}((u)) \overset{\cong}{\to} \bigoplus_{i=1}^{m} \boldsymbol{\mathcal{E}}^{\boldsymbol{c}_i/u} \otimes (\mathscr{R}_i, \nabla_i)$$

where $(\mathscr{R}_i, \nabla_i)$ are meromorphic bundles with connections with regular singularities and $\boldsymbol{c}_1, \ldots, \boldsymbol{c}_m \in \mathbb{C}$ denote the distinct eigenvalues of A_{-2}.

ncdR(ii) A pair (M, ∇), where M is an algebraic vector bundle on $\mathbb{A}^1 - \{0\}$ and ∇ is a connection on M. These data should satisfy the following

Property ncdR(ii): M can be extended to an algebraic vector bundle $\widetilde{M}$ on $\mathbb{P}^1$, and

- with respect to this extension and appropriate local trivializations at zero and infinity we must have

$$\nabla = d + \left(\sum_{k \geq -2} A_k u^k \right) du \qquad \text{near } 0,$$

$$\nabla = d + \left(\sum_{k \geq -1} B_k u^{-k} \right) d(u^{-1}) \qquad \text{near } \infty.$$

In other words $\nabla : \widetilde{M} \to \widetilde{M} \otimes_{\mathcal{O}_{\mathbb{P}^1}} \Omega^1_{\mathbb{P}^1}(2 \cdot \{0\} + \{\infty\})$.

- There is a formal isomorphism over $\mathbb{C}((u))$:

$$(M, \nabla) \otimes_{\mathbb{C}[u,u^{-1}]} \mathbb{C}((u)) \overset{\cong}{\to} \bigoplus_{i=1}^{m} \boldsymbol{\mathcal{E}}^{\boldsymbol{c}_i/u} \otimes (\mathscr{R}_i, \nabla_i)$$

where $(\mathscr{R}_i, \nabla_i)$ are meromorphic bundles with connections with regular singularities and $\boldsymbol{c}_1, \ldots, \boldsymbol{c}_m \in \mathbb{C}$ denote the distinct eigenvalues of A_{-2}.

ncdR(iii) An algebraic holonomic $\mathcal{D}$-module $\boldsymbol{M}$ on $\mathbb{A}^1$. The $\mathcal{D}$-module $\boldsymbol{M}$ should also satisfy the following

Property ncdR(iii): $\boldsymbol{M}$ has regular singularities and $H^{\bullet}_{\mathrm{DR}}(\mathbb{A}^1, \boldsymbol{M}) = 0$.

The **nc**-de Rham data of types **ncdR(i)**, **ncdR(ii)**, and **ncdR(iii)** form obvious full subcategories in the categories of meromorphic connections over $\mathbb{C}\{u\}[u^{-1}]$, algebraic vector bundles with connections on $\mathbb{A}^1 - \{0\}$, and coherent algebraic $\mathcal{D}$-modules on $\mathbb{A}^1$, respectively. We have the following

LEMMA 2.26. *The categories of* **nc**-*de Rham data of types* **ncdR(i)**, **ncdR(ii)**, *and* **ncdR(iii)** *are all equivalent.*

Proof. In essence we have already discussed the equivalence **ncdR(i)** $\Longleftrightarrow$ **ncdR(ii)** in Remark 2.1. Explicitly we have $(\mathscr{M}, \nabla) = \mathfrak{G}_0((M, \nabla)) = (M \otimes_{\mathbb{C}[u,u^{-1}]} \mathbb{C}\{u\}[u^{-1}], \nabla)$.

We define a functor $\mathfrak{F}$: (data **(iii)**) $\to$ (data **(ii)**) as follows. Let $\boldsymbol{M}$ be a regular holonomic algebraic $\mathcal{D}$-module on $\mathbb{A}^1$ with trivial de Rham cohomology. Denote the coordinate on $\mathbb{A}^1$ by v. The vanishing of de Rham cohomology means that the action $\frac{d}{dv} : \boldsymbol{M} \to \boldsymbol{M}$ is an invertible operator. Consider the algebraic Fourier transform $\boldsymbol{\Phi M}$ which is the same vector space as $\boldsymbol{M}$ endowed with an action of the Weyl algebra defined by

$$\tilde{v} := \frac{d}{dv}$$

$$\frac{d}{d\tilde{v}} := -v$$

where $\tilde{v}$ is the coordinate on the dual line. By our assumptions $\boldsymbol{\Phi M}$ is a holonomic $\mathcal{D}$-module on which $\tilde{v}$ acts invertibly. Hence $\boldsymbol{\Phi M}$ is the direct image of a holonomic $\mathcal{D}$-module $\boldsymbol{\Phi M}'$ on $\mathbb{A}^1 - \{0\}$ under the embedding

$$(\mathbb{A}^1 - \{0\}) \hookrightarrow \mathbb{A}^1 = \operatorname{Spec}(\mathbb{C}[\tilde{v}])$$

Finally, making the change of coordinates $u = 1/\tilde{v}$ we obtain a $\mathcal{D}$-module M on $\mathbb{A}^1 - \{0\}$ with coordinate u.

We claim that $\mathfrak{F}(\boldsymbol{M}) := M$ obtained in this way satisfies the property **ncdR(ii)**, and that by this construction one obtains all such modules. It follows from the well-known properties of the Fourier transform that $\boldsymbol{\Phi M}$ has no singularities in $\mathbb{A}^1 - \{0\}$ and that its singularity at $\tilde{v} = 0$ is regular. Hence M is a vector bundle on $\mathbb{A}^1 - \{0\}$ endowed with a connection with regular singularity at ∞. It only remains to use the well-known fact (see e.g. [**Mal91**, Chapters IX-XI] or [**Kat90**, Theorem 2.10.16]) that the exponential type property for M is equivalent to the property of $\boldsymbol{M}$ to have only regular singularities. □

REMARK 2.27. The characterization of the exponential type property in terms of the Fourier transform can be stated more precisely (see [**Mal91**, Chapters IX-XI] or [**Kat90**, Theorem 2.10.16]): For an algebraic holonomic $\mathcal{D}$-module M on the complex affine line, the following two conditions are equivalent:

1) M has regular singularities;
2) the Fourier transform $\boldsymbol{\Phi}M$ of M has no singularities outside 0, its singularity at 0 is regular, and its singularity at infinity is of exponential type.

Explicitly $\boldsymbol{\Phi}M$ being of exponential type at infinity means that if x is a coordinate on $\mathbb{A}^1$ centered at 0, then after passing to the formal completion

$(\mathbf{\Phi} M) \otimes_{\mathbb{C}[x]} \mathbb{C}((x^{-1}))$ the resulting module will be isomorphic to a finite sum

$$\bigoplus_{i=1}^{m} \boldsymbol{\mathcal{E}}^{\boldsymbol{c}_i x} \otimes (\mathscr{R}_i, \nabla_i)$$

where $(\mathscr{R}_i, \nabla_i)$ are $\mathcal{D}$-modules with a regular singularity at infinity.

REMARK 2.28. Note that the de Rham data **ncdR(i)** are analytic in nature, whereas **ncdR(ii)** and **ncdR(iii)** are algebraic. In fact from the proof it is clear that **ncdR(ii)** and **ncdR(iii)** and their equivalence still make sense if we replace $\mathbb{C}$ with any field of characteristic zero.

2.3.2. **nc-*Betti data.*** The (rational) Betti part of an **nc**-Hodge structure of exponential type can be prescribed in four ways:

ncB(i) A (middle perversity) perverse sheaf $\mathscr{G}^\bullet$ of $\mathbb{Q}$-vector spaces on the Riemann surface $\mathbb{C}$ (taken with the analytic topology) satisfying the following

Property ncB(i): $R\Gamma(\mathbb{C}, \mathscr{G}^\bullet) = 0$.

ncB(ii) A constructible sheaf $\mathscr{F}$ of $\mathbb{Q}$-vector spaces on the Riemann surface $\mathbb{C}$ (taken with the analytic topology) satisfying the following

Property ncB(ii): $R\Gamma(\mathbb{C}, \mathscr{F}) = 0$.

ncB(iii) A finite collection of distinct points $S = \{\boldsymbol{c}_1, \ldots, \boldsymbol{c}_n\} \subset \mathbb{C}$, and

- a collection $U_1, U_2, \ldots, U_n$ of finite dimensional non-zero $\mathbb{Q}$-vector spaces,
- a collection of linear maps $T_{ij} : U_j \to U_i$, for all $i, j = 1, \ldots, n$,

satisfying the following

Property ncB(iii): $T_{ii} \in GL(U_i)$.

ncB(iv) A local system S of $\mathbb{Q}$-vector spaces on $\boldsymbol{S}^1$ equipped with a filtration $\{\mathsf{S}_{\leq\lambda}\}_{\lambda\in\mathbb{R}}$ by subsheaves of $\mathbb{Q}$-vector spaces, satisfying the following

Property ncB(iv): The filtration $\{\mathsf{S}_{\leq\lambda} \otimes \mathbb{C}\}_{\lambda\in\mathbb{R}}$ of $\mathsf{S} \otimes \mathbb{C}$ is a Deligne-Malgrange-Stokes filtration of exponential type. In other words, there exist complex numbers $\boldsymbol{c}_1, \ldots, \boldsymbol{c}_n \in \mathbb{C}$ so that:

- For every $\varphi \in \boldsymbol{S}^1$, the filtration $\{(\mathsf{S}_{\leq\lambda}\otimes\mathbb{C})_\varphi\}_{\lambda\in\mathbb{R}}$ of the stalk $(\mathsf{S}\otimes\mathbb{C})_\varphi$ jumps exactly at the real numbers $\{\mathrm{Re}\left(\boldsymbol{c}_k e^{-i\varphi}\right)\}_{k=1}^n$.
- The associated graded sheaves of $\mathsf{S}\otimes\mathbb{C}$ with respect to $\{\mathsf{S}_{\leq\lambda}\otimes\mathbb{C}\}_{\lambda\in\mathbb{R}}$ are local systems on $\boldsymbol{S}^1$.

Again there are natural equivalences of the different types of Betti data (for **ncB(iii)** the equivalence depends on certain choices of paths as one can see from the proof of Theorem 2.29 and the statement of Lemma 2.30). Consider the full subcategories (**ncB(i)**) and (**ncB(ii)**) of **nc**-Betti data of types **ncB(i)** and **ncB(ii)** in the category of perverse sheaves of $\mathbb{Q}$-vector spaces on $\mathbb{C}$ and in the category of constructible sheaves of $\mathbb{Q}$-vector spaces on $\mathbb{C}$, respectively. We have the following

THEOREM 2.29. *The categories of* **nc**-*Betti data of types* **ncB(i)** *and* **ncB(ii)** *are naturally equivalent. More precisely, the natural functors*

$$\mathcal{H}^{-1} : D^b_{\mathrm{constr}}(\mathbb{C}, \mathbb{Q}) \to \mathsf{Constr}(\mathbb{C}, \mathbb{Q}) \quad \textit{and} \quad [1] : \mathsf{Constr}(\mathbb{C}, \mathbb{Q}) \to D^b_{\mathrm{constr}}(\mathbb{C}, \mathbb{Q})$$

induce mutually inverse equivalences of the full subcategories $(\mathbf{ncB(i)}) \subset D^b_{\text{constr}}(\mathbb{C},\mathbb{Q})$ *and* $(\mathbf{ncB(ii)}) \subset \mathsf{Constr}(\mathbb{C},\mathbb{Q})$.

Proof. First we look at the data $\mathbf{ncB(i)}$ more closely. Suppose X is a complex analytic space underlying a complex quasi-projective variety. Recall (see e.g. [**BBD82, KS94, Dim04**]) that a bounded complex $\mathscr{G}^\bullet$ of sheaves of $\mathbb{C}$-vector spaces on X is called a (middle perversity) perverse sheaf if it has constructible cohomology sheaves $\mathcal{H}^k(\mathscr{G}^\bullet)$ and if

- for all $k \in \mathbb{Z}$, we have $\dim_{\mathbb{R}}\{x \in X | \ \mathcal{H}^{-k}(i_x^*\mathscr{G}^\bullet) \neq 0\} \leq 2k$,
- for all $k \in \mathbb{Z}$, we have $\dim_{\mathbb{R}}\{x \in X | \ \mathcal{H}^{k}(i_x^!\mathscr{G}^\bullet) \neq 0\} \leq 2k$.

Here $i_x : x \hookrightarrow X$ denotes the inclusion of the point x in X.

For future reference we will write $D^b_{\text{constr}}(X,\mathbb{Q})$ for the derived category of complexes of $\mathbb{Q}$-vector spaces on X with constructible cohomology, $\mathsf{Perv}(X,\mathbb{Q}) \subset D^b_{\text{constr}}(X,\mathbb{Q})$ for the full subcategory of middle perversity perverse sheaves, and $\mathsf{Constr}(X,\mathbb{Q}) \subset D^b_{\text{constr}}(X,\mathbb{Q})$ for the full subcategory of constructible sheaves.

From the definition it is clear that if $\mathscr{G}^\bullet$ is a perverse sheaf on $\mathbb{C}$, then $\mathscr{G}^\bullet$ has at most two non-trivial cohomology sheaves $\mathcal{H}^{-1}(\mathscr{G}^\bullet)$ and $\mathcal{H}^{0}(\mathscr{G}^\bullet)$. Moreover the support of $\mathcal{H}^{0}(\mathscr{G}^\bullet)$ has dimension ≤ 0. Now the cohomology $R\Gamma^\bullet(\mathscr{G}^\bullet) = \mathbb{H}^\bullet(\mathbb{C},\mathscr{G}^\bullet)$ can be computed via the hypercohomology spectral sequence

$$E_2^{pq} = H^p(\mathbb{C},\mathcal{H}^q(\mathscr{G}^\bullet)) \Rightarrow \mathbb{H}^{p+q}(\mathbb{C},\mathscr{G}^\bullet).$$

Since $\mathscr{G}^\bullet$ has only two cohomology sheaves, the E_2 level of this spectral sequence is

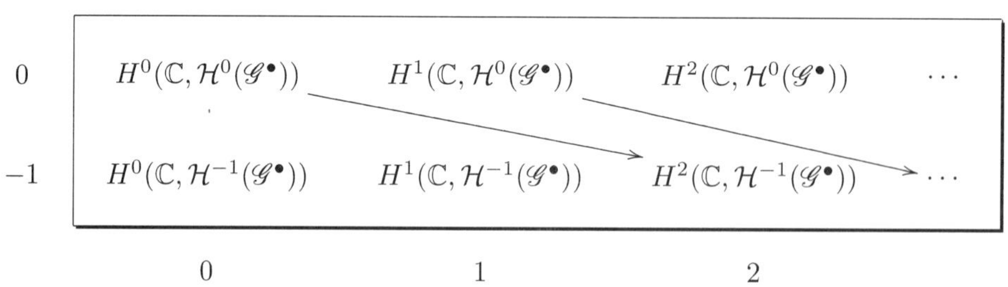

By Artin's vanishing theorem for constructible sheaves [**Art73**, Corollary 3.2] we have $H^p(\mathbb{C},\mathcal{H}^q(\mathscr{G}^\bullet)) = 0$ for all q and all $p > 1$. Furthermore, since $\mathcal{H}^0(\mathscr{G}^\bullet)$ has at most zero-dimensional support we have $H^1(\mathbb{C},\mathcal{H}^0(\mathscr{G}^\bullet)) = 0$. In particular the spectral sequence degenerates at E_2 and the only potentially non-trivial cohomology groups of $\mathscr{G}^\bullet$ are

$$\begin{aligned}\mathbb{H}^{-1}(\mathbb{C},\mathscr{G}^\bullet) &= H^0(\mathbb{C},\mathcal{H}^{-1}(\mathscr{G}^\bullet)), \text{ and}\\ \mathbb{H}^{0}(\mathbb{C},\mathscr{G}^\bullet) &= H^1(\mathbb{C},\mathcal{H}^{-1}(\mathscr{G}^\bullet)) \oplus H^0(\mathbb{C},\mathcal{H}^{0}(\mathscr{G}^\bullet)).\end{aligned}$$

Thus under the assumption that $R\Gamma(\mathscr{G}^\bullet) = 0$ we get that $H^0(\mathbb{C},\mathcal{H}^0(\mathscr{G}^\bullet)) = 0$, i.e. that $\mathcal{H}^0(\mathscr{G}^\bullet) = 0$. In other words $\mathscr{G}^\bullet = \mathscr{F}[1]$ for some constructible sheaf $\mathscr{F}$ with $R\Gamma(\mathscr{F}) = 0$.

To finish the proof of the theorem we need to show that for every constructible sheaf $\mathscr{F}$ with $R\Gamma(\mathscr{F}) = 0$, the object $\mathscr{F}[1]$ will be perverse (for the middle perversity). For this we will have to look more closely at constructible sheaves on the complex line.

Suppose $\mathscr{F}$ is a constructible sheaf of $\mathbb{Q}$-vector spaces on $\mathbb{C}$. Then there is a finite set $S = \{\boldsymbol{c}_1, \ldots, \boldsymbol{c}_n\}$ of points in $\mathbb{C}$ so that $\mathbb{C} - S$ is the maximal open set on which $\mathscr{F}$ restricts to a local system. Let $\mathbb{F} := \mathscr{F}_{|\mathbb{C}-S}$ denote this local system. Let $\mathbb{C} - S \overset{j}{\hookrightarrow} \mathbb{C} \overset{i}{\hookleftarrow} S$ be the natural inclusions and let $\varphi : \mathscr{F} \to j_* j^* \mathscr{F} = j_* \mathbb{F}$ be the adjunction homomorphism.

Before we can describe $\mathbb{F}$ and $\mathscr{F}$ via the quiver-like data of type **ncB(iii)** we will need to make some rigidifying choices. First we fix a base point $\boldsymbol{c}_0 \in \mathbb{C} - S$. For $i = 1, \ldots, n$ we choose a collection of small disjoint discs $\boldsymbol{D}_i \subset \mathbb{C}$, each $\boldsymbol{D}_i$ centered at $\boldsymbol{c}_i$. For each disc we fix a point $\boldsymbol{o}_i \in \partial \boldsymbol{D}_i$ and denote by $\mathfrak{l}_i$ the loop starting and ending at $\boldsymbol{o}_i$ and tracing $\partial \boldsymbol{D}_i$ once in the counterclockwise direction. We fix an ordered system of non-intersecting paths $\{a_i\}_{i=1}^n \subset \mathbb{C} - (\cup_{i=1}^n \boldsymbol{D}_i)$ which connect the base point $\boldsymbol{c}_0$ with each of the $\boldsymbol{o}_i$ as in Figure 2.

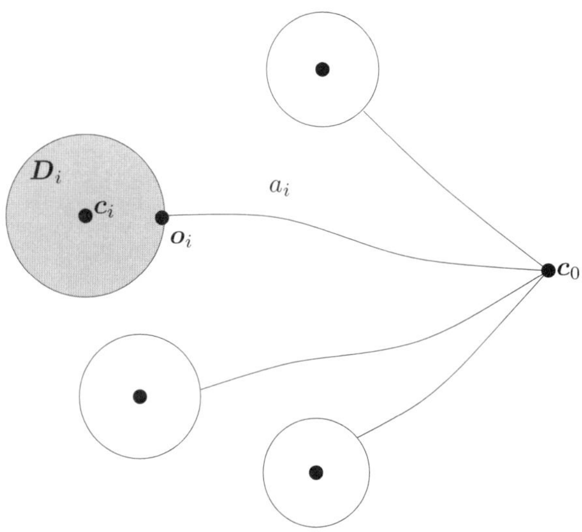

FIGURE 2. A system of paths for $S \subset \mathbb{C}$.

Let $\mathrm{mon}_{\mathfrak{l}_i} : \mathbb{F}_{\boldsymbol{o}_i} \to \mathbb{F}_{\boldsymbol{o}_i}$ be the monodromy operator associated with the local system $\mathbb{F}$ and the loop $\mathfrak{l}_i$. The stalk $(\boldsymbol{j}_*\mathbb{F})_{\boldsymbol{c}_i}$ of the constructible sheaf $\boldsymbol{j}_*\mathbb{F}$ at $\boldsymbol{c}_i$ can be naturally identified with the subspace $\mathbb{F}_{\boldsymbol{o}_i}^{\mathrm{mon}_{\mathfrak{l}_i}}$ of invariants for the local monodromy. Taking stalks at each $\boldsymbol{c}_i \in S$ we get $\mathbb{Q}$-vector spaces $\mathscr{F}_{\boldsymbol{c}_i}$ and the adjunction map $\varphi : \mathscr{F} \to \boldsymbol{j}_*\mathbb{F}$ induces linear maps

$$\varphi_{\boldsymbol{c}_i} : \mathscr{F}_{\boldsymbol{c}_i} \to \mathbb{F}_{\boldsymbol{o}_i}^{\mathrm{mon}_{\boldsymbol{c}_i}} \subset \mathbb{F}_{\boldsymbol{o}_i}.$$

Note that, by descent, specifying the constructible sheaf $\mathscr{F}$ is equivalent to specifying the collection of points $S \subset \mathbb{C}$, the local system $\mathbb{F}$ on $\mathbb{C} - S$, the collection of vector spaces $\{\mathscr{F}_{\boldsymbol{c}_i}\}_{i=1}^n$ and the collection of linear maps $\{\varphi_{\boldsymbol{c}_i}\}_{i=1}^n$. In particular, the compactly supported pullback of $\mathscr{F}[1]$ via the inclusion $\boldsymbol{i}_{\boldsymbol{c}_i} : \{\boldsymbol{c}_i\} \hookrightarrow \mathbb{C}$ can be computed in terms of these linear algebraic data and is given explicitly by the complex

$$\boldsymbol{i}_{\boldsymbol{c}_i}^!(\mathscr{F}[1]) = [\underset{-1}{\mathscr{F}_{\boldsymbol{c}_i}} \xrightarrow{\varphi_{\boldsymbol{c}_i}} \underset{0}{\mathbb{F}_{\boldsymbol{o}_i}} \xrightarrow{1-\mathrm{mon}_{\mathfrak{l}_i}} \underset{1}{\mathbb{F}_{\boldsymbol{o}_i}}].$$

By definition $\mathscr{F}[1]$ is a perverse sheaf iff for all $\boldsymbol{c}_i \in S$ the complex of vector spaces $\boldsymbol{i}^!_{\boldsymbol{c}_i}(\mathscr{F}[1])$ has no cohomology in strictly negative degrees, i.e. iff $\varphi_{\boldsymbol{c}_i}$ is injective for all $i = 1, \dots, n$.

Next we rewrite the condition $R\Gamma(\mathbb{C}, \mathscr{F}) = 0$ in terms of the descent data $(\mathbb{F}, \{\mathscr{F}_{\boldsymbol{c}_i}\}, \{\varphi_{\boldsymbol{c}_i}\})$. To simplify notation let $U := \mathscr{F}_{\boldsymbol{c}_0}$, $V_i = \mathscr{F}_{\boldsymbol{c}_i}$ for $i = 1, \dots, n$. Let $T_i : U \to U$ be the monodromy operator for the local system $\mathbb{F}$ and the $\boldsymbol{c}_0$-based loop γ_i obtained by first tracing the path a_i from $\boldsymbol{c}_0$ to $\boldsymbol{o}_i$, then tracing the loop $\mathfrak{l}_i$, and then tracing back a_i in the opposite direction. Similarly we have linear maps $\psi_i : V_i \to U^{T_i} \subset U$ obtained by conjugating $\varphi_{\boldsymbol{c}_i} : V_i \to \mathbb{F}_{\boldsymbol{o}_i}$ with the operator of parallel transport in $\mathbb{F}$ along the path a_i.

The descent data for $\mathscr{F}$ with respect to the open cover $\mathbb{C} = (\mathbb{C} - S) \cup (\cup_{i=1}^n \boldsymbol{D}_i)$ are now completely encoded in the linear algebraic data $(U, \{V_i\}_{i=1}^n, \{T_i\}_{i=1}^n, \{\psi_i\}_{i=1}^n)$. Cover $\mathbb{C}$ by the two opens $\mathbb{C} - S$ and $\cup_{i=1}^n \boldsymbol{D}_i$. The intersection of these two opens is the disjoint union of punctured discs $\coprod_{i=1}^n (\boldsymbol{D}_i - \boldsymbol{c}_i)$. The Mayer-Vietoris sequence for $\mathscr{F}$ and this cover identifies $R\Gamma(\mathbb{C}, \mathscr{F})$ with the complex:

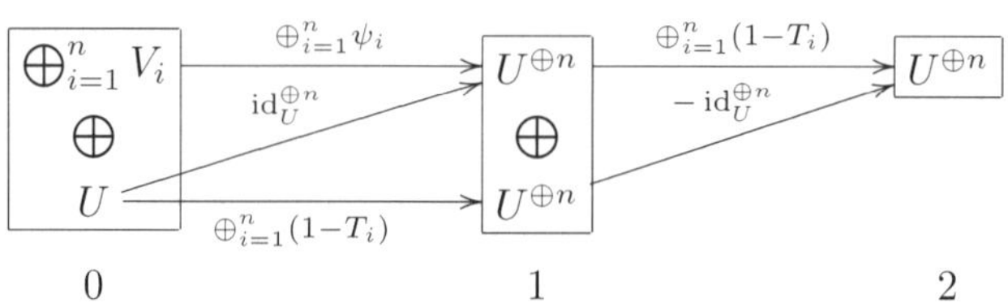

In other words we have a quasi-isomorphism of complexes of $\mathbb{Q}$-vector spaces:

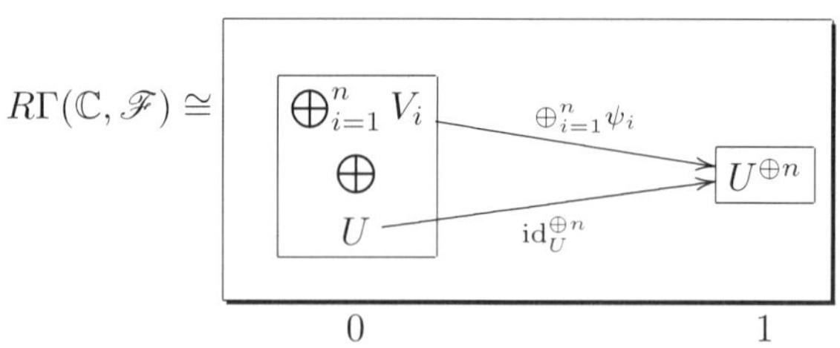

The acyclicity of this complex is equivalent to the conditions

(a) the maps $\psi_i : V_i \to U$ are injective for all $i = 1, \dots, n$, and
(b) the map $U \to \oplus_{i=1}^n U/V_i$ is an isomorphism.

Thus the acyclicity of $R\Gamma(\mathbb{C}, \mathscr{F})$ implies the perversity of $\mathscr{F}[1]$. The theorem is proven. □

The conditions (a) and (b) from the proof of Theorem 2.29 suggest a better way of recording the linear algebraic content of $\mathscr{F}$. Namely, if we set $U_i := U/V_i$, then we can use (b) to identify U with $\oplus_{i=1}^n U_i$, V_i with $\oplus_{j \neq i} U_j$ and the map $\psi_i : V_i \hookrightarrow U$ with the natural inclusion $\oplus_{j \neq i} U_j \subset \oplus_{i=1}^n U_i$. The only thing left is the data of the monodromy operators $T_i \in GL(U)$, $i = 1, \dots, n$. However for each i we have the embedding

$$V_i \overset{\psi_i}{\hookrightarrow} \operatorname{Ker}\left[U \overset{(1-T_i)}{\longrightarrow} U\right]$$

and so under the decomposition $U = \oplus_{i=1}^n U_i$ the automorphism T_i has a block form

$$T_i = \begin{pmatrix} 1 & 0 & \cdots & 0 & T_{1i} & 0 & \cdots & 0 \\ 0 & 1 & \cdots & 0 & T_{2i} & 0 & \cdots & 0 \\ \cdots & \cdots & \cdots & \cdots & \cdots & \cdots & \cdots & \cdots \\ 0 & 0 & \cdots & 0 & T_{ii} & 0 & \cdots & 0 \\ 0 & 0 & \cdots & 0 & T_{i+1,i} & 1 & \cdots & 0 \\ \cdots & \cdots & \cdots & \cdots & \cdots & \cdots & \cdots & \cdots \\ 0 & 0 & \cdots & 0 & T_{ni} & 0 & \cdots & 1 \end{pmatrix}$$

where $T_{i|U_i} = \sum_{j=1}^n T_{ji}$, and $T_{ji} : U_i \to U_j$. The linear maps T_{ji} are unconstrained except for the obvious condition that for all i the map T_i should be invertible, which is equivalent to $T_{ii} : U_i \to U_i$ being invertible for all $i = 1, \ldots, n$. Also since S was chosen to be such that $\mathbb{C} - S$ is the maximal open on which $\mathscr{F}$ is a local system, it follows that $U_i \neq \{0\}$ for all $i = 1, \ldots, n$.

In other words we have proven the following

LEMMA 2.30. *Fix the set of points* $S = \{\boldsymbol{c}_1, \ldots, \boldsymbol{c}_n\}$ *and choose the discs* $\{\boldsymbol{D}_i\}_{i=1}^n$ *and the system of paths* $\{a_i\}_{i=1}^n$. *The functor assigning to a constructible sheaf* $\mathscr{F}$ *with singularities at* S *the data* $(\{U_i\}_{i=1}^n, \{T_{ij}\})$ *establishes an equivalence between the groupoid of all data of type* **ncB(ii)** *with singularities exactly at* S *and all data of type* **ncB(iii)** *with the given* S.

The bridge between the **nc**- de Rham and Betti data is provided as usual by the Riemann-Hilbert correspondence. This is tautological but we record it for future reference:

LEMMA 2.31. *The de Rham functor:*

$$\boldsymbol{M} \to \text{cone}\left(\boldsymbol{M} \otimes_{\mathbb{C}[u]} \mathcal{O}_{\mathbb{A}^1}^{\text{an}} \xrightarrow{\frac{\partial}{\partial u}} \boldsymbol{M} \otimes_{\mathbb{C}[u]} \mathcal{O}_{\mathbb{A}^1}^{\text{an}} \right)$$

establishes an equivalence between the categories (**ncdR(iii)**) *and* (**ncB(i)**)$\otimes\mathbb{C}$.

Finally, note that Theorem 2.29, together with Lemma 2.31, and Deligne's classification [**BV85**, Theorem 4.7.3] of germs of irregular connections give immediately:

LEMMA 2.32. *The data* (**ncB(ii)**) *and* (**ncB(iv)**) *are equivalent.*

Proof. Let $\mathscr{F}$ be a constructible sheaf of $\mathbb{Q}$-vector spaces on $\mathbb{C}$. Define a local system S of $\mathbb{Q}$-vector spaces on $\boldsymbol{S}^1$ as the restriction of $\mathscr{F}$ to the circle "at infinity", i.e. define the stalk of S at $\varphi \in \boldsymbol{S}^1$ to be

$$\mathsf{S}_\varphi := \lim_{r \to +\infty} \mathscr{F}_{re^{i\varphi}}.$$

Next, for any $\lambda \in \mathbb{R}$ and any $\varphi \in \boldsymbol{S}^1$ consider the half-plane

$$\mathfrak{H}_{\varphi,\lambda} := (\lambda + \{u \in \mathbb{C} |\ \mathrm{Re}(u) \geq 0\}) \cdot e^{i\varphi},$$

as shown in Figure 3.

Now suppose that $R\Gamma(\mathbb{C}, \mathscr{F}) = 0$. By the long exact sequence for the cohomology of the pair $\mathfrak{H}_{\varphi,\lambda} \subset \mathbb{C}$ we get that $H^i(\mathbb{C}, \mathfrak{H}_{\varphi,\lambda}; \mathscr{F}) = 0$ unless $i = 1$. The Deligne-Malgrange-Stokes filtration on S is then given explicitly by

$$\mathsf{S}_{\varphi,\leq\lambda} := H^1(\mathbb{C}, \mathfrak{H}_{\varphi,\lambda}; \mathscr{F}) \subset \mathsf{S}_\varphi.$$

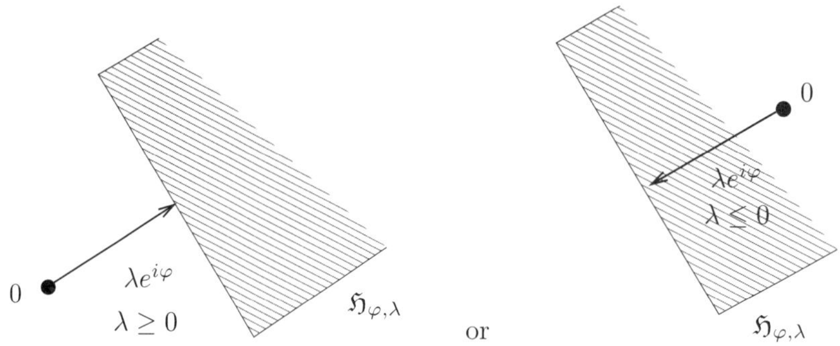

FIGURE 3. The half-plane $\mathfrak{H}_{\varphi,\lambda}$.

□

For the purposes of **nc**-Hodge theory all these statements can be summarized in the following

THEOREM 2.33. *There is a natural equivalence of categories*

$$\begin{pmatrix}\text{triples } (H, \mathscr{E}_B, \textsf{iso}) \text{ satisfying the } \textbf{(nc-filtration axiom)}^{\textsf{exp}} \text{ and the } \textbf{(}\mathbb{Q}\textbf{-structure axiom)}^{\textsf{exp}}\end{pmatrix} \leftrightarrow \begin{pmatrix}\text{quadruples } ((H,\nabla), \mathscr{F}_B, \boldsymbol{f}), \text{ where} \\ \bullet\ H \text{ is an algebraic } \mathbb{Z}/2\text{-graded vector bundle on } \mathbb{C} \text{ and } \nabla \text{ is a meromorphic connection on } H \text{ satisfying the } \textbf{(nc-filtration axiom)}^{\textsf{exp}}; \\ \bullet\ \mathscr{F}_B \in \mathsf{Constr}(\mathbb{C}, \mathbb{Q}), \text{ satisfying } R\Gamma(\mathbb{C}, \mathscr{F}_B) = 0; \\ \bullet\ \boldsymbol{f} \text{ is an isomorphism} \\ \boldsymbol{f} : \mathscr{F}_B \otimes \mathbb{C} \to DR\left(\boldsymbol{\Phi}\left[\iota_*\left((H,\nabla)_{|\mathbb{A}^1-\{0\}}\right)\right]\right) \\ \text{in } D^b_{\text{constr}}(\mathbb{C}, \mathbb{C})\end{pmatrix}$$

Here as before

DR is the de Rham complex functor from the derived category of regular holonomic $\mathbb{D}$-modules to the derived category of constructible sheaves,

ι is the inclusion map $\iota : \mathbb{A}^1 - \{0\} \hookrightarrow \mathbb{A}^1$ given by $\iota(v) = v^{-1}$, and

$\boldsymbol{\Phi}(\bullet)$ is the Fourier-Laplace transform for $\mathcal{D}$-modules on $\mathbb{A}^1$.

Proof. Follows immediately from previous equivalences. □

2.4. Structure results. In this section we collect a few results clarifying the structure properties of the **nc**-Hodge structures of exponential type.

2.4.1. ***A quiver description of* nc*-Betti data.*** Since the gluing data **ncB(iii)** are of essentially combinatorial nature, it is natural to look for a quiver interpretation of these data. To that end consider the algebra

$$\mathscr{A}_n := \left\langle \begin{matrix} \boldsymbol{p}_1, \ldots, \boldsymbol{p}_n \\ T, T_{11}^{-1}, \ldots, T_{nn}^{-1} \end{matrix} \;\middle|\; \begin{matrix} \boldsymbol{p}_1 + \boldsymbol{p}_2 + \cdots + \boldsymbol{p}_n = 1 \\ \boldsymbol{p}_i \boldsymbol{p}_j = \boldsymbol{p}_j \boldsymbol{p}_i \text{ for } i \neq j,\ \boldsymbol{p}_i^2 = \boldsymbol{p}_i \\ T_{ii}^{-1} \boldsymbol{p}_i T \boldsymbol{p}_i = \boldsymbol{p}_i T \boldsymbol{p}_i T_{ii}^{-1} = \boldsymbol{p}_i \end{matrix} \right\rangle \tag{2.4.1}$$

This is the path algebra of the complete quiver having n ordered vertices, $n^2 - n$ arrows connecting all pairs of distinct vertices, and $2n$-loops - two at each vertex, with the only relations being that the two loops at every given vertex are inverses of each other.

Note that our description of the gluing data **ncB(iii)** now immediately gives the following

LEMMA 2.34. *For a given set of points $S = \{\boldsymbol{c}_1, \ldots, \boldsymbol{c}_n\} \subset \mathbb{C}$, the category of gluing data* **ncB(iii)** *with singularities at S is equivalent to the category of finite dimensional representations of $\mathscr{A}_n$.*

In particular since the braid group $\boldsymbol{B}_n$ on n strands acts naturally on the data **ncB(iii)** we get a homomorphism $\boldsymbol{B}_n \to \mathrm{Aut}(\mathscr{A}_n)$ from the braid group to the group of algebra automorphisms of $\mathscr{A}_n$.

2.4.2. *Gluing of* **nc**-*Hodge structures.*

It is natural to expect that the usual classification of connections with second order poles in terms of formal regular type and Stokes multipliers can be promoted to a similar classification of **nc**-Hodge structures. The search for such a classification leads naturally to the following theorem:

THEOREM 2.35. *Let $\{(H, \mathscr{E}_B, \mathsf{iso})\}$ be an* **nc**-*Hodge structure of exponential type. Then specifying $\{(H, \mathscr{E}_B, \mathsf{iso})\}$ is equivalent to specifying the following data:*

- **(regular type):** *A finite set $S = \{\boldsymbol{c}_1, \ldots, \boldsymbol{c}_n\} \subset \mathbb{C}$ and a collection $\{((\mathscr{R}_i, \nabla_i), \mathscr{E}_{B,i}, \mathsf{iso}_i)\}_{i=1}^n$ of* **nc**-*Hodge structures with regular singularities.*
- **(gluing data):** *A base point $\boldsymbol{c}_0 \in \mathbb{C} - S$, a collection of discs $\{\boldsymbol{D}_i\}_{i=1}^n$ and paths $\{a_i\}_{i=1}^n$, chosen as in the proof of Theorem 2.29, and for every $i \neq j$, $i, j \in \{1, \ldots, n\}$ a map of rational vector spaces*

$$T_{ij} : (\mathscr{E}_{B,j})_{\boldsymbol{c}_0} \longrightarrow (\mathscr{E}_{B,i})_{\boldsymbol{c}_0}$$

Proof. It will be convenient to introduce formal counterparts to the de Rham parts of the **nc**-Hodge structures appearing in the statement of the theorem. We consider the following:

formal(a) A pair $(\mathscr{M}^{\mathrm{for}}, \nabla^{\mathrm{for}})$, where $\mathscr{M}^{\mathrm{for}}$ is a finite dimensional vector space over $\mathbb{C}((u))$ and ∇^{for} is a meromorphic connection on $\mathscr{M}^{\mathrm{for}}$ of exponential type.

formal(b) A finite set of points $S = \{\boldsymbol{c}_1, \ldots, \boldsymbol{c}_n\} \subset \mathbb{C}$ and a collection $\{(\mathscr{R}_i^{\mathrm{for}}, \nabla_i^{\mathrm{for}})\}_{i=1}^n$ where each $\mathscr{R}_i^{\mathrm{for}}$ is a non-zero finite dimensional vector space over $\mathbb{C}((u))$ and each ∇_i^{for} is a meromorphic connection on $\mathscr{R}_i^{\mathrm{for}}$ with a regular singularity.

formal(c) A finite collection of points $S = \{\boldsymbol{c}_1, \ldots, \boldsymbol{c}_n\} \subset \mathbb{C}$, and

- a collection $U_1, U_2, \ldots, U_n$ of finite dimensional non-zero $\mathbb{Q}$-vector spaces,
- a collection of linear maps $T_{ii} \in GL(U_i)$, for all $i = 1, \ldots, n$,

By Remark 2.13 the natural functor from the category of data **formal(b)** to the category of data **formal(a)**, which is given by

$$\begin{array}{ccc}(\mathbf{formal(b)}) & \longrightarrow & (\mathbf{formal(a)}) \\ \\ (S; \{(\mathscr{R}_i^{\text{for}}, \nabla_i^{\text{for}})\}_{i=1}^n) & \longrightarrow & \bigoplus_{i=1}^n \boldsymbol{\mathcal{E}}^{\boldsymbol{c}_i/u} \otimes (\mathscr{R}_i^{\text{for}}, \nabla_i^{\text{for}}) =: (\mathscr{M}^{\text{for}}, \nabla^{\text{for}})\end{array}$$

is an equivalence of categories.

Also we have the following

LEMMA 2.36. *The categories of data* **formal(b)** *and* **formal(c)** *are naturally equivalent.*

Proof. Indeed, consider the category $\mathcal{C}$ of all data consisting of a finite set of points $S = \{\boldsymbol{c}_1, \ldots, \boldsymbol{c}_n\} \subset \mathbb{C}$ and a collection $\{(\mathscr{R}_i, \nabla_i)\}_{i=1}^n$ where each $\mathscr{R}_i$ is a non-zero finite dimensional vector space over $\mathbb{C}\{u\}[u^{-1}]$ and each ∇_i is a meromorphic connection on $\mathscr{R}_i$ with a regular singularity and non-trivial monodromy. Then we have natural functors

$$\begin{array}{ccc} & & (\mathbf{formal(b)}) \\ & \overset{(\bullet)\otimes\mathbb{C}((u))}{\nearrow} & \\ \mathcal{C} & & \\ & \underset{\mathsf{mon}}{\searrow} & \\ & & (\mathbf{formal(c)}) \end{array}$$

where $(\bullet) \otimes \mathbb{C}((u))$ is the passage to a formal completion and mon is given by assigning to each $(\mathscr{R}_i, \nabla_i)$ the pair (U_i, T_i), where U_i is the fiber of the Birkhoff extension $\mathfrak{B}_0(\mathscr{R}_i, \nabla_i)$ of $(\mathscr{R}_i, \nabla_i)$ at $1 \in \mathbb{A}^1$, and T_i is the monodromy of $\mathfrak{B}_0(\mathscr{R}_i, \nabla_i)$ around the unit circle traced in the positive direction.

This proves the lemma since mon is an equivalence by the Riemann-Hilbert correspondence and $(\bullet) \otimes \mathbb{C}((u))$ is an equivalence by the formal decomposition theorem [**Sab02**, II.5.7]. □

Note that these equivalences are compatible with the corresponding equivalence of analytic de Rham data and Betti data. More precisely we have a commutative diagram of functors

$$\begin{array}{ccc}(\mathbf{ncdR(i)}) & \longrightarrow & (\mathbf{formal(a)}) \\ \updownarrow & & \updownarrow \\ (\mathbf{ncB(iii)}) & \longrightarrow & (\mathbf{formal(c)})\end{array} \tag{2.4.2}$$

Here the right vertical equivalence is the composition of the equivalences (**formal(a)**) $\cong$ (**formal(b)**) $\cong$ (**formal(c)**) that we just discussed. The left vertical equivalence is the composition of the equivalence (**nc**dR(i)) $\cong$ (**nc**dR(iii)) given in Lemma 2.26, the equivalence (**nc**dR(iii)) $\cong$ (**nc**B(i)) from Lemma 2.31, the equivalence (**nc**B(i)) $\cong$ (**nc**B(ii)) given in Theorem 2.29, and the equivalence (**nc**B(ii)) $\cong$ (**nc**B(iii)) from Lemma 2.30.

Horizontally we have the forgetful functors

$$\begin{array}{ccc} (\mathbf{ncdR(i)}) & \longrightarrow & (\mathbf{formal(a)}) \\ (\mathscr{M},\nabla) & \longmapsto & (\mathscr{M},\nabla)\otimes_{\mathbb{C}\{u\}[u^{-1}]}\mathbb{C}((u)), \end{array}$$

and

$$\begin{array}{ccc} (\mathbf{ncB(iii)}) & \longrightarrow & (\mathbf{formal(c)}) \\ \left(S;\{U_i\}_{i=1}^n,\{T_{ij}\}_{i,j=1}^n\right) & \longmapsto & \left(S;\{U_i\}_{i=1}^n,\{T_{ii}\}_{i=1}^n\right). \end{array}$$

Next we need the following

LEMMA 2.37. *Suppose that $(\mathscr{M},\nabla)$ is some de Rham data of type* **ncdR(i)** *and let*

$$(\mathscr{M}^{\text{for}},\nabla^{\text{for}}) = (\mathscr{M},\nabla)\otimes_{\mathbb{C}\{u\}[u^{-1}]}\mathbb{C}((u))$$

be the corresponding formal data. Then:

(a) *the map*

$$\begin{pmatrix}\mathbb{C}\{u\}\text{-submodules } \mathscr{H}\subset \\ \mathscr{M}\text{, on which } \nabla \text{ has a pole} \\ \text{of order} \leq 2\end{pmatrix} \xrightarrow{(\bullet)\otimes\mathbb{C}[[u]]} \begin{pmatrix}\mathbb{C}[[u]]\text{-submodules } \mathscr{H}^{\text{for}}\subset \\ \mathscr{M}^{\text{for}}\text{, on which } \nabla^{\text{for}} \text{ has a} \\ \text{pole of order} \leq 2\end{pmatrix},$$

is bijective.

(b) *If $\Psi : (\mathscr{M}^{\text{for}},\nabla^{\text{for}}) \to \bigoplus_{i=1}^n \boldsymbol{\mathcal{E}}^{c_i/u}\otimes(\mathscr{R}_i^{\text{for}},\nabla_i^{\text{for}})$ is a formal isomorphism, then the map*

$$\begin{pmatrix}\mathbb{C}[[u]]\text{-submodules } \mathscr{H}^{\text{for}}\subset \\ \mathscr{M}^{\text{for}}\text{, on which } \nabla^{\text{for}} \text{ has a} \\ \text{pole of order} \leq 2\end{pmatrix} \xleftarrow[\Psi]{} \begin{pmatrix}\mathbb{C}[[u]]\text{-submodules } \mathscr{H}_i^{\text{for}}\subset \\ \mathscr{R}_i^{\text{for}}\text{, for all } i=1,\ldots,n, \\ \text{on which } \nabla_i^{\text{for}} \text{ has a pole of} \\ \text{order} \leq 2\end{pmatrix},$$

is bijective.

Proof. **(a)** Pick some frame $\underline{e}$ of $\mathscr{M}$ over $\mathbb{C}\{u\}[u^{-1}]$ and let $\mathscr{H}^0 := \mathbb{C}\{u\}\cdot\underline{e}\subset\mathscr{M}$ be the submodule of all sections in $\mathscr{M}$ that are holomorphic in this frame. Now any $\mathbb{C}\{u\}$-submodule $\mathscr{H}\subset\mathscr{M}$ on which ∇ has a pole of order ≤ 2 will be a $\mathbb{C}\{u\}$-submodule of $\mathscr{M}$ which is commensurable with $\mathscr{H}^0$, i.e. we will have $u^N\mathscr{H}^0\subset\mathscr{H}\subset u^{-N}\mathscr{H}^0$ for $N\gg 1$. However the formal completion functor $(\bullet)\otimes_{\mathbb{C}\{u\}}\mathbb{C}[[u]]$ establishes an isomorphism between the Grassmannian $GL_r(\mathbb{C}\{u\}[u^{-1}])/GL_r(\mathbb{C}\{u\})$ and the affine Grassmannian $GL_r(\mathbb{C}((u))/GL_r(\mathbb{C}[[u]])$. But this map preserves the condition that a submodule $\mathscr{H}$ is invariant under $\nabla_{u^2d/du}$ which proves **(a)**.

(b) As already mentioned in Remark 2.13 this is proven in [**HS07**, Lemma 8.2]. Alternatively we can reason as in the proof of part **(a)**. Let $\mathscr{H}$ be a $\mathbb{C}[[u]]$-submodule in $\mathscr{M}^{\text{for}}$ which is commensurable with $\mathscr{H}^{0,\text{for}}$ and preserved by $\nabla_{u^2\frac{d}{du}}$. The operator $\nabla_{u^2\frac{d}{du}}$ acts on the infinite-dimensional topological complex vector space $\mathscr{M}^{\text{for}}$ with finitely many infinite Jordan blocks with eigenvalues $\{\boldsymbol{c}_1,\ldots,\boldsymbol{c}_n\}$. The corresponding generalized eigenspaces are exactly the modules $\boldsymbol{\mathcal{E}}^{c_i/u}\mathscr{R}_i^{\text{for}}$. Hence

$$\mathscr{H}^{\text{for}} = \oplus_i\left(\mathscr{H}^{\text{for}}\cap\boldsymbol{\mathcal{E}}^{c_i/u}\mathscr{R}_i^{\text{for}}\right)$$

Therefore we obtain extensions $\mathscr{R}_i^{\mathrm{for}}$ with second order poles and regular singularity.

Combining the previous lemma with the equivalences in diagram (2.4.2) and the description of **nc**-Hodge structures from Section 2.1.8 gives the theorem. □

2.5. Deformations of nc-spaces and gluing. In this section we will briefly examine how the gluing construction for **nc**-Hodge structures varies with parameters. In particular, we will look at deformations of **nc**-spaces and the way the gluing data for the **nc**-Hodge structures on the cohomology of these spaces interacts with the appearance of a curvature in the d$(\mathbb{Z}/2)$g algebra computing the sheaf theory of the space.

2.5.1. ***The cohomological Hochschild complex.*** Suppose $X = \mathbf{ncSpec}\, A$ is an **nc**-affine **nc**-space. Recall that the cohomological Hochschild complex is defined as

$$C^{\bullet}(A,A) := \prod_{n\geq 0} \mathrm{Hom}_{\mathbb{C}-\mathsf{Vect}}\left((\mathbf{\Pi}A)^{\otimes n}, A\right),$$

Its shift $\mathbf{\Pi}C^{\bullet}(A,A)$ is a Lie superalgebra with respect to the Gerstenhaber bracket [**Ger64**], and can be interpreted as the Lie algebra of continuous derivations of the free topological algebra $\prod_{n\geq 0}\left((\mathbf{\Pi}A)^{\otimes n}\right)^{\vee}$. The multiplication m_A and differential d_A of A combine into a cochain $\gamma_A := m_A + d_A \in C^{\bullet}(A,A)$ satisfying $[\gamma_A, \gamma_A] = 0$.

The formal deformation theory of X is controlled by a d$(\mathbb{Z}/2)$g Lie algebra structure $\mathbf{\Pi}C^{\bullet}(A,A)$ endowed with the differential $[\gamma_A, \bullet]$. It is convenient to consider also the reduced Hochschild complex

$$C^{\bullet}_{\mathrm{red}}(A,A) := \prod_{n\geq 0} \mathrm{Hom}_{\mathbb{C}-\mathsf{Vect}}\left(\left(\mathbf{\Pi}\left(A/\mathbb{C}\cdot 1_A\right)\right)^{\otimes n}, A\right),$$

which is naturally a subspace of $C^{\bullet}(A,A)$. The reduced complex is (after the parity change) a dg Lie subalgebra in $\mathbf{\Pi}C^{\bullet}(A,A)$. Moreover it is quasi-isomorphic to $\mathbf{\Pi}C^{\bullet}(A,A)$. Hence, for deformation theory purposes one can replace $\mathbf{\Pi}C^{\bullet}(A,A)$ by $\mathbf{\Pi}C^{\bullet}_{\mathrm{red}}(A,A)$.

Let $\gamma = \sum_{i\geq 1}\gamma_i t^i \in tC^{\mathrm{even}}_{\mathrm{red}}(A,A)[[t]]$ be a formal path consisting of solutions of the Maurer-Cartan equation, i.e.

$$d\gamma + \frac{1}{2}[\gamma,\gamma] = 0 \quad (\Leftrightarrow\ [\gamma+\gamma_A, \gamma+\gamma_A] = 0).$$

Such a solution defines so-called formal deformation of the d$(\mathbb{Z}/2)$g algebra A as a weak (or curved) A_∞-algebra (see e.g. [**LH03**] for the definition and [**Sch03**] for a more detailed analysis). We can use the cochain $\gamma + \gamma_A \in C^{\mathrm{even}}(A,A)[[t]]$ to twist the notion of an A-module. We will write A_γ for the (weak) A_∞-algebra over $\mathbb{C}[[t]]$ corresponding to A and $\gamma+\gamma_A$ and $(A_\gamma - \mathsf{mod})$ for the $\mathbb{C}[[t]]$-linear dg category of all modules over A_γ. By definition $(A_\gamma - \mathsf{mod})$ is the category of dg modules over a bar-type resolution of A_γ [**KS06b**]. As an algebra the relevant bar dg algebra is the completed tensor product

$$\prod_{n\geq 0}\left((\mathbf{\Pi}A)^{\otimes n}\right)^{\vee}\widehat{\otimes}\mathbb{C}[[t]] \tag{2.5.1}$$

where the algebra structure comes from the usual algebra structure on $\mathbb{C}[[t]]$ and the tensor algebra structure on $\prod_{n\geq 0}\left((\mathbf{\Pi}A)^{\otimes n}\right)^{\vee}$. Thus for every $\gamma \in tC^{\mathrm{even}}_{\mathrm{red}}(A,A)[[t]]$

which solves the Mauer-Cartan equation we get a differential $\gamma + \gamma_A$ on the graded algebra (2.5.1). The bar dg algebra of A_γ is now defined as the dg algebra

$$B_\gamma := \left(\prod_{n\geq 0} \left((\mathbf{\Pi} A)^{\otimes n}\right)^\vee \widehat{\otimes}\mathbb{C}[[t]], \gamma + \gamma_A \right).$$

The dg category $(A_\gamma - \mathsf{mod})$ is by definition the category of dg modules over B_γ which are topologically free as modules of the underlying algebra, i.e. after forgetting the differential, and also satisfying the condition of unitality at $t = 0$.

As before this category can be viewed as the category $C_{\mathbb{X}_\gamma} := (A_\gamma - \mathsf{mod})$ of quasi-coherent sheaves on an **nc**-affine **nc**-space $\mathbb{X}_\gamma \to \mathbb{D}$ defined over the formal disc $\mathbb{D} = \mathrm{Spf}(\mathbb{C}[[t]])$. More generally we will get an **nc**-space $\mathbb{X}$ over the formal scheme of solutions to the Maurer-Cartan equation and $\mathbb{X}_\gamma \to \mathbb{D}$ is the restriction of $\mathbb{X}$ to the formal path $\gamma + \gamma_A$ sitting inside that formal scheme.

Similarly we can use γ to twist the notion of a Hochschild cohomology class for A. Namely we can consider the Hochschild cohomology of the A_∞-algebra A_γ. It is given explicitly as the cohomology

$$HH^\bullet_\gamma(A) := H^\bullet\left(C^\bullet(A, A)[[t]], [\gamma + \gamma_A, \bullet]\right),$$

and is a commutative algebra with respect to the cup product. Note also that the algebra $HH^\bullet_\gamma(A)$ comes equipped with a unit $[1_A]$ and a distinguished even element $[\gamma + \gamma_A]$, i.e. a structure similar to the one discussed in Section 2.2.5.

REMARK 2.38. • If γ has no component of degree zero, i.e. if

$$\gamma \in tC^{\text{even}}_{\text{red},+}(A, A)[[t]], \text{ where } C^\bullet_{\text{red},+}(A, A) = \prod_{n\geq 1} \mathrm{Hom}_{\mathbb{C}-\mathsf{Vect}}\left((\mathbf{\Pi}\,(A/\mathbb{C}\cdot 1_A))^{\otimes n}, A\right),$$

then A_γ is an honest (strong) A_∞-algebra, and the category $(A_\gamma - \mathsf{mod})$ will typically have many interesting objects. Furthermore, in this case smoothness and compactness are stable under deformations. That is, if A is smooth (respectively compact) over $\mathbb{C}$, then A_γ is smooth (respectively compact) over $\mathbb{C}[[t]]$.

• If the $n = 0$ component of γ is non-trivial, i.e. if the corresponding A_∞ structure has a non-trivial m_0, then the category $(A_\gamma - \mathsf{mod})$ may have no non-zero objects. The basic example of this is when $A = \mathbb{C}$ and $\gamma = t \cdot 1_A$.

If the original algebra A has the degeneration property, then it is easy to see that the Hodge-to-de Rham spectral sequence will degenerate for the periodic cyclic homology of A_γ. In other words the formal **nc**-space $\mathbb{X}$ will give rise to a variation of **nc**-Hodge structures over the formal scheme of solutions of the Maurer-Cartan equation for A. When we have a non-trivial $n = 0$ component in γ this may lead to a paradoxical situation in which we have a family of **nc**-spaces over $\mathbb{D}$ which has no sheaves over the generic point but has non-trivial de Rham cohomology (i.e. periodic cyclic homology) generically. This suggests the following important

QUESTION 2.39. What is the geometrical meaning of $HH^\bullet_\gamma(A)$, $HH_\bullet(A_\gamma)$, $HH^-_\bullet(A_\gamma)$, and $HP_\bullet(A_\gamma)$, when γ has non-trivial $n = 0$ component and the objects of $(A_\gamma - \mathsf{mod})$ disappear over $\mathbb{D}^\times$?

REMARK 2.40. Note that if γ solves the Maurer-Cartan equation, then for any $c \in t\mathbb{C}[[t]]$, the cochain $\gamma + c \cdot 1_A$ will also solve the Maurer-Cartan equation[1]. So we have a natural mechanism for modifying formal paths of solutions of the Maurer-Cartan equation. We will exploit this mechanism in the next section.

2.5.2. ***Corrections by constants.*** The unpleasant phenomenon of having **nc**-spaces with no sheaves and non-trivial cohomology at the generic point is related to the gluing description for **nc**-Hodge structures. The idea is that the A-modules that disappear at the generic point of $\mathbb{D}$ may reappear again if we modify the weak A_∞-algebra A_γ appropriately. The periodic cyclic homologies of the different admissible modifications of A_γ then correspond to the regular pieces in the gluing description of the **nc**-de Rham data given by $HP_\bullet(A_\gamma)$. More precisely we have the following

CONJECTURE 2.41. *Suppose that A is a smooth and compact* $\mathrm{d}(\mathbb{Z}/2)\mathrm{g}$ *algebra. Let $\gamma \in tC^{\text{even}}_{\text{red}}(A,A)[[t]]$ be a formal even path of solutions of the Maurer-Cartan equation for A. Then the periodic cyclic homology $HP_\bullet(A_\gamma)$ carries a canonical functorial structure of a variation of $\mathbb{Q}$-**nc**-Hodge structures of exponential type over $\mathbb{D} = \operatorname{Spf}(\mathbb{C}[[t]])$. Furthermore there exists a positive integer N and a finite collection of pairwise distinct Puiseux series*

$$\mathbf{c}_i = \sum_{j \geq 1} c_{i,j} t^{\frac{j}{N}}, \quad c_{i,j} \in \mathbb{C}$$

such that:

- *The series $\mathbf{c}_i$ are the distinct eigenvalues of the operator of multiplication by the class $[\gamma + \gamma_A]$ in the supercommutative algebra $HH^\bullet_\gamma(A)\widehat{\otimes}_{\mathbb{C}[[t]]}\mathbb{C}((t))$.*
- *For each i the category $(A_{\gamma+\mathbf{c}_i \cdot 1_A} - \mathsf{mod})$ is a non-trivial $\mathbb{C}\left[\left[t^{1/N}\right]\right]$-linear* $\mathrm{d}(\mathbb{Z}/2)\mathrm{g}$ *category which are smooth and compact over $\mathbb{C}\left[\left[t^{1/N}\right]\right]$ and is computed by a $d(\mathbb{Z}/2)g$ algebra B_i defined over $\mathbb{C}\left[\left[t^{1/N}\right]\right]$ and quasi-isomorphic to the (weak) A_∞-algebra $A_{\gamma+\mathbf{c}_i \cdot 1_A}$.*
- *The Hochschild homologies $HH_\bullet(B_i)$ are flat $\mathbb{C}\left[\left[t^{1/N}\right]\right]$-modules and we have*

$$\sum_i \operatorname{rk}_{\mathbb{C}[[t^{1/N}]]}(HH_\bullet(B_i)) = \operatorname{rk}_{\mathbb{C}[[t^{1/N}]]} HH_\bullet(A_\gamma) = \dim_{\mathbb{C}} HH_\bullet(A).$$

- *The variation of **nc**-Hodge structures $HP_\bullet(A_\gamma)$ viewed as a variation over $\mathbb{C}\left[\left[t^{1/N}\right]\right]$ has as regular constituents the variations of **nc**-Hodge structures on $HP_\bullet(B_i)$ whose existence is predicted by Conjecture 2.24.*

In particular Conjecture 2.41 says that the categorical and Hodge theoretic content of the algebra A_γ consists of the following data:

(categories): A finite collection of smooth and compact $\mathbb{C}\left[\left[t^{1/N}\right]\right]$-linear $\mathrm{d}(\mathbb{Z}/2)\mathrm{g}$ categories $(B_i - \mathsf{mod})$.

[1]In fact this is the main reason for all the hassle with the unit and the reduced complex in this section.

(gluing): A finite collection of distinct Puiseux series $\boldsymbol{c}_i \in \mathbb{C}\left[\left[t^{1/N}\right]\right]$, and formal **nc**-gluing data which glues the variations of regular **nc**-Hodge structures on $HP_\bullet(B_i)$ into a variation of **nc**-Hodge structure of exponential type over $\mathbb{C}\left[\left[t^{1/N}\right]\right]$.

In the above discussion we have tacitly replaced the analytic setting from Section 2.3 by a formal setting. One can check that both the de Rham and Betti data make sense here, e.g. one can speak about homotopy classes of non-intersecting paths to points $\boldsymbol{c}_i$ thinking about t as a small real positive parameter.

REMARK 2.42. This situation is analogous to a well known setup in singularity theory. Namely, if we have a germ of an isolated hypersurface singularity given by an equation $f = 0$, and if we have a deformation of f which has several critical values, then the Milnor number of the original singularity is equal to the sum of the Milnor numbers of the simpler critical points of the deformed function. In fact, as we will see in Section 3.2 the singularity setup is a rigorous manifestation of the above conjectural picture.

2.5.3. ***Singular deformations.*** Suppose next that A is compact but not smooth (or smooth but non-compact) d($\mathbb{Z}/2$)g algebra and let again γ with $\gamma \in tC^{\text{even}}_{\text{red}}(A, A)[[t]]$ be a formal path of solutions of the Maurer-Cartan equation. We expect that the usual definition of smoothness and compactness can be modified to give a notion of smoothness together with compactness of A_γ at the generic point, i.e. over $\mathbb{C}((t))$, even when the objects in $(A_\gamma - \mathsf{mod})$ disappear over $\mathbb{C}((t))$.

In the case when A_γ is smooth and compact over $\mathbb{C}((t))$, i.e. when the deformation given by γ is a smoothing deformation, we also expect Conjecture 2.41 to hold at the generic point. More precisely, we expect to have Puiseux series $\boldsymbol{c}_i$ as above for which the associated categories $(A_{\gamma + \boldsymbol{c}_i 1_A} - \mathsf{mod})$ are non-trivial and smooth and compact over $\mathbb{C}\left(\left(t^{1/N}\right)\right)$. We also expect that the periodic cyclic homology $HP_\bullet\left(A_\gamma\right)$ is equipped with a variation of **nc**-Hodge structures of exponential type over $\mathbb{C}((t))$ so that the periodic cyclic homologies of the categories $(A_{\gamma + \boldsymbol{c}_i 1_A} - \mathsf{mod})$ are the regular pieces of this variation after we base change to $\mathbb{C}\left(\left(t^{1/N}\right)\right)$. Finally, the Puiseux series $\{\boldsymbol{c}_i\}$ should be the eigenvalues of the operator of multiplication by $[\gamma + \gamma_A] \in HH^\bullet\left(A_\gamma\right) \widehat{\otimes}_{\mathbb{C}[[t]]} \mathbb{C}((t))$.

3. Examples and relation to mirror symmetry

In this section we discuss examples of **nc**-Hodge structures arising from smooth and compact Calabi-Yau geometries and we study how these structures are affected by mirror symmetry. Specifically we look at a generalization of Homological Mirror Symmetry which relates categories of boundary topological field theories (or D-branes) associated with the following two types of geometric backgrounds:

A-model backgrounds: Pairs (X, ω), where X is a compact C^∞-manifold, and ω is a symplectic form on X satisfying a convergence property (see below).

B-model backgrounds: Pairs $\mathsf{w} : Y \to \text{disc} \subset \mathbb{C}$, where Y is a complex manifold with trivial canonical class, and w is a proper holomorphic map.

We will explain how each such background (both in the A and the B model) gives rise to the geometric and Hodge theoretic data described in Section 2.5.2. Namely we get:

- A finite collection $\{Z_i^{A/B}\}$ of smooth compact **nc**-spaces. In fact $\{Z_i^{A/B}\}$ will be (see Section 4.4.1 for the definition) odd/even Calabi-Yau **nc**-spaces of dimension $\left(\frac{\dim_{\mathbb{R}} X}{2} \mod 2\right) / \left(\dim_{\mathbb{C}} Y \mod 2\right)$.
- Complex numbers $\boldsymbol{c}_i^{A/B}$ and Betti gluing data $\left\{T_{ij}^{A/B}\right\}$ for the regular **nc**-Hodge structures on the periodic cyclic homology of $Z_i^{A/B}$.

In particular the data $\left(HC_\bullet^-\left(Z_i^A\right), \left\{\boldsymbol{c}_i^A\right\}, \left\{T_{ij}^A\right\}\right)$ and $\left(HC_\bullet^-\left(Z_i^B\right), \left\{\boldsymbol{c}_i^B\right\}, \left\{T_{ij}^B\right\}\right)$ each glue into a **nc**-Hodge structure of exponential type. The **generalized Homological Mirror Symmetry Conjecture** now asserts that if two A/B-model backgrounds $(X,\omega)/(Y,\mathsf{w})$ are mirror to each other, then the associated **nc**-geometry and **nc**-Hodge structure packages are isomorphic:

$$\left(Z_i^A, \left\{\boldsymbol{c}_i^A\right\}, \left\{T_{ij}^A\right\}\right) \cong \left(Z_i^B, \left\{\boldsymbol{c}_i^B\right\}, \left\{T_{ij}^B\right\}\right).$$

3.1. A-model Hodge structures: symplectic manifolds. Suppose (X,ω) is a compact symplectic manifold of dimension $\dim_{\mathbb{R}} X = 2d$. In the case when X is a Calabi-Yau variety (in particular $c_1(X) = 0$) one has a family of superconformal field theories attached to X in the large volume limit (i.e. after the rescaling $\omega \to \omega/\hbar$ where $0 < \hbar \ll 1$), and the A-twist gives a topological quantum field theory (see [**HKK$^+$03**]). In mathematical terms this means that we have Gromov-Witten invariants and a $\mathbb{Z}$-graded Fukaya category associated to $(X,\omega/\hbar)$. On the other side, Gromov-Witten invariants can be defined for an arbitrary compact symplectic manifold, not necessarily one with $c_1(X) = 0$. Our goal in this section is to describe what is an analog of the Fukaya category for general (X,ω).

Namely, it is expected that for (X,ω) of large volume the Fukaya category of (X,ω) is a weak $\mathbb{Z}/2$-graded A_∞-category which will satisfy the generalized smoothness and compactness properties conjectured in Section 2.5.3. Briefly this should work as follows. Following Fukaya-Oh-Ohta-Ono [**FOOO07**] consider a finite collection $\mathfrak{L} = \{L_i\}$ of transversal oriented spin Lagrangian submanifolds in X and form a "degenerate" version $\mathsf{Fuk}_{\mathfrak{L}}$ of Fukaya's category which only involves the L_i. More precisely we take $\mathsf{Ob}\left(\mathsf{Fuk}_{\mathfrak{L}}\right) = \{L_i\}$, and define

$$\mathrm{Hom}_{\mathsf{Fuk}_{\mathfrak{L}}}\left(L_i, L_j\right) = \begin{cases} \mathbb{C}^{L_i \cap L_j}, & i \neq j, \\ A^\bullet(L_i, \mathbb{C}), & i = j. \end{cases}$$

Here $\mathbb{C}^{L_i \cap L_j}$ is taken with the ordinary algebra structure but is put in degree equal to the Maslov grading mod 2, and $A^\bullet(L_i, \mathbb{C})$ is the dg algebra of C^∞ differential forms on L_i.

We consider a 1-parameter family of symplectic manifolds

$$\left(X, \frac{\omega}{\hbar}\right), \quad \hbar \in \mathbb{R}_{>0}, \quad \hbar \to 0. \tag{3.1.1}$$

It will be convenient to introduce a new parameter $q := \exp(-1/\hbar)$ (note that $q \to 0$ when $\hbar \to 0$). Denote by $\mathbb{C}_q$ the usual Novikov ring:

$$\mathbb{C}_q := \left\{ \sum_{i=0}^{\infty} a_i q^{E_i} \,\middle|\, \begin{array}{l} \text{formal series where } a_i \in \mathbb{C} \text{ and } E_i \in \mathbb{R} \\ \text{with } \lim_{i\to\infty} E_i = +\infty \end{array} \right\}$$

In the case $[\omega] \in H^2(X, \mathbb{Z})$ one can replace the Novikov ring $\mathbb{C}_q$ by the more familiar algebra $\mathbb{C}((q))$ of Laurent series. The three-point genus zero Gromov-Witten invariants of the symplectic family (3.1.1) give rise (see e.g. [**KM94, LT98, Sie99, CK99, FO01**]) to a $\mathbb{C}_q$-valued (small) quantum deformation of the cup product on $H^\bullet(X, \mathbb{C})$:

$$*_q : H^\bullet(X, \mathbb{C})^{\otimes 2} \to H^\bullet(X, \mathbb{C}) \otimes \mathbb{C}_q$$

Conjecturally the series for the quantum product is absolutely convergent for sufficiently small q.

What is constructed in [**FOOO07**] is a solution γ of the Maurer-Cartan equation in the cohomological Hochschild complex of $\mathsf{Fuk}_{\mathfrak{L}}$ with coefficients in the series in $\mathbb{C}_q$ with strictly positive exponents (equal to the areas of non-trivial pseudo-holomorphic discs). The meaning of the quantum product is the cup-product in the Hochschild cohomology of the deformed weak category.

The d$(\mathbb{Z}/2)$g category $\mathsf{Fuk}_{\mathfrak{L}}$ over $\mathbb{C}_q$ is compact but not smooth. If the collection $\mathfrak{L}$ is chosen to be big enough, i.e. if it generates the full Fukaya category, then $\mathsf{Fuk}_{\mathfrak{L}}$ is the large volume limit of $\mathsf{Fuk}(X, \omega)$, i.e. the limit in which all disc instantons for ω are supressed.

Now the formalism of Section 2.5.3 should associate with $\mathsf{Fuk}_{\mathfrak{L}} = (A - \mathsf{mod})$ and γ a finite collection $\{\boldsymbol{c}_i\}$ of formal series in positive powers of q and a collection $\{\mathsf{Fuk}_i\}$ of non-trivial smooth and compact modifications of the Fukaya category whose Hochschild homologies are the regular singularity constitutents of the Hochschild homology of the q-family of Fukaya categories near the large volume limit. In this geometric context, we expect that the $\{\boldsymbol{c}_i\}$ are the eigenvalues of the quantum multiplication operator $c_1(T_X) *_q (\bullet)$ acting on $H^\bullet(X, \mathbb{C}) \otimes \mathbb{C}[[u]]$. Some evidence for this comes from the observation that when $c_1(T_X)$ vanishes in $H^2(X, \mathbb{Z})$, then the Fukaya category is $\mathbb{Z}$-graded and thus is a fixed point of the renormalization group. There is also a more explicit direct argument identifying the class $c_1(T_X)$ with the infnitesimal generator of the renormalization group, but we will not discuss it here.

The formalism of Section 2.5.3 now predicts that the periodic cyclic homology of the Fukaya category, which additively should be the same as the de Rham cohomology of X, should carry a natural **nc**-Hodge structure satisfying the degeneration conjecture from Section 2.2.4. This expectation is supported by ample evidence coming from mirror symmetry for Calabi-Yau complete intersections. Here we present further evidence by describing a natural **nc**-Hodge structure on the de Rham cohomology of a symplectic manifold and by showing that as ω approaches the large volume limit this structure fits in a natural variation of **nc**-Hodge structures.

Using the quantum product $*_q$ we will attach to (X, ω) a variation $((\mathscr{H}, \nabla), \mathscr{E}_B,$ **iso**) of **nc**-Hodge structures over a small disc $\{q \in \mathbb{C} \,|\, |q| < r\}$ in the q-plane. First we describe the **nc**-Hodge filtration $(\mathscr{H}, \nabla)$ and its variation in the q-direction:

- $\mathscr{H} := H^\bullet(X, \mathbb{C}) \otimes \mathbb{C}\{u, q\}$ and

$$\mathscr{H}^0 := \left(\bigoplus_{k=d \mod 2} H^k(X, \mathbb{C}) \right) \otimes \mathbb{C}\{u, q\}$$

$$\mathscr{H}^1 := \left(\bigoplus_{k=d+1 \mod 2} H^k(X, \mathbb{C}) \right) \otimes \mathbb{C}\{u, q\}$$

- ∇ is a meromorphic connection on $\mathscr{H}$ with poles along the coordinate axes $u = 0$ and $q = 0$, given by

$$\nabla_{\frac{\partial}{\partial u}} := \frac{\partial}{\partial u} + u^{-2} (\kappa_X *_q \bullet) + u^{-1}\mathbf{Gr},$$

$$\nabla_{\frac{\partial}{\partial q}} := \frac{\partial}{\partial q} - q^{-1}u^{-1} ([\omega] *_q \bullet),$$

where:
$\kappa_X \in H^2(X, \mathbb{Z})$ denotes the first Chern class of the cotangent bundle of X computed w.r.t. any ω-compatible almost complex structure, and

$\mathbf{Gr} : \mathscr{H} \to \mathscr{H}$ is the grading operator, defined to be $\mathbf{Gr}_{|H^k(X,\mathbb{C})} := \frac{k-d}{2} \mathrm{id}_{H^k(X,\mathbb{C})}$.

The data $(\mathscr{H}, \nabla)$ define a q-variation of (the de Rham part of) **nc**-Hodge structures. Defining the $\mathbb{Q}$-structure is much more delicate. To gain some insight into the shape of the rational local system $\mathscr{E}_B$ one can look at the monodromy in the q direction of the algebraic bundle with connection

$$(H, \nabla)_{|(\mathbb{A}^1 - \{0\}) \times \{q \in \mathbb{C} | \ |q| < R\}}, \qquad (H, \nabla) = \mathfrak{B}_{\text{along } u}((\mathscr{H}, \nabla)).$$

In some cases the facts that $\mathscr{E}_B$ should be preserved by ∇ and that the Stokes filtration is rational with respect to $\mathscr{E}_B$ are enough to determine $\mathscr{E}_B$ completely:

PROPOSITION 3.1. *Let $X = \mathbb{CP}^{n-1}$ and let ω be the Fubini-Study form. Let (H, ∇) be the holomorphic bundle with meromorphic connection on $(\mathbb{A}^1 - \{0\}) \times \{q \in \mathbb{C} | \ |q| < R\}$ defined above. Let $\psi \in H$ be a holomorphic section which is covariantly constant with respect to ∇. Then*

(a) *For every $u \neq 0$, $\psi \neq 0$ the limit (in a sector of the q-plane)*

$$\psi_{\text{cl}}(u) = \lim_{q \to 0} \left(\exp \left(-\frac{\log(q)}{u} ([\omega] \wedge (\bullet)) \right) \right) \psi$$

exists. Furthermore, ψ_{cl} satisfies the differential equation

$$\left(\frac{d}{du} + u^{-2}\kappa_X \wedge + u^{-1}\mathbf{Gr} \right) \psi_{\text{cl}} = 0.$$

(b) *The vector*

$$\psi_{\text{const}}(u) := \exp(\log(u)\mathbf{Gr}) \exp \left(\frac{\log(u)}{u} \kappa_X \wedge (\bullet) \right) \psi_{\text{cl}} \in H^\bullet(X, \mathbb{C})$$

is independent of u.

(c) *Define the rational structure* $\mathscr{E}_B \subset H^\nabla$ *as the subsheaf of all covariantly constant sections* ψ *for which the vector* $\psi_{\mathrm{const}} \in H^\bullet(X, \mathbb{C})$ *belongs to the image of the map*

$$H^\bullet(X,\mathbb{Q}) \xrightarrow{\ \mathfrak{d}\ } H^\bullet(X,\mathbb{C}) \xrightarrow{\widehat{\Gamma}(X)\wedge(\bullet)} H^\bullet(X,\mathbb{C}),$$

where $\mathfrak{d} \in GL(H^\bullet(X,\mathbb{C}))$ *is the operator of multiplication by* $(2\pi i)^{k/2}$ *on* $H^k(X,\mathbb{C})$, *and* $\widehat{\Gamma}(X)$ *is a new characteristic class of* X *defined as*

$$\widehat{\Gamma}(X) := \exp\left(\boldsymbol{C}\,\mathrm{ch}_1(T_X) + \sum_{n\geq 2}\frac{\zeta(n)}{n}\mathrm{ch}_n(T_X)\right),$$

where

$$\boldsymbol{C} = \lim_{n\to\infty}\left(1 + \frac{1}{2} + \cdots + \frac{1}{n} - \ln(n)\right)$$

is Euler's constant, and $\zeta(s)$ *is Riemann's zeta function.*

Then the inclusion $\mathscr{E}_B \subset H^\nabla$ *is compatible with Stokes data, i.e. the rational structure* $\mathscr{E}_B$ *satisfies* **($\mathbb{Q}$-structure axiom)**$^{\boldsymbol{exp}}$.

The calculation presented below was known already to B.Dubrovin [**Dub98**, Section 4.2.1], where he also obtained a Taylor expansions of a power of a gamma function in quantum cohomology, although he did not identify it with a characteristic class.

Proof of Proposition 3.1. In the standard basis $\{1, h, h^2, \ldots, h^{n-1}\}$ of $H^\bullet(\mathbb{P}^{n-1},\mathbb{C})$ the connection ∇ on H is given by

$$\nabla_{\frac{\partial}{\partial u}} = \frac{\partial}{\partial u} + u^{-2}\begin{pmatrix} 0 & & & nq \\ n & 0 & & \\ & \ddots & \ddots & \\ & & n & 0 \end{pmatrix} + u^{-1}\begin{pmatrix} \frac{1-n}{2} & & & 0 \\ & \ddots & & \\ & & \ddots & \\ 0 & & & \frac{n-1}{2} \end{pmatrix}$$

$$\nabla_{\frac{\partial}{\partial q}} = \frac{\partial}{\partial q} - q^{-1}u^{-1}\begin{pmatrix} 0 & & & q \\ 1 & 0 & & \\ & \ddots & \ddots & \\ & & 1 & 0 \end{pmatrix}.$$

If $\psi = \sum_{i=1}^{n} \psi_i h^{i-1}$ is a local section of H, a straightforward check shows that the condition on ψ to be ∇-horizontal is solved by the following ansatz:

$$\psi_n = u^{\frac{1-n}{2}} \int\limits_{\Gamma_{u,q}} \exp(\mathcal{F}) \prod_{i=1}^{n-1} \frac{dz_i}{z_i}$$

$$\psi_{n-1} = \left(uq\frac{\partial}{\partial q} \right) \psi_n$$

$$\psi_{n-2} = \left(uq\frac{\partial}{\partial q} \right)^2 \psi_n$$

$$\cdots$$

$$\psi_1 = \left(uq\frac{\partial}{\partial q} \right)^{n-1} \psi_n.$$

Here $\mathcal{F}$ is the function on $(\mathbb{C}^{\times})^{n-1}$ with coordinates $z_1, \ldots, z_{n-1}$ depending on parameters $u, q \neq 0$ and given by

$$\mathcal{F}(z_1, z_2, \ldots, z_{n-1}; u, q) := u^{-1} \left(z_1 + z_2 + \cdots + z_{n-1} + \frac{q}{z_1 z_2 \cdots z_{n-1}} \right).$$

The integral is taken over some fixed $(n-1)$-dimensional semi-algebraic non-compact cycle $\Gamma_{u,q}$ in $(\mathbb{C}^{\times})^{n-1}$ (depending on the parameters u, q) which is going to infinity in directions where $\operatorname{Re}(\mathcal{F}) \to -\infty$.

More generally, the domain of integration $\Gamma_{u,q}$ used for defining ψ_n can be taken to be an $(n-1)$-dimensional rapid decay homology chain in $(\mathbb{C}^{\times})^{n-1}$. The rapid decay homology cycles on smooth complex algebraic varieties are the natural domains of integration for periods of cohomology classes of irregular connections. The rapid decay homology was introduced and studied by Hien [**Hie07, Hie08**], following previous works of Sabbah [**Sab00**] and Bloch-Esnault [**BE04**]. In particular by recent work of Mochizuki [**Moc08a, Moc08b**] and Hien [**Hie08**] it follows that (after a birational base change) taking periods induces a perfect pairing between the de Rham cohomology of an irregular connection and the rapid decay homology. This powerful general theory is not really needed in our case where the manifold is the affine algebraic torus $(\mathbb{C}^{\times})^{n-1}$, but it does provide a useful perspective.

Explicitly the non-compact cycles that we will use to generate horizontal sections of (H, ∇) will be the $(n-1)$-dimensional relative cycles for a pair $(\boldsymbol{\mathcal{X}}, Z)$ constructed as follows. Start with a smooth projective compactification $\mathfrak{X}$ of $(\mathbb{C}^{\times})^{n-1}$ with a normal crossing boundary divisor D which is adapted to $\mathcal{F}$ in the sense that if u and q are nonzero, the divisors of zeroes and poles of $\mathcal{F}$ in $\mathfrak{X}$ do not intersect with each other, and locally at points of D the function $\mathcal{F}$ can be written as a product of an invertible holomorphic function and a monomial in the local coordinates. Let $\boldsymbol{\mathcal{X}}$ be the real oriented blow-up of $\mathfrak{X}$ along the divisor D. Now consider the real boundary $\partial\boldsymbol{\mathcal{X}}$ of $\boldsymbol{\mathcal{X}}$, i.e. the union of all the boundary divisors of the real oriented blow-up. The boundary $\partial\boldsymbol{\mathcal{X}}$ contains a natural open real semi-algebraic subset $Z \subset \partial\boldsymbol{\mathcal{X}}$ consisting of all points $b \in \partial\boldsymbol{\mathcal{X}}$, such that $|\mathcal{F}(z; u, q)| \to \infty$ when $z \to b$, and for points $z \in t(\mathbb{C}^{\times})^{n-1}$ near b the argument of $\mathcal{F}(z; u, q)$ lies strictly in the left half-plane of $\mathbb{C}$. Note that the real blow-up X has the same homotopy type as $\mathfrak{X} - D = (\mathbb{C}^\times)^{n-1}$ and so relative cycles on $(\boldsymbol{\mathcal{X}}, Z)$ can be thought of as non-compact cycles on $(\mathbb{C}^\times)^{n-1}$. Moreover since Z is defined by our condition on the argument of $\mathcal{F}$, it follows that relative cycles with boundaries in Z give rise to well defined integrals of $\exp(\mathcal{F}) \prod z_i^{-1} dz_i$.

Next observe that the integrals over relative cycles with integral coefficients, i.e. elements in $H_{n-1}(\boldsymbol{\mathcal{X}}, Z; \mathbb{Z})$, give rise to a covariantly constant integral lattice in the bundle (H, ∇). Furthermore the Deligne-Malgrange-Stokes filtration is integral with respect to this lattice. Indeed if we fix a real number λ, then whenever $\mathrm{Re}\,(\mathcal{F}) < \lambda \cdot |u|^{-1}$, it follows that $|\exp(\mathcal{F})| < \exp(\lambda \cdot |u|^{-1})$ when $u \to 0$. Hence the steps of the Deligne-Malgrange-Stokes filtration of (H, ∇) are easy to describe in this language: they correspond to periods of $\exp(\mathcal{F}) \prod z_i^{-1} dz_i$ on relative cycles on $(\boldsymbol{\mathcal{X}}, Z)$ whose boundary is contained in half-planes of the form $\mathrm{Re}(\mathcal{F}) <$ const. The periods over cycles with integral coefficients and the same boundary property then give a full integral lattice in each such step.

Now to finish the proof of the proposition we just have to calculate the limiting lattice (which is independent of u and q) consisting of vectors $\psi_{\mathrm{const}} \in H^\bullet(X, \mathbb{C})$ defined in terms of ψ by the formula in part (b) of the statement of the proposition.

For a general ∇-horizontal local section $\psi = \sum_{i=1}^n \psi_i h^{i-1}$ in a sector at 0 in the q-plane (for given $u \neq 0$) one has an asymptotic expansion of ψ at $q \to 0$ given by:

$$\psi_n = \sum_{i=0}^{n-1} a_i(u)(\log q)^i + \boldsymbol{O}(q(\log q)^n) + \cdots \tag{3.1.2}$$

Then we have that the "classical limit" (at $q \to 0$ where the quantum multiplication becomes classical) is given by

$$\psi_{\mathrm{cl}}(u) = \begin{pmatrix} (n-1)!u^{n-1}a_{n-1}(u) \\ (n-2)!u^{n-2}a_{n-2}(u) \\ \vdots \\ 0!u^0 a_0(u) \end{pmatrix}.$$

Now we restrict to the case where all variables are real, $u < 0$, $q > 0$ and the contour of integration is being the positive octant $\{(z_1, \ldots, z_n) \in \mathbb{C}^n |\, z_i > 0\ \forall i\}$.

The function $\psi_n = \psi_n(u, q)$ decays exponentially fast at $q \to +\infty$ for a given $u < 0$, hence one can extract its asymptotic expansion at $q \to 0$ through the Mellin transform:

$$\int_0^{+\infty} \psi_n q^s \frac{dq}{q} = \sum_{i=0}^{\infty} a_i(u) \frac{i!(-1)^i}{s^{i+1}} + \boldsymbol{O}(1), \quad s \to 0.$$

This integral can be calculated explicitly

$$\int_0^{+\infty} \psi_n q^s \frac{dq}{q} = u^{\frac{1-n}{2}} \underbrace{\int_0^{+\infty} \cdots \int_0^{+\infty}}_{n \text{ times}} \frac{dq}{q} \prod_{i=1}^{n-1} \frac{dz_i}{z_i}$$

$$\cdot \exp\left(u^{-1}\left(z_1 + z_2 + \cdots + z_{n-1} + \frac{q}{z_1 z_2 \cdots z_{n-1}}\right)\right) q^s$$

$$= u^{\frac{1-n}{2}} \underbrace{\int_0^{+\infty} \cdots \int_0^{+\infty}}_{n-1 \text{ times}} \prod_{i=1}^{n-1} \frac{dz_i}{z_i} \exp\left(u^{-1} \sum_i z_i\right)$$

$$\cdot \underbrace{\int_0^{+\infty} \exp\left(\frac{q}{u z_1 z_2 \cdots z_{n-1}}\right) q^s \frac{dq}{q}}_{\substack{\| \\ \Gamma(s)(-u z_1 z_2 \cdots z_{n-1})^s}}$$

$$= u^{\frac{1-n}{2}} (-u)^s \Gamma(s) \int_0^{+\infty} \cdots \int_0^{+\infty} \prod_{i=1}^{n-1} \left(\frac{dz_i}{z_i} z_i^s \exp \frac{z_i}{u} \right)$$

$$= u^{\frac{1-n}{2}} (-u)^s \Gamma(s) \left((-u)^s \Gamma(s)\right)^{n-1}$$

$$= u^{\frac{1-n}{2}} (-u)^{ns} \Gamma(s)^n.$$

The conclusion is that the chosen branch $\psi_{\text{cl}}(u)$ is completely defined by the expansion

$$u^{\frac{1-n}{2}} (-u)^{ns} \Gamma(s)^n = \frac{\psi_{\text{cl},n}(u)}{(-u)^0 s} + \frac{\psi_{\text{cl},n-1}(u)}{(-u)^1 s^2} + \cdots + \frac{\psi_{\text{cl},1}(u)}{(-u)^{n-1} s^n} + \boldsymbol{O}(1), \quad s \to 0$$

Furthermore, all the other branches can be obtained by acting on the branch we know by the monodromy transformations (around $q = 0$)

$$\frac{(2\pi\sqrt{-1})^i}{u^i} \begin{pmatrix} 0 & & & 0 \\ 1 & 0 & & \\ & \ddots & \ddots & \\ & & 1 & 0 \end{pmatrix}^i,$$

for $i = 0, \ldots, n-1$.

Section ψ_{cl} satisfies the differential equation

$$\left(\frac{d}{du} + u^{-2} \kappa_X \wedge + u^{-1} \mathsf{Gr}\right) \psi_{\text{cl}} = 0.$$

which is the classical limit (at $q \to 0$) of the equation

$$\nabla_{\frac{\partial}{\partial u}} (\psi) = 0$$

One can check that the operator $\frac{d}{du} + u^{-2}\kappa_X \wedge + u^{-1}\mathsf{Gr}$ can be written as

$$\exp\left(-\frac{\log(u)}{u}\kappa_X \wedge (\bullet)\right) \exp(-\log(u)\mathsf{Gr}) \circ \frac{d}{du} \circ \exp(\log(u)\mathsf{Gr}) \exp\left(\frac{\log(u)}{u}\kappa_X \wedge (\bullet)\right)$$

This follows from the commutation relation

$$[\kappa_X \wedge (\bullet), \mathsf{Gr}] = -\kappa_X \wedge (\bullet)$$

Finally, in the above formulas one can replace $\log(u)$ by $\log(-u)$ (and also $u^{\frac{1-n}{2}}$ by $(-u)^{\frac{1-n}{2}}$) with principal values at the domain $u < 0$. Having this modification in mind, we conclude that the vector

$$\psi_{\text{const}} = \psi_{\text{const}}(u) := \exp(\log(-u)\mathsf{Gr}) \exp\left(\frac{\log(-u)}{u}\kappa_X \wedge (\bullet)\right) \psi_{\text{cl}} \in H^\bullet(X, \mathbb{C})$$

is independent of u, and in particular it coincides with $\psi_{\text{cl}}(-1)$, as for $u = -1$ the correction matrices relating $\psi_{\text{const}}(u)$ and $\psi_{\text{cl}}(u)$ are identity matrices. Therefore the vector ψ_{const} is given by Taylor coefficients

$$\psi_{\text{const},1} s^0 + \cdots + \psi_{\text{const},n} s^{n-1} = s^n \Gamma(s)^n + O(s^n) = \Gamma(1+s)^n + O(s^n)$$

We see that $\psi_{\text{const}} \in H^\bullet(X, \mathbb{C})$ (after rescaling by the operator $\mathfrak{d}$ from the Proposition) with the value of the multiplicative characteristic class associated to the series $\Gamma(1+s) = 1 + O(s) \in \mathbb{C}[[s]]$ and the tangent bundle T_X, because $[T_X] = n[\mathcal{O}(1)] - [\mathcal{O}]$ for $X = \mathbb{CP}^n$, and by the classical expansion

$$\log(\Gamma(1+s)) = Cs + \sum_{k \geq 2} \frac{\zeta(k)}{k} s^k$$

The action of the monodromy corresponds (up to torsion) to the multiplication by $\kappa_X \in H^\bullet(X, \mathbb{Z})$. □

The previous proposition suggests the following general definition:

DEFINITION 3.2. *The rational structure on (H, ∇) is the local subsystem $\mathcal{E}_B \subset H_{|\mathbb{A}^1 - \{0\}}$ of multivalued ∇-horizontal sections whose values at 1 belong to the image of*

$$H^\bullet(X, \mathbb{Q}) \xrightarrow{\ \mathfrak{d}\ } H^\bullet(X, \mathbb{C}) \xrightarrow{\widehat{\Gamma}(TX) \wedge (\bullet)} H^\bullet(X, \mathbb{C}),$$

where $\mathfrak{d} \in GL(H^\bullet(X, \mathbb{C}))$ is the operator of multiplication by $(2\pi i)^{k/2}$ on $H^k(X, \mathbb{C})$, and $\widehat{\Gamma}(TX)$ is a new characteristic class of X defined as

$$\widehat{\Gamma}(TX) := \prod_{i=1}^{d} \Gamma(1 + \lambda_i),$$

where $\Gamma(s)$ is the classical gamma function and λ_i are the Chern roots of T_X computed in any ω-admissible almost complex structure.

REMARK 3.3. Apart from the calculation in Proposition 3.1 there are a few other (loose) motivations for this definition:

- The class $\widehat{\Gamma}$ appears in the context of deformation quantization in the work of the second author [**Kon99**, Section 4.6].
- The number $\chi(X)\zeta(3)$ appears in the mirror formula for the quintic threefold.
- Golyshev's description [**Gol01, Gol07**] of the **nc**-motives associated with the Landau-Ginzburg mirror of a toric Fano involves similar hypergeometric series.

• The same class $\widehat{\Gamma}$ was derived and a definition similar to Definition 3.2 was proposed in the recent work of Iritani [**Iri07**] for the case of toric orbifolds by tracing out the mirror image of rational structure of the mirror Landau-Ginzburg model.

CONJECTURE 3.4. *The triple* $(H, \mathscr{E}_B, \mathsf{iso})$ *associated above with a symplectic manifold* (X, ω) *is a variation of* **nc**-*Hodge structures of exponential type.*

REMARK 3.5. **(i)** In general it is not clear if the **($\mathbb{Q}$-structure axiom)^{exp}** holds in this case. It does hold trivially in the graded case, i.e. when X is a Calabi-Yau.
(ii) At the moment the "exponential type" part of the conjecture is not supported by any evidence beyond the graded case in which the **nc**-Hodge structure is regular. It is possible that for non-Kähler symplectic manifolds the **nc**-Hodge structure on the de Rham cohomology is not of exponential type.

3.2. *B*-model Hodge structures: holomorphic Landau-Ginzburg models. Suppose we have an algebraic map $\mathsf{w} : Y \to \mathbb{C}$, where Y is a smooth quasi-projective manifold and w has a compact critical locus $\mathrm{crit}(\mathsf{w}) \subset Y$. Let $S = \{\boldsymbol{c}_1, \ldots, \boldsymbol{c}_m\} \subset \mathbb{C}$ denote the critical values of w.

A pair (Y, w) like that is called a ***holomorphic Landau-Ginzburg model*** and often arises (see e.g. [**HV00, HKK$^+$03**]) as the mirror of a symplectic manifold underlying a hypersurface, or a complete intersection in a toric variety. Remarkably the pair (Y, w) gives rise to a natural **nc**-space $\mathbf{nc}(Y, \mathsf{w})$. The category $C_{\mathbf{nc}(Y,\mathsf{w})}$ can be described in two equivalent ways (in fact these descriptions are valid even if the critical locus of w is not compact). First note that it is enough to define $\mathsf{Perf}_{C_{\mathbf{nc}(Y,\mathsf{w})}}$ since the category $C_{\mathbf{nc}(Y,\mathsf{w})}$ can be thought of as the homotopy colimit completion of $\mathsf{Perf}_{C_{\mathbf{nc}(Y,\mathsf{w})}}$. For the latter we have two models:

$\mathsf{Perf}_{C_{\mathbf{nc}(Y,\mathsf{w})}}$ **as a category of matrix factorizations:** This model was proposed originally by the second author as a mathematical description of the category of D-branes and was subsequently studied extensively in the physics and mathematics literature, see [**KL03, KL04**] and [**Orl04, Orl05b, Orl05a**].

A matrix factorization on (Y, w) is a pair $\left(E = E^0 \oplus E^1, d_E \in \mathrm{End}(E)^{\mathrm{opp}}\right)$, where

E is a $\mathbb{Z}/2$-graded algebraic vector bundle on Y, and
d_E is an odd endomorphism satisfying $d_E^2 = \mathsf{w} \cdot \mathrm{id}_E$.

In the case when Y is affine the $\mathbb{Z}/2$-graded complex $\underline{\mathrm{Hom}}((E, d_E), (F, d_F))$ of homomorphisms between two matrix factorizations is defined as $\underline{\mathrm{Hom}}((E, d_E), (F, d_F)) := (\mathrm{Hom}(E, F), d)$ where for a $\varphi : E \to F$ we have $d\varphi := \varphi \circ d_E - d_F \circ \varphi$. For general Y the same definition works if we replace $\mathrm{Hom}(E, F)$ by some acyclic model, e.g. if we use the Dolbeault resolution. The resulting category $\mathsf{MF}(Y, \mathsf{w})$ of matrix factorizations is a $\mathbb{C}$-linear $\mathrm{d}(\mathbb{Z}/2)\mathrm{g}$ category. We define $\mathsf{Perf}_{C_{\mathbf{nc}(Y,\mathsf{w})}}$ to be the derived category $D^b(\mathsf{MF}(Y, \mathsf{w}))$ of the category of matrix factorizations.

To construct $D^b(\mathsf{MF}(Y, \mathsf{w}))$ one notes that in addition to being a $\mathrm{d}(\mathbb{Z}/2)\mathrm{g}$ category $\mathsf{MF}(Y, \mathsf{w})$ can also be viewed as a curved $\mathrm{d}(\mathbb{Z}/2)\mathrm{g}$ category with central curvature w (see e.g. [**PP05**] for the definition) or as a $\mathbb{Z}/2$-graded weak A_∞-category, i.e. an A_∞ category with an m_0-operation given by w

(see e.g. [**Sch03, LH03**] for the definition). In particular we can form the associated homotopy category (in the A_∞-sense) which by definition will be the derived category of matrix factorizations.

Alternatively, one can use the following two step construction proposed by Orlov. First we pass to the homotopy category of $\mathsf{MF}(Y, \mathsf{w})$, i.e. we consider the category whose objects are matrix factorizations and whose morphisms are given by the quotient of $\underline{\mathrm{Hom}}((E, d_E), (F, d_F))$ by homotopy equivalences. Next (following the standard wisdom) we need to quotient $\mathrm{Ho}(\mathsf{MF}(Y, \mathsf{w}))$ by the subcategory of acyclic factorizations. Since the matrix factorizations are not complexes, they do not have cohomology and so we can not define acyclicity in the usual way. But there is another point of view on acyclicity. If we have a short exact sequence of usual complexes, then the total complex of this diagram will be an acyclic complex. So we define ***acyclic*** matrix factorizations as the total matrix factorization of an exact sequence of factorizations. With this definition we get a thick subcategory in the homotopy category $\mathrm{Ho}(\mathsf{MF}(Y, \mathsf{w}))$ of matrix factorizations and then we can pass to the Serre quotient of $\mathrm{Ho}(\mathsf{MF}(Y, \mathsf{w}))$ by this thick subcategory. We set $D^b(\mathsf{MF}(Y, \mathsf{w}))$ to be this Serre quotient.

$\mathsf{Perf}_{C_{\mathsf{nc}(Y,\mathsf{w})}}$ **as a category of singularities:** This model was proposed originally by D. Orlov as an alternative to the matrix factorization description which is localized near the critical set of w. Orlov proved the equivalence of the two models, various versions of the localization theorem, and proved several duality statements relating derived categories of singularities to other familiar categories [**Orl04, Orl05b, Orl05a**].

Suppose Z is a quasi-projective complex scheme. The ***derived category*** $D^b_{\mathrm{Sing}}(Z)$ ***of singularities*** of Z is defined as the quotient

$$D^b_{\mathrm{Sing}}(Z) := D^b(\mathsf{Coh}(Z))/\,\mathsf{Perf}_Z$$

of the (dg enhancement of the) bounded derived category $D^b(\mathsf{Coh}(Z))$ of coherent sheaves on Z by the thick subcategory of perfect complexes on Z. The syzygy theorem implies that $D^b_{\mathrm{Sing}}(Z) = 0$ whenever Z is smooth and so $D^b(\mathsf{Coh}(Z))$ can be thought of as an invariant of the singularities of Z.

If now $\mathsf{w} : Y \to \mathbb{C}$ is a holomorphic Landau-Ginzburg model we write Y_c for the fiber $\mathsf{w}^{-1}(c)$ and set

$$\mathsf{Perf}_{C_{\mathsf{nc}(Y,\mathsf{w})}} := D^b_{\mathrm{Sing}}(Y_0).$$

Note that if $0 \in \mathbb{A}^1$ is not a critical value of w, then with this definition we will get $\mathsf{Perf}_{C_{\mathsf{nc}(Y,\mathsf{w})}} = 0$. In order to get non-trivial categories we will use the critical values $S = \{\boldsymbol{c}_1, \ldots, \boldsymbol{c}_n\}$ to shift the potential $\mathsf{w} \rightsquigarrow \mathsf{w} - \boldsymbol{c}_i$ and associate with $\mathsf{nc}(Y, \mathsf{w})$ honest categories $\mathsf{Perf}_i := \mathsf{Perf}_{C_{\mathsf{nc}(Y,\mathsf{w}-\boldsymbol{c}_i)}} = D^b_{\mathrm{Sing}}(Y_{\boldsymbol{c}_i})$. Conjecturally, these categories are smooth and compact.

Mirror symmetry suggests that the nc-space $\mathsf{nc}(Y, \mathsf{w})$ gives rise to the B-model geometric and Hodge theoretic data described in Section 2.5.2, and in particular that the periodic cyclic homology of $C_{\mathsf{nc}(Y,\mathsf{w})}$ carries a canonical nc-Hodge structure. In fact we have already described the geometric part of the data, namely the numbers

$\{c_i\}$ and the categories $\{\mathsf{Perf}_i\}$. These data of course fix the regular type (in the sense of Theorem 2.35) of the **nc**-Hodge structure but we are still missing the gluing data. Here we propose a construction of the Hodge structure on the periodic cyclic homology of $C_{\mathsf{nc}(Y,\mathsf{w})}$ but similarly to the A-model we have to rely on the actual geometry of (Y, w) in order to produce the gluing data. At present it is not clear if the gluing data can be reconstructed from the category $C_{\mathsf{nc}(Y,\mathsf{w})}$ or more generally from its one-parameter deformation.

First we discuss the appropriate cohomologies of the Landau-Ginzburg model. Let

$$\begin{aligned}\mathscr{H}^\bullet_{\text{for}} &:= H^\bullet_{\text{DR}}((Y,\mathsf{w});\mathbb{C})\\ &= \mathbb{H}^{\bullet \text{ mod } 2}_{\text{Zar}}(Y, (\Omega^\bullet_Y[[u]], ud_{\text{DR}} + d\mathsf{w}\wedge))\end{aligned}$$

be the $\mathbb{Z}/2$-graded $\mathbb{C}[[u]]$-module of algebraic de Rham cohomology of the potential w. In the case when $\text{crit}(\mathsf{w})$ is compact, the $\mathbb{C}[[u]]$-module $\mathscr{H}^\bullet_{\text{for}}$ is known to be free by the work of Barannikov and the second author (unpublished), Sabbah [**Sab99**], or Ogus-Vologodsky [**OV05**]. This implies the following

LEMMA 3.6. *Assume that Y is quasi-projective and the critical locus of w is compact. Then we have:*

(i) *The fiber of $\mathscr{H}^\bullet_{\text{for}}$ at $u = 0$ is the algebraic Dolbeault cohomology*

$$\mathbb{H}^\bullet_{Zar}(Y, (\Omega^\bullet_Y, d\mathsf{w}\wedge)) \cong \mathbb{H}^\bullet_{an}(Y, (\Omega^\bullet_Y, d\mathsf{w}\wedge))$$

of the potential w.

(ii) *There is a canonical isomorphism*

$$\mathbb{H}^\bullet_{Zar}(Y, (\Omega^\bullet_Y[[u]], ud_{\text{DR}} + d\mathsf{w}\wedge)) \cong \mathbb{H}^\bullet_{an}(Y, (\Omega^\bullet_Y[[u]], ud_{\text{DR}} + d\mathsf{w}\wedge))$$

(iii) *If the map w is proper then $\mathscr{H}^\bullet_{\text{for}}$ is the formal germ at $u = 0$ of an algebraic vector bundle on the affine line*

$$\mathscr{H}^\bullet_{\text{alg}} := \mathbb{H}^{\bullet \text{ mod } 2}_{Zar}(Y, (\Omega^\bullet_Y[u], ud_{\text{DR}} + d\mathsf{w}\wedge))$$

Proof. The cohomology sheaves of the complex $(\Omega^\bullet_Y, d\mathsf{w}\wedge)$ are supported on the critical locus of w and so, by our compactness assumption, must be coherent sheaves on Y both in the analytic and in the Zariski topology. The hypercohomology spectral sequence then implies that the hypercohomology of the complex $(\Omega^\bullet_Y, d\mathsf{w}\wedge)$ is finite dimensional and the spectral sequence associated with the filtration induced by multiplication by u implies that $\mathbb{H}^\bullet_{\text{Zar/an}}(Y, (\Omega^\bullet_Y[[u]], ud_{\text{DR}} + d\mathsf{w}\wedge))$ is a finite rank $\mathbb{C}[[u]]$-module. Furthermore, the same spectral sequence implies that

$$\dim_{\mathbb{C}((u))} \mathbb{H}^\bullet_{\text{Zar/an}}(Y, (\Omega^\bullet_Y((u)), ud_{\text{DR}} + d\mathsf{w}\wedge)) \leq \dim_{\mathbb{C}} \mathbb{H}^\bullet_{\text{Zar/an}}(Y, (\Omega^\bullet_Y, d\mathsf{w}\wedge)).$$

The freeness statement of Barannikov and the second author (see e.g. [**Sab99**]) now gives that these two dimensions are equal and so $\mathbb{H}^\bullet_{\text{Zar}}(Y, (\Omega^\bullet_Y[[u]], ud_{\text{DR}} + d\mathsf{w}\wedge))$ is a free finite rank module over $\mathbb{C}[[u]]$. This proves part **(i)** of the lemma.

For part **(ii)** we only need to notice that the two spaces in question are computed by spectral sequences associated with the filtrations by the powers of u and that these spectral sequences have E_2-levels whose entries are finite sums of copies of $\mathbb{H}^\bullet_{\text{Zar}}(Y, (\Omega^\bullet_Y, d\mathsf{w}\wedge))$ and $\mathbb{H}^\bullet_{\text{an}}(Y, (\Omega^\bullet_Y, d\mathsf{w}\wedge))$, respectively. Each of these can in turn be computed from the hypercohomology spectral sequence for the complex $(\Omega^\bullet_Y, d\mathsf{w}\wedge)$ of (Zariski or analytic) coherent sheaves. But the cohomology sheaves of this complex are supported on the zero locus of $d\mathsf{w}$ which by assumption is projective. Hence by GAGA the Zariski and analytic cohomologies of this complex are naturally isomorphic. This gives isomorphisms of the hypercohomology and filtration spectral sequences in the Zariski and the analytic setup, respectively, and so the two types of hypercohomologies are isomorphic.

Finally, part **(iii)** was also proven by Barannikov and the second author, and by Sabbah [**Sab99**]. □

REMARK 3.7. The isomorphism in part **(ii)** of the previous lemma is not convergent for $u \to 0$ in general. Indeed if $u \neq 0$ is a complex number, then the complex vector space $\mathbb{H}^{\bullet}_{\mathrm{an}}(Y, (\Omega^{\bullet}_Y, ud_{\mathrm{DR}} + d\mathsf{w}\wedge))$ is the same as the usual de Rham cohomology $H^{\bullet}_{\mathrm{DR}}(Y, \mathbb{C})$ of Y. Indeed, for such a fixed $u \neq 0$, the complex $(\Omega^{\bullet}_Y, ud_{\mathrm{DR}} + d\mathsf{w}\wedge) \cong (\Omega^{\bullet}_Y, d_{\mathrm{DR}} + u^{-1}d\mathsf{w}\wedge)$ is the holomorphic de Rham complex of the local system $(\mathcal{O}_Y, d_{\mathrm{DR}} + u^{-1}d\mathsf{w})$. But the multiplication by $\exp(-u^{-1}\mathsf{w})$ is an analytic automorphism of the line bundle $\mathcal{O}_Y$ which gauge transforms the connection $d_{\mathrm{DR}} + u^{-1}d\mathsf{w}$ into the trivial connection d_{DR}. Hence $\exp(-u^{-1}\mathsf{w})$ identifies $(\Omega^{\bullet}_Y, ud_{\mathrm{DR}} + d\mathsf{w}\wedge)$ with the holomorphic de Rham complex $(\Omega^{\bullet}_Y, d_{\mathrm{DR}})$ and $\mathbb{H}^{\bullet}_{\mathrm{an}}(Y, (\Omega^{\bullet}_Y, ud_{\mathrm{DR}} + d\mathsf{w}\wedge))$ with $H^{\bullet}_{\mathrm{DR}}(Y, \mathbb{C})$. On the other hand, the space $\mathbb{H}^{\bullet}_{\mathrm{Zar}}(Y, (\Omega^{\bullet}_Y, ud_{\mathrm{DR}} + d\mathsf{w}\wedge))$ depends on the potential in an essential way. For instance, if $\mathsf{w} : Y \to \mathbb{A}^1$ is a Lefschetz fibration, then the complex $(\Omega^{\bullet}_Y, d\mathsf{w}\wedge)$ is just the Koszul complex associated with the regular section $d\mathsf{w} \in \Omega^1_Y$. In particular the space $\mathbb{H}^{\bullet}_{\mathrm{Zar}}(Y, (\Omega^{\bullet}_Y, d_{\mathrm{DR}} + d\mathsf{w}\wedge)) \cong \mathbb{H}^{\bullet}_{\mathrm{Zar}}(Y, (\Omega^{\bullet}_Y, d\mathsf{w}\wedge))$ has dimension equal to the number of critical points of w. More generally $\mathbb{H}^{\bullet}_{\mathrm{Zar}}(Y, (\Omega^{\bullet}_Y, d_{\mathrm{DR}} + d\mathsf{w}\wedge))$ can be identified (see e.g. [**Kap91**]) with the cohomology of the perverse sheaf of vanishing cycles of w.

REMARK 3.8. Under our assumptions, the algebraic de Rham and Dolbeault cohomologies $H^{\bullet}_{\mathrm{DR}}((Y, \mathsf{w}); \mathbb{C})$ and $H^{\bullet}_{\mathrm{Dol}}((Y, \mathsf{w}); \mathbb{C})$ of the potential w can be identified respectively with the periodic cyclic and Hochschild homologies $HP_{\bullet}(C_{\mathsf{nc}(Y,\mathsf{w})})$ and $HH_{\bullet}(C_{\mathsf{nc}(Y,\mathsf{w})})$ of the **nc**-space $C_{\mathsf{nc}(Y,\mathsf{w})}$ (more precisely, of the collection of categories Perf_i labeled by numbers $\{c_i\}$). This can be done, e.g. by choosing strong generators $\mathcal{E}_i$ of Perf_i, and then identifying $HP_{\bullet}(C_{\mathsf{nc}(Y,\mathsf{w})})$ and $HH_{\bullet}(C_{\mathsf{nc}(Y,\mathsf{w})})$ with the periodic cyclic and Hochschild homologies of the curved d$(\mathbb{Z}/2)$g algebra, which consists of the d$(\mathbb{Z}/2)$g algebra $R\mathrm{Hom}(\mathcal{E}, \mathcal{E})$ and a central curvature given by w. A detailed proof of the comparison theorem giving the identifications $H^{\bullet}_{\mathrm{DR}}((Y, \mathsf{w}); \mathbb{C}) \cong HP_{\bullet}(C_{\mathsf{nc}(Y,\mathsf{w})})$ and $H^{\bullet}_{\mathrm{Dol}}((Y, \mathsf{w}); \mathbb{C}) \cong HH_{\bullet}(C_{\mathsf{nc}(Y,\mathsf{w})})$ can be found in the recent work of Junwu Tu [**Tu08**].

We will construct an **nc**-Hodge structure on $H^{\bullet}_{\mathrm{DR}}((Y, \mathsf{w}); \mathbb{C})$ by using the dual description of **nc**-Hodge structures given in Theorem 2.35. Here we will assume that we choose an open subset (in the analytic topology) $Y' \subset Y$ such that

- $\mathrm{crit}(\mathsf{w}) \subset Y'$,
- $\mathsf{w}(Y')$ is an open disc in $\mathbb{C}$,
- the closure $\overline{Y}'$ of Y' is a manifold with corners,
- the restriction of w to the part of the boundary of $\overline{Y}'$ lying over $\mathsf{w}(Y')$ is a smooth fibration.

In the case when w is already proper one can choose Y' to be the pre-image under w of an open disc in $\mathbb{C}$ containing all the critical values $\boldsymbol{c}_i$.

Label the critical values of w: $S = \{\boldsymbol{c}_1, \ldots, \boldsymbol{c}_n\}$, and let $\boldsymbol{c}_0 \in \mathsf{w}(Y') - S$. Choose a system of paths $\{a_i\}_{i=1}^n$ and discs $\{\boldsymbol{D}\}_{i=1}^n$ as in the proof of Theorem 2.35. Choose $\boldsymbol{c}_0$-based loops $\gamma_1, \ldots, \gamma_n$, so that γ_i goes once around c_i in the counterclockwise direction, all γ_i intersect only at $\boldsymbol{c}_0$, and each γ_i encloses the path a_i and the disc $\boldsymbol{D}_i$ (see Figure 4). Let Γ_i denote the closed region in $\mathbb{C}$ enclosed by γ_i. Adjusting if necessary the choice of the γ_i we can ensure also that each Γ_i is convex. From now on we will always assume that this is the case.

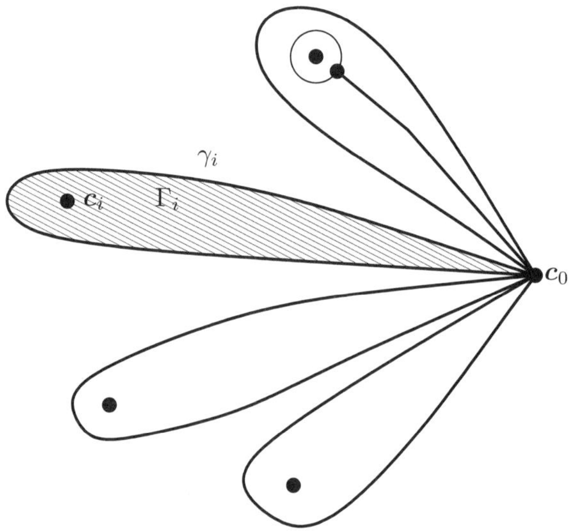

FIGURE 4. A system of thickened loops for $S \subset \mathbb{C}$.

For $i = 1, \ldots, n$ set $Y_i := \mathsf{w}^{-1}(\Gamma_i) \cap Y'$ and consider the $\mathbb{Q}$-vector spaces of relative cohomology

$$U_i := H^\bullet(Y_i, Y_{\boldsymbol{c}_0}; \mathbb{Q}),$$

and

$$\begin{aligned} U &:= \oplus_{i=1}^n U_i \\ &= H^\bullet(\mathsf{w}^{-1}(\cup_{i=1}^n \Gamma_i), Y_{\boldsymbol{c}_0}; \mathbb{Q}) \\ &= H^\bullet(Y, Y_{\boldsymbol{c}_0}; \mathbb{Q}). \end{aligned}$$

Let $T_i : U \to U$ be the monodromy along γ_i. By definition T_i satisfies

$$(T_i - 1)_{|\oplus_{j \neq i} U_j} = 0$$

and so we get operators $T_{ji} : U_i \to U_j$, such that $T_{i|U_i} = \sum_{j=1}^n T_{ji}$. By construction the operator T_{ii} is the monodromy along γ_i of the local system on Γ_i of local relative cohomology, i.e. the local system of $\mathbb{Q}$-vector spaces whose fiber at $\boldsymbol{c} \in \Gamma_i$ is $H^\bullet(Y_i, Y_{\boldsymbol{c}}; \mathbb{Q})$. Hence T_{ii} is an isomorphism, and so the data $(S, \{U_i\}_{i=1}^n, \{T_{ij}\})$ are **nc**-Betti data of type **ncB(iii)**.

REMARK 3.9. (a) By Lemma 2.30 the data $(S, \{U_i\}_{i=1}^n, \{T_{ij}\})$ are the same thing as a constructible sheaf $\mathscr{F}$ of $\mathbb{Q}$-vector spaces on $\mathbb{C}$, satisfying $R\Gamma(\mathbb{C}, \mathscr{F}) = 0$.

The sheaf $\mathscr{F}$ can be described directly in terms of the geometry of (Y, w): for a $\boldsymbol{c} \in \mathbb{C}$ the stalk $\mathscr{F}_{\boldsymbol{c}}$ of $\mathscr{F}$ at $\boldsymbol{c}$ is the relative cohomology $H^\bullet(Y, Y_{\boldsymbol{c}}; \mathbb{Q})$.

(b) The geometric construction of $\mathscr{F}$ makes sense for every cohomology theory K. Indeed for every such K we can form a constructible sheaf of abelian groups ${}^{\mathsf{K}}\mathscr{F}$ whose stalk at $\boldsymbol{c} \in \mathbb{C}$ is $\mathsf{K}(Y, Y_{\boldsymbol{c}})$ and which again satisfies $R\Gamma(\mathbb{C}, {}^{\mathsf{K}}\mathscr{F}) = 0$. The vanishing of cohomology here is not obvious but can be proven as follows. Given a disk $\boldsymbol{D} \subset \mathsf{w}(Y') \subset \mathbb{C}$ such that $\partial \boldsymbol{D} \cap S = \varnothing$, and given any point $\boldsymbol{c} \in \partial \boldsymbol{D}$, consider the abelian group $A(\boldsymbol{D}, \boldsymbol{c}) := \mathsf{K}(\mathsf{w}^{-1}(\boldsymbol{D}), Y_{\boldsymbol{c}})$. The collection of abelian groups $A(\boldsymbol{D}, \boldsymbol{c})$ satisfies:

- $A(\boldsymbol{D}, \boldsymbol{c})$ are locally constant under small perturbations of $(\boldsymbol{D}, \boldsymbol{c})$, and
- for every decomposition $(\boldsymbol{D}, \boldsymbol{c}) = (\boldsymbol{D}_1, \boldsymbol{c}) \cup (\boldsymbol{D}_2, \boldsymbol{c})$ of $\boldsymbol{D}$ obtained by cutting D along a chord starting at c, we have $A(\boldsymbol{D}, \boldsymbol{c}) = A(\boldsymbol{D}_1, \boldsymbol{c}) \oplus A(\boldsymbol{D}_2, \boldsymbol{c})$.

This immediately gives us an equivalent description of ${}^{\mathsf{K}}\mathscr{F}$ via data of type **ncB(iii)**, which in turn yields the vanishing of cohomology of ${}^{\mathsf{K}}\mathscr{F}$.

Next, in order to complete the data **ncB(iii)** to a full-fledged **nc**-Hodge structure of exponential type, we need to construct:

- a collection $\{(\mathscr{R}_i, \nabla_i)\}_{i=1}^m$ of holomorphic bundles $\mathscr{R}_i$ over $\mathbb{C}\{u\}$ equipped with meromorphic connections ∇_i with at most second order poles and regular singularities, and
- for each $i = 1, \ldots, m$, an isomorphism $\mathfrak{f}_i$ between the local system on $\boldsymbol{S}^1$ induced from $(\mathscr{R}_i, \nabla_i)$ and the local system on $\boldsymbol{S}^1$ corresponding to the vector space $U_i \otimes \mathbb{C}$ and the monodromy operator T_{ii}.

As explained above the local system on the circle corresponding to the vector space $U_i \otimes \mathbb{C}$ and the monodromy operator T_{ii} can be described geometrically as the sheaf of complex vector spaces on the loop γ_i, whose stalk at $\boldsymbol{c} \in \gamma_i$ is $H^\bullet((Y_i, Y_{\boldsymbol{c}}); \mathbb{C})$. We will exploit this geometric picture to produce $(\mathscr{R}_i, \nabla_i)$ and the isomorphism $\mathfrak{f}_i$. The most convenient way to define the ∇_i is by using a Betti-to-de Rham cohomology isomorphism given by oscillating integrals.

Fix $i \in \{1, \ldots, m\}$ and let $Z := Y_i$, $\boldsymbol{\Delta} := \Gamma_i - \boldsymbol{c}_i \subset \mathbb{C}$, $\boldsymbol{f} := \mathsf{w} - \boldsymbol{c}_i$. By construction we have:

Z is a C^∞-manifold with boundary which is the closure of an open (in the classical topology) subset in the quasiprojective complex manifold Y.

$\boldsymbol{\Delta} \subset \mathbb{C}$ is a closed disc containing zero.

$\boldsymbol{f} : Z \to \boldsymbol{\Delta}$ is an analytic surjective map whose only critical value is zero and whose critical locus $\mathrm{crit}(\boldsymbol{f}) \subset Z$ is compact.

Consider now the $\mathbb{Z}/2$-graded $\mathbb{C}[[u]]$-module $H^\bullet_{\mathrm{DR}}((Z, \boldsymbol{f}); \mathbb{C})$ of de Rham cohomology of $(Z, \boldsymbol{f})$. By Lemma 3.6 we know that $H^\bullet_{\mathrm{DR}}((Z, \boldsymbol{f}); \mathbb{C})$ is a free $\mathbb{C}[[u]]$-module which can be computed as the cohomology of the complex $(\mathcal{A}^\bullet(Z)[[u]], d_{\mathrm{tot}})$, where $\mathcal{A}^\bullet(Z)[[u]]$ are the global C^∞ complex-valued differential forms on Z, and

$d_{\text{tot}} := \bar{\partial} + u\partial + d\boldsymbol{f}\wedge$. The $\mathbb{C}[[u]]$-module $H^{\bullet}_{\text{DR}}((Z,\boldsymbol{f});\mathbb{C})$ carries a natural meromorphic connection ∇ differentiating in the u-direction and having a second order pole at $u = 0$. This connection is induced from a connection $\boldsymbol{\nabla}$ on the $\mathbb{C}[[u]]$-module $\mathcal{A}^{\bullet}(Z)[[u]]$ which also has a second order pole and is defined by the formula

$$\boldsymbol{\nabla}_{u^2\frac{d}{du}} := u^2\frac{d}{du} - \boldsymbol{f}\cdot(\bullet) + u\mathsf{Gr} : \mathcal{A}^{\bullet}(Z)[[u]] \longrightarrow \mathcal{A}^{\bullet}(Z)[[u]],$$

where

$$\mathsf{Gr}_{|\mathcal{A}^{p,q}(Z)[[u]]} := \frac{q-p}{2}\cdot \mathrm{id}_{\mathcal{A}^{p,q}(Z)[[u]]}$$

is the grading operator coming from **nc**-geometry (compare with 2.1.7).

With this definition we have

LEMMA 3.10. *The operator* $\boldsymbol{\nabla}_{u^2\frac{d}{du}}$ *satisfies:*

(a) $\left[\boldsymbol{\nabla}_{u^2\frac{d}{du}}, d_{\text{tot}}\right] = \frac{u}{2}\cdot d_{\text{tot}}$.

(b) $\boldsymbol{\nabla}_{u^2\frac{d}{du}}$ *preserves* $\ker(d_{\text{tot}})$ *and* $\operatorname{im}(d_{\text{tot}})$ *and so induces a meromorphic connection* ∇ *with a second order pole on the* $\mathbb{C}[[u]]$*-module* $H^{\bullet}_{\text{DR}}((Z,\boldsymbol{f});\mathbb{C})$.

Proof. We compute

$$\begin{aligned}
\left[\boldsymbol{\nabla}_{u^2\frac{d}{du}}, d_{\text{tot}}\right] &= \left[u^2\frac{d}{du} - \boldsymbol{f} + u\mathsf{Gr}, \bar{\partial} + u\partial + d\boldsymbol{f}\wedge\right] \\
&= \left[u^2\frac{d}{du}, u\partial\right] - [\boldsymbol{f}, u\partial] + \left[u\mathsf{Gr}, \bar{\partial} + u\partial + d\boldsymbol{f}\wedge\right] \\
&= u^2\partial + ud\boldsymbol{f}\wedge + \frac{u\bar{\partial}}{2} - \frac{ud\boldsymbol{f}\wedge}{2} - \frac{u^2\partial}{2} \\
&= \frac{u}{2}\cdot d_{\text{tot}}.
\end{aligned}$$

Part **(b)** follows immediately from **(a)** □

Suppose now that $\alpha = \alpha_0 + \alpha_1 u + \alpha_2 u^2 + \cdots \in \mathcal{A}^{\bullet}(Z)[[u]]$, $\alpha_i = \sum \alpha_i^{p,q}$, $\alpha_i^{p,q} \in \mathcal{A}^{p,q}(Z)$ is a d_{tot}-cocycle. Then the differential $d + u^{-1}d\boldsymbol{f}\wedge = \bar{\partial} + \partial + u^{-1}d\boldsymbol{f}\wedge = u^{-1/2}u^{\mathsf{Gr}}d_{\text{tot}}u^{-\mathsf{Gr}}$ will kill the element

$$u^{\mathsf{Gr}}\alpha := \sum_{\substack{i\geq 0 \\ 0\leq p,q\leq \dim Z}} \alpha_i^{p,q}u^{i+\frac{q-p}{2}} \in \mathcal{A}^{\bullet}(Z)((u^{1/2})).$$

Therefore the expression $e^{\frac{\boldsymbol{f}}{u}}u^{\mathsf{Gr}}\alpha$ satisfies formally

$$d\left(e^{\frac{\boldsymbol{f}}{u}}u^{\mathsf{Gr}}\alpha\right) = 0,$$

i.e. is d-closed. Moreover, the action of the operator $\boldsymbol{\nabla}_{u^2\frac{d}{du}}$ on α translates to the action of $u^2\frac{d}{du}$ on the above expression modulo formally exact forms.

Consider now a closed connected arc $\delta \subset \partial\boldsymbol{\Delta} = \gamma_i$ and let $\operatorname{Sec}(\delta) \subset \boldsymbol{\Delta}$ be the corresponding open sector (see Figure 5) with vertex at $0 \in \boldsymbol{\Delta}$, and boundary made out of the arc δ and the segments connecting 0 with the end points of δ. Note that

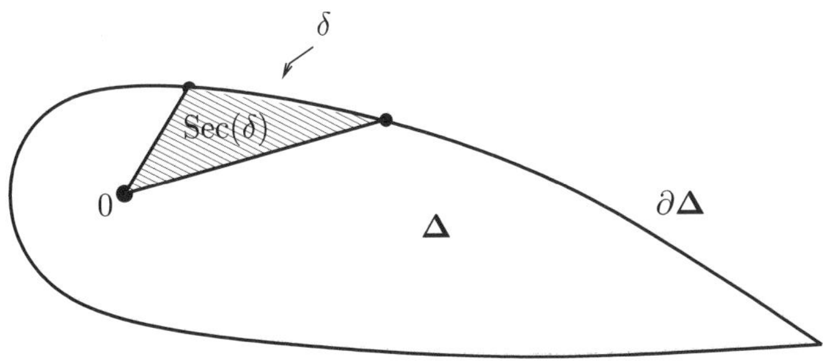

FIGURE 5. A sector in $\boldsymbol{\Delta}$.

the convexity of $\boldsymbol{\Delta}$ assures that $\mathrm{Sec}(\delta) \subset \boldsymbol{\Delta}$. Denote by $\mathrm{Sec}(\delta)^\vee \subset \mathbb{C}$ the dual angle sector consisting of $u \in \mathbb{C}$ such that $\mathrm{Re}(w/u) < 0$ for all $w \in \mathrm{Sec}(\delta)$.

Clearly, for each class in the relative integral homology $H_\bullet(Z, \boldsymbol{f}^{-1}(\delta); \mathbb{Z})$ we can choose a relative chain $\mathfrak{c}$ representing it, so that $\mathfrak{c}$ satisfies:

(†)
- $\mathfrak{c}$ is piece-wise real analytic;
- $\boldsymbol{f}(\mathrm{supp}(\mathfrak{c})) \subset \mathrm{Sec}(\delta)$;
- $\boldsymbol{f}(\mathrm{supp}(\partial\mathfrak{c})) \subset \delta$.

For every such relative chain $\mathfrak{c}$ we now have:

LEMMA 3.11. *For every d_{tot}-closed formal power series of forms $\alpha \in \mathcal{A}_Z^\bullet(Z)[[u]]$ and every relative chain $\mathfrak{c} \in C_\bullet(Z, \boldsymbol{f}^{-1}(\delta); \mathbb{Z})$ satisfying (†) the oscillating integral*

$$\int_{\mathfrak{c}} e^{\frac{\boldsymbol{f}}{u}} u^{\mathsf{Gr}} \alpha$$

is well defined as an asymptotic series in $u^{\mathbb{Q}}(\log u)^{\mathbb{N}}$ in the sector $\mathrm{Sec}(\delta)^\vee$.

Proof. Let $N \geq 0$ be a non-negative integer. Clearly the expression

$$e^{\boldsymbol{f}/u} u^{\mathsf{Gr}} \left(\sum_{0 \leq i \leq N} \alpha_i u^i \right)$$

is a well defined analytic function on $Z \times \mathrm{Sec}(\delta)^\vee$. Using the fact that $(d + u^{-1} d\boldsymbol{f} \wedge) u^{\mathsf{Gr}} \alpha = 0$ and the Malgrange-Sibuya theory of asymptotic sectorial solutions to analytic differential equations, we get that

$$\int_{\mathfrak{c}} e^{\boldsymbol{f}/u} u^{\mathsf{Gr}} \left(\sum_{0 \leq i \leq N} \alpha_i u^i \right) \simeq \sum_{j \in \mathbb{Q}, k \in \mathbb{N}} c_{j,k} u^j (\log(u))^k \tag{3.2.1}$$

is asymptotic to a series in $u^{\mathbb{Q}}(\log u)^{\mathbb{N}}$ in which the logarithms enter with bounded powers. Thus the limit of (3.2.1) as $N \to \infty$ is asymptotic to a series in $u^{\mathbb{Q}}(\log u)^{\mathbb{N}}$ on $\mathrm{Sec}(\delta)^\vee$. □

The previous lemma shows that the $\mathbb{C}[[u]]$-module with connection $(H^\bullet_{\mathrm{DR}}((Z, \boldsymbol{f}); \mathbb{C}), \nabla)$ is formally isomorphic to a meromorphic local system of the form $\boldsymbol{\mathcal{E}}^{\boldsymbol{f}/u} \otimes (\mathscr{R}_i, \nabla_i)$, where $\mathscr{R}_i$ is a free $\mathbb{C}[[u]]$-module, and ∇_i has regular singularities. Furthermore the lemma shows that the oscillating integrals above identify the local system on γ_i given by $(\boldsymbol{c} \in \gamma_i) \mapsto H^\bullet((Y_i, Y_{\boldsymbol{c}}), \mathbb{Q})$ with a rational structure on $(\mathscr{R}_i \otimes_{\mathbb{C}[[u]]} \mathbb{C}((u)), \nabla_i)$. In particular the data $\{(\mathscr{R}_i, \nabla_i)\}_{i=1}^m$ and $(S, \{U_i\}, \{T_{ij}\})$ constitute the regular type and gluing data (in the sense of Theorem 2.35) of an **nc**-Hodge structure of exponential type.

Usually if one tries to make a Landau-Ginzburg model with proper map w from non-proper examples above, one gets new parasitic critical points. Choosing an appropriate domain $Y' \subset Y$ one can define the gluing data for the relevant critical points.

3.3. Mirror symmetry examples. Finally, in order to give a general idea of the mirror correspondence, we briefly discuss three examples of Landau-Ginzburg models mirror dual to symplectic manifolds of positive, vanishing, and negative anti-canonical class respectively.

- For $X = \mathbb{CP}^n$ one of the possible mirror dual Landau-Ginzburg models is given by $Y = (\mathbb{C}^\times)^n$ endowed with potential
$$\mathsf{w}(z_1, \ldots, z_n) = z_1 + \cdots + z_n + \frac{q}{z_1 \cdots z_n}$$
where $q \in \mathbb{C}^\times$ is a parameter. In this model the map w is not proper. This can be repaired by compactifying the fibers of w to $(n-1)$-dimensional projective Calabi-Yau varieties. The compactification is not unique, it depends on combinatorial data, but the compactified space has the same critical points as Y. In general, for symplectic manifolds (X, ω) with ω representing the anticanonical class, one can combine equations for the connection in the q and u directions and get a beautiful variation of Hodge structures with strong arithmetic properties as predicted by our considerations in Section 3.1 (see also Golyshev's work [**Gol01, Gol07**]).
- For a smooth projective Calabi-Yau variety X one can take for Y the product $(X^\vee \times \mathbb{A}^{2N}, \mathsf{w})$ where $X^\vee$ is a Calabi-Yau variety mirror dual to X, $N \geq 1$ is an arbitrary integer and w is the pullback from $\mathbb{A}^{2N}$ of a non-degenerate quadratic form. In general, the complex dimension of the Landau-Ginzburg model is equal to half of the real dimension of X modulo 2.
- For X being a complex curve of genus $g \geq 2$ (considered as a symplectic manifold), the first author proposed several years ago a mirror Landau-Ginzburg model (Y, w) which is a complex algebraic 3-dimensional manifold with non-vanishing algebraic volume element, such that locally (in the analytic topology) near each point the pair (Y, w) is isomorphic to
$$\mathsf{w} : \mathbb{C}^3 \to \mathbb{C}, \qquad (x, y, z) \mapsto xyz$$
The set of critical points of w is the union of $3g - 3$ copies of $\mathbb{CP}^1$ glued along points $0, \infty$ meeting 3 curves at a point. The graph obtained by contracting each copy of $\mathbb{C}^\times$ to an edge is a connected 3-valent graph with g loops, representing a maximal degeneration point in the Deligne-Mumford moduli stack of stable genus g curves.

4. Generalized Tian-Todorov theorems and canonical coordinates

In this section we will examine more closely the other direction of the mirror symmetry correspondence, i.e. the situation in which symplectic Landau-Ginzburg models appear as mirrors of complex manifolds with a fixed anti-canonical section. In order to understand the Hodge theoretic implications of this process we first revisit a classical concept in the subject: the notion of canonical coordinates.

4.1. Canonical coordinates for Calabi-Yau variations of nc-Hodge structures.

4.1.1. ***Variations over supermanifolds.*** We begin with a reformulation of the definition of variations of **nc**-Hodge structures (Definition 2.7) to allow for bases that are supermanifolds:

DEFINITION 4.1. *For a complex analytic supermanifold S, a* ***variation of* nc-*Hodge structures over*** *S (respectively a* ***variation of* nc-*Hodge structures over*** *S* ***of exponential type****) is a triple $(H, \mathscr{E}_B, \mathsf{iso})$, where*

- *H is a holomorphic $\mathbb{Z}/2$-graded vector bundle on $\mathbb{A}^1 \times S$ which is algebraic in the $\mathbb{A}^1$-direction;*
- *$\mathscr{E}_B$ is a local system of $\mathbb{Z}/2$-graded $\mathbb{Q}$-vector spaces on $(\mathbb{A}^1 - \{0\}) \times S$;*
- **iso** *is an analytic isomorphism of holomorphic vector bundles*

$$\mathsf{iso} : \mathscr{E}_B \otimes \mathcal{O}_{(\mathbb{A}^1-\{0\})\times S} \xrightarrow{\cong} H_{|(\mathbb{A}^1-\{0\})\times S};$$

so that:

◊ *the induced meromorphic connection ∇ on $H_{|(\mathbb{A}^1-\{0\})\times S}$ satisfies: locally on S, for every section ξ of T_S, the operators $\nabla_{u^2 \frac{\partial}{\partial u}}$, $\nabla_{u\xi}$ extend to operators on $\mathbb{A}^1 \times S$, and*

◊ *the triple $(H, \mathscr{E}_B, \mathsf{iso})$ satisfies the* **($\mathbb{Q}$-structure axiom)** *and the* **(Opposedness axiom)** *(respectively (H, ∇) is of exponential type and $(H, \mathscr{E}_B,$* **iso***) satisfies the* **($\mathbb{Q}$-structure axiom)^{exp}** *and the* **(Opposedness axiom)^{exp}***).*

REMARK 4.2. From now on we will suppress the $\mathbb{Q}$-structure and the opposedness axioms since they will not play any special role in our analysis. At any given stage of the discussion they can be added without any harm or alteration to the arguments.

4.1.2. ***Calabi-Yau variations.*** Suppose now that $(H, \mathscr{E}_B, \mathsf{iso})$ is a variation of **nc**-Hodge structures over a supermanifold S. For any point $x \in S$ let $H_{0,x}$ denote the fiber of H at $(0, x) \in \mathbb{A}^1 \times S$. We get a canonical map

$$\mu_x : T_x S \to \operatorname{End}(H_{0,x}),$$

defined as follows: Extend the tangent vector $v \in T_x S$ to some analytic vector field ξ defined in a neighborhood of x. Consider the holomorphic first order differential operator $\nabla_{u\xi} : H \to H$. By construction this operator has symbol $(u\xi) \otimes \mathrm{id}_H$. In particular, the restriction of $\nabla_{u\xi}$ to the slice $\{0\} \times S \subset \mathbb{A}^1 \times S$ will have zero symbol, and so will be an $\mathcal{O}$-linear endomorphism of $H_{|\{0\}\times S}$. We define $\mu_x(v)$ to be the action of this $\mathcal{O}$-linear map on the fiber $H_{(0,x)}$. It is straightforward to check that this action is independent of the extension ξ and depends only on v.

DEFINITION 4.3. *Let S be a complex analytic supermanifold. We say that a variation $(H, \mathscr{E}_B, \mathsf{iso})$ of* **nc**-*Hodge structures on S is of* ***Calabi-Yau type at a point*** *$x \in S$ if there exists an (even or odd) vector $h \in H_{(0,x)}$, so that the linear map*

$$\begin{aligned} T_x S &\longrightarrow H_{(0,x)} \\ v &\longmapsto (\mu_x(v))(h) \end{aligned}$$

is an isomorphism. Such a vector h will be called a ***generating vector for*** *H* ***at*** *x.*

It follows from the definition that if S is the base of a variation of **nc**-Hodge structures which is of Calabi-Yau type at a point $x \in S$, then the tangent space $T_x S$ is a unital commutative associative algebra acting on $H_{0,x}$ via the map μ_x and such that $H_{0,x}$ is a free module of rank one. The condition on a variation to have a Calabi-Yau type (even or odd) is an open condition on $x \in S$. Variations of **nc**-Hodge structures of Calabi-Yau type should arise naturally on the periodic cyclic homology of smooth and compact d$(\mathbb{Z}/2)$g categories which are Calabi-Yau in the sense of [**KS06b**]. The basic geometric example of a Calabi-Yau variation is an extension of the setup we discussed in section 3.1:

EXAMPLE 4.4. Let (X, ω) be a compact symplectic manifold with $\dim_{\mathbb{R}} X = 2d$. Conjecturally there exists a non-empty open subset $S \subset H^\bullet(X, \mathbb{C})$ so that the big quantum product $*_x$ is absolutely convergent for all $x \in S$ (the product is given by a formula similar to one on page 78). The manifold S has a natural structure of a supermanifold, being an open subset in the affine superspace $H^\bullet(X, \mathbb{C})$. As in Section 3.1 we define a variation of **nc**-Hodge structures $(H, \mathscr{E}_B, \mathsf{iso})$ on S by taking H to be the trivial vector bundle on $\mathbb{A}^1 \times S$ with fiber $H^\bullet(X, \mathbb{C})$, and defining the connection ∇ on H by the formulas:

$$\nabla_{\frac{\partial}{\partial u}} := \frac{\partial}{\partial u} + u^{-2}(\kappa_X *_x \bullet) + u^{-1}\mathsf{Gr},$$

$$\nabla_{\frac{\partial}{\partial t^i}} := \frac{\partial}{\partial t^i} - q^{-1}u^{-1}(t_i *_x \bullet),$$

where the (t_i) form a basis on $H^\bullet(X, \mathbb{C})$, and (t^i) are the dual linear coordinates.

Clearly, if we restrict (H, ∇) to $S \cap H^2(X, \mathbb{C})$ we will get back the bundle with connection we defined in section 3.1. We now define the integral lattice $\mathscr{E}_B$ and isomorphism **iso** on S as the ∇-horizontal extensions of the integral lattice and isomorphism we had defined on $S \cap H^2(X, \mathbb{C})$. Finally, in order to match the framework of **nc**-geometry, we should change the parity of the bundle H in the case $d \equiv 1 \bmod 2$.

4.1.3. ***Decorated Calabi-Yau variations.*** The variations of **nc**-Hodge structures of Calabi-Yau type need to be decorated by a few additional pieces of data before we can extract canonical coordinates from them. To motivate our choice of such data we first recall the Deligne-Malgrange classification of logarithmic holomorphic extensions of regular connections.

Let S be a complex analytic supermanifold, let $\boldsymbol{D}$ be a one-dimensional complex disc, and let $\mathscr{E}$ be a complex local system on $(\boldsymbol{D} - \{\mathrm{pt}\}) \times S$ and let $(\mathscr{E}, \nabla)$ be the associated holomorphic bundle $E := \mathscr{E} \otimes \mathcal{O}_{(D-\{\mathrm{pt}\})\times S}$ on $(D - \{\mathrm{pt}\}) \times S$ with the induced flat connection ∇. Suppose that $\widetilde{E}$ is a holomorphic bundle on $\boldsymbol{D}\times S$ which extends E and on which ∇ has a logarithmic pole. The restriction $\widetilde{E}_{|\{\mathrm{pt}\}\times S}$ is a holomorphic bundle on S and ∇ induces: a holomorphic connection ${}^{\widetilde{E}}\nabla$ and an $\mathcal{O}_S$-linear residue endomorphism $\mathrm{Res}_{\widetilde{E}}(\nabla)$ on $\widetilde{E}_{|\{\mathrm{pt}\}\times S}$. Furthermore the integrability of ∇ on $(\boldsymbol{D}-\{\mathrm{pt}\})\times S$ implies that ${}^{\widetilde{E}}\nabla$ is also integrable and that the endomorphism $\mathrm{Res}_{\widetilde{E}}(\nabla)$ is covariantly constant with respect to ${}^{\widetilde{E}}\nabla$ [**Sab02**, Section 0.14b].

Recall next that by Deligne's extension theorem (see e.g. [**Del70**, Chapter II.5] or [**Sab02**, Corollary II.2.21]) meromorphic bundles with connections with regular singularities always admit functorial holomorphic extensions across the pole divisor. Deligne's extension procedure is not unique and depends on the choice of a set-theoretic section of the quotient map $\mathbb{C} \to \mathbb{C}/\mathbb{Z}$. We fix $\mathcal{V}$ to be the unique Deligne extension of E for which ∇ has a logarithmic pole at $\{\mathrm{pt}\} \times S$ and a residue with eigenvalues whose real parts are in the interval $(-1, 0]$. Now the classification theorem of Deligne-Malgrange [**Sab02**, Theorem III.1.1] asserts that there is a natural equivalence of categories

$$\begin{pmatrix}\text{Holomorphic extensions of } E \text{ to} \\ \boldsymbol{D}\times S \text{ for which } \nabla \text{ has a logarith-} \\ \text{mic singularity along } \{\mathrm{pt}\} \times S\end{pmatrix} \longleftrightarrow \begin{pmatrix}\text{Decreasing filtrations of } \mathscr{E} \text{ by } \mathbb{C}\text{-} \\ \text{local subsystems on } (\boldsymbol{D}-\{\mathrm{pt}\}) \times \\ S\end{pmatrix}.$$

The equivalence depends on the chosen Deligne extension and is explicitly given as follows. Let t be a complex coordinate on $\boldsymbol{D}$ which vanishes at $\{\mathrm{pt}\} \in \boldsymbol{D}$. Consider the restriction $\mathcal{V}/t\mathcal{V}$ of $\mathcal{V}$ to $\{\mathrm{pt}\} \times S$. This is a holomorphic bundle on S equipped as above with the holomorphic connection ${}^{\mathcal{V}}\nabla$ and the covariantly constant residue endomorphism $\mathrm{Res}_{\mathcal{V}}(\nabla)$. Suppose now that $\widetilde{E}$ is another holomorphic bundle on $\boldsymbol{D} \times S$ which extends E and on which ∇ has a logarithmic pole. For any $k \in \mathbb{Z}$ we define a subbundle $(\mathcal{V}/t\mathcal{V})^k \subset \mathcal{V}/t\mathcal{V}$ by setting

$$(\mathcal{V}/t\mathcal{V})^k := \frac{\mathcal{V}\cap t^k\widetilde{E}}{t\mathcal{V}\cap t^k\widetilde{E}}$$

where $\mathcal{V}$ and $\widetilde{E}$ are viewed as subsheaves in the meromorphic bundle E.

By construction the subbundles $(\mathcal{V}/t\mathcal{V})^k$ are preserved both by ${}^{\mathcal{V}}\nabla$ and by the residue endomorphism $\mathrm{Res}_{\mathcal{V}}(\nabla)$ and so give rise to ∇-covariantly constant meromorphic subbundles of E, or equivalently to $\mathbb{C}$-local subsystems of $\mathscr{E}$.

Alternatively we can use a more intrinsic description of holomorphic extensions of $(\mathscr{E}, \nabla)$ which is beter adapted to our examples and in particular to Example 4.8. Namely, instead of relying on the Deligne extension and the induced filtration we can use decreasing filtrations $\mathscr{E}_{\leq\lambda}$ of $\mathscr{E}$ labeled by real numbers $\lambda \in \mathbb{R}$ and such that on the associated graded pieces the monodromy on $\boldsymbol{D} - \{\mathrm{pt}\}$ has eigenvalues in $\mathbb{R}_+ \times \exp(2\pi i\lambda)$.

We can now introduce the additional data that one needs for the canonical coordinates

DEFINITION 4.5. *Let S be a complex supermanifold and let $(H, \mathscr{E}_B, \mathsf{iso})$ be a variation of* **nc***-Hodge structures of Calabi-Yau type on S. A* ***decoration*** *on $(H, \mathscr{E}_B, \mathsf{iso})$ is a pair $(\widetilde{H}, \psi)$ where:*

$\widetilde{H}$ is an extension of H to a $(\mathbb{Z}/2)$-graded vector bundle on $\mathbb{P}^1 \times S$ so that ∇ has a regular singularity at $\{\infty\} \times S$.

ψ is a ${}^{\widetilde{H}}\nabla$-covariantly constant section of $\widetilde{H}_{\{\infty\}\times S}$.

*A decoration is called **rational** iff the $\mathbb{R}$-filtration on the local system $\mathscr{E}_B \otimes \mathbb{C}$ is compatible with the rational structure, and if the vector $\psi(x) \in \widetilde{H}_{\{\infty\}\times\{x\}} = \mathrm{gr}(\mathscr{E}_B \otimes \mathbb{C})_x$ is rational, i.e. if $\psi(x) \in \mathrm{gr}(\mathscr{E}_B)_x$.*

The previous discussion applied to the local system $\mathscr{E}_B \otimes \mathbb{C}$, the disc $\boldsymbol{D} = \{|u| > 1\} \cup \{\infty\}$ and the point $\mathrm{pt} = \infty$ shows that the data of a decoration are equivalent to the data $((\mathscr{E}_B \otimes \mathbb{C})_{\leq \bullet}, \psi)$, where $(\mathscr{E}_B \otimes \mathbb{C})_{\leq \bullet}$ is a decreasing filtration of $\mathscr{E}_B \otimes \mathbb{C}$ (labelled by real numbers) and ψ is a covariantly constant section (along S) of the corresponding logarithmic holomorphic extension of H. We will freely go back and forth between these two points of view.

Any decorated variation $(H, \mathscr{E}, \mathsf{iso}; \widetilde{H}, \psi)$ of **nc**-Hodge structures of Calabi-Yau type gives rise to a natural open domain $U \subset S$ defined by

$$U := \left\{ x \in S \;\middle|\; \begin{array}{l} \widetilde{H}_{\mathbb{P}^1\times\{x\}} \text{ is holomorphically trivial and if } s \in \Gamma\left(\mathbb{P}^1 \times \{x\}\right) \\ \text{is such that } s_x(\infty) = \psi(\infty, x), \text{ then } s_x(0) \text{ is a generating} \\ \text{vector for } (H, \mathscr{E}, \mathsf{iso}). \end{array} \right\}$$

Furthermore for every $x \in U$ we get a natural map $\mathsf{can}_x : T_x S \to \widetilde{H}_{\infty,x}$ defined as the composition

$$\mathsf{can}_x : T_x S \xrightarrow{\mu_x(\bullet)(s_x(0))} H_{0,x} \xrightarrow{\mathrm{ev}_{(0,x)}^{-1}} \Gamma\left(\widetilde{H}_{|\mathbb{P}^1\times\{x\}}\right) \xrightarrow{\mathrm{ev}_{(\infty,x)}} \widetilde{H}_{\infty,x}.$$

where $\mathrm{ev}_{(t,x)} : \Gamma\left(\mathbb{P}^1, \widetilde{H}_{|\mathbb{P}^1\times\{x\}}\right) \to \widetilde{H}_{t,x}$ denotes the natural evaluation of sections, which is invertible by the triviality assumption on $\widetilde{H}_{|\mathbb{P}^1\times\{x\}}$.

The pullback of the flat connection ${}^{\widetilde{H}}\nabla$ by the map can induces a flat connection on $TS_{|U}$. The canonical coordinates on S come from the following easy claim whose proof we omit:

CLAIM 4.6. *The flat connection $\mathsf{can}^*\left({}^{\widetilde{H}}\nabla\right)$ on $TS_{|U}$ is torsion-free and so gives rise to a natural affine structure and affine coordinates on U. If the decoration is rational then the tangent bundle $TS_{|U}$ carries a natural rational structure.*

REMARK 4.7. **(i)** The canonical coordinates on U corresponding to a decorated **nc**-variation of Hodge structures are only affine coordinates and are defined only up to a translation.

(ii) For any $u \in \mathbb{A}^1 - \{0\}$ we can introduce another affine structure which is a **vector structure**. In fact, we get an analytic isomorphism between U and a domain in $H_{u,\bullet} = (\mathscr{E}_B)_{u,\bullet} \otimes \mathbb{C}$:

$$x \in U \mapsto \mathrm{ev}_{(u,x)}\, \mathrm{ev}_{(\infty,x)}^{-1}(\psi(x)) \in H_{(u,x)}.$$

One can use this to show that the local Torelli theorem holds for decorated Calabi-Yau variations of **nc**-Hodge structures.

EXAMPLE 4.8. The setup of Example 4.4 gives not only a variation of **nc**-Hodge structures but in fact gives a rationally decorated **nc**-Hodge structure of Calabi-Yau type. Indeed by definition the fibers of H are identified with $\mathbf{\Pi}^d H^\bullet(X,\mathbb{C})$. The monodromy of the connection around $\infty \in \mathbb{P}^1$ is the operator acting by $(-1)^{i+d}\exp(\kappa_X \wedge (\bullet))$ on $H^i(X,\mathbb{C})$. Consider the monodromy invariant filtration on $H^\bullet(X,\mathbb{C})$ whose component in degree $\frac{d-i}{2}$ is $H^{\geq i}(X,\mathbb{C})$. Let $\widetilde{H}$ be the corresponding logarithmic extension of H and let ψ be the section of $\widetilde{H}$ corresponding to the image of $1 \in H^0(X,\mathbb{C}) \subset H^\bullet(X,\mathbb{C})$. The bundle $\widetilde{H}_{|\{\infty\}\times S}$ is trivialized and $\nabla_{\frac{\partial}{\partial t^i}} = \frac{\partial}{\partial t^i}$ in this trivialization. This gives the desired decoration $(\widetilde{H},\psi)$ and the associated canonical coordinates are the standard canonical coordinates in Gromov-Witten theory.

4.1.4. ***Generalized decorations.*** The notion of a decorated Calabi-Yau variation of **nc**-Hodge structures can be generalized in various ways. For instance, instead of specifying a covariantly constant filtration on H giving the extension $\widetilde{H}$ we can start with any holomorphic bundle H' defined on $\{u \in \mathbb{P}^1 | \, |u| \geq R\}$, and an identification of C^∞-bundles

$$p_1^*\left(H'_{|\{|u|=R\}}\right) \cong (\mathscr{E}_B \otimes \mathbb{C})_{\{|u|=R\}\times S},$$

where $p_1 : \{|u| = R\} \times S \to \{|u| = R\}$ is the projection on the first factor.

Furthermore (locally in S) the holomorphic bundle $p_1^* H'$ on $\{u \in \mathbb{P}^1 | \, |u| \geq R\} \times S$ carries a flat connection defined along S only. We can use the above identification to glue this together with H along $\{|u| = R\} \times S$ to get a bundle $\widetilde{H}$ on $\mathbb{P}^1 \times S$ equipped with a flat connection $\nabla_{/S}$ along S. This generalizes the first part of the decoration. For the second part we will take a $\nabla_{/S}$-covariantly constant section ψ of $\widetilde{H}_{|\{\infty\}\times S}$. Now the same definition of the set U and the canonical map can make sense in this context. The resulting connection on $TS_{|U}$ is again torsion-free.

4.1.5. ***Formal variations of Calabi-Yau type.*** The notion of a Calabi-Yau variation extends readily to the formal context. Suppose $S = \operatorname{Spf}\mathbb{C}[[x_1,\ldots,x_N, \xi_1,\ldots,\xi_M]]$ be a formal algebraic supermanifold, where x_i are even and ξ_j are odd formal variables. The de Rham part of a formal variation of **nc**-Hodge structures on S is a pair (H,∇) where H is a $(\mathbb{Z}/2)$-graded algebraic vector bundle over $\mathbb{D} \times S$, where $\mathbb{D}$ is the one-dimensional formal disc $\mathbb{D} := \operatorname{Spf}(\mathbb{C}[[u]])$. Here ∇ is a meromorphic connection on H such that $\nabla_{u^2\frac{\partial}{\partial u}}$, $\nabla_{u\frac{\partial}{\partial x_i}}$, $\nabla_{u\frac{\partial}{\partial \xi_j}}$ are regular differential operators on H.

We say that such a pair (H,∇) has the Calabi-Yau property if we can find a vector $h \in H_{0,0}$, so that the natural linear map $T_0 S \to H_{0,0}$, $v \mapsto \mu_0(v)(h)$ is an isomorphism.

Finally a decoration of a formal Calabi-Yau de Rham data (H,∇) is a pair $(\boldsymbol{e},h)$, where $\boldsymbol{e}$ is a trivialization $\boldsymbol{e} : H_{|\mathbb{D}\times\{0\}} \to H_{0,0} \otimes \mathcal{O}_{\mathbb{D}\times\{0\}}$, and $h \in H_{0,0}$ is a generating vector for the Calabi-Yau property.

Again a decorated de Rham data of Calabi-Yau type gives an affine structure and canonical formal coordinates on S.

4.2. Algebraic framework: dg Batalin-Vilkovisky algebras. In this section we discuss the aspects of algebraic deformation theory relevant to the study of **nc**-Hodge structures. We will work over $\mathbb{C}$ but all algebraic considerations in this section make sense over any field of characteristic zero.

4.2.1. ***Preliminaries on L_∞ algebras.*** Our main objects of interest here will be differential $\mathbb{Z}/2$-graded algebras over $\mathbb{C}$ or more generally $\mathbb{Z}/2$-graded L_∞-algebras over $\mathbb{C}$. We begin with a definition:

DEFINITION 4.9. *A complex differential $\mathbb{Z}/2$-graded Lie algebra $\mathfrak{g}$ (or a $\mathbb{Z}/2$-graded L_∞-algebra) is called **homotopy abelian** if it is L_∞ quasi-isomorphic to an abelian $d(\mathbb{Z}/2)g$ Lie algebra.*

REMARK 4.10. Homotopy abelian differential $\mathbb{Z}/2$-graded Lie algebras can be characterized in a variety of ways. In particular we have the following statements that follow readily from the definition:

- A differential $\mathbb{Z}/2$-graded Lie algebra $\mathfrak{g}$ is homotopy abelian if and only if all the higher operations m_n vanish on its L_∞ minimal model $\mathfrak{g}^{\min} = H^\bullet(\mathfrak{g}, d_\mathfrak{g})$, i.e. $m_n = 0$ for $n \geq 1$.
- A differential $\mathbb{Z}/2$-graded Lie algebra $\mathfrak{g}$ is homotopy abelian if and only if there exist d($\mathbb{Z}/2$)g Lie algebras $\mathfrak{g}_1$ and $\mathfrak{g}_2$, and morphisms of d($\mathbb{Z}/2$)g Lie algebras:

$$\begin{array}{ccc} & \mathfrak{g}_1 & \\ {\scriptstyle\cong}\swarrow & & \searrow{\scriptstyle\cong} \\ \mathfrak{g} & & \mathfrak{g}_2 \end{array}$$

  so that $\mathfrak{g}_2$ is an abelian d($\mathbb{Z}/2$)g Lie algebra, and the morphisms $\mathfrak{g}_1 \to \mathfrak{g}$ and $\mathfrak{g}_1 \to \mathfrak{g}_2$ are quasi-isomorphisms.
- A differential $\mathbb{Z}/2$-graded Lie algebra $\mathfrak{g}$ is homotopy abelian if and only if the Lie algebra cohomology algebra $H^\bullet(\mathfrak{g}, \mathbb{C})$ is free, i.e. is isomorphic to the algebra of formal power series on some (possibly infinitely many) supervariables. Here the Lie algebra cohomology is defined as

$$H^\bullet(\mathfrak{g}, \mathbb{C}) := H^\bullet\left(\prod_{n\geq 0} \operatorname{Hom}_{(\mathbb{C}-\mathsf{Vect})}(\operatorname{Sym}^n \mathbf{\Pi}\mathfrak{g}, \mathbb{C})^\bullet, d\right)$$

  where d is the cochain Cartan-Eilenberg differential.

After the pioneering work of Deligne and Drinfeld in the 80's, it is by now common wisdom (see e.g. [**Man99a**, Chapter III.9]) that dg Lie algebras give rise to solutions of moduli problems. In particular a homotopy abelian d($\mathbb{Z}/2$)g Lie algebra $\mathfrak{g}$ gives rise to a moduli space - the formal supermanifold $\mathbb{M}\mathsf{od}_{(\mathfrak{g}, d_\mathfrak{g})} := \operatorname{Spf}(H^\bullet(\mathfrak{g}, \mathbb{C}))$.

The property of being homotopy abelian is preserved by suitably non-degenerate deformations and various other natural operations:

PROPOSITION 4.11. **(i)** *Let $\mathfrak{g}$ be a flat family of $d(\mathbb{Z}/2)g$ Lie algebras (or $(\mathbb{Z}/2)$-graded L_∞ algebras) over $\mathbb{C}[[u]]$. That is, $\mathfrak{g}$ is a flat $(\mathbb{Z}/2)$-graded $\mathbb{C}[[u]]$-module, and the Lie bracket and differential on $\mathfrak{g}$ are $\mathbb{C}[[u]]$-linear. Assume further that*

(A) $\mathfrak{g}_{\text{gen}} := \mathfrak{g} \otimes_{\mathbb{C}[[u]]} \mathbb{C}((u))$ *is homotopy abelian over* $\mathbb{C}((u))$*, and*
(B) $H^\bullet(\mathfrak{g}, d_\mathfrak{g})$ *is a flat* $\mathbb{C}[[u]]$*-module.*

Then the special fiber $\mathfrak{g}_0 := \mathfrak{g}\widehat{\otimes}_{\mathbb{C}[[u]]}\mathbb{C}$ *is also a homotopy abelian* $\mathrm{d}(\mathbb{Z}/2)\mathrm{g}$ *Lie algebra over* $\mathbb{C}$.

(ii) *If $\mathfrak{g}$ is a homotopy abelian $\mathrm{d}(\mathbb{Z}/2)\mathrm{g}$ Lie algebra over $\mathbb{C}$, and $\mathfrak{g}_1 \to \mathfrak{g}$ is a morphism of L_∞-algebras inducing a monomorphism $H^\bullet(\mathfrak{g}_1, d_{\mathfrak{g}_1}) \hookrightarrow H^\bullet(\mathfrak{g}, d_\mathfrak{g})$, then $\mathfrak{g}_1$ is homotopy abelian as well.*

(iii) *If $\mathfrak{g}$ is a homotopy abelian $\mathrm{d}(\mathbb{Z}/2)\mathrm{g}$ Lie algebra over $\mathbb{C}$, and $\mathfrak{g} \to \mathfrak{g}_2$ is a morphism of L_∞-algebras inducing an epimorphism $H^\bullet(\mathfrak{g}, d_\mathfrak{g}) \twoheadrightarrow H^\bullet(\mathfrak{g}_2, d_{\mathfrak{g}_2})$, then $\mathfrak{g}_2$ is homotopy abelian as well.*

Proof. The proof is standard so we only mention some of the highlights of the argument. First note that parts **(ii)** and **(iii)** follow immediately by passing to minimal models. For part **(i)** we note first that the assumption *(B)* implies (and is in fact equivalent to) the existence of $\mathbb{C}[[u]]$-linear quasi-isomorphisms p_1, p_2 of complexes:

$$(H^\bullet(\mathfrak{g}_0, d_{\mathfrak{g}_0})[[u]], 0) \quad \cong \quad (H^\bullet(\mathfrak{g}, d_\mathfrak{g}), 0) \underset{p_2}{\overset{p_1}{\rightleftarrows}} (\mathfrak{g}, d_\mathfrak{g}),$$

and a $\mathbb{C}[[u]]$-linear homotopy h so that

$$\begin{aligned} p_2 \circ p_1 &= \mathrm{id} \\ p_1 \circ p_2 &= \mathrm{id} + [d_\mathfrak{g}, h]. \end{aligned}$$

Next note that the homological perturbation theory of [**KS01**] carries over verbatim to the L_∞-context and gives explicit expressions for the higher products m_n on $(H^\bullet(\mathfrak{g}_0, d_{\mathfrak{g}_0})[[u]], 0)$ as a polynomial expression in p_1, p_2 and h. In particular the operations m_n are all $\mathbb{C}[[u]]$-linear and are given by universal expressions. But by assumption *(A)* we know that the higher operations are zero after tensoring with $\otimes_{\mathbb{C}[[u]]}\mathbb{C}((u))$ and so $m_n = 0$ as formal power series in u for all $n \geq 1$. This implies that $m_{n|u=0} = 0$ for all $n \geq 1$ and so the proposition is proven. □

4.2.2. ***DG Batalin-Vilkovisky algebras.*** Recall [**Man99a**, Chapter III.10] the notion of a dg BV algebra:

DEFINITION 4.12. *A differential $\mathbb{Z}/2$-graded Batalin-Vilkovisky algebra over $\mathbb{C}$ is the data (A, d, Δ), where A is a $\mathbb{Z}/2$-graded supercommutative associative unital algebra, and $d : A \to A$, $\Delta : A \to A$ are odd $\mathbb{C}$-linear maps satisfying:*

- $d(1) = \Delta(1) = 0$,
- *d is a differential operator of order ≤ 1 on A,*
- *Δ is a differential operator of order ≤ 2 on A,*
- $d^2 = \Delta^2 = d\Delta + \Delta d = 0$.

Note that the first two properties in the definition imply that d is a derivation of A. Also $\mathfrak{g} := \mathbf{\Pi} A$ together with $[a,b] := \Delta(ab) - \Delta(a)b - (-1)^{\deg(a)} a\Delta(b)$ is a Lie superalgebra with two anti-commuting differentials d and Δ.

DEFINITION 4.13. *We will say that a* d$(\mathbb{Z}/2)$g *Batalin-Vilkovisky algebra* (A, d, Δ) *has the* ***degeneration property*** *if for every* $N \geq 1$ *we have that* $H^\bullet(A[u]/(u^N), d + u\Delta)$ *is a free* $\mathbb{C}[u]/(u^N)$*-module.*

Equivalently (A, d, Δ) has the degeneration property iff $H^\bullet(A[[u]], d + u\Delta)$ is a topologically free (flat) $\mathbb{C}[[u]]$-module. This in turn is equivalent to the existence of a (non-unique) isomorphism of topological $\mathbb{C}[[u]]$-modules:

$$T : \ H^\bullet(A[[u]], d + u\Delta) \xrightarrow{\cong} H^\bullet(A, d)[[u]]. \tag{4.2.1}$$

In this situation we will always normalize T so that $T_{|u=0} = \mathrm{id}_{H^\bullet(A,d)}$.

The degeneration property for dg Batalin-Vilkovisky algebras defined above is weaker than the $\partial\bar{\partial}$-lemma used by Barannikov and the second author in [**BK98**] and by Manin in [**Man99a, Man99b**]. In particular it has potentially a wider scope of applications – a feature that we will exploit next. We begin with a general smoothness result which was also proven by J.Terilla [**Ter07**].

THEOREM 4.14. *Suppose* (A, d, Δ) *is a* d$(\mathbb{Z}/2)$g *Batalin-Vilkovisky algebra which has the degeneration property. Let* $\mathfrak{g} := \mathbf{\Pi} A$ *be the associated super Lie algebra with anti-commuting differentials* d *and* Δ*. Then:*

(1) *The* d$(\mathbb{Z}/2)$g *Lie algebra* $(\mathfrak{g}, d)$ *is homotopy abelian, i.e. is quasi-isomorphic to* $H^\bullet(\mathfrak{g}, d)$ *endowed with the trivial bracket and the zero differential. In particular the associated moduli space* $\mathsf{Mod}_{(\mathfrak{g},d)}$ *is (non-canonically) isomorphic to a formal neighborhood of* 0 *in the superaffine space* $\mathbf{\Pi} H^\bullet(\mathfrak{g}, d)$*.*

(2) *Every choice of a normalized degeneration isomorphism* T *as in equation* (4.2.1) *gives an identification of formal manifolds*

$$\Phi_T : \ \mathsf{Mod}_{(\mathfrak{g},d)} \xrightarrow{\cong} \begin{pmatrix} \text{formal neighborhood of } 0 \\ \text{in } \mathbf{\Pi} H^\bullet(\mathfrak{g}, d) \end{pmatrix}$$

Proof. Part **(1)** of the theorem follows immediately from

LEMMA 4.15. *The* d$(\mathbb{Z}/2)$g *Lie algebra* $(\mathfrak{g}((u)), d + u\Delta)$ *is homotopy abelian over* $\mathbb{C}((u))$*.*

Proof. Consider the formal completion at zero $\widehat{A}$ of the vector superspace underlying $A = \mathbf{\Pi}\mathfrak{g}$ as an algebraic supermanifold, and let as before $\mathbb{D} = \mathrm{Spf}(\mathbb{C}[[u]])$ be the formal one-dimensional disc. The d$(\mathbb{Z}/2)$g Lie algebra structure on $\mathfrak{g}[[u]]$ is encoded in an odd vector field $\boldsymbol{\xi} \in \Gamma(\widehat{A} \times \mathbb{D}, T)$ on the supermanifold $\widehat{A} \times \mathbb{D}$, defined by

$$\dot{a} := \boldsymbol{\xi}(a) = da + u\Delta a + \frac{1}{2}[a, a].$$

There is a natural automorphism (i.e. a formal change of coordinates) $F : \widehat{A} \times \mathbb{D}^\times \to \widehat{A} \times \mathbb{D}^\times$ on the formal supermanifold $\widehat{A} \times \mathbb{D}^\times$ given by

$$F(a) := u\left(\exp\left(\frac{a}{u}\right) - 1\right) = a + \frac{1}{u}\frac{1}{2!}a^2 + \frac{1}{u^2}\frac{1}{3!}a^3 + \cdots,$$

and in the new coordinates $b = F(a)$ the vector field $\boldsymbol{\xi}$ is linear:

$$\begin{aligned}
\dot{b} &= \dot{a} \cdot \exp\left(\frac{a}{u}\right) \\
&= \left(da + u\Delta a + \frac{1}{2}[a, a]\right) \cdot \exp\left(\frac{a}{u}\right) \\
&= u \cdot \left(\frac{da}{u} + u\Delta\left(\frac{a}{u}\right) + u\frac{1}{2}\left[\frac{a}{u}, \frac{a}{u}\right]\right) \cdot \exp\left(\frac{a}{u}\right) \\
&= u \cdot (d + u\Delta) \exp\left(\frac{a}{u}\right) = (d + u\Delta)b.
\end{aligned}$$

So, in the b-coordinates, the vector field $\boldsymbol{\xi}$ depends only on the differential $d + u\Delta$ and does not depend on any higher operations. Passing to the minimal model we see that $(\mathfrak{g}((u)), d + u\Delta)$ is homotopy abelian, which proves the lemma. □

The lemma implies that the hypothesis *(A)* of Proposition 4.11 **(i)** holds. On the other hand the hypothesis (B) holds by the degeneration assumption. Therefore by Proposition 4.11 **(i)** we conclude that $(\mathfrak{g}, d)$ is homotopy abelian. This proves part **(1)** of the theorem.

Next we construct the identification Φ_T. Given a formal path in $\mathsf{Mod}_{(\mathfrak{g},d)}$, i.e. a family of solutions (up to gauge equivalence)

$$\begin{aligned}
a(\varepsilon) &= a_1\varepsilon + a_2\varepsilon^2 + a_3\varepsilon^3 + \cdots \in \varepsilon A[[\varepsilon]] \\
d(a(\varepsilon)) &+ \frac{1}{2}[a(\varepsilon), a(\varepsilon)] = 0
\end{aligned}$$

of the Maurer-Cartan equation in $(\mathfrak{g}, d)$, we have to construct the corresponding formal path through the origin in $H^\bullet(\mathfrak{g}, d)$.

As a first step choose a lift of the formal arc $a(\varepsilon)$ to a formal series in two variables $\tilde{a}(\varepsilon, u) \in \varepsilon A[[\varepsilon, u]]$ such that

$$\begin{aligned}
(d{+}u\Delta)\tilde{a} + \frac{1}{2}[\tilde{a}, \tilde{a}] &= 0, \\
a(\varepsilon, 0) &= a(\varepsilon).
\end{aligned}$$

Consider the reparameterization

$$\tilde{b} = F(\tilde{a}) = u\left(\exp\left(\frac{\tilde{a}}{u}\right) - 1\right) \in \varepsilon A((u))[[\varepsilon]].$$

Arguing as before we see that $\tilde{b}$ satisfies $(d + u\Delta)\tilde{b} = 0$. So if we expand

$$\tilde{b} = \tilde{b}_1\varepsilon + \tilde{b}_2\varepsilon^2 + \cdots, \qquad \text{where } \tilde{b}_n \in A((u)) \text{ satisfy } (d + u\Delta)\tilde{b}_n = 0,$$

we can define cohomology classes $\left[\tilde{b}_n\right] \in H^\bullet(A((u)), d + u\Delta)$. We can now apply the isomorphism $T \otimes_{\mathbb{C}[[u]]} \mathbb{C}((u))$ to the series

$$\sum_{n \geq 1} \left[\tilde{b}_n\right] \varepsilon^n \in \varepsilon H^\bullet(A((u)), d + u\Delta)[[\varepsilon]],$$

to obtain an element

$$T\left(\sum_{n\geq 1}\left[\tilde{b}_n\right]\varepsilon^n\right)\in \varepsilon H^\bullet(A,d)((u))[[\varepsilon]].$$

In fact one has the following lemma whose proof we will skip since it is a somewhat tedious application of homological perturbation theory:

LEMMA 4.16. *There exists a lift* $\tilde{a}(\varepsilon,u)$ *of* $a(\varepsilon)$ *such that the associated class* $T\left(\sum_{n\geq 1}\left[\tilde{b}_n\right]\right)$ *belongs to* $\varepsilon H^\bullet(A,d)[[\varepsilon]]\subset \varepsilon H^\bullet(A,d)((u))[[\varepsilon]]$. *Any such lift* $\widetilde{a}$ *produces the same class* $T\left(\sum_{n\geq 1}\left[\tilde{b}_n\right]\right)$ *and this class depends only on the gauge equivalence class of the original arc* a, *i.e. on the image* $\underline{a}(\varepsilon)$ *of* $a(\varepsilon)$ *in* $\mathbb{M}\mathsf{od}_{(\mathfrak{g},d)}$.

Now by definition the map Φ_T assigns the class $T\left(\sum_{n\geq 1}\left[\tilde{b}_n\right]\right)\subset \varepsilon H^\bullet(A,d)[[\varepsilon]]$ to the formal arc $\underline{a}(\varepsilon)$. □

4.2.3. ***Geometric interpretation.*** The previous discussion can be repackaged geometrically as follows. A $(\mathbb{Z}/2)$-graded Batalin-Vilkovisky algebra (A,d,Δ) gives rise to a family $\mathscr{M}\to\mathbb{D}=\operatorname{Spf}(\mathbb{C}[[u]])$ of formal manifolds over the one-dimensional formal disc. The family $\mathscr{M}$ is the total space of the relative moduli space $\mathbb{M}\mathsf{od}_{(\mathfrak{g},d+u\Delta)}$ over $\mathbb{C}[[u]]$. If (A,d,Δ) has the degeneration property, then by Lemma 4.15 we have an affine structure on the generic fiber $\mathscr{M}^{\mathrm{gen}}:=\mathscr{M}\otimes_{\mathbb{C}[[u]]}\mathbb{C}((u))$ of the family (see Figure 6) given by the map F.

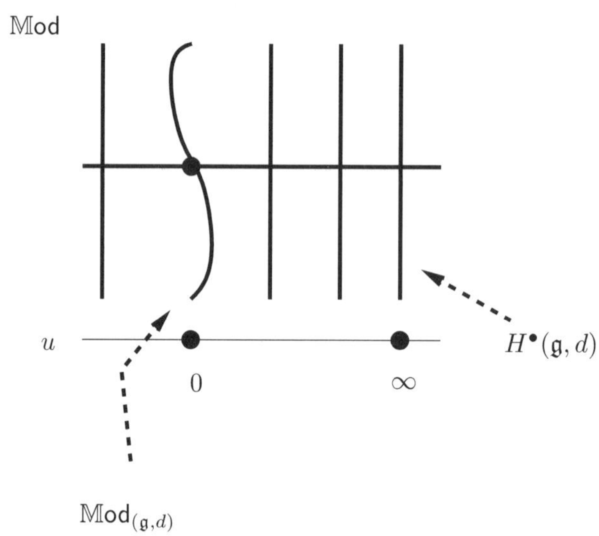

FIGURE 6. The relative moduli $\mathbb{M}\mathsf{od}\to\mathbb{P}^1$.

Furthermore the map T can be viewed as an extension of the affine bundle $\mathscr{M}^{\mathrm{gen}}\to\mathbb{D}^\times$ to a trivial bundle on $\mathbb{P}^1-\{0\}$ of formal superaffine spaces, where the fiber at ∞ is the superaffine space $H^\bullet(\mathfrak{g},d)$. This results in a family $\mathbb{M}\mathsf{od}\to\mathbb{P}^1$ of formal supermanifolds, which is a trivial vector bundle outside of zero but has a non-linear fiber at $0 \in \mathbb{P}^1$. Moreover by picking the closed point in each fiber we get a section of $\mathbb{M}\mathsf{od} \to \mathbb{P}^1$, which is just the zero section of the vector bundle $\mathbb{M}\mathsf{od}_{|\mathbb{P}^1-\{0\}} \to \mathbb{P}^1 - \{0\}$. The normal bundle to this section in $\mathbb{M}\mathsf{od}$ is trivial (hence $\mathbb{M}\mathsf{od}$ is trivial as a non-linear bundle over $\mathbb{P}^1$), and the map Φ_T gives a (non-linear) trivialization of $\mathbb{M}\mathsf{od}$ over $\mathbb{P}^1$. This type of geometry was already discussed in [**CKS05**].

4.2.4. ***Relation to Calabi-Yau variations of* nc*-Hodge structures.*** Suppose (A, d, Δ) is a d($\mathbb{Z}/2$)g Batalin-Vilkovisky algebra which has the degeneration property. In this generality one does not expect to find a natural connection on $H^\bullet(A, d+u\Delta)$ along u, i.e. one does not expect to have a general formal analogue of an **nc**-Hodge structure.

However, a natural connection along the u-line may exist if we specify some additional data on (A, d, Δ). Following the analogy with the **nc**-Hodge structure associated with a symplectic manifold and the Gromov-Witten invariants, it is sufficient to specify:

- an even element $\kappa \in A$, with $d\kappa = 0$, and
- a grading operator $\mathsf{Gr} : A \to A$,

so that if we consider $\Gamma_{-1} := \mathsf{Gr} : A \to A$, and $\Gamma_{-2} : A \to A$ (the operator of multiplication by κ), then we have the commutation relations:

$$
\begin{aligned}
[\Gamma_{-1}, \Delta] &= -\frac{1}{2}\Delta \\
[\Gamma_{-2}, d] &= 0 \\
d &= [\Gamma_{-1}, d] + [\Gamma_{-2}, \Delta] .
\end{aligned}
$$

These commutation relations imply the identity

$$
\left[u\frac{\partial}{\partial u} + u^{-1}\Gamma_{-2} + \Gamma_{-1}, d + u\Delta\right] = \frac{1}{2}(d + u\Delta),
$$

which is consistent with the general formulas from Section 2.2.5. In particular, we can define a connection on $H^\bullet(A, d+u\Delta)$ along the u-line by setting

$$
\nabla_{\frac{\partial}{\partial u}} := \frac{\partial}{\partial u} + u^{-2}\Gamma_{-2} + u^{-1}\Gamma_{-1}.
$$

EXAMPLE 4.17. Let Y be a (possibly non-compact) d-dimensional Calabi-Yau manifold with a fixed holomorphic volume form $\mathbf{\Omega}_Y$. Let $\mathsf{w} : Y \to \mathbb{C}$ be a proper holomorphic function. This geometry gives rise to a natural dg Batalin-Vilkovisky algebra:

$$
\begin{aligned}
A &:= \Gamma_{C^\infty}\left(Y, \wedge^\bullet T_Y^{1,0} \otimes \wedge^\bullet A_Y^{0,1}\right), \\
d &:= \bar\partial + \iota_{d\mathsf{w}}, \\
\Delta &:= \mathsf{div}_{\mathbf{\Omega}_Y} = \iota_{\mathbf{\Omega}_Y}^{-1} \circ \partial \circ \iota_{\mathbf{\Omega}_Y},
\end{aligned}
$$

where $\iota_{\mathbf{\Omega}_Y} : \wedge^\bullet T_Y^{1,0} \to \wedge^{d-\bullet}\Omega_Y^{1,0}$ denotes the contraction with $\mathbf{\Omega}_Y$.

As discussed in Section 3.2 in this situation we get a connection along u which conjecturally defines an **nc**-Hodge structure. The connection is defined by the above formula with Γ_{-2} = the operator of multiplication by $-\mathsf{w}$, and $\Gamma_{-1} = \mathbf{Gr} : A \to A$, the grading operator which is equal to $\frac{q+p-d}{2} \cdot \mathrm{id}$ on $\Gamma_{C^\infty}\left(Y, \wedge^p T_Y^{1,0} \otimes \wedge^q A_Y^{0,1}\right)$.

We will elaborate on this geometric picture in the next section.

4.3. B-model framework: manifolds with anticanonical sections.

4.3.1. ***The classical Tian-Todorov theorem.*** Let X be a compact Kähler manifold. By Kodaira-Spencer theory we know that the deformations of X are controlled by the dg Lie algebra

$$\left(\mathfrak{g}^{(1)}, d_{\mathfrak{g}^{(1)}}\right) := \left(\Gamma_{C^\infty}\left(X, T_X^{1,0} \otimes_{C_X^\infty} A_X^{0,\bullet}\right), \bar{\partial}\right).$$

The classical Tian-Todorov theorem [**Tia87, Tod89**] can be formulated as follows:

THEOREM 4.18. *If X is a compact Kähler manifold with $c_1(X) = 0 \in \mathrm{Pic}(X)$, then $\left(\mathfrak{g}^{(1)}, d_{\mathfrak{g}^{(1)}}\right)$ is homotopy abelian. In particular the formal moduli space of X is smooth.*

Proof. Since $c_1(X) = 0 \in \mathrm{Pic}(X)$, up to scale we can find a unique holomorphic volume form $\mathbf{\Omega}_X$ on X. As in example 4.17 the pair $(X, \mathbf{\Omega}_X)$ gives rise to a dg Batalin-Vilkovisky algebra (A, d, Δ):

$$\begin{aligned} A &:= \Gamma_{C^\infty}\left(X, \wedge^\bullet T_X^{1,0} \otimes \wedge^\bullet A_X^{0,1}\right) \\ d &:= \bar{\partial} \\ \Delta &:= \mathsf{div}_{\mathbf{\Omega}_X} = \iota_{\mathbf{\Omega}_X}^{-1} \circ \partial \circ \iota_{\mathbf{\Omega}_X}. \end{aligned}$$

Consider the associated dg Lie algebra $(\mathfrak{g}, d_{\mathfrak{g}}) := (\mathbf{\Pi} A, d)$. We have a natural inclusion of dg Lie algebras

$$\begin{array}{ccc} \left(\mathfrak{g}^{(1)}, d_{\mathfrak{g}^{(1)}}\right) & \hookrightarrow & (\mathfrak{g}, d_{\mathfrak{g}}) \\ \| & & \| \\ \left(\Gamma_{C^\infty}\left(X, T_X^{1,0} \otimes_{C_X^\infty} A_X^{0,\bullet}\right), \bar{\partial}\right) & \hookrightarrow & \Gamma_{C^\infty}\left(X, \wedge^\bullet T_X^{1,0} \otimes \wedge^\bullet A_X^{0,1}\right) \end{array}$$

which embeds $\left(\mathfrak{g}^{(1)}, d_{\mathfrak{g}^{(1)}}\right)$ as a direct summand in $(\mathfrak{g}, d_{\mathfrak{g}})$, and so induces an embedding $H^\bullet\left(\mathfrak{g}^{(1)}, d_{\mathfrak{g}^{(1)}}\right) \subset H^\bullet(\mathfrak{g}, d_{\mathfrak{g}})$ in cohomology. So by Proposition 4.11 it suffices to check that $(\mathfrak{g}, d_{\mathfrak{g}})$ is homotopy abelian.

On the other hand the contraction map $\iota_{\mathbf{\Omega}_X}$ gives an isomorphism of bicomplexes between the dg Batalin-Vilkovisky algebra (A, d, Δ) and the Dolbeault bicomplex $(A^\bullet(X), \bar{\partial}, \partial)$. Since X is assumed compact and Kähler, the Hodge-to-de Rham spectral sequence degenerates on X, which is equivalent to the equality $\dim H^k_{\mathrm{dR}}(X, \mathbb{C}) = \dim(\oplus_{p+q=k} H^p(X, \Omega_X^q))$ which implies that the Dolbeault bicomplex $(A^\bullet(X), \bar{\partial}, \partial)$ has the degeneration property. Thus by Theorem 4.14 **(1)** it follows that $(\mathfrak{g}, d_{\mathfrak{g}})$ is homotopy abelian. The theorem is proven. □

4.3.2. ***Canonical coordinates on the moduli of Calabi-Yau manifolds.*** Let X be a Calabi-Yau manifold, i.e. a d-dimensional compact Kähler manifold with $c_1(X) = 0$ in $\mathrm{Pic}(X)$. Let (A, d, Δ) be the dg Batalin-Vilkovisky algebra defined in Section 4.3.1. The contraction map $\iota_{\mathbf{\Omega}_X}$ identifies the $\mathbb{C}[[u]]$-module $H^\bullet(\mathfrak{g}[[u]], d + u\Delta)$ with the Rees module of the **nc**-Hodge filtration on $H^\bullet_{\mathrm{dR}}(X, \mathbb{C})$

for which $H^{p,q}(X) \subset F^{\frac{p-q}{2}}$. Now choose one of the following equivalent pieces of data:

- a filtration $G^\bullet$ on $H^\bullet_{\mathrm{dR}}(X,\mathbb{C})$ which is opposed to the **nc**-Hodge filtration,
- a splitting of the **nc**-Hodge filtration,
- an extension of the associated **nc**-Hodge structure to a trivial bundle on $\mathbb{P}^1$ such that the connection has at most a first order pole at infinity.

Each such choice gives rise to an affine structure on $\mathbb{M}\mathsf{od}_{(\mathfrak{g},d_\mathfrak{g})}$. This affine structure is the same as the one described in Section 4.1.3 corresponding to the **nc**-Hodge structure above and the decoration ψ given by the class $[\boldsymbol{\Omega}_X]$ in the associated graded $\mathsf{gr}_{G^\bullet} H^\bullet_{\mathrm{dR}}(X,\mathbb{C})$.

In mirror symmetry considerations a choice of this type arises naturally when X is a Calabi-Yau manifold near a large complex structure limit point. Concretely, suppose $X = X_\mathfrak{z}$ is a member in a holomorphic family $\{X_z\}$ of compact d-dimensional Calabi-Yau manifolds parameterized by z in a polydisc $\prod_{i=1}^M\{z_i \in \mathbb{C}\,|\,0<|z_i|\ll 1\}$, and such that:

- $M = \dim_\mathbb{C} H^1\left(X_z, T_{X_z}\right)$;
- for each $i=1,\ldots,M$ the monodromy operator $t_i \in GL\left(H^1\left(X_\mathfrak{z}, T_{X_\mathfrak{z}}\right)\right)$ assigned to the circle (traced counterclockwise)
$$\gamma_i = \left\{ z \,\middle|\, \begin{array}{l} z_j = \mathfrak{z}_j,\ j\neq i, \\ |z_i| = |\mathfrak{z}_i| \end{array} \right\}$$
is unipotent of order d.

In this setup, the filtration $G^\bullet$ of $H^\bullet\left(X_\mathfrak{z},\mathbb{C}\right)$ invariant under all unipotent operators $\prod_{i=1}^M t_i^{a_i}$, $a_i \in \mathbb{Z}_{>0}$ will be opposed to the Hodge filtration and will thus give us canonical coordinates on the polydisc. This affine structure corresponds to a rational decoration of a Calabi-Yau variation of **nc**-Hodge structures.

4.3.3. ***Generalizations.*** Here we generalize the previous discussion to the case of varieties with divisors.

(i) Let X be a d-dimensional smooth projective variety over $\mathbb{C}$, and let $D\subset X$ be a normal crossings anti-canonical divisor, i.e $\mathcal{O}_X(D) = K_X^{-1} \in \mathrm{Pic}(X)$. Typically such an X will be a Fano or a quasi-Fano. If D is smooth, then by adjunction D will be a Calabi-Yau. Specifying such a divisor is equivalent to specifying a logarithmic volume form on X. This is up to scale a unique n-form $\boldsymbol{\Omega}_{X\log D} \in \Gamma\left(X, \Omega^d_X(\log D)\right)$ on X which has a first order pole along D and does not vanish anywhere on $X-D$.

Let $T_{X,D}$ be the subsheaf of T_X of holomorphic vector fields on X which at the points of D are tangent to D. This is a locally free subsheaf of T_X of rank d which controls the deformation theory of the pair (X,D). The relevant dg Batalin-Vilkovisky algebra (A,d,Δ) is an obvious generalization of the one in the absolute case:

$$\begin{aligned} A &:= \Gamma_{C^\infty}\left(X, \wedge^\bullet T_{X,D} \otimes_{\mathcal{C}^\infty_X} \wedge^\bullet A^{0,1}_X\right) \\ d &:= \bar\partial \\ \Delta &:= \mathsf{div}_{\boldsymbol{\Omega}_{X\log D}} = \iota^{-1}_{\boldsymbol{\Omega}_{X\log D}} \circ \partial \circ \iota_{\boldsymbol{\Omega}_{X\log D}}, \end{aligned}$$

where $\iota_{\boldsymbol{\Omega}_{X\log D}} : \wedge^\bullet T_{X,D} \to \Omega^{d-\bullet}_X(\log D)$ is the isomorphism given by contraction with $\boldsymbol{\Omega}_{X\log D}$.

Again, the map $\iota_{\mathbf{\Omega}_{X \log D}}$ identifies (A, d, Δ) with the logarithmic Dolbeault bicomplex $\left(A^{\bullet,\bullet}(\log D), \bar{\partial}, \partial\right)$. In particular, for all $u \neq 0$ we get an identification of the cohomology of the complex $(A, d+u\Delta)$ with the cohomology of the total complex of the double complex $\left(\Omega_X^{\bullet,\bullet}(\log D), \bar{\partial}, \partial\right)$, which is equal [**Voi03**, Section 6.1] to the cohomology of the open variety $X - D$. In other words for all $u \neq 0$ we have an isomorphism

$$H^\bullet(A, d + u\Delta) \cong H^\bullet_{\mathrm{dR}}(X - D, \mathbb{C}). \tag{4.3.1}$$

Now mixed Hodge theory implies the following

LEMMA 4.19. *The logarithmic dg Batalin-Vilkovisky algebra (A, d, Δ) has the degeneration property. In particular the formal moduli space of the pair (X, D) is smooth.*

We will return to the proof of this lemma in section 4.3.4 but first we will discuss a couple of variants of this geometric setup.

(ii) Suppose X is a smooth projective d-dimensional Calabi-Yau manifold. Let as before $\mathbf{\Omega}_X$ be the holomorphic volume form on X. Let $D \subset X$ be a normal crossings divisor. Typically if D is smooth, it will be a variety of general type.

Consider the dg Batalin-Vilkovisky algebra (A, d, Δ) given by

$$\begin{aligned} A &:= \Gamma_{C^\infty}\left(X, \wedge^\bullet T_{X,D} \otimes_{C^\infty_X} \wedge^\bullet A_X^{0,1}\right) \\ d &:= \bar{\partial} \\ \Delta &:= \mathsf{div}_{\mathbf{\Omega}_X} = \iota_{\mathbf{\Omega}_X}^{-1} \circ \partial \circ \iota_{\mathbf{\Omega}_X}, \end{aligned}$$

The contraction $\iota_{\mathbf{\Omega}_X}$ identifies this algebra with the dg Batalin-Vilkovisky algebra

$$\left(\Gamma_{C^\infty}\left(X, \Omega_X^\bullet(\mathrm{rel}\, D) \otimes_{C^\infty_X} \wedge^\bullet A_X^{0,1}\right), \bar{\partial}, \partial\right),$$

where $\Omega_X^k(\mathrm{rel}\, D) \subset \Omega_X^k$ denotes the subsheaf of all holomorphic k-forms that restrict to $0 \in \Omega^k_{D-\mathrm{sing}(D)}$. The cohomology of the total complex associated with this double complex is the de Rham cohomology of the pair (X, D), and so again we get an identification

$$H^\bullet(A, d + u\Delta) \cong H^\bullet_{\mathrm{dR}}(X, D; \mathbb{C}) \tag{4.3.2}$$

valid for all fixed $u \neq 0$. Again using this identification and mixed Hodge theory one deduces the following

LEMMA 4.20. *The dg Batalin-Vilkovisky algebra (A, d, Δ) has the degeneration property and hence the formal moduli space of the pair (X, D) is smooth.*

(iii) The setups **(i)** and **(ii)** have a natural common generalization. Fix a smooth projective complex variety of dimension d, a normal crossings divisor $D = \cup_{i \in I} D_i \subset X$, and a collection of weights $\{a_i\}_{i \in I} \subset [0, 1] \cap \mathbb{Q}$, so that

$$\sum_{i \in I} a_i [D_i] = -K_X \in \mathrm{Pic}(X) \otimes \mathbb{Q}.$$

Represent the a_i's by reduced fractions, take $N \geq 1$ to be the least common multiple of the denominators of these fractions and such that

$$\sum_{i \in I} (N a_i)[D_i] = -N K_X \in \mathrm{Pic}(X),$$

and set $n_i := a_i N$. In particular up to scale we have a unique section $\widetilde{\mathbf{\Omega}}_X \in \Gamma\left(X, K_X^{\otimes(-N)}\right)$ whose divisor is $\sum_{i\in I} n_i D_i$. In this situation we can again promote the Dolbeault dg Lie algebra which computes the deformation theory of (X, D) to a dg Batalin-Vilkovisky algebra (A, d, Δ), where

$$\begin{aligned} A &:= \Gamma_{C^\infty}\left(X, \wedge^\bullet T_{X,D} \otimes_{C^\infty_X} \wedge^\bullet A_X^{0,1}\right) \\ d &:= \bar{\partial} \\ \Delta &:= \mathsf{div}_{\widetilde{\mathbf{\Omega}}_X} . \end{aligned}$$

The divergence operator $\mathsf{div}_{\widetilde{\mathbf{\Omega}}_X}$ is defined as follows. Restricting the section $\widetilde{\mathbf{\Omega}}_X$ to $X - D$ we get a nowhere vanishing section of $K_{X-D}^{\otimes(-N)}$, i.e. a flat holomorphic connection on K_{X-D}. If $U \subset X - D$ is a simply connected open, then we can choose $\mathbf{\Omega}_U$ a holomorphic volume form on U which is covariantly constant for this flat connection, and define the associated divergence operator $\mathsf{div}_{\mathbf{\Omega}_U} := \iota_{\mathbf{\Omega}_U}^{-1} \circ \partial \circ \iota_{\mathbf{\Omega}_U}$. But by the flatness of the connection it follows that any other covariantly constant volume form on U will be proportional to $\mathbf{\Omega}_U$ with a constant proportionality coefficient. Since by definition $\mathsf{div}_{c\mathbf{\Omega}_U} = \mathsf{div}_{\mathbf{\Omega}_U}$ for any constant c we get a well defined divergence operator on $X - U$. Furthermore, locally this divergence operator is given by a holomorphic volume form which is a branch of $\left(\widetilde{\mathbf{\Omega}}_X\right)^{-1/N}$ and so by continuity it gives a well defined map of locally free sheaves $\mathsf{div}_{\widetilde{\mathbf{\Omega}}_X} : \wedge^i T_{X,D} \to \wedge^{i-1} T_{X,D}$.

Again, we claim that

LEMMA 4.21. *The dg Batalin-Vilkovisky algebra (A, d, Δ) has the degeneration property and the formal moduli space of the pair (X, D) is smooth.*

Proof. The proof of this lemma again reduces to mixed Hodge theory via a map similar to the isomorphisms (4.3.1) and (4.3.2). However constructing this map is a bit more involved than the arguments we used to construct (4.3.1) and (4.3.2).

Consider the root stack $Z = X\left\langle \left\{\frac{D_i}{N}\right\}_{i\in I}\right\rangle$ as defined in e.g. [**MO05, IS07**]. By construction Z is a smooth proper Deligne-Mumford stack, equipped with a finite and flat morphism $\pi : Z \to X$.

Conceptually the best way to define the stack Z is as a moduli stack classifying (special) log structures associated with X, the divisor D and the number N (see [**MO05**] for the details). Étale locally on X the stack Z can be described easily as a quotient stack. Indeed, choose étale locally an identification of X with a neighborhood of zero in $\mathbb{A}^d$ with coordinates $z_1, \ldots, z_d$, so that $D = D_1 \cup \cdots \cup D_r$ and D_i is identified with the hyperplane $z_i = 0$. Then the corresponding étale local patch in Z is canonically isomorphic to the stack quotient

$$[\mathbb{A}^d / \underbrace{\boldsymbol{\mu}_N \times \cdots \times \boldsymbol{\mu}_N}_{r\text{-times}}],$$

where $\boldsymbol{\mu}_N \subset \mathbb{C}^\times$ is the group of N-th roots of unity, and $(\zeta_1, \ldots, \zeta_r) \in \boldsymbol{\mu}_N^{\times r}$ acts as $(z_1, \ldots, z_r, z_{r+1}, \ldots, z_d) \mapsto (\zeta_1 z_1, \ldots, \zeta_r z_r, z_{r+1}, \ldots, z_d)$.

In particular, this description shows (see [**MO05**, Theorem 4.1]) that:

- The map π is an isomorphism over $X - D$ and in general identifies X with the coarse moduli space of Z;
- There is a strict normal crossings divisor $\widetilde{D} = \cup_{i\in I} \widetilde{D}_i \subset Z$, such that
$$\mathcal{O}_Z\left(-N\widetilde{D}_i\right) = \pi^*\mathcal{O}_X(-D_i)$$
as ideal subsheaves of $\mathcal{O}_Z$;
- For all j we have the Hurwitz formula $\Omega^j_Z(\log \widetilde{D}) = \pi^*\Omega^j_X(\log D)$.

In particular we have canonical isomorphisms

$$\pi^* K_X \cong \mathcal{O}_Z\left(-\sum_{i\in I} n_i \widetilde{D}_i\right)$$

$$\pi^* K_X \cong K_Z \otimes \mathcal{O}_Z\left((1-N)\sum_{i\in I} \widetilde{D}_i\right)$$

the first given by the section $\pi^*\widetilde{\boldsymbol{\Omega}}_X$ and the second coming from the Hurwitz formula.

There is a natural complex local system of rank one on $X - D$ with monodromy in $\boldsymbol{\mu}_N$ associated with the choices of N-th root of the section $\widetilde{\boldsymbol{\Omega}}_X$. It is easy to see that the pullback of this local system admits a canonical extension (as a local system) to Z, which we denote by $\boldsymbol{\Xi}$. Moreover, we have a canonical meromorphic section $\boldsymbol{\Omega}_Z$ of $K_Z \otimes_{\mathbb{C}} \boldsymbol{\Xi}$ with divisor $\sum_{i\in I}(N-1-n_i)\widetilde{D}_i$. It is easy to check locally by using the étale local description of Z as a quotient stack that the contraction $\iota_{\boldsymbol{\Omega}_Z}$ gives a well defined isomorphism of locally free sheaves:

$$\iota_{\boldsymbol{\Omega}_Z} : \quad \wedge^j T_{Z,\widetilde{D}} \xrightarrow{\cong} \Omega^{d-j}_Z\left(\log \widetilde{D}_{(1)}, \mathrm{rel}\, \widetilde{D}_{(0)}\right) \otimes_{\mathbb{C}} \boldsymbol{\Xi}.$$

Here

$$\widetilde{D}_{(0)} := \cup_{i\in I_0} \widetilde{D}_i \qquad I_0 = \{i \in I | a_i = 0\}$$
$$\widetilde{D}_{(1)} := \cup_{i\in I_1} \widetilde{D}_i \qquad I_1 = \{i \in I | a_i = 1\}.$$

Now taking into account the Hurwitz isomorphism $\wedge^j T_{Z,\widetilde{D}} \cong \pi^* \wedge^j T_{X,D}$ and using adjunction, we can view $\iota_{\boldsymbol{\Omega}_Z}$ as an isomorphism

$$\wedge^j T_{X,D} \xrightarrow{\cong} \left(\pi_*\Omega^{d-j}_Z\left(\log \widetilde{D}_{(1)}, \mathrm{rel}\, \widetilde{D}_{(0)}\right) \otimes_{\mathbb{C}} \boldsymbol{\Xi}\right) \tag{4.3.3}$$

It is immediate from the definition that the isomorphism (4.3.3) (taken for all j) identifies the dg Batalin-Vilkovisky algebra (A, d, Δ) with the Dolbeault bicomplex

$$\left(\Gamma_{C^\infty}\left(X, \left(\pi_*\Omega^{\bullet}_Z\left(\log \widetilde{D}_{(1)}, \mathrm{rel}\, \widetilde{D}_{(0)}\right) \otimes_{\mathbb{C}} \boldsymbol{\Xi}\right) \otimes_{C^\infty_X} A^{0,\bullet}_X\right), \bar\partial, \partial\right).$$

But the above complex equipped with the differential $\partial + \bar\partial$ is the Dolbeault resolution of the complex of sheaves $\pi_*\left(\Omega^{\bullet}_Z\left(\log \widetilde{D}_{(1)}, \mathrm{rel}\, \widetilde{D}_{(0)}\right) \otimes_{\mathbb{C}} \boldsymbol{\Xi}, \partial\right)$ which is equal to the derived direct image $R\pi_*\left(\Omega^{\bullet}_Z\left(\log \widetilde{D}_{(1)}, \mathrm{rel}\, \widetilde{D}_{(0)}\right) \otimes_{\mathbb{C}} \boldsymbol{\Xi}, \partial\right)$ since π is finite. Now combined with the Leray spectral sequence for π this gives, for all $u \neq 0$, an isomorphism

$$H^{\bullet}(A, d + u\Delta) \cong H^{\bullet}_{\mathrm{dR}}\left(Z - \widetilde{D}_{(1)}, \widetilde{D}_{(0)} - \widetilde{D}_{(1)}; \boldsymbol{\Xi}\right), \tag{4.3.4}$$

which specializes to both isomorphisms (4.3.1) and (4.3.2).

Now the fact that Z is a smooth and proper Deligne-Mumford stack and mixed Hodge theory (see 4.3.4) for $\left(Z - \widetilde{D}_{(1)}, \widetilde{D}_{(0)} - \widetilde{D}_{(1)}\right)$ endowed with the local system Ξ imply that (A, d, Δ) has the degeneration property. □

REMARK 4.22. The fact that the root stack in the previous proof can be viewed as the moduli stack of special log structures is very interesting. It suggests that the setup we just discussed may fit naturally in the recent approach of Gross-Siebert [**GS06, GS07**] to mirror symmetry and instanton corrections via log degenerations of toric Fano manifolds (see also [**KS06a, KS01**]). The relationship between these two setups is certainly worth studying and we plan to return to it in the future.

(iv) Yet another generalization of the previous picture arises when we take the variety X to be a normal-crossings Calabi-Yau. More precisely, assume that X is a strict normal crossings variety with irreducible components $X = \cup_{i\in I} X_i$ equipped with a holomorphic volume form $\boldsymbol{\Omega}_X$ on $X - X^{\text{sing}}$ such that the restriction of $\boldsymbol{\Omega}_X$ on each X_i has a logarithmic pole along $X_i \cap (\cup_{j\neq i} X_j)$ and the residues of these restricted forms cancel along each $X_i \cup X_j$. Taking a colimit along the projective system of all finite intersections of components of X we get again a dg Batalin-Vilkovisky algebra $A_{\text{tot}}(X) = \mathsf{colim}_{J\subset I}\, A\left(\cap_{i\in J} X_i\right)$ and again by using mixed Hodge theory we can check that this algebra has the degeneration property.

4.3.4. ***Mixed Hodge theory in a nutshell.*** In this section we briefly recall the basic arguments from Deligne's mixed Hodge theory [**Del74**] that are necessary for proving the degeneration property of the dg Batalin-Vilkovisky algebras in section 4.3.3 **(i)**-**(iv)**.

Suppose we are given:

- a finite ordered collection (X_α) of smooth complex projective varieties;
- for every α a choice of a $\mathbb{Z}\times\mathbb{Z}$-graded complex of sheaves of differential forms which are either C^∞ or are $C^{-\infty}$ (i.e. currents) and constrained so that their wave-front (singular support) is contained in a given conical Lagrangian in $T^\vee X_\alpha$ which is the conormal bundle to a normal crossing divisor in X_α;
- a collection of integers $n_\alpha \in \mathbb{Z}$.

Consider the complex $C^{\text{tot}} = \oplus_\alpha C^\bullet_\alpha[n_\alpha]$ equipped with three differentials ∂, $\bar\partial$, δ, where $\delta = \sum_{\alpha<\beta} \delta_{\alpha\beta}$, and the $\delta_{\alpha\beta}$ come from pullbacks and pushforwards for some maps $X_\beta \hookrightarrow X_\alpha$ or $X_\alpha \hookrightarrow X_\beta$. The statement we need now can be formulated as follows:

CLAIM 4.23. *For every $k \geq 1$ the cohomology*

$$H^\bullet\left(C^{\text{tot}}[u]/(u^k), \bar\partial + \delta + u\partial\right)$$

is a free $\mathbb{C}[u]/(u^k)$-module.

Proof. If X is smooth projective over $\mathbb{C}$ and if $\left(A^\bullet(X), \bar\partial\right)$ is the $\bar\partial$-complex of (either C^∞ or $C^{-\infty}$) differential forms on X, then the inclusion

$$\left(\ker\partial, \bar\partial\right) \hookrightarrow \left(A^\bullet(X), \bar\partial\right)$$

is a quasi-isomorphism.

This implies that the horizontal arrows in the diagram of complexes

$$\begin{array}{ccc} \left(\ker \partial[u]/(u^k), \bar{\partial} + \delta + u\partial\right) & \longrightarrow & \left(C^{\text{tot}}[u]/(u^k), \bar{\partial} + \delta + u\partial\right) \\ \| & & \\ \left(\ker \partial[u]/(u^k), \bar{\partial} + \delta\right) & \longrightarrow & \left(C^{\text{tot}}, \bar{\partial} + \delta\right)[u]/(u^k), \end{array}$$

are quasi-isomorphisms. Indeed, this follows by noticing that there are natural filtrations on both sides (by the powers of u and the index α) which give rise to convergent spectral sequences and induce the quasi-isomorphic inclusion $\left(\ker \partial, \bar{\partial}\right) \hookrightarrow \left(C^{\text{tot}}, \bar{\partial}\right)$ on the associated graded. This proves the claim. □

REMARK 4.24. • Note that the same reasoning implies that the natural map

$$\left(\ker \partial, \bar{\partial} + \delta\right) \twoheadrightarrow \left(\ker \partial / \operatorname{im} \partial, \bar{\partial} + \delta\right) = \left(H^\bullet(X_\alpha), \delta\right),$$

is also a quasi-isomorphism, which reduces the problem of computing $H^\bullet\left(C^{\text{tot}}[u]/(u^k), \bar{\partial} + \delta + u\partial\right)$ to a homological algebra question on a complex of finite dimensional vector spaces.

• There is a useful variant of the theory, also discussed in [**Del74**]: the previous discussion immediately generalizes to the case of cochain complexes of a collection of projective manifolds with coefficients in some unitary local systems.

Next we discuss a few examples and applications of the geometric setup from section 4.3.3.

4.3.5. ***The moduli stack of Fano varieties.*** As a consequence of section 4.3.3 **(iii)** we get a new proof and a refinement of the following result of Ran [**Ran92, Kaw92**]:

THEOREM 4.25. *Let X be a complex Fano manifold, that is, let X be a smooth proper $\mathbb{C}$-variety for which K_X^{-1} is ample. Then the versal deformations of X are unobstructed.*

Proof: Choose $N > 1$ so that $K_X^{\otimes(-N)}$ is very ample and all the higher cohomology groups $H^k\left(X, K_X^{\otimes(-N)}\right)$ vanish for $k \geq 1$. Choose a generic section $\widetilde{\mathbf{\Omega}}_X \in H^0\left(X, K_X^{\otimes(-N)}\right) = 0$ whose zero locus is a smooth and connected divisor $D \subset X$.

Consider now $\mathfrak{g} = \mathbf{\Pi} R\Gamma\left(X, \wedge^\bullet T_{X,D}\right)$ with the Schouten bracket. By Lemma 4.21 this d$(\mathbb{Z}/2)$g Lie algebra is homotopy abelian and so as in the proof of Theorem 4.18 we conclude that $\mathfrak{g}^{(1)} = R\Gamma\left(X, T_{X,D}\right)$ is homotopy abelian. Since this d$(\mathbb{Z}/2)$g Lie algebra governs the deformation theory of (X, D) as a variety with a divisor, it follows that the formal germ of the deformation space of the pair (X, D) is smooth. Next we will need the following simple

LEMMA 4.26. *Suppose (X', D') is a small deformation of (X, D) as a variety with divisor. Then X' is still a Fano with $K_{X'}^{\otimes(-N)}$ very ample and $D' \in \left|K_{X'}^{\otimes(-N)}\right|$.*

Proof: The condition of $K_X^{\otimes(-N)}$ being very ample is open in the moduli of X. Furthermore by definition $K_X^{\otimes(-N)} \otimes \mathcal{O}_X(-D) = \mathcal{O}_X$ and so by the small deformation hypothesis it follows that $K_{X'}^{\otimes(-N)} \otimes \mathcal{O}_{X'}(-D)$ is in the connected component of the identity of $\mathrm{Pic}(X')$. But X' is a Fano and so $\mathrm{Lie}(\mathrm{Pic}^0(X')) = H^1(X', \mathcal{O}_{X'}) = 0$. Hence $K_{X'}^{\otimes(-N)} \otimes \mathcal{O}_{X'}(-D) = \mathcal{O}_X$ as well. □

The theorem now follows easily. The versal deformation space of smooth connected D's for a given X is smooth and isomorphic to a domain in $\mathbb{P}^{h^0\left(X, K_{X'}^{\otimes(-N)}\right)-1}$. Since the dimension of these projective spaces is locally constant in X by Riemann-Roch and vanishing of the higher cohomologies, it follows that the map from the versal deformation space of the pairs (X, D) to the versal deformation stack of X is smooth. In other words the versal deformation stack of X has a presentation in the smooth topology with a smooth atlas – the versal deformation space for (X, D). Hence the versal deformations of X form a smooth stack. □

4.3.6. ***Algebras for the Landau-Ginzburg model.*** Consider again the setup of a holomorphic Landau-Ginzburg model. Suppose Y is smooth and quasiprojective over $\mathbb{C}$ and of dimension $\dim Y = d$. Suppose there exists a nowhere vanishing algebraic volume form $\mathbf{\Omega}_Y \in \Gamma(Y, K_Y)$, and let $\mathsf{w} : Y \to \mathbb{A}^1$ be a regular function with compact critical locus.

This data gives a dg Batalin-Vilkovisky algebra (A, d, Δ) where

$$\begin{aligned} A &:= \Gamma_{C^\infty}\left(Y, \wedge^\bullet T_Y^{1,0} \otimes_{C_Y^\infty} \wedge^\bullet A_Y^{0,1}\right) \\ d &:= \bar{\partial} + \iota_{d\mathsf{w}} \\ \Delta &:= \mathsf{div}_{\tilde{\mathbf{\Omega}}_Y} . \end{aligned}$$

Again the contraction $\iota_{\mathbf{\Omega}_Y}$ identifies (A, d, Δ) with the twisted Dolbeault bicomplex $\left(A^\bullet(Y), \bar{\partial} + d\mathsf{w}\wedge, \partial\right)$. The latter satisfies the degeneration property by the work of Barannikov and the second author, Sabbah [**Sab99**], or Ogus-Vologodsky [**OV05**].

REMARK 4.27. It will be interesting to combine the previous discussion with the discussion in section 4.3.3 **(iii)** or with the broken Calabi-Yau geometry from section 4.3.3 **(iv)**. Suppose we have a quasiprojective smooth complex Y, a regular function $\mathsf{w} : Y \to \mathbb{A}^1$ with compact critical locus, and suppose we are given a normal crossings divisor $D = \cup_{i\in I} D_i$ and a system of weights $\{a_i\}_{i\in I}$ as in Section 4.3.3 **(iii)**. Then we can write the w-twisted version of the dg Batalin-Vilkovisky algebra for (Y, D) which by general nonsense will compute the deformation theory of the data (Y, D, w). Similarly we can add a potential to Y which itself is a normal-crossings Calabi-Yau, as in Section 4.3.3 **(iv)**. We expect that the resulting algebras will again have the degeneration property, but we have not investigated this question.

4.4. Categorical framework: spherical functors.

In this section we briefly discuss some algebraic aspects of the deformation theory of **nc**-spaces (see section 2.2.1). For simplicity we will discuss the $\mathbb{Z}$-graded case but in fact all definitions and statements readily generalize to the $\mathbb{Z}/2$ case.

4.4.1. ***Calabi-Yau* nc*-spaces.*** Suppose $X = \mathbf{ncSpec}(A)$ is a graded **nc**-affine **nc**-space over $\mathbb{C}$. If X is smooth, then $A \in \mathsf{Perf}_{X\times X^{\mathrm{op}}} = \mathsf{Perf}\,(A \otimes A^{\mathrm{op}} - \mathsf{mod})$ and we define the ***smooth dual*** of A to be $A^! := \mathrm{Hom}_{A\otimes A^{\mathrm{op}}}(A, A \otimes A)$. Similarly if X is compact, then $A \in \mathsf{Perf}_{\mathrm{pt}}$ and we define the ***compact dual*** of A to be $A^* := \mathrm{Hom}_{\mathbb{C}}(A, \mathbb{C}) \in (A \otimes A^{\mathrm{op}} - \mathsf{mod})$.

If X is both a smooth and compact **nc**-space, then we have isomorphisms

$$A^! \otimes_A A^* \cong A^* \otimes_A A^! \cong A$$

in the category $(A \otimes A^{\mathrm{op}} - \mathsf{mod})$. The endofunctor $S_X : C_X \to C_X$ given by the A-bimodule A^* is called ***the Serre functor*** of X. It is an autoequivalence of C_X which is central (i.e. commutes with all autoequivalences). Moreover, for any two objects $\mathcal{E}, \mathcal{F} \in \mathsf{Perf}_X$ there is a functorial identification

$$\mathrm{Hom}_X(\mathcal{E}, \mathcal{F})^\vee \cong \mathrm{Hom}_X(\mathcal{F}, S_X \mathcal{E}).$$

With this notation we have the following definition (see also [**KS06b**]):

DEFINITION 4.28. *We say that a smooth graded* **nc**-*affine* **nc**-*space* $X =$ **ncSpec**(A) *is a* ***Calabi-Yau of dimension*** $d \in \mathbb{Z}$ *if* $A^! \cong A[-d]$ *in* $(A \otimes A^{\mathrm{op}} - \mathsf{mod})$. *We say that a compact* **nc**-*affine* **nc**-*space* $X =$ **ncSpec**(A) *is a* ***Calabi-Yau of dimension*** $d \in \mathbb{Z}$ *if* $A^* \cong A[d]$ *in* $(A \otimes A^{\mathrm{op}} - \mathsf{mod})$.

The definition works also in the $\mathbb{Z}/2$-graded case, where the dimension d is understood as an element of $\mathbb{Z}/2$.

For an **nc**-space which is both smooth and compact the two conditions are equivalent and are equivalent to having an isomorphism of endofunctors $S_X \cong [d]$.

REMARK 4.29. This definition of a Calabi-Yau structure on a smooth compact **nc**-space is somewhat simplistic and should be taken with a grain of salt. The true definition (see [**KS06b**]) implies the isomorphism of functors $S_X \cong [d]$ but also involves higher homotopical data which are encoded in a cyclic category structure on C_X. We will suppress the cyclic structure here in order to simplify the discussion.

We are interested in **nc**-space analogues of the Tian-Todorov theorem. The unobstructedness of graded smooth and compact **nc**-Calabi-Yau spaces was recently analyzed by Pandit [**Pan08**] via the T^1-lifting property of Ran [**Ran92**] and Kawamata [**Kaw92**]. Here we formulate the following general

THEOREM 4.30. *Suppose that X is a smooth and compact* **nc**-*Calabi-Yau space of dimension $d \in \mathbb{Z}$ (or of dimension $d \in \mathbb{Z}/2$ in the $\mathbb{Z}/2$-graded case). Assume that X satisfies the degeneration conjecture (see Section 2.2.4). Then:*

- *the Hochschild cochain algebra $C^\bullet(X)$ of X is a homotopy abelian L_∞ algebra;*
- *the formal moduli space $\mathbb{M}\mathsf{od}_X$ of X is a formal supermanifold, i.e.*

$$\mathbb{M}\mathsf{od}_X := \mathbb{M}\mathsf{od}_{C^\bullet(A,A)} \cong \mathrm{Spf}\, \mathbb{C}[[x_1, \ldots, x_N, \xi_1, \ldots, \xi_M]];$$

- *the negative cyclic homology of the universal family over $\mathbb{M}\mathsf{od}_X$ gives a vector bundle $H \to \mathbb{M}\mathsf{od}_X \times \mathbb{D}$ which is equipped with a flat meromorphic connection ∇ so that $\nabla_{u\partial/\partial x_i}$, $\nabla_{u\partial/\partial \xi_j}$, and $\nabla_{u^2\partial/\partial u}$ are regular;*
- *(H, ∇) is the de Rham part of a Calabi-Yau variation of* **nc**-*Hodge structures.*

We will only sketch some of the highlights of the proof of this theorem here since going into full detail would take us too far afield. The proof is based on a mildly generalized version of Deligne's conjecture (see e.g. [**KS00, Tam03**]) which states that the Hochschild cochain complex of an affine **nc**-space is also an algebra over the operad of chains of the little discs operad. The first step is to show that under the Calabi-Yau assumption the Hochschild cochain complex $C^\bullet(X)$ is also naturally an algebra over the cyclic operad of chains of the framed little discs operad (i.e. the operad of little discs with a marked point on the boundary). Next one shows that the validity of the degeneration conjecture for X implies that the induced S^1-action on the cochain complex is homotopically trivial. Finally by a topological argument one deduces from this the fact that all the higher L_∞ operations on $C^\bullet(X)$ must vanish.

REMARK 4.31. It seems certain[2] that from deformation quantization it follows that if X is a smooth and projective Calabi-Yau variety, then the data described in the above theorem are canonically isomorphic to the formal completion of the variation of **nc**-Hodge structures described in section 4.3.2.

For a general smooth and compact **nc**-Calabi-Yau space we expect that the formal variation of **nc**-de Rham data in Theorem 4.30 converges to give analytic de Rham data which contain a compatible **nc**-Betti data $\mathscr{E}_B$ and so extend to an honest variation of **nc**-Hodge structures.

4.4.2. ***Spherical functors.*** In this section we introduce a special version of the general notion of a spherical functor [**Ann07**] which is tailored to the Calabi-Yau condition. We begin with a definition:

DEFINITION 4.32. *Let X and Y be two graded **nc**-spaces. A **morphism** $f : X \to Y$ is a triple of functors*

$$\begin{array}{c} C_X \\ f^*\big\uparrow \;\; \big\downarrow f_* \;\; \big\uparrow f^! \\ C_Y \end{array}$$

so that (f^*, f_*) and $(f_*, f^!)$ are (left, right) pairs of adjoint functors.

Suppose now X, Y are smooth and compact graded **nc**-spaces and let Y be an **nc**-Calabi-Yau of dimension d.

DEFINITION 4.33. *A morphism $f : X \to Y$ is **spherical** if:*

(a) *the cone of the natural adjunction morphism $\mathrm{id}_{C_X} \to f^! \circ f_*$ is isomorphic to the shifted Serre functor of X: $\mathrm{cone}\left(\mathrm{id}_{C_X} \to f^! \circ f_*\right) \cong S_X[1-d]$,*

(b) *the natural map $f^! \to S_X[1-d] \circ f^*$, induced from the isomorphism in* **(a)** *and the adjunction $f^! \to f^! \circ f_* \circ f^*$, is an isomorphism of functors.*

REMARK 4.34. **(a)** If f is spherical, then the associated ***reflection functor*** $\mathcal{R}_f := \mathrm{cone}\left(f_* \circ f^! \to \mathrm{id}_{C_Y}\right)$ is an auto-equivalence of C_Y [**Ann07**].
(b) Similarly to the definition of a Calabi-Yau structure the above notion of a spherical functor should be viewed as a weak preliminary version of a stronger more refined notion which has to involve higher homotopical data and has yet to be defined carefully.

EXAMPLE 4.35. **(i)** Let $X = \mathrm{pt}$ and Y be a d-dimensional smooth and compact **nc**-Calabi-Yau and let $\mathcal{E} \in C_Y$ be a spherical object, i.e. an object for which the complex of $\mathbb{C}$-vector spaces $\mathrm{Hom}_Y(\mathcal{E}, \mathcal{E})$ is quasi-isomorphic to $(H^\bullet(S^d, \mathbb{C}), 0)$. The

[2] We borrowed this delightful expression from [**Kap91**].

morphism of **nc**-spaces $f : \mathrm{pt} \to Y$ given by $f_*(V) = \mathcal{E} \otimes V$ for any $V \in C_{\mathrm{pt}} = (\mathsf{Vect}_{\mathbb{C}})$ is spherical.
(ii) Let X be smooth and projective of dimension $d+1$, and let $i : Y \hookrightarrow X$ be a smooth anti-canonical divisor in X. Then Y is a d-dimensional Calabi-Yau and we have a natural spherical **nc**-morphism $f : X \to Y$ given by $f_* := i^*$, $f^! := i_*$, etc.
(iii) Let Y be a smooth projective d-dimensional Calabi-Yau. Let $i : X \hookrightarrow Y$ be a smooth hypersurface. Then we have a natural spherical **nc**-morphism $f : X \to Y$ given by $f_* = i_*$, $f^! = i^!$, and $f^* = i^*$.

REMARK 4.36. The geometry of Example 4.35 **(ii)**, where X is taken to be a smooth projective Fano, and $i : Y \hookrightarrow X$ is a smooth anti-canonical divisor, can be encoded algebraically in the categories $C_X = D(\mathsf{Qcoh}(X))$, $C_Y = D(\mathsf{Qcoh}(Y))$, the functor $f_* = i^*$, and another natural triple of categories:

- the compact category $D_{\substack{\text{compact}\\ \text{support}}}(\mathsf{Qcoh}(X - Y)) = \ker(f_*)$,
- the compact category $D_{\mathrm{supp}\ Y}(\mathsf{Qcoh}(X)) =$ the subcategory in $D(\mathsf{Qcoh}(X))$ generated by $i_* D(\mathsf{Qcoh}(Y))$,
- the smooth category $D(\mathsf{Qcoh}(X - Y)) =$ the quotient $D(\mathsf{Qcoh}(X))/D_{\mathrm{supp}\ Y}(\mathsf{Qcoh}(X))$.

There is a similar triple of categories for the setup in Example 4.35 **(iii)**. It will be very interesting to describe the categorical data that encode anti-canonical divisors with normal crossings or more generally the fractional anti-canonical divisor setup from section 4.3.3 **(iii)**. It seems likely that in this situation one gets a system of nested categories and functors with a "spherical" condition imposed on the whole system rather than on individual functors. This is a very interesting question that we plan to investigate in the future.

REMARK 4.37. It is clear from the examples above that spherical functors give a unifying framework for handling different types of geometric pairs.

Suppose that X and Y are smooth and compact **nc**-spaces, Y is an **nc**-Calabi-Yau, $f : X \to Y$ is a spherical map, and the degeneration conjecture holds for both X and Y. In this situation we expect that the deformation theory of $f : X \to Y$ is controlled by a homotopy abelian d$(\mathbb{Z}/2)$g Lie algebra which is L_∞-quasi-isomorphic to

$$\mathrm{cone}\left(C_\bullet(Y) \xrightarrow{f^!} C_\bullet(X) \right)[1-d]. \tag{4.4.1}$$

Moreover, using f_* (or $f^!$) we can build a new **nc**-space Z by taking C_Z to be the semi-orthogonal extension $C_Z = \langle C_X, C_Y \rangle$, where we set

$$\mathrm{Hom}_Z(C_Y, C_X) := 0$$

$$\mathrm{Hom}_Z(\mathcal{E}, \mathcal{F}) := \mathrm{Hom}_Y(f_*\mathcal{E}, \mathcal{F}) \quad \text{for all } \mathcal{E} \in C_X, \mathcal{F} \in C_Y.$$

We expect that the deformation theory of $f : X \to Y$ is equivalent to the deformation theory of Z and in particular that the L_∞ algebra $C^\bullet(Z)$ is quasi-isomorphic to the algebra (4.4.1).

REMARK 4.38. We should point out that even though deformation quantization provides a conceptual bridge between the categorical framework and the geometric framework of the previous section, the actual connection between the two frameworks is tenuous at best. The source of the problem lies in the fact that the deformation quantization of general Poisson maps can be obstructed [**Wil07**].

4.5. A-model framework: symplectic Landau-Ginzburg models. We already noted in Examples 4.4 and 4.8 that there are natural canonical coordinates and a Calabi-Yau variation of **nc**-Hodge structures that one can attach to the A-model on a compact symplectic manifold. An interesting open problem is to find an algebraic description of these coordinates and variations in terms of some d($\mathbb{Z}/2$)g Batalin-Vilkovisky algebra that is naturally attached to the Fukaya category. This question is hard and we will not study it directly here. Instead we will look at the question of finding canonical coordinates and variations in another symplectic context, i.e. for symplectic Landau-Ginzburg models, and try to get an insight into a possible algebraic formulation in that case. It will be interesting to compare our formalism with the recent work of Fan-Jarvis-Ruan [**FJR07**] on the symplectic geometry of quasi-homogeneous Landau-Ginzburg potentials with isolated singularities but at the moment we do not see a direct relationship.

4.5.1. ***Symplectic geometry with potentials.*** The objects we would like to understand are triples (Y, w, ω), where

- Y is a C^∞-manifold and ω is a C^∞-symplectic form on Y.
- $\mathsf{w} : Y \to \mathbb{C}$ is a proper C^∞-map such that there exists an $R > 0$ so that over $\{z \in \mathbb{C} | \, |z| \geq R\}$ the map w is a smooth fibration with fibers symplectic submanifolds in (Y, ω).

Similarly to the case of compact symplectic manifolds one can associate Gromov-Witten invariants to such a geometry. Specifically, if we fix $n \geq 1$, $g \geq 0$, and $\beta \in H_2(Y, \mathbb{Z})$, then we can use stable pseudo-holomorphic pointed curves in Y to define a natural linear (correlator) map

$$I^{(1)}_{g,\beta,n-1} : \quad H^\bullet(Y, \mathbb{Q})^{\otimes(n-1)} \otimes H^\bullet\left(\overline{\mathcal{M}}_{g,n}, \mathbb{Q}\right) \longrightarrow H^\bullet(Y, \mathbb{Q}).$$

Indeed, note that Poincaré duality gives an identification

$$H^\bullet(Y, \mathbb{Q}) \cong H_\bullet(Y, Y_R; \mathbb{Q})[-\dim Y],$$

where $R > 0$ is as above and $Y_R = \mathsf{w}^{-1}(\{z \in \mathbb{C} | \, |z| \geq R\}) \subset Y$. Combining this identification with the isomorphism $(H^\bullet)^\vee = H_\bullet$ we see that $I^{(1)}_{g,\beta,n-1}$ will be given by a class in $H_\bullet(Y, \mathbb{Q})^{\otimes(n-1)} \otimes H_\bullet(Y, Y_R; \mathbb{Q}) \otimes H_\bullet(\overline{\mathcal{M}}_{g,n}, \mathbb{Q})$.

Next consider the usual moduli stack $\overline{\mathcal{M}}_{g,n}(Y, \beta)$ of stable pseudo-holomorphic maps. Here it will be convenient to assume that an almost-complex structure on Y tamed by ω is chosen in such a way that $\mathsf{w}_{|Y_R}$ is holomorphic. The stack $\overline{\mathcal{M}}_{g,n}(Y, \beta)$ is non-compact but near infinity it parameterizes only pseudo-holomorphic maps $\varphi : C \to Y$ such that $\mathsf{w} \circ \varphi : C \to \mathbb{C}$ is constant and $\mathsf{w} \circ \varphi(C) \in \mathbb{C}$ is close to infinity. Thus the virtual fundamental class of $\overline{\mathcal{M}}_{g,n}(Y, \beta)$ is well defined as a class in the relative homology

$$\begin{aligned}\left[\overline{\mathcal{M}}_{g,n}(Y, \beta)\right]_{\text{vir}} &\in H_\bullet\left(Y^n \times \overline{\mathcal{M}}_{g,n}, Y^{n-1} \times Y_R \times \overline{\mathcal{M}}_{g,n}; \mathbb{Q}\right) \\ &= H_\bullet(Y, \mathbb{Q})^{\otimes(n-1)} \otimes H_\bullet(Y, Y_R; \mathbb{Q}) \otimes H_\bullet(\overline{\mathcal{M}}_{g,n}, \mathbb{Q}).\end{aligned}$$

We define $I^{(1)}_{g,\beta,n-1}$ to be the map given by the relative virtual fundamental class $[\overline{\mathcal{M}}_{g,n}(Y,\beta)]_{\mathrm{vir}}$.

This collection of correlators satisfies analogues of the usual axioms of a cohomological field theory [**KM94**] but we will not discuss them here. Consider now a cohomology class

$$x = (x_2, x_{\neq 2}) \in H^\bullet(Y,\mathbb{C}) = H^2(Y,\mathbb{C}) \oplus H^{\neq 2}(Y,\mathbb{C}),$$

where $H^\bullet(Y,\mathbb{C})$ is viewed as a supermanifold over $\mathbb{C}$.

Now for every such x we define a quantum product

$$\bullet *_x \bullet : \quad H^\bullet(Y,\mathbb{C}) \otimes H^\bullet(Y,\mathbb{C}) \longrightarrow H^\bullet(Y,\mathbb{C})$$

by the formula

$$\begin{aligned} \alpha_1 *_x \alpha_2 \quad := & \sum_{m\geq 0} \sum_{\beta \in H_2(Y,\mathbb{Z})} \exp\left(\langle \beta, x_2\rangle\right) \\ & \cdot \frac{1}{m!} I^{(1)}_{g,\beta,m+1}\big((\alpha_1 \otimes \alpha_2 \otimes \underbrace{x_{\neq 2} \otimes \cdots \otimes x_{\neq 2}}_{m \text{ times}}) \otimes 1_{\overline{\mathcal{M}}_{0,m+1}}\big). \end{aligned}$$

Now this quantum multiplication together with the usual formulas (see Examples 4.4 and 4.8) can be used to define a decorated variation of **nc**-Hodge structures over the (conjecturally non-empty) domain in $H^\bullet(Y,\mathbb{C})$ where the series defining $*_x$ is absolutely convergent.

REMARK 4.39. There are some interesting variants of this construction. For instance we can take a symplectic manifold (Y,ω) with no potential and a pseudoconvex boundary. In this situation $\overline{\mathscr{M}}_{g,n}(Y,\beta)$ is already compact, as long as $\beta \neq 0$. Also, in a symplectic Landau-Ginzburg model (Y,ω,w) we can allow for w to be non-proper and instead require that its fibers have pseudoconvex boundary. Finally one can consider a symplectic Y equipped with a proper map $Y \to \mathbb{C}^k$, holomorphic at infinity and with $k \geq 2$.

4.5.2. ***Categories of branes.*** Let (Y,ω,w) be a symplectic geometry with a proper potential. There are two natural categories that we can attach to this geometry: the Fukaya category of the general fiber of w, and the Fukaya-Seidel category of w. Understanding the structural properties of these categories or even defining them properly is a difficult task which requires a lot of effort and hard work. We will not explain any of these intricate details but will rather use the Fukaya and Fukaya-Seidel categories as conceptual entities. For details of the definitions and a rigorous development of the theory we refer the reader to the main sources [**FOOO07, FO01**], [**Sei07b, Sei07a**]. The categories that we are interested in are:

(1) The Fukaya-Seidel category $\mathsf{FS}(Y,\omega,\mathsf{w})$ of the potential w has objects which are unitary local systems $\mathbb{V}$ on (graded) ω-Lagrangian submanifolds $L \subset Y$ such that:

- $\mathsf{w}(L) \subset (\text{compact}) \cup \mathbb{R}_{\leq 0}$;
- The restriction of L over the ray $\mathbb{R}_{\leq 0}$ is a fibration on $\mathbb{R}_{\leq -R}$ and when $z \in \mathbb{R}_{\leq 0}$, and $z \to -\infty$, we have that the fiber $L_z \subset Y_z$ is a Lagrangian submanifold in the symplectic manifold $(Y_z, \omega_{|Y_z})$.

The morphisms between two objects $(L_1, \mathbb{V}_1)$ and $(L_2, \mathbb{V}_2)$ are defined as homomorphisms between the fibers of the local systems at the intersection points of the two Lagrangians. As usual, to make this work one has to perturb one of the Lagrangians, say L_2, by a Hamiltonian isotopy to ensure transversality of the intersection. A new feature of this setup (compared to the situation of symplectic manifolds with no potential) is that the allowable isotopies are tightly controlled – they correspond to small wiggling, see Figure 7, of the tail of the tadpole-like image of the Lagrangian in $\mathbb{C}$ and a lift of this wiggling to Y given by a non-linear symplectic connection identifying the fibers of Y.

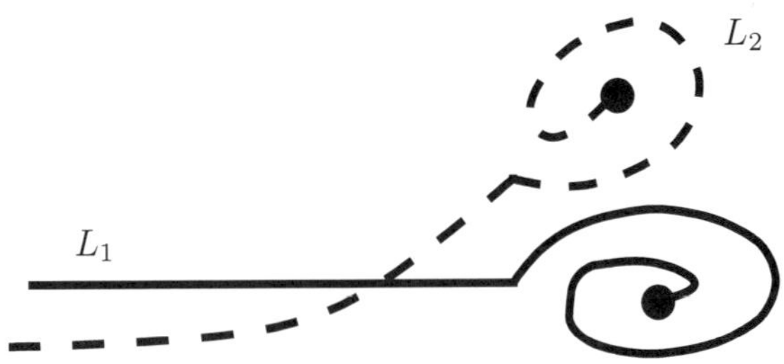

FIGURE 7. Tadpole-like w-images of two Lagrangian submanifolds.

The compositions of morphisms are given by correlators counting pseudo-holomorphic discs whose boundary is contained in the given Lagrangian submanifolds.

(2) The Fukaya category $\mathsf{Fuk}(Y_z)$ of a fiber $\left(Y_z, \omega_{|Y_z}\right)$ over a point $z \in \mathbb{C}$ which is not a critical value for w. The objects in this category are again pairs consisting of (graded) Lagrangian submanifolds in Y_z equipped with unitary local systems, and morphisms and compositions are defined again by maps between the fibers of the local systems at the intersection points and by counting discs. The parallel transport w.r.t. a non-linear symplectic connection on $\mathsf{w} : Y \to \mathbb{C}$ identifies symplectically all fibers $\left(Y_z, \omega_{|Y_z}\right)$ over points $z \in \mathbb{R}_{\leq 0}$ when $z \to -\infty$. We will denote any one such fiber by $(Y_{-\infty}, \omega_{-\infty})$.

Now observe that by intersecting a Lagrangian $L \subset Y$ with the fiber $Y_{-\infty}$ we get an assignment $L \mapsto L_{-\infty} := L \cap Y_{-\infty}$. We expect that this assignment can be promoted to a spherical functor (see also [**Sei07a**] for a similar discussion)

$$F : \quad \mathsf{FS}(Y, \mathsf{w}, \omega) \longrightarrow \mathsf{Fuk}(Y_{-\infty}, \omega_{-\infty})$$

so that the associated spherical twist $\mathcal{R}_F : \mathsf{Fuk}(Y_{-\infty}, \omega_{-\infty}) \to \mathsf{Fuk}(Y_{-\infty}, \omega_{-\infty})$ categorifies the monodromy around the circle $\{z \in \mathbb{C} |\ |z| = R\}$.

In this situation one can also define relative Gromov-Witten invariants

$$J^{(1)}_{g,\beta,n-2} : \quad H^\bullet(Y, Y_{-\infty}; \mathbb{Q}) \otimes H^\bullet(Y, \mathbb{Q})^{\otimes(n-2)} \otimes H^\bullet\left(\overline{\mathcal{M}}_{g,n}, \mathbb{Q}\right) \longrightarrow H^\bullet(Y, Y_{-\infty}; \mathbb{Q}).$$

For this we again use the duality $(H^\bullet)^\vee \cong H_\bullet$ and the Poincaré duality $H^\bullet(Y, Y_{-\infty}; \mathbb{Q}) \cong H_\bullet(Y, Y_{+\infty}; \mathbb{Q})$ to rewrite $J^{(1)}_{g,\beta,n-2}$ as a class in

$$H_\bullet(Y, Y_{-\infty}; \mathbb{Q}) \otimes H_\bullet(Y, Y_{+\infty}; \mathbb{Q}) \otimes H_\bullet(Y, \mathbb{Q})^{\otimes(n-2)} \otimes H_\bullet\left(\overline{\mathcal{M}}_{g,n}, \mathbb{Q}\right).$$

This class can again be defined as a virtual fundamental class space $\overline{\mathcal{M}}_{g,n}(Y,\beta)$ of stable pseudo-holomorphic maps. Again we can interpret the virtual class as a relative homology class:

$$\begin{aligned}\left[\overline{\mathcal{M}}_{g,n}(Y,\beta)\right]_{\text{vir}} &\in H_\bullet\left(Y^n \times \overline{\mathcal{M}}_{g,n},\, Y^{n-2} \times \left((Y^-_{R,\varepsilon} \times Y) \cup (Y \times Y^+_{R,\varepsilon})\right) \times \overline{\mathcal{M}}_{g,n}; \mathbb{Q}\right) \\ &= H_\bullet(Y, Y^-_{R,\varepsilon}; \mathbb{Q}) \otimes H_\bullet(Y, Y^+_{R,\varepsilon}; \mathbb{Q}) \otimes H_\bullet(Y,\mathbb{Q})^{\otimes(n-2)} \otimes H_\bullet\left(\overline{\mathcal{M}}_{g,n}, \mathbb{Q}\right) \\ &= H_\bullet(Y, Y_{-\infty}; \mathbb{Q}) \otimes H_\bullet(Y, Y_{+\infty}; \mathbb{Q}) \otimes H_\bullet(Y,\mathbb{Q})^{\otimes(n-2)} \otimes H_\bullet\left(\overline{\mathcal{M}}_{g,n}, \mathbb{Q}\right),\end{aligned}$$

and so it gives the desired map $J^{(1)}_{g,\beta,n-2}$.

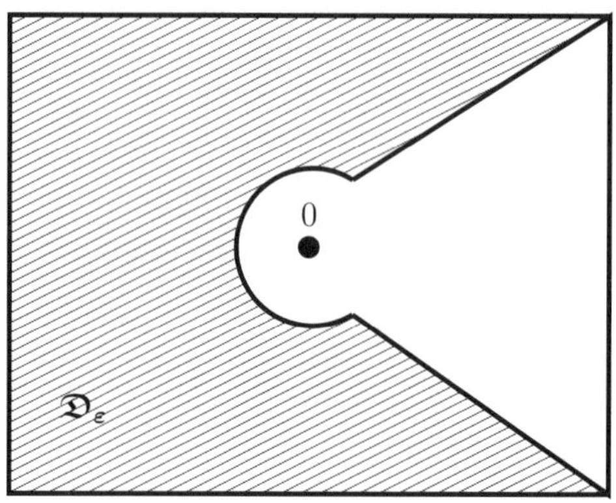

FIGURE 8. The domain $\mathfrak{D}_\varepsilon$.

Here $1 \gg \varepsilon > 0$, and $Y^{\pm}_{R,\varepsilon} = \mathsf{w}^{-1}(\pm\mathfrak{D}_\varepsilon)$, where $\mathfrak{D}_\varepsilon \subset \mathbb{C}$ is the domain given by (see Figure 8)

$$\mathfrak{D}_\varepsilon := \left\{ z \in \mathbb{C} \;\middle|\; |z| \geq R \text{ and } \operatorname{Arg} z \in \left(\frac{\pi}{2} - \varepsilon, \frac{3\pi}{2} + \varepsilon\right) \right\}.$$

Again the relative invariants $J^{(1)}_{g,\beta,n-2}$ give rise to a quantum multiplication and through the usual formulas from Examples 4.4 and 4.8 we again get a decorated variation of **nc**-Hodge structures over a (conjecturally non-empty) domain in $H^\bullet(Y,\mathbb{C})$ with fiber $H^\bullet(Y, Y_{-\infty})$.

4.5.3. ***Mirror symmetry.*** In conclusion we systematize (see also [**Aur07**]) all the objects introduced above in a mirror table describing the corresponding A and B-model entities in parallel:

invariants	A-model	B-model
geometry	a triple (Y, w, ω) where: $\mathsf{w} : Y \to \mathbb{C}$ is a proper C^∞-map, and (Y, ω) is symplectic with $c_1(T_Y) = 0$	a pair $Z \subset X$ where: X is smooth projective, and $Z \subset X$ is a smooth anticanonical divisor
cohomology	$\left.\begin{matrix} H^\bullet(Y, \mathbb{C}) \\ H^\bullet(Y, Y_{-\infty}; \mathbb{C}) \\ H^\bullet(Y_{-\infty}, \mathbb{C}) \end{matrix}\right\}$ variations of **nc**HS over a domain in $H^\bullet(Y, \mathbb{C})$	$\left.\begin{matrix} H^\bullet(X-Z, \mathbb{C}) \\ H^\bullet(X, \mathbb{C}) \\ H^\bullet(Z, \mathbb{C}) \end{matrix}\right\}$ variations of **nc**HS over a domain in $H^\bullet(X-Z, \mathbb{C})$
categories	$\left.\begin{matrix} \mathsf{Fuk}(Y_{-\infty}) \\ \uparrow F \\ \mathsf{FS}(Y) \end{matrix}\right\}$ $\mathsf{Fuk}(Y_{-\infty})$ is a CY category and F is a spherical functor	$\left.\begin{matrix} D(Z) \\ \uparrow F \\ D(X) \end{matrix}\right\}$ $D(Z)$ is a CY category and F is a spherical functor
	The part of $\mathsf{FS}(Y)$ consisting of Lagrangians fibered over 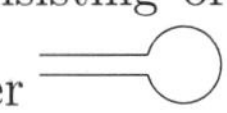 where the circle is of radius $R \gg 0$.	$D_{\mathrm{supp}\, Z}(X) \left\{ \begin{matrix} \text{a full compact} \\ \text{(non smooth)} \\ \text{subcategory in} \\ D(X) \end{matrix} \right.$
	The part of $\mathsf{FS}(Y)$ consisting of compact Lagrangian sumanifolds in Y.	$D_{\substack{\text{compact} \\ \text{support}}}(X - Z) \left\{ \begin{matrix} \text{a full compact (non smooth)} \\ \text{subcategory in} \\ D(X) \end{matrix} \right.$
	The wrapped FS category: the Hom space between $(L_1, \mathbb{V}_1)$ and $(L_2, \mathbb{V}_2)$ is the sum of $\mathrm{Hom}(\mathbb{V}_1, \mathbb{V}_2)_x$, $x \in L_1 \cap L_2$, and L_2 is deformed so that $\mathsf{w}(L_2)$ becomes a spiral:  wrapped infinitely many times	$D(X - Z) \left\{ \begin{matrix} \text{a smooth (non} \\ \text{compact) cate-} \\ \text{gory} \end{matrix} \right.$

References

[AKO04] D. Auroux, L. Katzarkov, and D. Orlov. Mirror symmetry for weighted projective planes and their noncommutative deformations, 2004, arXiv:math.AG/0404281. To appear in Annals of Math.

[AKO06] D. Auroux, L. Katzarkov, and D. Orlov. Mirror symmetry for del Pezzo surfaces: vanishing cycles and coherent sheaves. *Invent. Math.*, 166(3):537–582, 2006.

[Ann07] R. Anno. Spherical functors, 2007, arXiv.org:0711.4409.

[Art73] M. Artin. Théorème de finitude pour un morphisme propre; dimension cohomologique des schémas algébriques affines. In *Théorie de topos and cohomologie étale des schémas, SGA 4, tome 3, exposé XIV*, pages 145–167. Springer, LNM 305, 1973.

[Aur07] D. Auroux. Mirror symmetry and T-duality in the complement of an anticanonical divisor. *J. Gökova Geom. Topol.*, 1:51–91, 2007.

[Bar01] S. Barannikov. Quantum periods. I. Semi-infinite variations of Hodge structures. *Internat. Math. Res. Notices*, 23:1243–1264, 2001.

[Bar02a] S. Barannikov. Non-commutative periods and mirror symmetry in higher dimensions. *Comm. Math. Phys.*, 228(2):281–325, 2002.

[Bar02b] S. Barannikov. Semi-infinite variations of Hodge structures and integrable hierarchies of KdV type. *Int. Math. Res. Not.*, 19:973–990, 2002.

[BBD82] A. Beĭlinson, J. Bernstein, and P. Deligne. Faisceaux pervers. In *Analysis and topology on singular spaces, I (Luminy, 1981)*, volume 100 of *Astérisque*, pages 5–171. Soc. Math. France, Paris, 1982.

[BCOV94] M. Bershadsky, S. Cecotti, H. Ooguri, and C. Vafa. Kodaira-Spencer theory of gravity and exact results for quantum string amplitudes. *Comm. Math. Phys.*, 165(2):311–427, 1994.

[BE04] S. Bloch and H. Esnault. Homology for irregular connections. *J. Théor. Nombres Bordeaux*, 16(2):357–371, 2004.

[BK91] A. Bondal and M. Kapranov. Enhanced triangulated categories. *Math. USSR Sb.*, 70:93–107, 1991.

[BK98] S. Barannikov and M. Kontsevich. Frobenius manifolds and formality of Lie algebras of polyvector fields. *Internat. Math. Res. Notices*, 4:201–215, 1998.

[BN93] M. Bökstedt and A. Neeman. Homotopy limits in triangulated categories. *Compositio Math.*, 86(2):209–234, 1993.

[BO01] A. Bondal and D. Orlov. Reconstruction of a variety from the derived category and groups of autoequivalences. *Compositio Math.*, 125(3):327–344, 2001.

[BO02] A. Bondal and D. Orlov. Derived categories of coherent sheaves. In *Proceedings of the International Congress of Mathematicians, Vol. II (Beijing, 2002)*, pages 47–56, Beijing, 2002. Higher Ed. Press.

[Bon93] A. Bondal. Non-commutative deformations and Poisson brackets on projective spaces. Preprint no. 67, Max-Planck-Institut, Bonn 1993.

[BV85] D. Babbitt and V.S. Varadarajan. Deformations of nilpotent matrices over rings and reduction of analytic families of meromorphic differential equations. *Mem. Amer. Math. Soc.*, 55(325):iv+147, 1985.

[BvdB03] A. Bondal and M. van den Bergh. Generators and representability of functors in commutative and noncommutative geometry. *Mosc. Math. J.*, 3(1):1–36, 258, 2003.

[CK99] D. Cox and S. Katz. *Mirror symmetry and algebraic geometry*, volume 68 of *Mathematical Surveys and Monographs.* American Mathematical Society, Providence, RI, 1999.

[CK05] P. Caldero and B. Keller. From triangulated categories to cluster algebras, *Invent. Math.*, 172(1):169-211, 2008.

[CK06] P. Caldero and B. Keller. From triangulated categories to cluster algebras. II. *Ann. Sci. École Norm. Sup. (4)*, 39(6):983–1009, 2006.

[CKS05] Y. Chen, M. Kontsevich, and A. Schwarz. Symmetries of the WDVV equations. *Nucl.Phys. B*, 730:352–363, 2005.

[Cla08] P. Clarke. Duality for toric Landau-Ginzburg models, 2008, arXiv.org:0803.0447.

[Cos05] K. Costello. The Gromov-Witten potential associated to a TCFT, 2005, arXiv.org:math/0509264.

[Cos07a] K. Costello. Renormalisation and the Batalin-Vilkovisky formalism, 2007, arXiv.org:0706.1533.

[Cos07b] K. Costello. Topological conformal field theories and Calabi-Yau categories. *Adv. Math.*, 210(1):165–214, 2007.

[Cos07c] K. Costello. Topological conformal field theories and gauge theories. *Geom. Topol.*, 11:1539–1579, 2007.

[CV91] S. Cecotti and C. Vafa. Topological–anti-topological fusion. *Nuclear Phys. B*, 367(2):359–461, 1991.

[CV93a] S. Cecotti and C. Vafa. Ising model and $N = 2$ supersymmetric theories. *Comm. Math. Phys.*, 157(1):139–178, 1993.

[CV93b] S. Cecotti and C. Vafa. On classification of $N = 2$ supersymmetric theories. *Comm. Math. Phys.*, 158(3):569–644, 1993.

[Del70] P. Deligne. *Équations différentielles à points singuliers réguliers.* Springer-Verlag, Berlin, 1970. Lecture Notes in Mathematics, Vol. 163.

[Del71] P. Deligne. Théorie de Hodge. II. *Inst. Hautes Études Sci. Publ. Math.*, 40:5–57, 1971.

[Del74] P. Deligne. Théorie de Hodge. III. *Inst. Hautes Études Sci. Publ. Math.*, 44:5–77, 1974.

[Dim04] A. Dimca. *Sheaves in topology.* Universitext. Springer-Verlag, Berlin, 2004.

[Dub98] B. Dubrovin. Geometry and analytic theory of Frobenius manifolds. In *Proceedings of the International Congress of Mathematicians, Vol. II (Berlin, 1998)*, Extra Vol. II, pages 315–326 (electronic), 1998.

[FJR07] H. Fan, T. Jarvis, and Y. Ruan. The Witten equation and its virtual fundamental cycle, 2007, preprint, 99 pp., arXiv.org:0712.4025.

[FO01] K. Fukaya and K. Ono. Floer homology and Gromov-Witten invariants over the integers of general symplectic manifolds—summary. In *Taniguchi Conference on Mathematics Nara '98*, volume 31 of *Adv. Stud. Pure Math.*, pages 75–91. Math. Soc. Japan, Tokyo, 2001.

[FOOO07] K. Fukaya, Y.-G. Oh, H. Ohta, and K. Ono. Lagrangian intersection Floer theory - anomaly and obstruction, 2007, available at http://www.math.kyoto-u.ac.jp/~fukaya/fukaya.html.

[FW01] E. Friedlander and M. Walker. Comparing K-theories for complex varieties. *Amer. J. Math.*, 123(5):779–810, 2001.

[FW05] E. Friedlander and M. Walker. Semi-topological K-theory. In *Handbook of K-theory. Vol. 1, 2*, pages 877–924. Springer, Berlin, 2005.

[Ger64] M. Gerstenhaber. On the deformation of rings and algebras. *Ann. of Math. (2)*, 79:59–103, 1964.

[Get93] E. Getzler. Cartan homotopy formulas and the Gauss-Manin connection in cyclic homology. In *Quantum deformations of algebras and their representations (Ramat-Gan, 1991/1992; Rehovot, 1991/1992)*, volume 7 of *Israel Math. Conf. Proc.*, pages 65–78. Bar-Ilan Univ., Ramat Gan, 1993.

[Gol01] V. Golyshev. Riemann-Roch variations. *Izv. Ross. Akad. Nauk Ser. Mat.*, 65(5):3–32, 2001.

[Gol07] V. Golyshev. Classification problems and mirror duality. In *Surveys in geometry and number theory: reports on contemporary Russian mathematics*, volume 338 of *London Math. Soc. Lecture Note Ser.*, pages 88–121. Cambridge Univ. Press, Cambridge, 2007.

[GS06] M. Gross and B. Siebert. Mirror symmetry via logarithmic degeneration data. I. *J. Differential Geom.*, 72(2):169–338, 2006.

[GS07] M. Gross and B. Siebert. Mirror symmetry via logarithmic degeneration data II, 2007, arXiv.org:0709.2290.

[Her03] C. Hertling. tt^* geometry, Frobenius manifolds, their connections, and the construction for singularities. *J. Reine Angew. Math.*, 555:77–161, 2003.

[Her06] C. Hertling. tt^* geometry and mixed Hodge structures. In *Singularity theory and its applications*, volume 43 of *Adv. Stud. Pure Math.*, pages 73–84, Tokyo, 2006. Math. Soc. Japan.

[Hie07] M. Hien. Periods for irregular singular connections on surfaces. *Math. Ann.*, 337(3):631–669, 2007.

[Hie08] M. Hien. Periods for flat algebraic connections, 2008, arXiv.org:0803.3463.

[HKK+03] K. Hori, S. Katz, A. Klemm, R. Pandharipande, R. Thomas, C. Vafa, R. Vakil, and E. Zaslow. *Mirror symmetry*, volume 1 of *Clay Mathematics Monographs.* American Mathematical Society, Providence, RI, 2003.

[HS07] C. Hertling and C. Sevenheck. Nilpotent orbits of a generalization of Hodge structures. *J. Reine Angew. Math.*, 609:23-80, 2007.

[HV00] K. Hori and C. Vafa. Mirror symmetry, 2000, hep-th/0002222.

[Iri07] H. Iritani. Real and integral structures in quantum cohomology I: toric orbifolds, 2007, arXiv.org:0712.2204.

[IS07] J. Iyer and C. Simpson. A relation between the parabolic Chern characters of the de Rham bundles. *Math. Ann.*, 338(2):347–383, 2007.

[Kal06] D. Kaledin. Non-commutative Hodge-to-de Rham degeneration via the method of Deligne-Illusie, 2006, arXiv.org:math/0611623. To appear in F.Bogomolov's 60th anniversary volume of Pure and Applied Mathematics Quaterly.

[Kal07a] D. Kaledin. Cartier isomorphism and Hodge theory in the non-commutative case, 2007, arXiv.org:0708.1574.

[Kal07b] D. Kaledin. Cyclic homology with coefficients, 2007, arXiv.org:math/0702068. To appear in Yu.I. Manin Festschrift.

[Kap91] M. Kapranov. On DG-modules over the de Rham complex and the vanishing cycles functor. In *Algebraic geometry (Chicago, IL, 1989)*, volume 1479 of *Lecture Notes in Math.*, pages 57–86. Springer, Berlin, 1991.

[Kat90] N. Katz. *Exponential sums and differential equations*, volume 124 of *Annals of Mathematics Studies*. Princeton University Press, Princeton, NJ, 1990.

[Kaw92] Y. Kawamata. Unobstructed deformations. A remark on a paper of Z. Ran: "Deformations of manifolds with torsion or negative canonical bundle" [J. Algebraic Geom. **1** (1992), no. 2, 279–291; MR1144440 (93e:14015)]. *J. Algebraic Geom.*, 1(2):183–190, 1992.

[Kel05] B. Keller. On triangulated orbit categories. *Doc. Math.*, 10:551–581, 2005.

[KL03] A. Kapustin and Y. Li. D-branes in Landau-Ginzburg models and algebraic geometry. *J. High Energy Phys.*, 12:005, 2003.

[KL04] A. Kapustin and Y. Li. D-branes in topological minimal models: the Landau-Ginzburg approach. *J. High Energy Phys.*, 7:045, 2004.

[KM94] M. Kontsevich and Yu. Manin. Gromov-Witten classes, quantum cohomology, and enumerative geometry. *Comm. Math. Phys.*, 164(3):525–562, 1994.

[Kon99] M. Kontsevich. Operads and motives in deformation quantization. *Lett. Math. Phys.*, 48(1):35–72, 1999.

[Kon08] M. Kontsevich. XI Solomon Lefschetz memorial lecture series: Hodge structures in non-commutative geometry. Notes by Ernesto Lupercio, 2008, arXiv.org:0801.4760.

[KS94] M. Kashiwara and P. Schapira. *Sheaves on manifolds*, volume 292 of *Grundlehren der Mathematischen Wissenschaften [Fundamental Principles of Mathematical Sciences]*. Springer-Verlag, Berlin, 1994. With a chapter in French by Christian Houzel, Corrected reprint of the 1990 original.

[KS00] M. Kontsevich and Y. Soibelman. Deformations of algebras over operads and the Deligne conjecture. In *Conférence Moshé Flato 1999, Vol. I (Dijon)*, volume 21 of *Math. Phys. Stud.*, pages 255–307. Kluwer Acad. Publ., Dordrecht, 2000.

[KS01] M. Kontsevich and Y. Soibelman. Homological mirror symmetry and torus fibrations. In *Symplectic geometry and mirror symmetry (Seoul, 2000)*, pages 203–263. World Sci. Publ., River Edge, NJ, 2001.

[KS06a] M. Kontsevich and Y. Soibelman. Affine structures and non-Archimedean analytic spaces. In *The unity of mathematics*, volume 244 of *Progr. Math.*, pages 321–385. Birkhäuser Boston, Boston, MA, 2006.

[KS06b] M. Kontsevich and Y. Soibelman. Notes on A_∞ algebras, A_∞ categories and non-commutative geometry. I, 2006, arXiv.org:math/0606241.

[Kuz05a] A. Kuznetsov. Derived categories of quadric fibrations and intersections of quadrics, 2005, arXiv.org:math/0510670. To appear in Adv. in Math.

[Kuz05b] A. Kuznetsov. Homological projective duality, *Publ. Math. de IHES*, 105:157–220, 2007.

[Kuz06] A. Kuznetsov. Homological projective duality for Grassmannians of lines, 2006, arXiv.org:math/0610957.

[LH03] K. Lefévre-Hasegawa. *Sur le A_∞ catègories*. PhD thesis, Université Denis Diderot - Paris 7 2, place Jussieu, 2003, available at http://www.math.jussieu.fr/~lefevre/publ.html.

[LT98] J. Li and G. Tian. Virtual moduli cycles and Gromov-Witten invariants of general symplectic manifolds. In *Topics in symplectic 4-manifolds (Irvine, CA, 1996)*, First Int. Press Lect. Ser., I, pages 47–83. Internat. Press, Cambridge, MA, 1998.

[Mal79] B. Malgrange. Sur la réduction formelle des équations differentielles à singularités irréguliers. Preprint, Institute Fourier, Grenoble, 1979.

[Mal83] B. Malgrange. La classification des connexions irrégulières à une variable. In *Mathematics and physics (Paris, 1979/1982)*, volume 37 of *Progr. Math.*, pages 381–399. Birkhäuser Boston, Boston, MA, 1983.

[Mal91] B. Malgrange. *Équations différentielles à coefficients polynomiaux*, volume 96 of *Progress in Mathematics*. Birkhäuser Boston Inc., Boston, MA, 1991.

[Man99a] Yu. Manin. *Frobenius manifolds, quantum cohomology, and moduli spaces*, volume 47 of *American Mathematical Society Colloquium Publications*. American Mathematical Society, Providence, RI, 1999.

[Man99b] Yu. Manin. Three constructions of Frobenius manifolds: a comparative study. *Asian J. Math.*, 3(1):179–220, 1999. Sir Michael Atiyah: a great mathematician of the twentieth century.

[MO05] K. Matsuki and M. Olsson. Kawamata-Viehweg vanishing as Kodaira vanishing for stacks. *Math. Res. Lett.*, 12(2-3):207–217, 2005.

[Moc06a] T. Mochizuki. Asymptotic behavior of tame harmonic bundles and pure twistor D-modules. *Sūgaku*, 58(4):337–363, 2006.

[Moc06b] T. Mochizuki. Kobayashi-Hitchin correspondence for tame harmonic bundles and an application. *Astérisque*, 309:viii+117, 2006.

[Moc07a] T. Mochizuki. Asymptotic behaviour of tame harmonic bundles and an application to pure twistor D-modules. I. *Mem. Amer. Math. Soc.*, 185(869):xii+324, 2007.

[Moc07b] T. Mochizuki. Asymptotic behaviour of tame harmonic bundles and an application to pure twistor D-modules. II. *Mem. Amer. Math. Soc.*, 185(870):xii+565, 2007.

[Moc08a] T. Mochizuki. Good formal structure for meromorphic flat connections on smooth projective surfaces, 2008, arXiv.org:0803.1346.

[Moc08b] T. Mochizuki. Wild harmonic bundles and wild pure twistor D-modules, 2008, arXiv.org:0803.1344.

[Orl04] D. Orlov. Triangulated categories of singularities and D-branes in Landau-Ginzburg models. *Tr. Mat. Inst. Steklova*, 246(Algebr. Geom. Metody, Svyazi i Prilozh.):240–262, 2004.

[Orl05a] D. Orlov. Derived categories of coherent sheaves and triangulated categories of singularities, 2005, arXiv.org:math/0503632.

[Orl05b] D. Orlov. Triangulated categories of singularities and equivalences between Landau-Ginzburg models, 2005, arXiv.org:math/0503630.

[OV05] A. Ogus and V. Vologodsky. Nonabelian Hodge theory in characteristic p, *Publ. Math. IHES*, 106:1–138, 2007.

[Pan08] P. Pandit. Unobstructedness of graded Calabi-Yau categories. In preparation, 2008.

[PP05] A. Polishchuk and L. Positselski. *Quadratic algebras*, volume 37 of *University Lecture Series*. American Mathematical Society, Providence, RI, 2005.

[Ran92] Z. Ran. Deformations of manifolds with torsion or negative canonical bundle. *J. Algebraic Geom.*, 1(2):279–291, 1992.

[Sab99] C. Sabbah. On a twisted de Rham complex. *Tohoku Math. J. (2)*, 51(1):125–140, 1999.

[Sab00] C. Sabbah. Équations différentielles à points singuliers irréguliers et phénomène de Stokes en dimension 2. *Astérisque*, 263:viii+190, 2000.

[Sab02] C. Sabbah. *Déformations isomonodromiques et variétés de Frobenius.* Savoirs Actuels (Les Ulis). EDP Sciences, Les Ulis, 2002. Mathématiques (Les Ulis).

[Sab05a] C. Sabbah. Fourier-Laplace transform of a variation of polarized complex Hodge structure, 2005, arXiv.org:math/0508551.

[Sab05b] C. Sabbah. Polarizable twistor $\mathscr{D}$-modules. *Astérisque*, 300:vi+208, 2005.

[Sai83] K. Saito. Period mapping associated to a primitive form. *Publ. Res. Inst. Math. Sci.*, 19(3):1231–1264, 1983.

[Sai98a] K. Saito. Around the theory of the generalized weight system: relations with singularity theory, the generalized Weyl group and its invariant theory, etc. In *Selected papers on harmonic analysis, groups, and invariants*, volume 183 of *Amer. Math. Soc. Transl. Ser. 2*, pages 101–143. Amer. Math. Soc., Providence, RI, 1998.

[Sai98b] K. Saito. Duality for regular systems of weights: a précis. In *Topological field theory, primitive forms and related topics (Kyoto, 1996)*, volume 160 of *Progr. Math.*, pages 379–426. Birkhäuser Boston, Boston, MA, 1998.

[Sch03] A. Schwarz. Noncommutative supergeometry, duality and deformations. *Nuclear Physics B*, 650:475–496, 2003.

[Sei07a] P. Seidel. Directed A_∞-subcategories and natural transformations, 2007, arXiv.org:math/0701778.

[Sei07b] P. Seidel. Fukaya categories and Picard-Lefschetz theory, 2007, available at http://www-math.mit.edu/~seidel/revision.pdf.

[Shk08] D. Shklyarov. *Hirzebruch-Riemann-Roch theorem for DG algebras.* PhD thesis, Kansas State University, 2008.

[Sie99] B. Siebert. Symplectic Gromov-Witten invariants. In *New trends in algebraic geometry (Warwick, 1996)*, volume 264 of *London Math. Soc. Lecture Note Ser.*, pages 375–424. Cambridge Univ. Press, Cambridge, 1999.

[Sim97a] C. Simpson. The Hodge filtration on nonabelian cohomology. In *Algebraic geometry—Santa Cruz 1995*, volume 62 of *Proc. Sympos. Pure Math.*, pages 217–281. Amer. Math. Soc., Providence, RI, 1997.

[Sim97b] C. Simpson. Mixed twistor structures, 1997, alg-geom/9705006.

[Sza07] S. Szabo. Algebraic interpretation of Nahm transform for integrable connections, 2007, arXiv.org:0704.2744.

[Tam03] D. Tamarkin. Formality of chain operad of little discs. *Lett. Math. Phys.*, 66(1-2):65–72, 2003.

[Ter07] J. Terilla. Smoothness theorem for differential BV algebras, 2007, preprint, arXiv:0707.1290.

[Tia87] G. Tian. Smoothness of the universal deformation space of compact Calabi-Yau manifolds and its Petersson-Weil metric. In *Mathematical aspects of string theory (San Diego, Calif., 1986)*, volume 1 of *Adv. Ser. Math. Phys.*, pages 629–646. World Sci. Publishing, Singapore, 1987.

[Tod89] A. Todorov. The Weil-Petersson geometry of the moduli space of SU($n \geq 3$) (Calabi-Yau) manifolds. I. *Comm. Math. Phys.*, 126(2):325–346, 1989.

[Toë07a] B. Toën. The homotopy theory of *dg*-categories and derived Morita theory. *Invent. Math.*, 167(3):615–667, 2007.

[Toë07b] B. Toën. Letter to the authors, May 10, 2007.

[Toë07c] B. Toën. Letter to the authors, September 14, 2007.

[Tsy07] B. Tsygan. On the Gauss-Manin connection in cyclic homology. *Methods Funct. Anal. Topology*, 13(1):83–94, 2007.

[Tu08] J. Tu. The Hochschild cohomology of Landau-Ginzburg models, 2008, preprint University of Wisconsin.

[TV05] B. Toën and M. Vaquie. Moduli of objects in dg-categories, 2005, arXiv.org:math/0503269. To appear in Annales de l'ENS.

[Voi03] C. Voisin. *Hodge theory and complex algebraic geometry. II*, volume 77 of *Cambridge Studies in Advanced Mathematics*. Cambridge University Press, Cambridge, 2003. Translated from the French by Leila Schneps.

[Wei96] C. Weibel. Cyclic homology for schemes. *Proc. Amer. Math. Soc.*, 124(6):1655–1662, 1996.

[Wil07] T. Willwacher. A counterexample to the quantizability of modules. *Lett. Math. Phys.*, 81(3):265–280, 2007.

Department of Mathematics, University of Miami, Coral Gables, Florida 33124, USA

E-mail address: `l.katzarkov@math.miami.edu`

IHES, Le Bois-Marie 35, route de Chartres, 91440 Bures-sur-Yvette, France, and Department of Mathematics, University of Miami, Coral Gables, Florida 33124, USA

E-mail address: `maxim@ihes.fr`

Department of Mathematics, University of Pennsylvania, DRL 209 South 33rd Street, Philadelphia, PA 19104-6395, USA

E-mail address: `tpantev@math.upenn.edu`

Proceedings of Symposia in Pure Mathematics
Volume **78**, 2008

A weight two phenomenon for the moduli of rank one local systems on open varieties

Carlos Simpson

Abstract. The twistor space of representations on an open variety maps to a weight two space of local monodromy transformations around a divisor component at infinty. The space of σ-invariant sections of this slope-two bundle over the twistor line is a real 3-dimensional space whose parameters correspond to the complex residue of the Higgs field and the real parabolic weight of a harmonic bundle.

1. Introduction

Let X be a smooth projective variety and $D \subset X$ a reduced effective divisor with simple normal crossings. We would like to define a Deligne glueing for the Hitchin twistor space of the moduli of local systems over $X - D$. Making the construction presents new difficulties which are not present in the case of a compact base, so we only treat the case of local systems of rank 1. Every local system comes from a vector bundle on X with connection logarithmic along D; however one can make local meromorphic gauge transformations near components of D, and this changes the structure of the bundle as well as the eigenvalues of the residue of the connection. The change in eigenvalues is by subtracting an integer. There is no reasonable algebraic quotient by such an operation: for our main example in §4, that would amount to taking the quotient of $\mathbb{A}^1$ by the translation action of $\mathbb{Z}$. Hence, we are tempted to look at the moduli space of logarithmic connections [**43**] and accept the fact that the Riemann-Hilbert correspondence from there to the moduli space of local systems is many-to-one.

We first concentrate on looking at the simplest case, which is when $X := \mathbb{P}^1$ and $D := \{0, \infty\}$ and the local systems have rank 1. In this case, much as in Goldman and Xia [**23**], one can explicitly write down everything, in particular we can write down a model. This will allow observation of the weight two phenomenon which is new in the noncompact case.

The residue of a connection takes values in a space which represents the local monodromy around a puncture. As might be expected, this space has weight two,

2000 *Mathematics Subject Classification.* Primary 14D21, 32J25; Secondary 14C30, 14F35, 32L25.

Key words and phrases. Connection, Fundamental group, Higgs bundle, Parabolic structure, Quasiprojective variety, Representation, Twistor space.

so when we do the Deligne glueing we get a bundle of the form $\mathcal{O}_{\mathbb{P}^1}(2)$. There is an antipodal involution σ on this bundle, and the preferred sections corresponding to harmonic bundles are σ-invariant. The space of σ-invariant sections of $\mathcal{O}_{\mathbb{P}^1}(2)$ is $\mathbb{R}^3$, in particular it doesn't map isomorphically to a fiber over one point of $\mathbb{P}^1$. The kernel of the map to the fiber is the parabolic weight parameter. Remarkably, the parabolic structure appears "out of nowhere", as a result of the holomorphic structure of the Deligne-Hitchin twistor space constructed only using the notion of logarithmic λ-connections.

After §4, treating in detail the case of $\mathbb{P}^1 - \{0, \infty\}$, we look in §5 more closely at the bundle $\mathcal{O}_{\mathbb{P}^1}(2)$ which occurs: it is the *Tate twistor structure*, and is also seen as a twist of the tangent bundle $T\mathbb{P}^1$. Then §6 concerns the case of rank one local systems when X has arbitrary dimension. In §7 we state a conjecture about strictness which should follow from a full mixed theory as we are suggesting here.

Since we are considering rank one local systems, the tangent space is Deligne's mixed Hodge structure on $H^1(X - D, \mathbb{C})$ (see Theorem 6.3). However, a number of authors, such as Pridham [**45, 46**] and Brylinski-Foth [**7, 21**] have already constructed and studied a mixed Hodge structure on the deformation space of representations of rank $r > 1$ over an open variety. These structures should amount to the local version of what we are looking for in the higher rank case, and motivate the present paper. They might also allow a direct proof of the infinitesimal version of the strictness conjecture 7.1.

In the higher rank case, there are a number of problems blocking a direct generalization of what we do here. These are mostly related to monodromy operators which are not regular semisimple group elements. In a certain sense, the local structure of a connection with diagonalizable monodromy operators is like the direct sum of rank 1 pieces. However, the action of the gauge group contracts to a trivial action at $\lambda = 0$, so there is no easy way to cut out an open substack corresponding only to regular semisimple values. We leave this generalization as a problem for future study. This will necessitate using contributions from other works in the subject, such as Inaba-Iwasaki-Saito [**28, 29**] and Gukov-Witten [**24**].

This paper corresponds to my talk in the conference "Interactions with Algebraic Geometry" in Florence (May 30th-June 2nd 2007), just a week after the Augsburg conference. Sections 5–7 were added later. We hope that the observation we make here can contribute to some understanding of this subject, which is related to a number of other works such as the notion of tt^*-geometry [**25, 48**], geometric Langlands theory [**24**], Deligne cohomology [**20, 22**], parabolic bundles [**2, 3, 6, 30, 33, 34, 35, 38, 39, 40, 41, 44, 49, 50, 55, 56, 58**], harmonic bundles [**4**] [**31**] [**39**] and twistor $\mathcal{D}$-modules [**47**], Painlevé equations [**5**] [**28, 29**], and the theory of rank one local systems on open varieties [**1, 8, 14, 15, 16, 36**].

I would like to thank the referees for many useful corrections and improvements.

2. Preliminary definitions

It is useful to follow Deligne's way of not choosing a square root of -1. This serves as a guide to making constructions more canonically, which in turn serves to avoid encountering unnecessary choices later. We do this because one of the goals below is to understand in a natural way the Tate twistor structure $\mathbb{T}(-1)$. In particular, this has served as a useful guide for finding the explanation given in §5.1 for the sign change necessary in the logarithmic version $\mathbb{T}(-1, \log)$. We have tried, when possible, to explain the motivation for various other minus signs too. *Caution:* there may remain sign errors, especially towards the end.

Let $\mathbb{C}$ be an algebraic closure of $\mathbb{R}$, but without a chosen $\sqrt{-1}$. Nevertheless, occasional explanations using a choice of $i = \sqrt{-1} \in \mathbb{C}$ are admitted so as not to leave things too abstruse.

2.1. Complex manifolds. There is a notion of $\mathbb{C}$-linear complex manifold M. This means that at each point $m \in M$ there should be an action of $\mathbb{C}$ on the real tangent space $T_{\mathbb{R}}(M)$. Holomorphic functions are functions $M \to \mathbb{C}$ whose 1-jets are compatible with this action. Usual Hodge theory still goes through without referring to a choice of $i \in \mathbb{C}$. We get the spaces $A^{p,q}(M)$ of forms on M, and the operators ∂ and $\overline{\partial}$.

Let $\mathbb{R}^{\perp}$ denote the imaginary line in $\mathbb{C}$. This is what Deligne would call $\mathbb{R}(1)$; however we don't divide by 2π.

If h is a metric on M, there is a naturally associated two-form $\omega \in A^2(M, \mathbb{R}^{\perp})$. The Kähler class is $[\omega] \in H^2(X, \mathbb{R}^{\perp}) = H^2(X, \mathbb{R}(1))$. Classically this is brought back to a real-valued 2-form by multiplying by a choice of $\sqrt{-1}$, but we shouldn't do that here. Then, the operators L and Λ are defined independently of $\sqrt{-1}$, but they take values in $\mathbb{R}^{\perp}$. The Kähler identities now hold without $\sqrt{-1}$ appearing, but it is left to the reader to establish a convention for the signs.

Note that M may not be canonically oriented. If $Q = \{\pm\sqrt{-1}\}$ as below, then the orientation of M is canonically defined in the n-th power $Q^n \subset \mathbb{C}$, where $n = \dim_{\mathbb{C}} M$. In particular, the orientation in codimension 1 is always ill-defined. If D is a divisor, this means that $[D] \in H^2(M, \mathbb{R}^{\perp})$. This agrees with what happens with the Kähler metric. Similarly, if L is a line bundle then $c_1(L) \in H^2(M, \mathbb{R}^{\perp})$.

If X is a quasiprojective variety over $\mathbb{C}$ then $X(\mathbb{C})$ has a natural topology. Denote this topological space by X^{top}. It is the topological space underlying a structure of complex analytic space. In the present paper, we don't distinguish too much between algebraic and analytic varieties, so we use the same letter X to denote the analytic space.

Let $\overline{X}$ denote the conjugate variety, where the structural map is composed with the complex conjugation $\text{Spec}(\mathbb{C}) \to \text{Spec}(\mathbb{C})$. In terms of coordinates, $\overline{X}$ is given by equations whose coefficients are the complex conjugates of the coefficients of the equations of X. There is a natural isomorphism $\varphi : X^{\text{top}} \stackrel{\cong}{\to} \overline{X}^{\text{top}}$, which in terms of equations is given by $x \mapsto \overline{x}$ conjugating the coordinates of each point.

2.2. The imaginary scheme of a group. Let $Q \subset \mathbb{C}$ be the zero set of the polynomial $x^2 + 1$; in other words $Q = \{\pm\sqrt{-1}\}$. Multiplication by -1 is equal to multiplicative inversion, which is equal to complex conjugation, and these all define an involution

$$c_Q : Q \to Q.$$

Suppose Y is a set provided with an involution τ_Y. Then we define a new set denoted $Y^{\perp}$ starting from $\text{Hom}(Q, Y)$ with its two involutions

$$f \mapsto \tau_Y \circ f, \quad f \mapsto f \circ c_Q.$$

Let $Y^{\perp}$ be the equalizer of these two involutions; in other words

$$Y^{\perp} := \{f \in \text{Hom}(Q, Y), \ \tau_Y \circ f = f \circ c_Q\}.$$

Thus, an element of $Y^{\perp}$ is a function $\gamma : q \mapsto \gamma(q)$ such that $\gamma(-q) = \tau_Y(\gamma(q))$.

The two equal involutions will be denoted $\tau_{Y^\perp}$.

If we choose $i = \sqrt{-1} \in \mathbb{C}$, then $Y^\perp$ becomes identified with Y via $\gamma \mapsto \gamma(i)$. For the opposite choice of i this isomorphism gets composed with τ_Y.

If G is a scheme defined over $\mathbb{R}$ then $G^\perp$ is also defined over $\mathbb{R}$. For example, if $G = \mathbb{R}$ with involution $x \mapsto -x$ then $G^\perp$ is the imaginary line $\mathbb{R}^\perp$ defined above.

Using the involution $x \mapsto -x$ we could also define $\mathbb{C}^\perp$. However there is a natural isomorphism $\mathbb{C} \cong \mathbb{C}^\perp$ sending a to the function $\gamma : q \mapsto qa$. In view of this, and in order to lighten notation, we don't distinguish between $\mathbb{C}$ and $\mathbb{C}^\perp$ even in places where that might be natural, for example throughout §6.4.

If G is an algebraic group over $\mathbb{C}$, it has an involution $g \mapsto g^{-1}$, which doesn't preserve the group structure unless G is abelian. Using this involution yields a scheme, denoted by $G^\perp$. If G is abelian then $G^\perp$ has a natural group structure. In general there is a natural action of G on $G^\perp$ by conjugation: if $g \in G$ and $q \mapsto \gamma(q)$ is an element of $G^\perp$ then the element $q \mapsto g\gamma(q)g^{-1}$ is again an element of $G^\perp$.

Since we will be looking mostly at rank one local systems, we are particularly interested in the case $G = \mathbb{G}_m$. Then

$$\mathbb{G}_m^\perp = \{(x, y) \in \mathbb{C}^2, \ \ x^2 + y^2 = 1\}.$$

The equality is given as follows: to an element $\gamma : q \mapsto \gamma(q)$ of $\mathbb{G}_m^\perp$, associate the point (x, y) given by

$$x := \frac{1}{2} \sum_{q \in Q} \gamma(q),$$

$$y := \frac{1}{2} \sum_{q \in Q} q^{-1}\gamma(q).$$

Call (x, y) the *circular coordinates* on $\mathbb{G}_m^\perp$.

It is well-known that the exponential should really be considered as a map $\exp : \mathbb{C}(1) \to \mathbb{G}_m$. Alternatively, we can view the exponential as a map

$$\exp^\perp : \mathbb{C} \to \mathbb{G}_m^\perp.$$

given in circular coordinates by

$$\exp^\perp(\theta) := (\cos(2\pi\theta), \sin(2\pi\theta)).$$

It is useful to include 2π here because of the relationship with residues; see below. The kernel of $\exp^\perp$ is the usual $\mathbb{Z} \subset \mathbb{C}$. We call θ a *circular logarithm* of its image point.

2.3. Logarithmic connections. Suppose X is a smooth projective variety and $D \subset X$ is a normal crossings divisor. Let $U := X - D$ and $j : U \hookrightarrow X$ be the inclusion. Recall that the sheaf of *logarithmic forms* on (X, D), denoted $\Omega^1_X(\log D)$, is the locally free sheaf, a subsheaf of $j_*\Omega^1_U$, which is generated in local coordinates by $d\log z_1, \ldots, d\log z_k, dz_{k+1}, \ldots, dz_n$ whenever $(z_1, \ldots, z_n)$ is a system of local coordinates in which D is given by $z_1 \cdots z_k = 0$.

A *logarithmic connection* ∇ on a vector bundle E over X is a morphism of sheaves

$$\nabla : E \to E \otimes_{\mathcal{O}_X} \Omega^1_X(\log D)$$

such that $\nabla(af) = a\nabla(f) + da \cdot f$. More generally, for $\lambda \in \mathbb{C}$ a *logarithmic λ-connection* is a map ∇ as above such that $\nabla(af) = a\nabla(f) + \lambda da \cdot f$. For $\lambda = 1$

this is a usual connection, and for any $\lambda \neq 0$ we get a usual connection $\lambda^{-1}\nabla$. For $\lambda = 0$ it is a Higgs field.

The *Riemann-Hilbert correspondence* takes a vector bundle with logarithmic connection (E, ∇) to its monodromy representation ρ. This is well-defined independently of the choice of $\sqrt{-1} \in \mathbb{C}$. In the compact case, it is an equivalence of categories between vector bundles with connection and representations up to conjugacy. However, in our open case there are many possible choices of (E, ∇) which give the same representation ρ, because of the possibility of making *meromorphic gauge transformations* along the components of the divisor D; see §2.5 below. For any $\lambda \neq 0$, the *monodromy representation* of a λ-connection is by definition that of the normalized connection $\lambda^{-1}\nabla$.

2.4. Local monodromy. The reason for introducing the imaginary scheme $G^{\perp}$ was to discuss local monodromy. Keep the notation that (X, D) is a smooth variety with a normal crossings divisor. For each component D_i of D, choose a point x_i near D_i. Choose a local coordinate system $(z_1, \ldots, z_n)$ for X near a smooth point of D_i, such that D_i is given by $z_1 = 0$ and x_i is the point $(\epsilon, 0, \ldots, 0)$. We get a map from Q to $\pi_1(X - D, x_i)$ as follows: for $q \in Q$, consider the path $t \mapsto (\epsilon \cdot e^{2\pi q t}, 0, \ldots, 0)$. For $-q$ we get the inverse path, in other words we really have an element of $\pi_1(X - D, x_i)^{\perp}$. Conjugating by a choice of path from x to x_i, we get an element

$$\gamma_{D_i} \in \pi_1(X - D, x)^{\perp}.$$

It is well-defined up to the conjugation action of $\pi_1(X - D, x)$.

If $\rho : \pi_1(X - D, x) \to G$ is a representation, we obtain by functoriality of the construction $(\ )^{\perp}$ a map

$$\rho^{\perp} : \pi_1(X - D, x)^{\perp} \to G^{\perp},$$

so we get the *local monodromy element*

$$\mathrm{mon}(\rho, D_i) := \rho^{\perp}(\gamma_{D_i}) \in G^{\perp},$$

which is well-defined up to the conjugation action of G. If G is abelian, such as $G = \mathbb{G}_m$, then the local monodromy element is well-defined.

2.5. Meromorphic gauge group. Since we will mostly be working with line bundles, we describe the meromorphic gauge group only in this case. It is much easier than in general. Decompose $D = D_1 + \ldots + D_k$ into a union of smooth irreducible components. The gauge group is just

$$\mathcal{G} := \mathbb{Z}^k,$$

acting as follows. Suppose (L, ∇) is a line bundle with logarithmic λ-connection on (X, D) and $g = (g_1, \ldots, g_k)$ is an element of $\mathcal{G}$. Then the new line bundle is defined by

$$L^g := L(g_1 D_1 + \ldots + g_k D_k),$$

and ∇^g is the unique logarithmic λ-connection on L^g which coincides with ∇ over the open set U via the canonical isomorphism $L^g|_U \cong L|_U$.

The gauge transformation affects the first Chern class:

$$c_1(L^g) = c_1(L) + g_1[D_1] + \ldots + g_k[D_k], \tag{2.1}$$

and the residue:

$$\mathrm{res}(\nabla^g; D_i) = \mathrm{res}(\nabla; D_i) - \lambda g_i. \tag{2.2}$$

For convenience, here is the proof of (2.2). If u is a nonvanishing holomorphic section of L near a point of D_i (but not near the other divisor components), then u may also be considered as a meromorphic section of L^g, but it has a zero of order g_i along D_i. Hence, $u' := z_i^{-g_i} u$ is a nonvanishing holomorphic section of L^g near our point of D_i.

Let $R_i := \mathrm{res}(\nabla; D_i)$, so

$$\nabla(au) = \lambda d(a)u + R_i \frac{dz_i}{z_i} au + \cdots .$$

Generically, ∇ and ∇^g are the same connection. However, a section of L^g is written in terms of the unit section u' as $au' = az_i^{-g_i}u$, so

$$\nabla^g(au') = \nabla(az_i^{-g_i}u) = \lambda d(a)u' - \lambda g_i \frac{dz_i}{z_i} au' + R_i \frac{dz_i}{z_i} au + \cdots .$$

The residue of ∇^g is $\mathrm{res}(\nabla^g; D_i) = R_i - \lambda g_i$ as claimed in (2.2).

The restrictions to the open set are isomorphic:

$$(L^g, \nabla^g)|_U \cong (L, \nabla)|_U,$$

hence the monodromy representations are the same in the case $\lambda \neq 0$. Conversely, again in the case $\lambda \neq 0$, given (L, ∇) and (L', ∇') two logarithmic λ-connections with the same monodromy representations, there is a unique meromorphic gauge transformation $g \in \mathcal{G}$ such that $(L', \nabla') \cong (L^g, \nabla^g)$.

Throughout the paper, make the convention that spaces and maps are in the complex analytic category. The reader will notice which parts of these analytic spaces have natural algebraic structures, for example the Betti spaces or the charts $M_{\mathrm{Hod}}(X, \log D)$. Often these algebraic charts will be divided by a group action or glued to other charts in an analytic way, so the result only has a structure of analytic space.

3. The Deligne glueing in the compact case

In this section we recall the Deligne glueing construction for the twistor space, in the case of a compact base variety X, that is, $D = \emptyset$. The hyperkähler structure on the moduli space was constructed by Hitchin [**26**], who also considered the Penrose twistor space associated to the quaternionic structure. Deligne in [**12**] proposed a construction of the twistor space using a deformation called the space of *λ-connections* closely related to the Hodge filtration, plus the Riemann-Hilbert correspondence relating connections on X and the conjugate variety $\overline{X}$. Apparently Witten contributed something too because Deligne's letter [**12**] starts off:

> "As I understand, Hitchin's understanding of why one has a hyperkähler structure—as explained to me by Witten—works in your case. ...".

The twistor space structure is related to the notion of tt^* geometry [**9, 25, 48**]. The idea of a deformation relating de Rham and Dolbeault cohomology goes back further, to the theory of Γ-factors [**13**], Esnault's notion of τ-connection [**19**], Dolbeault homotopy theory [**42**], to the relation between cyclic and Hochschild cohomology [**10, 32**], and to singular perturbation theory [**57**].

3.1. Moduli spaces. Fix a basepoint $x \in X$. Complex conjugation provides a map of topological spaces $\varphi : X^{\text{top}} \to \overline{X}^{\text{top}}$, which is antiholomorphic for the complex structures. In particular, we get

$$\varphi_* : \pi_1(X, x) \xrightarrow{\cong} \pi_1(\overline{X}, \overline{x}).$$

Recall the following moduli spaces, which are universal categorical quotients of the corresponding moduli stacks. Most of what we say holds also for the moduli stacks. We are fixing the target group as $GL(r, \mathbb{C})$ which will be left out of the notation. In this section we let r be arbitrary, although the next sections will specialize to $r = 1$, in which case the moduli stacks would be $\mathbb{G}_m$-gerbs over the spaces.

Write $M_{\text{Hod}}(X) \to \mathbb{A}^1$ for the moduli space of semistable vector bundles of rank r with λ-connection with vanishing Chern classes. The fibers over 0 and 1 are denoted respectively by $M_{\text{Dol}}(X)$ and $M_{\text{DR}}(X)$. The group $\mathbb{G}_m$ acts, and over $\mathbb{G}_m \subset \mathbb{A}^1$ this action provides an isomorphism

$$\mathbb{G}_m \times_{\mathbb{A}^1} M_{\text{Hod}}(X) \cong \mathbb{G}_m \times M_{\text{DR}}(X).$$

These natural constructions applied to the conjugate variety give conjugate varieties:

$$M_{\text{Hod}}(\overline{X}) \cong \overline{M_{\text{Hod}}(X)}, \quad M_{\text{DR}}(\overline{X}) \cong \overline{M_{\text{DR}}(X)}, \quad M_{\text{Dol}}(\overline{X}) \cong \overline{M_{\text{Dol}}(X)}.$$

We also have the Betti moduli space [**37**], a universal categorical quotient

$$M_{\text{B}}(X) = \frac{\text{Hom}(\pi_1(X, x), GL(r, \mathbb{C}))}{GL(r, \mathbb{C})}$$

whereas the moduli stack would be the stacky quotient. The Riemann-Hilbert correspondence gives an isomorphism of analytic spaces or stacks

$$M_{\text{DR}}(X)^{\text{an}} \cong M_B(X)^{\text{an}}.$$

It doesn't depend on a choice of square root of -1.

3.2. Glueing. The *Deligne glueing* is an isomorphism of complex analytic spaces

$$\mathbf{d} : \mathbb{G}_m \times_{\mathbb{A}^1} M_{\text{Hod}}(X) \cong \mathbb{G}_m \times_{\mathbb{A}^1} M_{\text{Hod}}(\overline{X}).$$

It is defined as follows. A point in the source is a triple (λ, E, ∇) with $\lambda \in \mathbb{G}_m \subset \mathbb{A}^1$, where E is a vector bundle on X, and ∇ is a λ-connection on E. This corresponds to the point $(\lambda, (E, \lambda^{-1}\nabla))$ in $\mathbb{G}_m \times M_{\text{DR}}(X)$. Let $\rho(\lambda^{-1}\nabla)$ denote the monodromy representation of $\pi_1(X, x)$ corresponding to the connection $\lambda^{-1}\nabla$. Then $\rho(\lambda^{-1}\nabla) \circ \varphi_*^{-1}$ is a representation of $\pi_1(\overline{X}, \overline{x})$. It corresponds to a vector bundle with connection (F, Φ) on $\overline{X}$.

This vector bundle with connection may be characterized as follows:
—we have a natural identification $F_{\overline{x}} \cong E_x$; and
—the monodromy of (F, Φ) around a loop γ in $\pi_1(\overline{X}, \overline{x})$ is equal, via this identification, to the monodromy of $(E, \lambda^{-1}\nabla)$ around the loop $\varphi^{-1}(\gamma)$ in $\pi_1(X, x)$.

To continue with the definition of $\mathbf{d}$, choose the point $\mu = \lambda^{-1} \in \mathbb{G}_m$, and look at the point

$$(\mu, (F, \Phi)) \in \mathbb{G}_m \times M_{\text{DR}}(\overline{X}).$$

It corresponds to a point

$$(\mu, F, \mu\Phi) \in M_{\text{Hod}}(\overline{X}) \times_{\mathbb{A}^1} \mathbb{G}_m.$$

We set

$$\mathbf{d}(\lambda, E, \nabla) := (\mu, F, \mu\Phi).$$

Note that by definition $\mathbf{d}$ covers the map $\mathbb{G}_m \to \mathbb{G}_m$ given by $\lambda \mapsto \mu := \lambda^{-1}$.

This isomorphism can now be used to glue together the two analytic spaces $M_{\mathrm{Hod}}(X)^{\mathrm{an}}$ and $M_{\mathrm{Hod}}(\overline{X})^{\mathrm{an}}$ along their open sets which are the source and target of $\mathbf{d}$. The resulting space is denoted $M_{\mathrm{DH}}(X)$ for *Deligne-Hitchin*. It is Hitchin's twistor space [**26**], constructed as suggested by Deligne [**12**].

Interpreting $\mathbb{P}^1$ as obtained by glueing two copies of $\mathbb{A}^1$ along the map $\mu = \lambda^{-1}$, we get a map $M_{\mathrm{DH}}(X) \to \mathbb{P}^1$.

It is essential to make some remarks on the choices above. The space $M_{\mathrm{Hod}}(X)$ and its conjugate counterpart admit numerous natural automorphisms, for example multiplication by any element of $\mathbb{G}_m$, but also taking the dual of an object. In particular, it would have been possible to insert these operations in the middle of the definition of $\mathbf{d}$. They would extend to automorphisms of either of the two sides being glued, so the resulting space would be isomorphic. We feel that it is reasonable at each step of the way to use the simplest choice. This will nonetheless result in more complicated choices in the definition of preferred sections later.

Note that we have not used any choice of $\sqrt{-1} \in \mathbb{C}$ in the construction, so $M_{\mathrm{DH}}(X)$ is independent of that.

According to the construction, notice that we have two inclusions

$$u : M_{\mathrm{Hod}}(X) \hookrightarrow M_{\mathrm{DH}}(X), \quad v : M_{\mathrm{Hod}}(\overline{X}) \hookrightarrow M_{\mathrm{DH}}(X).$$

We have $u(\lambda, E, \nabla) = v(\mu, F, \mu\Phi)$ exactly when $\mathbf{d}(\lambda, E, \nabla) = (\mu, F, \mu\Phi)$ as constructed above.

We leave to the reader the problem of comparison of $M_{\mathrm{DH}}(X)$ and $M_{\mathrm{DH}}(\overline{X})$.

3.3. The antipodal involution. A crucial part of the structure is an antilinear involution $\sigma : M_{\mathrm{DH}}(X) \to M_{\mathrm{DH}}(X)$, covering the antipodal involution of $\mathbb{P}^1$. The antipodal involution exchanges the two charts $\mathbb{A}^1$ of $\mathbb{P}^1$. Thus, in order to define σ it suffices to define an antiholomorphic map

$$\sigma_{\mathrm{Hod},X} : M_{\mathrm{Hod}}(X) \to M_{\mathrm{Hod}}(\overline{X})$$

which is an antilinear isomorphism, and involutive, that is $\sigma_{\mathrm{Hod},\overline{X}} \circ \sigma_{\mathrm{Hod},X} = \mathrm{Id}$.

Suppose we have a point (λ, E, ∇). Taking the complex conjugate of everything gives a point $(\overline{\lambda}, \overline{E}, \overline{\nabla}) \in M_{\mathrm{Hod}}(\overline{X})$. This gives an antiholomorphic map, denoted

$$C_{\mathrm{Hod},X} : M_{\mathrm{Hod}}(X) \to M_{\mathrm{Hod}}(\overline{X}).$$

We need to show that it is compatible with the glueing $\mathbf{d}$ in the sense that

$$C_{\mathrm{Hod},\overline{X}} \circ \mathbf{d} = \mathbf{d}^{-1} \circ C_{\mathrm{Hod},X}. \tag{3.1}$$

First of all $C_{\mathrm{Hod},X}$ and $C_{\mathrm{Hod},\overline{X}}$ intertwine the multiplication action of $\mathbb{G}_m$ with the complex conjugation $\mathbb{G}_m \cong \overline{\mathbb{G}}_m$. Also, C and $\mathbf{d}$ both fix the de Rham fiber over $\lambda = 1$. Hence, to verify the compatibility (3.1), it suffices to verify it over $\lambda = 1$. Here

$$C_{\mathrm{DR},X} : M_{\mathrm{DR}}(X) \to M_{\mathrm{DR}}(\overline{X}), \quad (E, \nabla) \mapsto (\overline{E}, \overline{\nabla})$$

and composing with the isomorphism $\mathbf{d}^{-1}$ which comes from $\pi_1(X, x) \cong \pi_1(\overline{X}, \overline{x})$ we get an antilinear automorphism of $M_{\mathrm{DR}}(X)$. It is easy to see that, in terms of the isomorphism with $M_{\mathrm{B}}(X)$, this antilinear automorphism is just the complex conjugation action on representations, $\rho \mapsto \overline{\rho}$ where $\overline{\rho}(\gamma) := \overline{\rho(\gamma)}$. Similarly,

$C_{\mathrm{Hod},\overline{X}} \circ \mathbf{d}$ is also seen to be the same automorphism $\rho \mapsto \overline{\rho}$. This proves the equality (3.1).

With this compatibility, $C_{\mathrm{Hod},X}$ and $C_{\mathrm{Hod},\overline{X}}$ glue to give an antiholomorphic involution

$$C : M_{\mathrm{DH}}(X) \to M_{\mathrm{DH}}(X)$$

covering the involution $\lambda \mapsto \overline{\lambda}^{-1}$ of $\mathbb{P}^1$. As described above, on the fiber over $\lambda = 1$ which is $M_{\mathrm{DR}}(X) \cong M_{\mathrm{B}}(X)$, the involution is $C(\rho) = \overline{\rho}$.

The dual of a vector bundle with λ-connection (E, ∇) is again a vector bundle with λ-connection, denoted (E^*, ∇^*). This operation is compatible with multiplication by $\mathbb{G}_m$, and with the operation of taking the dual of a local system via the Riemann-Hilbert correspondence. Therefore, it is compatible with the glueing $\mathbf{d}$ and gives an involution, holomorphic this time, denoted

$$D : M_{\mathrm{DH}}(X) \to M_{\mathrm{DH}}(X)$$

which covers the identity of $\mathbb{P}^1$.

Finally, multiplication by $-1 \in \mathbb{G}_m$ gives an involution of $M_{\mathrm{DH}}(X)$, denoted by N, covering the involution $\lambda \mapsto -\lambda$ of $\mathbb{P}^1$.

LEMMA 3.1. *The involutions C, D and N commute. Their product is an involution σ of $M_{\mathrm{DH}}(X)$ covering the antipodal involution $\lambda \mapsto -\overline{\lambda}^{-1}$ of $\mathbb{P}^1$.*

Proof: The complex conjugate of the dual of a vector bundle is naturally isomorphic to the dual of the complex conjugate. These also clearly commute with the operation of multiplying the connection by -1. Hence, the three involutions commute, which implies that the product CDN is again an involution. It is antilinear because C is antilinear whereas D and N are $\mathbb{C}$-linear. Since C, D and N cover respectively the involutions $\lambda \mapsto \overline{\lambda}^{-1}$, $\lambda \mapsto \lambda$ and $\lambda \mapsto -\lambda$, their product covers the product of these three, which is the antipodal involution. □

3.4. Preferred sections and the twistor property. Deligne proposed to construct a family of "preferred sections" of the glued space $M_{\mathrm{DH}}(X)$, one for each harmonic bundle on X.

PROPOSITION 3.2 ([**26, 12, 53**]). *Suppose $(E, \partial, \overline{\partial}, \theta, \overline{\theta})$ is a harmonic bundle on X. Then it leads to a section $\mathcal{P} : \mathbb{P}^1 \to M_{\mathrm{DH}}(X)$ which is σ-invariant and which sends $\lambda \in \mathbb{A}^1$ to the holomorphic bundle $(E, \overline{\partial} + \lambda\overline{\theta})$ with λ-connection $\nabla = \lambda\partial + \theta$.*

Proof: We first define the value of the section at $\lambda \in \mathbb{A}^1$. On the $\mathcal{C}^\infty$ bundle E, consider the holomorphic structure

$$\overline{\partial}_\lambda := \overline{\partial} + \lambda\overline{\theta}.$$

The holomorphic bundle $E^\lambda := (E, \overline{\partial}_\lambda)$ admits a λ-connection operator $\nabla_\lambda := \lambda\partial + \theta$. This gives a point $(E^\lambda, \nabla_\lambda)$ in $M_{\mathrm{Hod}}(X)_\lambda$. One checks that over $\mathbb{G}_m \subset \mathbb{A}^1$, this section is invariant under the antipodal involution operator. Hence, taking the image of the graph of our section already defined over $\mathbb{A}^1$, by σ gives the section over the other chart $\mathbb{A}^1$ at infinity, and over $\mathbb{G}_m$ these glue together. One can also define directly the value of the section on the complex conjugate chart; see for example [**54**, pp 20-24]. □

In Hitchin's original point of view [27, 26], the twistor space $M_{\rm DH}(X)$ came from the Penrose construction for the quaternionic structure on $M(X)$ whose different complex structures were those of $M_{\rm Dol}$ and $M_{\rm DR}$. The Penrose twistor space has a natural product structure of the form $\mathbb{P}^1 \times M(X)$.

In Deligne's reinterpretation [12] we can first construct the space $M_{\rm DH}(X)$ using the notion of λ-connection, complex conjugation and the Riemann-Hilbert correspondence as described above. The product structure is obtained by considering the family of preferred sections as described in the previous proposition. This leads back to the quaternionic structure by looking at the tangent space near a preferred section. The key to this beautiful procedure is the observation that the relative tangent space, or equivalently the normal bundle, along a preferred section is a semistable bundle of slope 1 on $\mathbb{P}^1$, which is to say it is isomorphic to $\mathcal{O}_{\mathbb{P}^1}(1)^{\oplus a}$. This weight one property is equivalent to having a quaternionic structure, as was observed in [27].

There is an equivalence of categories between quaternionic vector spaces and vector bundles of slope 1 over $\mathbb{P}^1$ with involution σ covering the antipodal involution. If $V = \mathcal{O}_{\mathbb{P}^1}(1)^d$ is a slope one bundle, the space of sections is $H^0(\mathbb{P}^1, V) \cong \mathbb{C}^{2d}$. If σ is an antipodal involution, the space of σ-invariant sections is a real form of the space of sections; thus

$$H^0(\mathbb{P}^1, V)^\sigma \cong \mathbb{R}^{2d},$$

and the twistor property says that the map from here to any of the fibers V_λ is an isomorphism. This is what provides the single real vector space $\mathbb{R}^{2d}$ with a whole sphere of different complex structures.

One can show directly the weight 1 property for the normal bundle of $M_{\rm DH}(X)$ along a preferred section, given the glueing construction and the definition of preferred sections in Proposition 3.2, see [53]. It then follows that the deformation space of a preferred section in the world of σ-invariant sections of the fibration $M_{\rm DH}(X) \to \mathbb{P}^1$ maps isomorphically to the tangent space of any of the moduli space fibers (for example $M_{\rm DR}(X)$ over $\lambda = 1$ or $M_{\rm Dol}(X)$ over $\lambda = 0$). It implies that, locally, there is a unique σ-invariant section going through any point, and gives an alternative proof of Hitchin's theorem that the moduli space has a quaternionic structure. For rank one bundles, this property can be globalized:

LEMMA 3.3. *For bundles of rank* 1 *on a compact* X, *the evaluation morphism at any point* $p \in \mathbb{P}^1$

$$\Gamma(\mathbb{P}^1, M_{\rm DH}(X))^\sigma \to M_{\rm DH}(X)_p$$

is an isomorphism.

PROOF. In the rank one case, the moduli space is a Lie group so we can use its exponential exact sequence. The tangent at the identity preferred section is purely semistable of slope 1. There is a lattice $A = H^1(X, \mathbb{Z}) \cong \mathbb{Z}^a$ and a finite group $B = H^2(X, \mathbb{Z})^{\rm tors} = NS(X)^{\rm tors}$ such that we have an exact sequence

$$0 \to A \to A \otimes \mathcal{O}_{\mathbb{P}^1}(1) \to M_{\rm DH}(X) \to B \to 0.$$

We have $H^1(\mathbb{P}^1, A) = H^1(\mathbb{P}^1, A \otimes \mathcal{O}_{\mathbb{P}^1}) = H^1(\mathbb{P}^1, B) = 0$, so taking sections gives an exact sequence. The subgroups of σ-invariants again form an exact sequence. The weight one property, equivalent to the quaternionic structure, says

$$\Gamma(\mathbb{P}^1, A \otimes \mathcal{O}_{\mathbb{P}^1}(1))^\sigma \stackrel{\cong}{\to} A \otimes \mathcal{O}_{\mathbb{P}^1}(1)_p.$$

Comparing with the exact sequence of values at p gives the result for $M_{\rm DH}(X)$. □

Deligne gave the construction of a quaternionic structure associated to a weight 1 real Hodge structure in [**12**]. Given a vector space V with two filtrations F and $\overline{F}$, we can form a bundle $\xi(V, F, \overline{F}) \to \mathbb{P}^1$ and this bundle has slope 1 if and only if the two filtrations are 1-opposed, i.e. they define a Hodge structure of weight 1. In this sense, the twistor property is analogous to saying that a Hodge structure has weight 1. These slightly different points of view are compatible for a preferred section which comes from a variation of Hodge structure, where the tangent space to the moduli space has a natural weight 1 Hodge structure.

The bundles of the form $\xi(V, F, \overline{F})$ are the slope 1 bundles together with additional structure of an action of $\mathbb{G}_m$. A somewhat different collection of additional structures involving a connection is used in the notion of tt^* geometry [**9, 25**], which also has its physical roots in Hitchin's twistor space. Schmid mentioned, during his courses on Hodge theory, the similarity between the equations governing the local structure of variations of Hodge structure, and the monopole or Nahm's equations. It is interesting that these objects from physics are so closely related to variations of Hodge structure, the analytic incarnation of the idea of motives. This suggests a relationship between physics and motives which might be philosophically compelling.

4. The twistor space for $\mathbb{P}^1 - \{0, \infty\}$

We would now like to mimic the Deligne-Hitchin construction for a quasiprojective curve. For simplicity of calculation, let us take the easiest case, which is $X := \mathbb{P}^1$ and $D := \{0, \infty\}$. Let $U := X - D$ and fix $x = 1$ as basepoint in X or U. Then $\pi_1(U, x) \cong \mathbb{Z}^\perp$. A choice of $q = \sqrt{-1} \in Q \subset \mathbb{C}$ yields a choice of generator $\gamma_0(q) \in \pi_1(U, x)$ going once around the origin, counterclockwise if 1 is pictured to the right of the origin and q is pictured above the origin. Changing the choice of q changes the generator to its inverse, which is why we get $\mathbb{Z}^\perp$ rather than $\mathbb{Z}$. For the local monodromy transformations, this yields

$$\pi_1(U, x)^\perp \cong \mathbb{Z}$$

with a distinguished generator, denoted γ_0.

Let z denote the standard coordinate on X. A logarithmic λ-connection on the trivial bundle $E := \mathcal{O}_X$ is of the form

$$\nabla = \lambda d + \alpha \frac{dz}{z}.$$

In particular, we can write

$$M_{\mathrm{Hod}}(X, \log D) = \mathbb{A}^1 \times \mathbb{A}^1 = \{(\lambda, \alpha)\}.$$

The first coordinate is the parameter λ and the second the residue parameter α.

For $\lambda \neq 0$ a point (λ, α) corresponds to the 1-connection $d + \lambda^{-1}\alpha(dz/z)$. Let $\rho : \pi_1(U, x) \to \mathbb{G}_m$ be its monodromy representation.

The local monodromy transformation at the origin (see §2.4 above) is

$$\rho^\perp(\gamma_0) = \exp^\perp(\alpha/\lambda) = (\cos(2\pi\alpha/\lambda), \sin(2\pi\alpha/\lambda)) \in \mathbb{G}_m^\perp.$$

In order to write the global monodromy, note that any loop $\gamma \in \pi_1(U, x)$ can be expressed as a unique function

$$\gamma : [0, 1] \to U$$

such that $|\gamma(t)| = 1$ and γ proceeds at a uniform speed, i.e. $\left|\frac{d\gamma}{dt}\right|$ is constant. The function γ is then real analytic and extends by analytic continuation to a unique map $\gamma : \mathbb{C} \to U$. In usual terms, choosing $i = \sqrt{-1} \in \mathbb{C}$, the path γ is written as $t \mapsto e^{2\pi i t}$ and this expression is valid for any $t \in \mathbb{C}$.

The global monodromy of our differential equation $d + \lambda^{-1}\alpha(dz/z)$ can now be expressed by the formula

$$\rho(\gamma) = \gamma(\alpha/\lambda).$$

Next, note that $\overline{X} = \mathbb{P}^1 - \{0, \infty\}$ too, and $\overline{x} = 1$ still, so we can write $(\overline{X}, \overline{D}, \overline{x}) = (X, D, x)$. In terms of this equality, φ is just the geometric operation of complex conjugation on U^{top}. Hence, for any $\gamma \in \pi_1(U, x)$, the complex conjugation map φ takes γ to the loop

$$\varphi(\gamma) = \overline{\gamma} = \gamma^{-1}.$$

4.1. Computation of the Deligne glueing. We also have

$$M_{\mathrm{Hod}}(\overline{X}, \log \overline{D}) = \mathbb{A}^1 \times \mathbb{A}^1 = \{(\lambda, \alpha)\}$$

via the identification between X and $\overline{X}$. In order to compute the Deligne glueing map $\mathbf{d}$, suppose we start with a point (λ, α) in $M_{\mathrm{Hod}}(X, \log D)$. This corresponds to a monodromy representation $\gamma \mapsto \gamma(\alpha/\lambda)$ as explained above. The fact that φ interchanges γ and γ^{-1} means that, after re-identifying $\overline{X}$ with X, the image of this representation by φ^* is

$$\gamma \mapsto \gamma^{-1}(\alpha/\lambda).$$

There is a unique way to lift this to a logarithmic connection, if we want to send the point $\alpha = 0$ to the point $\alpha = 0$ and keep everything continuous: it is the connection $\Phi = d - \lambda^{-1}\alpha(dz/z)$. Finally, in the prescription for the Deligne glueing we set $\mu := \lambda^{-1}$ and transform this to a μ-connection $\mu\Phi$. This yields the point

$$\mathbf{d}(\lambda, \alpha) = (\mu, \beta) = (\lambda^{-1}, -\lambda^{-2}\alpha).$$

We can now glue the two charts to get the Deligne-Hodge twistor space:

$$M_{\mathrm{DH}}(X, \log D) := M_{\mathrm{Hod}}(X, \log D) \sqcup^{\mathbf{d}} M_{\mathrm{Hod}}(\overline{X}, \log \overline{D}).$$

4.2. The weight two property. Notice that the glueing map is linear in α, so in this case the result is a vector bundle over $\mathbb{P}^1$, in fact it is clearly the bundle $\mathcal{O}_{\mathbb{P}^1}(2)$ with glueing function $-\lambda^{-2}$. The minus sign will have an effect on the antilinear involution below.

Suppose $\lambda \mapsto P(\lambda)$ is a polynomial considered as a section of $M_{\mathrm{Hod}}(X, \log D)$. Then its graph is the set of points $(\lambda, P(\lambda))$ and these correspond to points of the form $(\lambda^{-1}, -\lambda^{-2}P(\lambda))$ in $M_{\mathrm{Hod}}(\overline{X}, \log \overline{D})$ for λ invertible. Taking the closure over $\mu = \lambda^{-1} \to 0$ yields the set of points of the form $(\mu, -\mu^2 P(\mu^{-1}))$. This is a holomorphic section in the μ chart if and only if P is a polynomial of degree ≤ 2. This is one way to see that $M_{\mathrm{DH}}(X, \log D) \cong \mathcal{O}_{\mathbb{P}^1}(2)$. The global sections are those which, in the standard chart $M_{\mathrm{Hod}}(X, \log D)$, are polynomials of degree ≤ 2.

The main point of the title of this paper is that, since this bundle has slope 2, it corresponds in some sense to a Hodge structure of weight 2. That contrasts with the normal bundle of a preferred section in the compact case (§3.4), which has slope 1. The weight two behavior is to be expected in the present situation, by analogy with the usual mixed Hodge theory of open varieties, where $H^1(\mathbb{P}^1 - \{0, \infty\})$ has a pure Hodge structure of weight two.

In the present case, the Deligne-Hitchin space $M_{\rm DH}(X, \log D)$ will again have an involution σ to be calculated below. Given that it is a line bundle of slope 2, the space of sections has dimension 3:

$$\Gamma(\mathbb{P}^1, M_{\rm DH}(X, \log D)) \cong \Gamma(\mathbb{P}^1, \mathcal{O}_{\mathbb{P}^1}(2)) \cong \mathbb{C}^3,$$

and the σ-invariant sections are a real form

$$\Gamma(\mathbb{P}^1, M_{\rm DH}(X, \log D))^{\sigma} \cong \mathbb{R}^3.$$

The map from here to any one of the fibers over $\lambda \in \mathbb{P}^1$ will be surjective but have a real one-dimensional kernel. It turns out that this real kernel corresponds to the real parameter involved in a parabolic structure, even though we have seen the existence of this additional real parameter without refering *a priori* to the notion of parabolic structure.

4.3. The antipodal involution. We now calculate explicitly the involution σ. Recall that it is a product of the three involutions C, D and N. The duality involution is trivial on the underlying bundles because we are using the trivial bundle: $E^* = \mathcal{O}^* = \mathcal{O} = E$. The connection on $E \otimes E^*$ should be trivial, so we see that for $\nabla = d + \alpha(dz/z)$ the dual connection is $\nabla^* = d - \alpha(dz/z)$. Thus

$$D(\lambda, \alpha) = (\lambda, -\alpha).$$

Similarly, by definition

$$N(\lambda, \alpha) = (-\lambda, -\alpha).$$

Putting these together gives $DN(\lambda, \alpha) = (-\lambda, \alpha)$. These are expressed within a single chart $M_{\rm Hod}(X, \log D)$. On the other hand, the involution C goes from the chart $M_{\rm Hod}(X, \log D)$ to the chart $M_{\rm Hod}(\overline{X}, \log \overline{D})$, and with respect to these charts it is given by

$$C(\lambda, \alpha) = (\overline{\lambda}, \overline{\alpha}).$$

For λ invertible we would like to put this back in the original chart. Recall that a point of the form (μ, β) in the chart $M_{\rm Hod}(\overline{X}, \log \overline{D})$ corresponds to $(\mu^{-1}, -\mu^{-2}\beta)$ in the chart $M_{\rm Hod}(X, \log D)$. Thus, within the same chart $M_{\rm Hod}(X, \log D)$ and for λ invertible, the involution C can be expressed as

$$C(\lambda, \alpha) = (\overline{\lambda}^{-1}, -\overline{\lambda}^{-2}\overline{\alpha}).$$

Putting these together gives our expression for $\sigma = CDN$ again within the original chart and for λ invertible:

$$\sigma(\lambda, \alpha) = (-\overline{\lambda}^{-1}, -\overline{\lambda}^{-2}\overline{\alpha}).$$

We would now like to calculate which are the σ-invariant sections. Recall from §4.1 that a global section of $M_{\rm DH}(X, \log D)$ is, in the first chart, a polynomial P of order ≤ 2. Thus we can write our section as

$$P : \lambda \mapsto (\lambda, a_0 + a_1\lambda + a_2\lambda^2).$$

Its graph is the set of image points. The transformed section P^{σ} has a graph which is the closure of the set of points of the form

$$\sigma P(\lambda) = \left(-\overline{\lambda}^{-1}, -\overline{\lambda}^{-2}\overline{(a_0 + a_1\lambda + a_2\lambda^2)}\right) = \left(-\overline{\lambda}^{-1}, -\overline{a_2} - \overline{a_1}\overline{\lambda}^{-1} - \overline{a_0}\overline{\lambda}^{-2}\right).$$

Substituting in the above expression $-\overline{\lambda}^{-1}$ by t, the graph becomes the set of points of the form

$$(t, -\overline{a_2} + \overline{a_1}t - \overline{a_0}t^2).$$

This is the graph of the polynomial $t \mapsto -\overline{a_2} + \overline{a_1}t - \overline{a_0}t^2$. Thus, writing our polynomials generically with a variable u we can write

$$(4.1) \qquad \left(a_0 + a_1 u + a_2 u^2\right)^\sigma = \left(-\overline{a_2} + \overline{a_1}u - \overline{a_0}u^2\right).$$

The σ-invariant sections are the polynomials with $a_2 = -\overline{a_0}$ and $a_1 = \overline{a_1}$. In other words, an invariant section corresponds to a pair $(a, \alpha) \in \mathbb{R} \times \mathbb{C} \cong \mathbb{R}^3$ with the formula

$$(4.2) \qquad P(\lambda) = \psi(a, \alpha)(\lambda) := \alpha - a\lambda - \overline{\alpha}\lambda^2.$$

The reader will recognize this as the formula from Mochizuki [**39**, 2.1.7, p. 25].

4.4. Gauge transformations. A logarithmic connection is not uniquely determined by its monodromy representation. This situation is complicated in higher rank, but is understood easily in our case from the fact that the monodromy associated to a λ-connection (λ, α) is $\exp^\perp(\alpha/\lambda)$, or with a choice of $i = \sqrt{-1}$ it is $e^{2\pi i\alpha/\lambda}$. If we replace α by $\alpha - k\lambda$ for $k \in \mathbb{Z}$ we get the same monodromy representation. This process may be viewed as making the meromorphic gauge transformation $v \mapsto z^{-k}v$ on the bundle $E = \mathcal{O}_X$, or equivalently changing the bundle E to $E(kD_0 - kD_\infty)$, where $D_0 = \{0\}$ and $D_\infty = \{\infty\}$. The zeros or poles of the gauge transformation at points of D change the residues of the λ-connection by integer multiples of λ. The sign here and in the definition (4.2) of $\psi(a, \alpha)$ comes from the formula (2.2).

In terms of our space $M_{\rm Hod}(X, \log D)$ we have an action of $\mathbb{Z}$ obtained by letting $k \in \mathbb{Z}$ act as $(\lambda, \alpha) \mapsto (\lambda, \alpha - \lambda k)$. This action extends to the other chart, and gives an action of $\mathbb{Z}$ on $M_{\rm Hod}(X, \log D)$. Over $\mathbb{G}_m \subset \mathbb{P}^1$ the action is discrete and the quotient is $\mathbb{G}_m \times \mathbb{C}/\mathbb{Z} = \mathbb{G}_m \times M_{\rm B}(U)$. Note that the action degenerates to a trivial action on the fibers over 0 and ∞, and the stacky quotients by these actions are trivial $B\mathbb{Z}$-gerbs over $M_{\rm Dol}(X, \log D)$ and $M_{\rm Dol}(\overline{X}, \log \overline{D})$.

The $\mathbb{Z}$ action respects the involution σ so it gives an action on the space of σ-invariant sections. In terms of the previous formulae, this clearly acts on the degree 1 term in the polynomials, or in terms of the coordinates $(a, \alpha) \in \mathbb{R} \times \mathbb{C}$ it acts with generator $(1, 0)$. Thus, we can write

$$\frac{\Gamma(\mathbb{P}^1, M_{\rm DH}(X, \log D))^\sigma}{\mathbb{Z}} \cong \frac{\mathbb{R} \times \mathbb{C}}{(1, 0) \cdot \mathbb{Z}}.$$

Given a harmonic bundle on U we get a σ-invariant preferred section, where a is the parabolic weight of the Higgs bundle and α is the residue of the Higgs field; see Theorem 6.7 below.

We recover in this way the space of possible residues of parabolic λ-connections, with the residue of the λ-connection being given by the previous formula (4.2). Note that the action of $\mathbb{Z}$ is by meromorphic gauge transformations, so moving the parabolic index once around the circle induces an elementary transformation of the bundle.

5. The Tate twistor structure

Before getting to the general rank one case, we investigate the structures associated to the bundle $\mathcal{O}_{\mathbb{P}^1}(2)$ which occurs above. Recall that $T\mathbb{P}^1 \cong \mathcal{O}_{\mathbb{P}^1}(2)$. Furthermore, the sign $-\lambda^2$ which occurs in the glueing function for residues of points in $M_{\rm DH}(X, \log D)$ is the same as in the glueing function for $T\mathbb{P}^1$. The Tate motive (denoted by a twist of -1 following Deligne) is a pure Hodge structure of type $(1,1)$, hence weight 2. In view of this, we define the *additive Tate twistor structure* to be the bundle

$$\mathbb{T}(-1) := T\mathbb{P}^1,$$

with its natural antilinear involution

$$\sigma_{\mathbb{T}(-1)} := \sigma_{T\mathbb{P}^1}.$$

Here $\sigma_{T\mathbb{P}^1}$ is the real differential of σ considered as an involution of $T\mathbb{P}^1$ covering σ, via the identification of real bundles $T\mathbb{P}^1 \cong T_{\mathbb{R}}\mathbb{P}^1$; it is also the composite of the $\mathbb{C}$-linear bundle map $d\sigma : \sigma^* T\mathbb{P}^1 \to \overline{T\mathbb{P}^1}$ with complex conjugation $\overline{T\mathbb{P}^1} \to T\mathbb{P}^1$. See also Mochizuki [**39**, §3.10.2] and Sabbah [**47**, §2.1.3].

On the other hand, we define the *logarithmic Tate twistor structure* to be the same bundle $\mathbb{T}(-1, \log) := T\mathbb{P}^1$, but here the antilinear involution should have a sign change with respect to the natural one on $T\mathbb{P}^1$,

$$\sigma_{\mathbb{T}(-1,\log)} := -\sigma_{T\mathbb{P}^1}.$$

The reason for the difference between these two will be explained below*.

5.1. Integer subgroups. The action of $\mathbb{G}_m$ on $\mathbb{P}^1$ preserving 0 and ∞ gives an action of $\mathbb{G}_m$ on $T\mathbb{P}^1$. The derivative of this action is a section of the tangent bundle, defining the integer subgroup $\mathbb{Z} \times \mathbb{P}^1 \to T\mathbb{P}^1$. This section is antipreserved by the standard involution $\sigma_{T\mathbb{P}^1}$. For the additive Tate twistor structure, we therefore use the imaginary version of this integer subgroup, the set of integer multiples of $\pm(2\pi\sqrt{-1})^{-1}$, denoted

$$\mathbb{Z}(-1) \cong \mathbb{Z}^{\perp} \subset \Gamma(\mathbb{P}^1, \mathbb{T}(-1))^{\sigma}.$$

For the logarithmic Tate twistor structure, we use the integer subgroup itself

$$\mathbb{Z}(-1, \log) := \mathbb{Z} \subset \Gamma(\mathbb{P}^1, \mathbb{T}(-1, \log))^{\sigma}.$$

Take for generator of $\mathbb{Z}(-1, \log)$ the vector field $-\lambda\frac{\partial}{\partial\lambda}$ in the standard chart $\mathbb{A}^1$. This minus sign comes from the formula (2.2); see also §4.4 above. If we go into the other chart then the generator changes sign once again. See the paragraph above equation (6.3) for a reflection of this sign change in the action of the gauge group on the complex conjugate chart.

The sign change for σ on $\mathbb{T}(-1, \log)$ guarantees that the integer sections are preserved by σ.

The action of $\mathbb{G}_m$ on the additive Tate twistor structure $\mathbb{T}(-1)$ corresponds to the usual Tate Hodge structure of type $(1,1)$ with its integral subgroup $\mathbb{Z}(-1)$.

Define the *multiplicative Tate twistor structure* to be

$$\mathbb{G}_m(-1) := \mathbb{T}(-1, \log)/\mathbb{Z}(-1, \log). \tag{5.1}$$

The fiber over $\lambda = 1$, the de Rham version, is naturally identified with $\mathbb{C}/\mathbb{Z}$. The exponential map gives the isomorphism

$$\exp^{\perp} : \mathbb{C}/\mathbb{Z} \xrightarrow{\cong} \mathbb{G}_m^{\perp}.$$

This explains why we needed to change the sign of σ for the logarithmic Tate structure: the exponential map interchanges objects before and after $(\ )^{\perp}$. Thus, to get a Deligne-Tate type twist on the multiplicative group corresponding to the

*The referee suggests that one could instead adopt the point of view of looking at σ-antiinvariant sections of $\mathbb{T}(-1)$, which would fit naturally with antisymmetric polarization pairings.

local monodromy operator of a connection, we need to undo this twist which occurs naturally in $\mathbb{T}(-1) = T\mathbb{P}^1$.

Over $\lambda = 0$ and $\lambda = \infty$, the quotient defining $\mathbb{G}_m(-1)$ is to be taken in the stack sense. Hence,

$$\mathbb{G}_m(-1)_{\mathrm{Dol}} = \mathbb{C} \times B\mathbb{Z}.$$

5.2. The antipodal involution in the additive case. Start by computing the natural antipodal involution $\sigma_{T\mathbb{P}^1}$ of the tangent bundle or equivalently the additive Tate structure $\mathbb{T}(-1)$. For this subsection, the notation σ represents $\sigma_{T\mathbb{P}^1} = \sigma_{\mathbb{T}(-1)}$.

The vector field $\lambda\frac{\partial}{\partial\lambda}$ goes radially outward from 0 towards ∞. Up to a scalar it is the unique vector field with zeros at 0 and ∞, and this property is preserved by σ. Geometrically we see that the antipodal involution changes the sign of this radial vector field. Acting on this section considered as a section of $T\mathbb{P}^1$ we get

$$\sigma^*\lambda\frac{\partial}{\partial\lambda} = -\lambda\frac{\partial}{\partial\lambda}.$$

Similar geometric consideration shows that σ interchanges the vector fields $\frac{\partial}{\partial\lambda}$ and $\lambda^2\frac{\partial}{\partial\lambda}$, this time with no sign change. Thus

$$\sigma^*\frac{\partial}{\partial\lambda} = \lambda^2\frac{\partial}{\partial\lambda}$$

and vice versa.

A point of the total space of the tangent bundle, in the first standard chart, has the form $\left(\lambda, v\frac{\partial}{\partial\lambda}\right)$. We know that σ acts on the first coordinate by sending λ to $-\overline{\lambda}^{-1}$. Thinking of the above sections as corresponding to their graphs which are sets of points, and noting that σ is antilinear in the coordinate v, the formula for σ on points of the total bundle is

$$\sigma\left(\lambda, v\frac{\partial}{\partial\lambda}\right) = \left(-\overline{\lambda}^{-1}, \overline{\lambda}^{-2}\overline{v}\frac{\partial}{\partial\lambda}\right).$$

5.3. The antipodal involution in the logarithmic or multiplicative case. Recall that $\sigma_{\mathbb{T}(-1,\log)} = -\sigma_{T\mathbb{P}^1}$, with the minus sign acting only in the bundle fiber direction. Hence, for $\sigma = \sigma_{\mathbb{T}(-1,\log)}$ the formulae from the previous section become

$$\sigma\left(\lambda, v\frac{\partial}{\partial\lambda}\right) = \left(-\overline{\lambda}^{-1}, -\overline{\lambda}^{-2}\overline{v}\frac{\partial}{\partial\lambda}\right),$$

and

$$\sigma^*(u + v\lambda + w\lambda^2)\frac{\partial}{\partial\lambda} = (-\overline{w} + \overline{v}\lambda - \overline{u}\lambda^2)\frac{\partial}{\partial\lambda}.$$

This fits with the formula (4.2): a σ-invariant section has the form

$$\psi(a, \alpha) = \lambda \mapsto (\alpha - a\lambda - \overline{\alpha}\lambda^2)\frac{\partial}{\partial\lambda}$$

with $\alpha \in \mathbb{C}$ and $a \in \mathbb{R}$.

On the quotient (5.1) $\mathbb{G}_m(-1) = \mathbb{T}(-1, \log)/\mathbb{Z}(-1, \log)$ we get the involution $\sigma_{\mathbb{G}_m(-1)}$.

5.4. The space of invariant sections. The space of σ-invariant sections of $\mathbb{T}(-1,\log)$ inherits some canonical structure. For any point $p \in \mathbb{P}^1$ we get a distinguished σ-invariant direction in $\Gamma(\mathbb{P}^1, \mathbb{T}(-1,\log))$: the sections having simple zeros at p and $\sigma(p)$.

This space $\Gamma(\mathbb{P}^1, \mathbb{T}(-1,\log)(-p-\sigma(p)))^\sigma$ is naturally identified with the Lie algebra of the one-parameter group of σ-antipreserving homotheties of $\mathbb{P}^1$ which fix p and $\sigma(p)$. One must say "antipreserving" here because of the sign change in $\sigma_{\mathbb{T}(-1,\log)}$.

The elements of this group are radial homotheties; the group is isomorphic to $\mathbb{G}_m(\mathbb{R})$ and its Lie algebra is isomorphic to $\mathbb{R}$. In particular, there is a distinguished generator which is the vector field going inward from $\sigma(p)$ towards p attaining speed 1 at the equator between the two fixed points. Equivalently, we can require that the expansion factor at p be equal to -1. This is normalized so that when $p=0$ it gives the generator of $\mathbb{Z}(-1,\log)$. The expansion factor of a vector field with a zero is a well-defined complex number: it is dual to the residue, and can be defined as the value of the vector field on the differential form $\frac{dz}{z}$.

Let $\nu_p \in \Gamma(\mathbb{P}^1, \mathbb{T}(-1,\log)(-p-\sigma(p)))^\sigma$ denote the generator normalized to have expansion factor -1 at p. We get a canonical isomorphism

$$\Gamma(\mathbb{P}^1, \mathbb{T}(-1,\log)(-p-\sigma(p)))^\sigma \cong \mathbb{R}, \quad \nu_p \mapsto 1.$$

Evaluating at p gives a map ev_p from the space of sections to the fiber $\mathbb{T}(-1,\log)_p$.

LEMMA 5.1. *These maps fit into a canonical exact sequence depending on $p \in \mathbb{P}^1$,*

$$0 \to \mathbb{R} \overset{\nu_p}{\to} \Gamma(\mathbb{P}^1, \mathbb{T}(-1,\log))^\sigma \overset{\mathrm{ev}_p}{\to} \mathbb{T}(-1,\log)_p \to 0.$$

PROOF. It is exact in the middle because $\mathbb{R}\cdot\nu_p$ is exactly the space of sections vanishing at p. Exactness on the left and right follow by dimension count. □

5.5. The residue evaluation. The standard translation action of $\mathbb{G}_a$ on $\mathbb{P}^1$ fixing the point ∞ gives a trivialization

$$\mathbb{T}(-1,\log)|_{\mathbb{A}^1} \cong \mathcal{O}.$$

For any point $p \in \mathbb{A}^1$, let res_p denote the composition of this trivialization at p with the evaluation map ev_p. Then the exact sequence 5.1 can be written

$$0 \to \mathbb{R} \overset{\nu_p}{\to} \Gamma(\mathbb{P}^1, \mathbb{T}(-1,\log))^\sigma \overset{\mathrm{res}_p}{\to} \mathbb{C} \to 0. \tag{5.2}$$

As calculated above, the σ-invariant sections are identified with the polynomials of the form

$$\psi(a,\alpha) = \left(\alpha - a\lambda - \overline{\alpha}\lambda^2\right)\frac{\partial}{\partial\lambda}$$

with $a \in \mathbb{R}$ and $\alpha \in \mathbb{C}$. The residue evaluation at p is

$$\mathrm{res}_p(\psi(a,\alpha)) = \alpha - ap - \overline{\alpha}p^2. \tag{5.3}$$

Notice that $\mathrm{res}_p(\psi(a_p,\alpha_p)) = 0$, since ν_p is a section vanishing at p.

5.6. The generator ν_p. Suppose given a point $p \in \mathbb{A}^1$. In coordinates, $\sigma(p) = -\overline{p}^{-1}$. The line of σ-invariant sections which vanish to first order at p is

$$\alpha - ap - \overline{\alpha}p^2 = 0.$$

Vanishing at $\sigma(p)$ is a consequence, because a σ-invariant section vanishing at p also has to vanish at $\sigma(p)$. Recall that $a \in \mathbb{R}$. We get a real one-dimensional space of solutions, generated for example by (a, α), with

$$a = 1 - |p|^2, \quad \alpha = p.$$

Then

$$\psi(a, \alpha)(\lambda) = \left(p + (|p|^2 - 1)\lambda - \overline{p}\lambda^2\right) \frac{\partial}{\partial \lambda}.$$

Let us calculate the expansion factor at p of the vector field corresponding to our section $\psi(a, \alpha)$. For this, express the vector field in the form

$$\psi(a, \alpha) = \eta \cdot (\lambda - p) \frac{\partial}{\partial \lambda} + \cdots$$

where the $\cdots$ signify higher order terms at p. The constant η is the expansion factor. We have

$$\begin{aligned} \eta &= \frac{d}{d\lambda} \left(p + (|p|^2 - 1)\lambda - \overline{p}\lambda^2\right)\big|_{\lambda = p} \\ &= (|p|^2 - 1 - 2\overline{p}\lambda)|_{\lambda=p} = -(1 + |p|^2). \end{aligned}$$

We can normalize to get the canonical generator $\nu_p = \psi(a_p, \alpha_p)$ whose expansion factor is -1:

$$a_p = \frac{1 - |p|^2}{1 + |p|^2}, \quad \alpha_p = \frac{p}{1 + |p|^2}. \tag{5.4}$$

For $p = 1$ it is the generator $(1, 0)$ of $\mathbb{Z}(-1, \log)$.

5.7. A natural inner product. For all the above vectors (5.4), we have $a_p^2 + 4|\alpha_p|^2 = 1$. The equation $a^2 + 4|\alpha|^2 = 1$ defines an $S^2 \subset \mathbb{R} \times \mathbb{C}$, and the function $p \mapsto (a_p, \alpha_p)$ provides an isomorphism between $\mathbb{P}^1$ and this S^2. This is the unit sphere for the inner product

$$(a, \alpha) \cdot (b, \beta) := ab + 2(\alpha\overline{\beta} + \overline{\alpha}\beta). \tag{5.5}$$

LEMMA 5.2. *The group $SO(3)$ acts naturally as the group of σ-intertwining complex automorphisms of $\mathbb{P}^1$, so it acts on the space of sections $\Gamma(\mathbb{P}^1, \mathbb{T}(-1, \log))^{\sigma}$. The above inner product* (5.5) *is the $SO(3)$-invariant one, unique up to a scalar.*

PROOF. Any σ-intertwining automorphism f of $\mathbb{P}^1$ acts on $\mathbb{T}(-1, \log)$ because that bundle is naturally defined as the tangent bundle with a twisted σ. Hence it acts on the space of σ-invariant sections. The section ν_p is canonically defined depending on the point p and the involution σ. Hence f_* takes ν_p to $\nu_{f(p)}$. It follows that the action of f preserves the sphere S^2 image of the map ν. Therefore f is in the orthogonal group $O(3)$ for this scalar product. It has determinant 1 because of holomorphicity. We get $f \in SO(3)$, the special orthogonal group for the scalar product (5.5). □

5.8. The parabolic weight function. The space of σ-invariant sections of $\mathbb{T}(-1,\log)$ is an $\mathbb{R}^3$, and at any $p \in \mathbb{P}^1$ it is naturally an extension of $\mathbb{T}(-1,\log)_p$ by $\mathbb{R}$ (Lemma 5.1). The quotient $\mathbb{T}(-1,\log)_p$ represents the residue of a λ-connection. The extra real parameter corresponds to the real parabolic weight of a parabolic structure. However, we need to discuss the normalization of this identification splitting the exact sequence.

We use the coordinates a, α for the set of invariant sections denoted $\psi(a,\alpha)$, giving an isomorphism between this space and $\mathbb{R} \times \mathbb{C}$.

The point $(1,0) \in \mathbb{R} \times \mathbb{C}$ is the generator of the subgroup $\mathbb{Z}(-1,\log)$. For each p we have the point $\nu_p = \psi(a_p, \alpha_p)$ given by (5.4). The quotient of $\mathbb{R} \times \mathbb{C}$ by the line generated by (a_p, α_p) is the space of residues at p.

We would like to define a *parabolic weight function* $\varpi_p : \mathbb{R} \times \mathbb{C} \to \mathbb{R}$, depending on the point p, such that $\varpi_p(1,0) = 1$ for compatibility with local gauge transformations; and $\varpi_p(a_p, \alpha_p) \neq 0$ so that the local residue map is an isomorphism between $\ker(\varpi_p)$ and $\mathbb{C}$.

Using the inner product $(a,\alpha)\cdot(a',\alpha') = aa' + 2(\alpha\overline{\alpha}' + \overline{\alpha}\alpha')$, the simplest thing to do is to let ϖ_p be given by the inner product with the average of the two vectors $(1,0)$ and (a_p, α_p), then normalize to get $\varpi_p(1,0) = 1$. This is

$$
\begin{aligned}
\varpi_p(a,\alpha) &:= \frac{(a,\alpha)\cdot(1,0) + (a,\alpha)\cdot(a_p,\alpha_p)}{(1,0)\cdot(1,0) + (1,0)\cdot(a_p,\alpha_p)} \\
&= \frac{(1+|p|^2)a + (1-|p|^2)a + 2(\alpha\overline{p} + \overline{\alpha}p)}{(1+|p|^2) + (1-|p|^2)} \\
&= a + \alpha\overline{p} + \overline{\alpha}p. \qquad (5.6)
\end{aligned}
$$

The parabolic weight and residue functions are the same as Mochizuki's functions $\mathfrak{p}$ and $\mathfrak{e}$ of [**39**, §2.1.7]; however we have preferred to motivate their introduction independently above.

PROPOSITION 5.3. *For any point $p \in \mathbb{A}^1$, the parabolic weight function and the residue give an isomorphism*

$$(\varpi_p, \mathrm{res}_p) : \Gamma(\mathbb{P}^1, \mathbb{T}(-1,\log))^\sigma \overset{\cong}{\to} \mathbb{R} \times \mathbb{C}.$$

Let $\mathbb{Z} = \mathbb{Z}(-1,\log)$ act on $\mathbb{R} \times \mathbb{C}$ with generator $(1,-p)$ in keeping with (2.2). *Then the above isomorphism descends to the quotient to give*

$$(\varpi_p, \mathrm{res}_p) : \frac{\Gamma(\mathbb{P}^1, \mathbb{T}(-1,\log))^\sigma}{\mathbb{Z}(-1,\log)} \overset{\cong}{\to} \frac{\mathbb{R} \times \mathbb{C}}{(1,-p)\mathbb{Z}}.$$

In terms of the coordinates (a,α) given by the construction ψ, we have

$$(\varpi_p, \mathrm{res}_p)(\psi(a,\alpha)) = (a + \alpha\overline{p} + \overline{\alpha}p, \alpha - ap - \overline{\alpha}p^2).$$

PROOF. We have chosen ϖ_p so that $\varpi_p(\nu_p) = 1$. From the exact sequence of Lemma 5.1, this implies that ϖ_p and res_p are linearly independent so by dimension count we get the first isomorphism. For the second part, it suffices to recall that the subgroup $\mathbb{Z}(-1,\log) \subset \Gamma(\mathbb{P}^1, \mathbb{T}(-1,\log))^\sigma$ is given by the generator $\psi(1,0)$, and to note that

$$(\varpi_p, \mathrm{res}_p)(\psi(1,0)) = (1,-p).$$

The formula in terms of coordinates just recalls our calculations above. □

5.9. Some questions. The residue of a logarithmic λ-connection at a singular point lies in the fiber $\mathbb{T}(-1,\log)_\lambda$. This was seen by direct calculation: using the standard frame for $\mathbb{T}(-1,\log)$ over $\mathbb{A}^1$ and the expression of the residue as a well-defined complex number, we obtain this identification over the chart $\mathbb{A}^1$. Then, by calculation, it is compatible with the corresponding identification for $\overline{X}$ over the chart at infinity, using the Riemann-Hilbert correspondence. This passage by an explicit calculation is unsatisfactory but I haven't seen any way of improving it, so we formulate a question:

QUESTION 5.4. *Is there some more natural geometric way of identifying the residue of a logarithmic λ-connection at a singular point with a tangent vector to the λ-line?*

One possible approach would be to give a geometric description of the meaning of points in the twistor line $\mathbb{P}^1$.

Similarly, we picked the definition of the parabolic weight function "out of a hat". Taking the scalar product with the average of the two vectors, then normalizing, is certainly the easiest way to make sure that the function takes on nonzero values on the two vectors; furthermore the resulting formula for ϖ_p is relatively simple. Nonetheless, it would be better to have a more motivated reason for this choice.

QUESTION 5.5. *Is there a geometric interpretation of the meaning of the parabolic weight function, preferably going with the geometric interpretation we are looking for in Question 5.4?*

Another direction of questions is the relationship with $SO(3)$. The group of σ-invariant automorphisms of $\mathbb{P}^1$ is the group of metric automorphisms of S^2, in other words it is $SO(3)$ by Lemma 5.2. The space of σ-invariant sections of $\mathbb{T}(-1,\log)$, which are the σ-antiinvariant sections of the tangent bundle, may be identified with the perpendicular of the Lie algebra $\mathfrak{so}(3)^\perp$. By naturality, $\mathbb{T}(-1,\log)$ also has an action of $SO(3)$, corresponding to the adjoint action on the Lie algebra.

QUESTION 5.6. *What is the significance of this action of $SO(3)$ and the identification of elements of the space of parabolic weights and residues, with vectors in the Lie algebra?*

This seems to be one of the subjects of Gukov and Witten's paper [**24**].

6. The general rank one case

Consider now the following situation: X is a smooth projective variety, and $D \subset X$ is a reduced strict normal crossings divisor written as $D = D_1 + \cdots + D_k$ where D_i are its distinct smooth connected irreducible components. Let $U := X - D$.

6.1. The Hodge moduli space. Let $M_{\mathrm{Hod}}(X, \log D)$ denote the moduli space [**43, 53**] of triples (λ, L, ∇) where $\lambda \in \mathbb{A}^1$, L is a line bundle on X such that

$$c_1(L)_{\mathbb{Q}} \in \mathbb{Q}\cdot[D_1] + \cdots + \mathbb{Q}\cdot[D_k] \subset H^2(X, \mathbb{Q}^\perp), \tag{6.1}$$

and

$$\nabla : L \to L \otimes_{\mathcal{O}_X} \Omega^1_X(\log D)$$

is an integrable logarithmic λ-connection on L. The first coordinate is a map

$$\lambda : M_{\mathrm{Hod}}(X, \log D) \to \mathbb{A}^1.$$

Let $\mathrm{res}(\nabla; D_i) \in \mathbb{C}$ denote the residue of ∇ along D_i. Recall that the residue is a locally constant function and D_i is connected so it is a complex scalar.

For $\lambda \neq 0$, we have

$$\lambda c_1(L) = -\sum_i \mathrm{res}(\nabla; D_i) \cdot [D_i] \text{ in } H^2(X, \mathbb{C}^{\perp}), \tag{6.2}$$

as can be calibrated by comparing with the gauge transformation formulae (2.1) and (2.2). So the condition about $c_1(L)$ in the definition of $M_{\mathrm{Hod}}(X, \log D)$ is automatically satisfied when $\lambda \neq 0$; however for $\lambda = 0$ this condition is nontrivial.

Tensor product gives $M_{\mathrm{Hod}}(X, \log D)$ a structure of abelian group scheme relative to $\mathbb{A}^1$. We need to use the condition about Chern classes to prove that it is smooth over $\mathbb{A}^1$, otherwise there would be additional irreducible components lying over $\lambda = 0$.

LEMMA 6.1. *The morphism $M_{\mathrm{Hod}}(X, \log D) \to \mathbb{A}^1$ is smooth.*

Proof: Suppose $\phi : Y \to M_{\mathrm{Hod}}(X, \log D)$ is a morphism from an artinian scheme. Suppose $Y \subset Y'$ is an artinian extension provided with a morphism $\lambda' : Y' \to \mathbb{A}^1$. We need to extend to $Y' \to M_{\mathrm{Hod}}(X, \log D)$ lifting λ'. The map ϕ corresponds to a line bundle with integrable λ-connection (L, ∇) on $X \times Y$. By smoothness of the Picard scheme of X, this extends to a line bundle L' on $X \times Y'$. The condition about the Chern class of L implies that there exists some logarithmic connection $\nabla_{1,y}$ on L_y where $y \in Y$ denotes the closed point and L_y is the restriction of L to the fiber over y. By smoothness of $M_{\mathrm{DR}}(X, \log D)$, which follows because of its group structure under tensor product, we can extend $\nabla_{1,y}$ to some integrable connection ∇_1' on L'. Then $\lambda'\nabla_1'$ is an integrable λ'-connection on L'. Restricted to $X \times Y$, we can write

$$\nabla = \lambda'\nabla_1'|_{X\times Y} + A$$

where

$$A \in H^0(X \times Y, \Omega^1_X(\log D) \otimes_{\mathcal{O}_X} \mathcal{O}_{X\times Y}) \cong H^0(X, \Omega^1_X(\log D)) \otimes_{\mathbb{C}} \mathcal{O}_Y.$$

Now extend A in any way to a section

$$A' \in H^0(X \times Y', \Omega^1_X(\log D) \otimes_{\mathcal{O}_X} \mathcal{O}_{X\times Y'}) \cong H^0(X, \Omega^1_X(\log D)) \otimes_{\mathbb{C}} \mathcal{O}_{Y'}$$

and set

$$\nabla' := \lambda'\nabla_1' + A'.$$

This provides the required extension. □

6.2. Gauge group action. Recall the action of the local meromorphic gauge group $\mathcal{G} := \mathbb{Z}^k$ on $M_{\mathrm{Hod}}(X, \log D)$. A vector $g = (g_1, \ldots, g_k)$ sends (λ, L, ∇) to $(\lambda, L(g_1D_1 + \cdots + g_kD_k), \nabla^{\alpha})$ where ∇^{α} is the logarithmic λ-connection on $L(\alpha_1 D_1 + \cdots + \alpha_k D_k)$ which coincides with ∇ over U, via the isomorphism

$$L(g_1D_1 + \cdots + g_kD_k)|_U \cong L|_U.$$

We have (2.2)

$$\mathrm{res}(\nabla^g; D_i) = \mathrm{res}(\nabla; D_i) - \lambda g_i.$$

The vector of residues, viewed in the standard framing $\frac{\partial}{\partial\lambda}$ for $\mathbb{T}(-1,\log)$, provides a morphism

$$R : M_{\mathrm{Hod}}(X, \log D) \to \mathbb{T}(-1,\log)^k$$

where the i-th coordinate of $R(\lambda, L, \nabla)$ is by definition $\mathrm{res}(\nabla; D_i)\cdot\frac{\partial}{\partial\lambda}$.

The morphism R is compatible with the action of $\mathcal{G} = \mathbb{Z}^k = \mathbb{Z}(-1,\log)^k$, where $g \in \mathbb{Z}^k$ acts on $\mathbb{T}(-1,\log)^k$ using our standard generator of $\mathbb{Z}(-1,\log)$

$$(v_1,\ldots,v_k) \mapsto (v_1 - g_1\lambda\frac{\partial}{\partial\lambda},\ldots,v_k - g_1\lambda\frac{\partial}{\partial\lambda}).$$

Let

$$M_{\mathrm{Hod}}(X,\log D)_{\mathbb{G}_m} := \mathbb{G}_m \times_{\mathbb{A}^1} M_{\mathrm{Hod}}(X,\log D).$$

Then $\mathcal{G}$ acts properly discontinuously on $M_{\mathrm{Hod}}(X,\log D)_{\mathbb{G}_m}$ because this action lies over the proper discontinuous action on $\mathbb{T}(-1,\log)^k_{\mathbb{G}_m}$ via the map R. In the analytic category, we can form the quotient, and the Riemann-Hilbert correspondence gives an isomorphism

$$M_{\mathrm{Hod}}(X,\log D)^{\mathrm{an}}_{\mathbb{G}_m}/\mathcal{G} \cong \mathbb{G}_m \times M_{\mathrm{B}}(U)$$

where $M_{\mathrm{B}}(U) := \mathrm{Hom}(\pi_1(U), \mathbb{G}_m)$.

In view of the Riemann-Hilbert correspondence, we define $M_{\mathrm{Hod}}(U)$ to be the stack-theoretical quotient

$$M_{\mathrm{Hod}}(U) := M_{\mathrm{Hod}}(X,\log D)/\mathcal{G},$$

and similarly for the fibers over $\lambda = 0, 1$:

$$M_{\mathrm{Dol}}(U) := M_{\mathrm{Dol}}(X,\log D)/\mathcal{G}, \quad M_{\mathrm{DR}}(U) := M_{\mathrm{DR}}(X,\log D)/\mathcal{G}.$$

Note that $\mathcal{G}$ acts trivially on $M_{\mathrm{Dol}}(X,\log D)$ so the quotient $M_{\mathrm{Dol}}(U)$ is a stack with $\mathcal{G}$ in the general stabilizer group. If we started with a stack version $\mathcal{M}_{\mathrm{Dol}}(X,\log D)$ then the general stabilizer also contains the automorphism group $\mathbb{G}_m$ of a rank one Higgs bundle. In the rank one case, the stabilizer groups are all the same. So the full stabilizer group of any point of $\mathcal{M}_{\mathrm{Dol}}(U)$ would be $\mathbb{G}_m \times \mathcal{G}$.

Using these definitions, the RH correspondence again says $M_{\mathrm{DR}}(U) \cong M_{\mathrm{B}}(U)$, and the Deligne glueing process applies as in §3.2 to give an analytic stack

$$M_{\mathrm{DH}}(U) \to \mathbb{P}^1$$

whose charts are $M_{\mathrm{Hod}}(U)$ and $M_{\mathrm{Hod}}(\overline{U})$. Note however that these charts don't have algebraic structures.

We would like to investigate how to lift to a Deligne-Hitchin glueing on the space of logarithmic connections to get an analytic space $M_{\mathrm{DH}}(X,\log D)$ which would have nicer geometric properties—its charts would be algebraic. We would then have a quotient expression

$$M_{\mathrm{DH}}(U) = M_{\mathrm{DH}}(X,\log D)/\mathcal{G} \to \mathbb{P}^1.$$

One way of looking at this question would be to calculate the fundamental group of $M_{\mathrm{DH}}(U) \to \mathbb{P}^1$ and see if it has a covering which resolves the stackiness over 0 and ∞. Instead, we construct directly the covering.

6.3. The Riemann-Hilbert correspondence and glueing. Our goal in this subsection is to define $M_{\mathrm{DH}}(X, \log D)$ by Deligne glueing of $M_{\mathrm{Hod}}(X, \log D)$ with $M_{\mathrm{Hod}}(\overline{X}, \log \overline{D})$.

It will be useful to have a Betti version of $M_{\mathrm{DR}}(X, \log D)$ to intervene in the glueing. Suppose $\rho \in M_{\mathrm{B}}(U)$. Recall from §2.4 that for each component D_i of the divisor, we get a well-defined local monodromy element $\rho^{\perp}(\gamma_{D_i}) \in \mathbb{G}_m^{\perp}$.

Consider the following diagram:

$$M_{\mathrm{B}}(U) \to (\mathbb{G}_m^{\perp})^k \stackrel{\exp^{\perp}}{\longleftarrow} \mathbb{C}^k,$$

where the k copies are for the k components of the divisor $D = D_1 + \ldots + D_k$, the first map sends ρ to its vector of local monodromy transformations, and

$$\exp^{\perp} : (a_1, \ldots, a_k) \mapsto ((\cos(2\pi a_1), \sin(2\pi a_1)), \ldots, (\cos(2\pi a_k), \sin(2\pi a_k)))\,.$$

Let $M_{\mathrm{B}}(X, \log D)$ denote the fiber product. Thus, a point in $M_{\mathrm{B}}(X, \log D)$ may be written $(\rho; a_1, \ldots, a_k)$ where ρ is a representation of rank one over U and $a_i \in \mathbb{C}$ are choices of circular logarithms of the monodromy operators $\rho^{\perp}(\gamma_{D_i}) \in \mathbb{G}_m^{\perp}$.

Define an action of the gauge group $\mathcal{G} = \mathbb{Z}^k$ on $M_{\mathrm{B}}(X, \log D)$ by

$$g = (g_1, \ldots, g_k) : (\rho; a_1, \ldots, a_k) \mapsto (\rho; a_1 - g_1, \ldots, a_k - g_k).$$

LEMMA 6.2. *The Riemann-Hilbert correspondence lifts to*

$$M_{\mathrm{DR}}(X, \log D) \cong M_{\mathrm{B}}(X, \log D).$$

Proof: Given a line bundle with integrable connection (L, ∇), associate the point

$$(\rho, \mathrm{res}(\nabla; D_1), \ldots, \mathrm{res}(\nabla; D_k)) \in M_{\mathrm{B}}(U) \times_{(\mathbb{G}_m^{\perp})^k} \mathbb{C}^k = M_{\mathrm{B}}(X, \log D).$$

This is a morphism of analytical groups with the same dimension, so it suffices to prove that it is injective and surjective. Before doing that, we verify compatibility with the gauge group action. Given $g = (g_1, \ldots, g_k) \in \mathcal{G}$, notice that the monodromy representation of (L^g, ∇^g) is the same as ρ. For the circular logarithms, the formula for residues

$$\mathrm{res}(\nabla^g; D_i) = \mathrm{res}(\nabla; D_i) - g_i$$

implies the required compatibility with the gauge group action.

For injectivity, suppose (L, ∇) and (L', ∇') are two line bundles with connection, with the same monodromy representation and the same residues. There is a unique isomrphism $\psi : L|_U \cong L'|_U$ compatible with the monodromy representation or equivalently with ∇ and ∇' on U. Then, the poles or zeros of ψ along a component D_i are determined by the difference between the residues of ∇ and ∇'. The condition that the residues are the same means that ψ has neither pole nor zero along each component D_i. Thus, ψ is an isomorphism of bundles over X.

For surjectivity, suppose $(\rho, a_1, \ldots, a_k)$ is a point in $M_{\mathrm{B}}(X, \log D)$. Choose a line bundle with logarithmic connection (L, ∇) inducing the monodromy representation ρ on U. Let $a'_1, \ldots, a'_k$ be the residues of ∇' along the D_i. We have $a_i = a'_i - g_i$ with $g_i \in \mathbb{Z}$. Now (L^g, ∇^g) maps to $(\rho, a_1, \ldots, a_k)$. □

We have the conjugation isomorphism $\varphi : U \cong \overline{X} - \overline{D}$. If ρ is a local system on U then we obtain $\varphi_*(\rho)$ a local system on $\overline{X} - \overline{D}$, defined by

$$\varphi_*(\rho)(\eta) := \rho(\varphi^{-1}\eta).$$

The divisor $\overline{D}$ breaks up into components $\overline{D}_1 + \cdots + \overline{D}_k$. Let $\gamma_{\overline{D}_i} \in \pi_1(\overline{U}, \overline{x})^{\perp}$ denote the local monodromy operator around $\overline{D}_i$. Then we have

$$\varphi^{-1}(\gamma_{\overline{D}_i}) = \gamma_{D_i}^{-1}$$

because φ reverses orientation. Thus,

$$\varphi_*(\rho)^{\perp}(\gamma_{\overline{D}_i}) = \rho^{\perp}(\gamma_{D_i})^{-1}.$$

Given a logarithm a_i of $\rho^{\perp}(\gamma_{D_i})$, its negative $-a_i$ is a logarithm of $\varphi_*(\rho)^{\perp}(\gamma_{\overline{D}_i})$. Therefore define the isomorphism

$$\varphi : M_{\mathrm{B}}(X, \log D) \stackrel{\cong}{\to} M_{\mathrm{B}}(\overline{X}, \log \overline{D})$$

by

$$\varphi(\rho; a_1, \ldots, a_k) := (\varphi_*(\rho); -a_1, \ldots, -a_k).$$

Using the Riemann-Hilbert correspondence of Lemma 6.2 we can do the Deligne glueing exactly as before to get a moduli space

$$M_{\mathrm{DH}}(X, \log D) \to \mathbb{P}^1.$$

The gauge group of meromorphic gauge transformations along the divisors $\mathcal{G} = \mathbb{Z}^k$ acts on $M_{\mathrm{DH}}(X, \log D)$ in the following way. It acts in the canonical way on the first chart $M_{\mathrm{Hod}}(X, \log D)$. On the other hand, a point $(g_1, \ldots, g_k)$ acts by the canonical action of $(-g_1, \ldots, -g_k)$ on the chart $M_{\mathrm{Hod}}(\overline{X}, \log \overline{D})$, because of the sign change in the definition of φ. The global quotient is the Deligne glueing considered previously:

$$M_{\mathrm{DH}}(U) = M_{\mathrm{DH}}(X, \log D)/\mathcal{G} \to \mathbb{P}^1. \tag{6.3}$$

Over $\mathbb{G}_m \subset \mathbb{P}^1$ this quotient is isomorphic to $M_{\mathrm{B}}(U) \times \mathbb{G}_m$ so it has a reasonable structure; however near the fibers over 0 and ∞ the quotient is analytically stacky, with $\mathcal{G}$ contributing to the stabilizer group.

We obtain an antilinear involution $\sigma : M_{\mathrm{DH}}(X, \log D) \to M_{\mathrm{DH}}(X, \log D)$ covering the antipodal involution $\sigma_{\mathbb{P}^1}$, by the construction of §3.3. To do this for the Betti case (the intermediate piece in the glueing) use the calculation of §4.3 to define the effect of the involution on the residual pieces at divisor components. The involution is compatible with the gauge group action so it descends to an involution of $M_{\mathrm{DH}}(U)$.

6.4. Exact sequences. Since we are treating the case $r = 1$, our moduli spaces are really just abelian cohomology groups, for example

$$M_{\mathrm{B}}(U) = H^1(U, \mathbb{G}_m).$$

This may also be interpreted as a Deligne cohomology group; see [**20, 22**] for example; we leave to the reader to make the link between our Hodge filtration and the Hodge filtration on Deligne cohomology.

The exponential exact sequence for $U = X - D$ is

$$0 \to H^1(U, \mathbb{Z}^{\perp}) \to H^1(U, \mathbb{C}) \to M_{\mathrm{B}}(U) \to H^2(U, \mathbb{Z}^{\perp}) \to H^2(U, \mathbb{C}).$$

The exact sequence for the gauge group action is

$$0 \to \mathcal{G} = \mathbb{Z}^k \to M_{\mathrm{B}}(X, \log D) \to M_{\mathrm{B}}(U) \to 1.$$

Let $W_1H^1(U) = H^1(X, \mathbb{Z}^\perp)$ denote the weight 1 piece of the weight filtration. We have an exact sequence

$$0 \to W_1H^1(U) = H^1(X, \mathbb{Z}^\perp) \to H^1(U, \mathbb{Z}^\perp) \to \mathcal{G} = \mathbb{Z}^k = H^2(X, U, \mathbb{Z}^\perp) \to$$

$$\to H^2(X, \mathbb{Z}^\perp) \to H^2(U, \mathbb{Z}^\perp) \to H^3(X, U, \mathbb{Z}^\perp) \to \dots$$

The exponential exact sequence lifts to an exact sequence for the logarithmic space

$$0 \to W_1H^1(U, \mathbb{Z}^\perp) \to H^1(U, \mathbb{C}) \to M_{\mathrm{B}}(X, \log D) \to H^2(X, \mathbb{Z}^\perp) \to H^2(X, \mathbb{C}).$$

The image of the connecting map in the first exponential exact sequence is the torsion subgroup of $H^2(U)$. A duality calculation relates $H^3(X, U, \mathbb{Z}^\perp)$ to the $H^1(D_i, \mathbb{Z}^\perp)$ so this is torsion-free. Hence, any torsion element in $H^2(U, \mathbb{Z}^\perp)$ comes from $H^2(X, \mathbb{Z}^\perp)$. This fits in with the fact that any element of $M_{\mathrm{B}}(U)$ admits a Deligne canonical extension to a line bundle with logarithmic connection on X.

These exact sequences all fit together into a diagram

$$\begin{array}{ccccccccc}
 & & & & & & 0 & \to & H^1(X, \mathbb{Z}^\perp) \\
 & & & & & & \downarrow & & \downarrow \\
 & & & & & & 0 & \to & H^1(U, \mathbb{Z}^\perp) \\
 & & & & & & \downarrow & & \downarrow \\
 & & 0 & \to & 0 & \to & \mathcal{G} = \mathbb{Z}^k & \overset{\equiv}{\to} & H^2(X, U, \mathbb{Z}^\perp) \\
 & & \downarrow & & \downarrow & & \downarrow & & \downarrow \\
0 \to & & H^1(X, \mathbb{Z}^\perp) & \to & H^1(U, \mathbb{C}) & \to & M_{\mathrm{B}}(X, \log D) & \to & H^2(X, \mathbb{Z}^\perp) \\
 & & \downarrow & & \downarrow & & \downarrow & & \downarrow \\
0 \to & & H^1(U, \mathbb{Z}^\perp) & \to & H^1(U, \mathbb{C}) & \to & M_{\mathrm{B}}(U) & \to & H^2(U, \mathbb{Z}^\perp) \\
 & & \downarrow & & \downarrow & & \downarrow & & \downarrow \\
\mathcal{G} & \overset{\equiv}{\to} & H^2(X, U, \mathbb{Z}^\perp) & \to & 0 & \to & 0 & \to & H^3(X, U, \mathbb{Z}^\perp).
\end{array}$$

There are also exact sequences for localization near the singular points. The main one is

$$0 \to M_{\mathrm{B}}(X) \to M_{\mathrm{B}}(X, \log D) \overset{\mathrm{res}}{\to} \mathbb{C}^k \to H^2(X, \mathbb{G}_m). \tag{6.4}$$

It fits with the exponential exact sequence to give a diagram

$$\begin{array}{cccccccc}
 & 0 & & 0 & & & & \\
 & \downarrow & & \downarrow & & & & \\
0 \to & H^1(X, \mathbb{Z}^\perp) & \to & H^1(X, \mathbb{Z}^\perp) & \to & 0 & \to & H^2(X, \mathbb{Z}^\perp) \\
 & \downarrow & & \downarrow & & \downarrow & & \downarrow \\
0 \to & H^1(X, \mathbb{C}) & \to & H^1(U, \mathbb{C}) & \to & \mathbb{C}^k & \to & H^2(X, \mathbb{C}) \\
 & \downarrow & & \downarrow & & \downarrow & & \downarrow \\
0 \to & M_{\mathrm{B}}(X) & \to & M_{\mathrm{B}}(X, \log D) & \to & \mathbb{C}^k & \to & H^2(X, \mathbb{G}_m) \\
 & \downarrow & & \downarrow & & \downarrow & & \downarrow \\
0 \to & H^2(X, \mathbb{Z}^\perp) & \to & H^2(X, \mathbb{Z}^\perp) & \to & 0 & \to & H^3(X, \mathbb{Z}^\perp).
\end{array}$$

Dividing by the gauge group gives the diagram

$$\begin{array}{ccccccccc} & & 0 & & 0 & & 0 & & \\ & & \downarrow & & \downarrow & & \downarrow & & \\ 0 \to & & H^1(X,\mathbb{Z}^\perp) & \to & H^1(U,\mathbb{Z}^\perp) & \to & \mathbb{Z}^k & \to & H^2(X,\mathbb{Z}^\perp) \\ & & \downarrow & & \downarrow & & \downarrow & & \downarrow \\ 0 \to & & H^1(X,\mathbb{C}) & \to & H^1(U,\mathbb{C}) & \to & \mathbb{C}^k & \to & H^2(X,\mathbb{C}) \\ & & \downarrow & & \downarrow & & \downarrow & & \downarrow \\ 0 \to & & M_{\mathrm{B}}(X) & \to & M_{\mathrm{B}}(U) & \to & (\mathbb{G}_m^\perp)^k & \to & H^2(X,\mathbb{G}_m) \\ & & \downarrow & & \downarrow & & \downarrow & & \downarrow \\ 0 \to & & H^2(X,\mathbb{Z}^\perp) & \to & H^2(U,\mathbb{Z}^\perp) & \to & 0 & \to & H^3(X,\mathbb{Z}^\perp). \end{array}$$

The connected component of $M_{\mathrm{B}}(U)$ containing the identity representation is a quotient:

$$M_{\mathrm{B}}(U)^o = H^1(U,\mathbb{C})/H^1(U,\mathbb{Z}^\perp).$$

This identification is valid in the analytic category. The logarithmic space is obtained by dividing out instead by $W_1H^1(X-D,\mathbb{Z}^\perp) = H^1(X,\mathbb{Z}^\perp)$:

$$M_{\mathrm{B}}(X,\log D)^o = H^1(U,\mathbb{C})/W_1H^1(U,\mathbb{Z}^\perp).$$

Let $\mathcal{G}^o := \ker\left(\mathbb{Z}^k \to H^2(X,\mathbb{Z}^\perp)\right)$. Then $\mathcal{G}^o$ acts on $M_{\mathrm{B}}(X,\log D)^o$ with quotient $M_{\mathrm{B}}(U)^o$. The above diagrams show that this is compatible with the quotient descriptions.

6.5. Compatibility with the Hodge filtration. The above diagrams can be replaced with the corresponding diagrams of twistor spaces over $\mathbb{P}^1$. This raises the question of showing that the maps preserve the twistor structure, another way of saying that they should be compatible with the Hodge filtrations.

Recall that the Hodge filtration and its complex conjugate for $H^1(U,\mathbb{C})$ lead to the twistor bundle $\xi(H^1(U,\mathbb{C}),F,\overline{F})$ over $\mathbb{P}^1$ see **[51]**. We can again take the quotient by the action of $H^1(U,\mathbb{Z})$ or $W_1H^1(U,\mathbb{Z})$. This gives an identification of the Deligne-Hitchin twistor space, at least for the connected component of the identity representation.

THEOREM 6.3. *Denote by a superscript $(\ )^o$ the connected component of the identity representation. We have identifications of analytic spaces over $\mathbb{P}^1$,*

$$M_{\mathrm{DH}}(X,\log D)^o \cong \xi(H^1(U,\mathbb{C}),F,\overline{F})/W_1H^1(U,\mathbb{Z}^\perp)$$

and

$$M_{\mathrm{DH}}(X,\log D)^o/\mathcal{G}^o \cong \xi(H^1(U,\mathbb{C}),F,\overline{F})/H^1(U,\mathbb{Z}^\perp).$$

Thus, the maps in the above big diagrams are compatible with the twistor structures.

PROOF. Use a cocycle description of $H^1(U,\mathbb{C})$. Suppose we are given an open analytic covering of X by open sets U_i, and let $U_{ij} := U_i \cap U_j$ etc. Recall Grothendieck's theorem

$$H^1(U,\mathbb{C}) = \mathbb{H}^1\left(\mathcal{O}_X \to \Omega^1_X(\log D) \to \Omega^2_X(\log D) \to \cdots\right).$$

An element here is given by a pair $(\{g_{ij}\},\{a_i\})$, where

$$g_{ij} \in \mathcal{O}_X(U_{ij}), \quad a_i \in \Omega^1_X(\log D)(U_i)$$

and these satisfy the cocycle condition $g_{ij}+g_{jk}+g_{ki} = 0$, the compatibility condition $d(g_{ij}) = a_i - a_j$, and $d(a_i) = 0$. The image of this pair in $M_{\mathrm{DR}}(X,\log D)$ is (L,∇) where L is the line bundle whose transition functions are $e^{g_{ij}}$, and $\nabla := d + a_i$ over

U_i, with d being the constant connection with respect to the trivialization $L|_{U_i} \cong \mathcal{O}_{U_i}$. This image is compatible with the exponential map $H^1(U,\mathbb{C}) \to M_{\mathrm{B}}(U)$ via the Riemann-Hilbert correspondence.

The Hodge filtration or Rees-bundle $\xi(H^1(U,\mathbb{C}),F) \to \mathbb{A}^1$ may also be described as the bundle of triples $(\lambda, \{g_{ij}\}, \{a_i\})$ subject to the conditions, analogues of the notion of λ-connection:

$$g_{ij} + g_{jk} + g_{ki} = 0, \quad \lambda d(g_{ij}) = a_i - a_j, \quad \lambda d(a_i) = 0.$$

Map this triple to (λ, L, ∇) where L is again given by transition functions $e^{g_{ij}}$, and $\nabla := \lambda d + a_i$ over U_i. This gives a map

$$\xi(H^1(U,\mathbb{C}),F) \to M_{\mathrm{Hod}}(X, \log D).$$

It is compatible with the action of $\mathbb{G}_m$, and is the same as the previous map in the fiber over $\lambda = 1$, so it is compatible with the exponential map on Betti cohomology.

Note that the complex conjugate of the Hodge filtration on $H^1(U,\mathbb{C})$ is the same as the pullback by $\varphi : U^{\mathrm{top}} \cong \overline{U}^{\mathrm{top}}$, of the Hodge filtration on $H^1(\overline{U},\mathbb{C})$. Indeed, φ is antiholomorphic, so the pullback by φ of a cohomology class containing at least a certain number of dz_i is a cohomology class containing at least that many $d\overline{z}_i$.

Using all of these things, our map glues together with the corresponding map for $\overline{U} = \overline{X} - \overline{D}$ to give a map of twistor spaces over $\mathbb{P}^1$,

$$\xi(H^1(U,\mathbb{C}),F,\overline{F}) \to M_{\mathrm{DH}}(X, \log D).$$

This is the required compatibility.

From the cocycle description, we get that the map is surjective to the connected component $M_{\mathrm{DH}}(X, \log D)^o$, even in the fibers over $\lambda = 0, \infty$. Using smoothness of both sides over $\mathbb{P}^1$ and a dimension count, we see that the kernel is discrete and flat over $\mathbb{P}^1$. In the general fiber it is $W_1 H^1(U,\mathbb{Z})$. The closure of the graph of this subgroup is again a subgroup of the form $W_1H^1(U,\mathbb{Z}) \times \mathbb{P}^1 \subset \xi(H^1(U,\mathbb{C}),F,\overline{F})$. Hence we have the isomorphism

$$M_{\mathrm{DH}}(X, \log D)^o \cong \xi(H^1(U,\mathbb{C}),F,\overline{F})/W_1H^1(U,\mathbb{Z}).$$

The other one is obtained by dividing out by the gauge group $\mathcal{G}^o$. □

Problem: Find a similar description for the twistor spaces of other connected components of $M_{\mathrm{B}}(U)$ corresponding to torsion elements in $H^2(U,\mathbb{Z})$. This should be doable using the fact that the points of finite order in $M_{\mathrm{B}}(U)$ occur in every connected component, as can be seen from the analogue of the exponential exact sequence

$$0 \to H^1(U,\mathbb{Z}) \to H^1(X_D,\mathbb{Q}) \to H^1(U,\mu_\infty) \to H^2(U,\mathbb{Z}) \to H^2(U,\mathbb{Q}).$$

6.6. Preferred sections. We now describe how a tame harmonic bundle of rank one on $U = X - D$ gives rise to a section of the fibration (6.3). In this discussion, we use the notion of parabolic structure and in particular Mochizuki's notion of KMS-spectrum [**39**]. See also Budur [**8**] for a discussion of the rank one case. In Theorem 6.4, the space of harmonic bundles will be identified with the space of σ-invariant sections of $M_{\mathrm{DH}}(U)$. The latter doesn't refer to the notion of parabolic structure, but the identification map between them does.

Fix a Kähler metric ω on X, which restricts to a Kähler metric on U. Recall that a tame harmonic bundle over U is a vector bundle E together with operators D' and D'' and a metric h, with respect to which these operators satisfy certain equations [**26, 11, 52**] plus the tameness condition [**4, 31, 39**]. Our preferred section will not depend on changes of h by multiplying by a positive constant. Decompose

$$D'' = \overline{\partial} + \theta, \quad D' = \partial + \overline{\theta}. \tag{6.5}$$

Use the notation $\mathcal{E} = (E, D', D'', h)$ for our harmonic bundle. Tameness for rank one means that θ extends to a holomorphic section of the logarithmic cotangent bundle [**39**].

Fix $\lambda \in \mathbb{P}^1$ and for now we suppose it is in the first standard chart $\mathbb{A}^1$, so we think of $\lambda \in \mathbb{C}$. Then we get a holomorphic structure $\overline{\partial} + \lambda\overline{\theta}$ on the bundle E, and a λ-connection $\lambda\partial + \theta$. Denote the holomorphic bundle with this holomorphic structure by $\mathcal{E}^\lambda$ and the λ-connection by ∇^λ. By [**39**], for any vector $a = (a_1, \ldots, a_k)$ of real numbers, we get an extension of $\mathcal{E}^\lambda$ to a holomorphic bundle denoted E^λ_a on X, and ∇^λ extends to a logarithmic λ-connection on E^λ_a.

If we pick λ_0 and any $i = 1, \ldots, k$, then there is a set of critical values of a_i which are the parabolic weight pieces of the *KMS-spectrum* [**39**]. The elements of the KMS-spectrum are pairs consisting of the residue eigenvalue of the λ_0-connection and the parabolic weight. For $a = (a_1, \ldots, a_k)$ with a_i not a parabolic weight in the KMS-spectrum at λ_0 for the divisor component D_i, there is a neighborhood $\lambda_0 \in L \subset \mathbb{P}^1$ such that for $\lambda \in L$, the bundles with logarithmic connection $(\mathcal{E}^\lambda_a, \nabla^\lambda)$ vary holomorphically in λ. For each divisor component and fixed λ_0, the KMS-spectrum is a $\mathbb{Z}$-translation orbit in $\mathbb{R}$, that is, it consists of everything of the form $a_i + u_i$ for $u_i \in \mathbb{Z}$. This is special to the rank one case, where there is only one KMS spectrum weight in $\mathbb{R}/\mathbb{Z}$.

The *KMS-critical locus* at λ_0 is the set of all a such that some a_i is a parabolic weight in the KMS-spectrum for λ_0 and D_i. This locus is a union of translates of the k coordinate hyperplanes. The translates included are all of the form $(a_1 + u_1, \ldots, a_k + u_k)$ where a_i are parabolic weights of the KMS-spectrum, and u_i are any integers. The *KMS-chambers* are the connected components of the complement of the KMS-critical locus. Note that $\mathbb{Z}^k$ acts simply transitively on the set of KMS-chambers for any λ_0. Furthermore, the set of KMS-chambers varies continuously with λ_0: a point which is well in the middle of a chamber for λ_1 will remain in a uniquely determined chamber for λ_1 nearby, or to put it another way, the KMS-critical locus varies continuously as a function of λ.

In particular, if for any λ_0 we choose a particular KMS-chamber, then by following this around it determines a KMS-chamber for all other $\lambda \in \mathbb{A}^1$.

For different values of a in the same KMS-chamber, the bundles with logarithmic connection $(\mathcal{E}^\lambda_a, \nabla^\lambda)$ are all canonically isomorphic. Hence, the choice of a KMS-chamber for a given λ_0 determines a choice of KMS-chamber for all $\lambda \in \mathbb{A}^1$; let $a(\lambda)$ denote a function taking values in this chamber for each λ. We thus get the collection of bundles with logarithmic connection depending on λ,

$$\lambda \mapsto (\mathcal{E}^\lambda_{a(\lambda)}, \nabla^\lambda).$$

This is our preferred section of $M_{\mathrm{DH}}(X, \log D)$ over $\mathbb{A}^1$. If we choose a different chamber to begin with, then the section is modified by the corresponding element of the local meromorphic gauge group $\mathcal{G} = \mathbb{Z}^k$. The projection to the quotient gives a uniquely defined section of the fibration (6.3), at least over $\mathbb{A}^1$.

This construction patches together with the corresponding construction on the other chart $\mathbb{A}^1$ at ∞. See [**39**, Chapter 11].

The construction we have described here is an isomorphism between harmonic bundles and σ-invariant sections of the fibration $M_{\mathrm{DH}}(X, \log D)/\mathcal{G} \to \mathbb{P}^1$.

THEOREM 6.4. *Let $M_{\mathrm{har}}(U)$ denote the group of tame harmonic line bundles on U. The map described above goes from here to the space of σ-invariant sections of $M_{\mathrm{DH}}(X, \log D)$ modulo the gauge group action:*

$$\mathcal{P} : M_{\mathrm{har}}(U) \to \Gamma(\mathbb{P}^1, M_{\mathrm{DH}}(X, \log D))^{\sigma}/\mathcal{G}.$$

This map is an isomorphism.

The map is given by the discussion above. The proof that it is an isomorphism, which requires techniques from the next subsections, will be given in Corollary 6.8 and §6.10 below.

This theorem, which is only in the rank one case, nevertheless suggests that in the correspondence between harmonic bundles and pure twistor $\mathcal{D}$-modules of [**40**] and [**47**] the parabolic weight should come out of the structure of the twistor $\mathcal{D}$-module, without having to impose an additional parabolic structure on the $\mathcal{D}$-module side. It isn't clear to me to what extent this statement may already be contained implicitly in [**40**] and [**47**] (see §6.11 below).

6.7. Residues and parabolic structures. We now get to one of the main observations in this article: that the three-dimensional space of σ-invariant sections of $\mathbb{T}(-1, \log)$ encodes the data of residues and parabolic weights for a harmonic bundle.

Fix a divisor component D_i, and a point $p \in \mathbb{A}^1$. The fiber $\mathbb{T}(-1, \log)_p$ is identified with $\mathbb{C}$ by the frame $\frac{\partial}{\partial \lambda}$. Hence the residue map can be composed with this identification to give

$$\mathrm{res}_{D_i,p} : M_{\mathrm{Hod}}(X, \log D)_p \to \mathbb{C} \cong \mathbb{T}(-1, \log)_p.$$

It sends a logarithmic p-connection (E, ∇) to $\mathrm{res}(\nabla; D_i)\frac{\partial}{\partial \lambda}(p)$.

The glueing function for residues of logarithmic λ-connections is $-\lambda^2$, the same as for $\mathbb{T}(-1, \log) = T\mathbb{P}^1$. Therefore, this map glues with the same map on the chart $M_{\mathrm{Hod}}(\overline{X}, \log \overline{D})$ to give a bundle map over $\mathbb{P}^1$,

$$\mathrm{res}^{\mathrm{DH}}_{D_i} : M_{\mathrm{DH}}(X, \log D) \to \mathbb{T}(-1, \log).$$

LEMMA 6.5. *The residue map $\mathrm{res}^{\mathrm{DH}}_{D_i}$ is compatible with the antipodal involutions on $M_{\mathrm{DH}}(X, \log D)$ and $\mathbb{T}(-1, \log)$, so it gives a map on σ-invariant sections*

$$\Gamma(\mathbb{P}^1, M_{\mathrm{DH}}(X, \log D))^{\sigma} \to \Gamma(\mathbb{P}^1, \mathbb{T}(-1, \log))^{\sigma}.$$

PROOF. The calculation for X near D is the same as that of §4.3. Comparing with the calculation of §5.3, we see that the residue is compatible with σ and it induces a map on σ-invariant sections. □

Next, consider the projection $\mathrm{pr}_i : \mathcal{G} \to \mathbb{Z}(-1, \log)$ which sends $(g_1, \ldots, g_k)$ to g_i.

LEMMA 6.6. *The residue map $\mathrm{res}^{\mathrm{DH}}_{D_i}$ is compatible with the action of the local meromorphic gauge group $\mathcal{G}$ via the projection pr_i composed with the morphism $\mathbb{Z}(-1, \log) \to \mathbb{T}(-1, \log)$, so it gives a map on quotients*

$$\Gamma(\mathbb{P}^1, M_{\mathrm{DH}}(X, \log D))^{\sigma}/\mathcal{G} \to \Gamma(\mathbb{P}^1, \mathbb{T}(-1, \log))^{\sigma}/\mathbb{Z}(-1, \log).$$

PROOF. At each point p, the action of the gauge group is compatible by equation (2.2) with the map pr_i via the standard morphism $\mathbb{Z}(-1,\log) \to \mathbb{T}(-1,\log)_p$ which sends the generator to $-p\frac{\partial}{\partial\lambda}$. This gives the compatibility on global sections. $\square$

For any $p \in \mathbb{A}^1 \subset \mathbb{P}^1$, we can compose the map of Lemma 6.5 with the isomorphism $(\varpi_p, \mathrm{res}_p)$ of Proposition 5.3, comprising the parabolic weight function ϖ_p and the residue or evaluation at p. This gives a map

$$(\varpi_p, \mathrm{res}_p)_{D_i} : \Gamma(\mathbb{P}^1, M_{\mathrm{DH}}(X, \log D))^{\sigma} \to \mathbb{R} \times \mathbb{C}. \tag{6.6}$$

Dividing by the action of the local meromorphic gauge group corresponds to dividing by the action of $\mathbb{Z}$ on $\mathbb{R} \times \mathbb{C}$ generated by $(\varpi_p, \mathrm{res}_p)(\psi(1,0)) = (1, -p)$. We get a quotient map

$$(\varpi_p, \mathrm{res}_p)^{\mathcal{G}}_{D_i} : \Gamma(\mathbb{P}^1, M_{\mathrm{DH}}(X, \log D))^{\sigma}/\mathcal{G} \to \frac{\mathbb{R} \times \mathbb{C}}{(1,-p)\cdot\mathbb{Z}}. \tag{6.7}$$

Compose with the preferred-sections map $\mathcal{P}$ of Theorem 6.4. Our main observation is that this encodes the parabolic weight and residue of a harmonic bundle. Following Seshadri for the case of unitary connections [**49**], these were defined for the case of curves, at $\lambda = 0$ and $\lambda = 1$, in [**50**]. They were defined in higher dimensions and for all λ in [**39**]. The parabolic weight parameter has been extensively studied in many places [**2, 3, 6, 33, 34, 35, 38, 41, 44, 55, 56, 58**].

Given a parabolic bundle $F = \{F_b\}$ filtered by bundles indexed in the increasing sense by $b \in \mathbb{R}^k$, suppose we have chosen E as one of these bundles. Define the *parabolic weight* to be the element $b = (b_1, \ldots, b_k)$ with b_i as small as possible so that $E = F_b$. Given a harmonic bundle $\mathcal{E} = (E, D', D'', h) \in M_{\mathrm{Har}}(U)$, we obtain for any λ a parabolic logarithmic λ-connection $\mathcal{E}^{\lambda}$ by [**39**]. Its underlying parabolic bundle has a parabolic weight as defined at the start of this paragraph, and the parabolic λ-connection on $\mathcal{E}^{\lambda}$ has a residue along each D_i. The parabolic weight of $\mathcal{E}^{\lambda}$ is determined by the rate of growth of the harmonic metric: if u is a unit section near a point of D_i, and if D_i is cut out by the equation $z = 0$, then $|u|_h \sim |z|^{-b_i}$ where b_i is the parabolic weight along D_i.

THEOREM 6.7. *Suppose D_i is a divisor component and $p \in \mathbb{A}^1 \subset \mathbb{P}^1$. Suppose $\mathcal{E} = (E, D', D'', h) \in M_{\mathrm{Har}}(U)$ is a rank one harmonic bundle on U. Then*

$$(\varpi_p, \mathrm{res}_p)^{\mathcal{G}}_{D_i}(\mathcal{P}(\mathcal{E})) \in \frac{\mathbb{R} \times \mathbb{C}}{(1,-p)\cdot\mathbb{Z}}$$

is the parabolic weight and residue of the parabolic logarithmic λ-connection $\mathcal{E}^{\lambda}$ at $\lambda = p$.

PROOF. Fix an extension of the logarithmic Higgs bundle $(\mathcal{E}^0, \nabla^0)$ to a line bundle over X. It then has a harmonic metric h. Let a' be the parabolic weight along D_i. Let α' be the residue of the Higgs field along D_i. Mochizuki defines functions $\mathfrak{p}(\lambda, a, \alpha)$ and $\mathfrak{e}(\lambda, a, \alpha)$ in [**39**, §2.1.7], and in Corollary 7.71, [**39**, §7.3.3] he points out that the rule obeyed by the KMS-spectrum of a harmonic bundle is given by the transformation $(\mathfrak{p}, \mathfrak{e})$. In the rank one case, the KMS-spectrum has only one element. Hence, the transformation rule [**39**, Cor. 7.71] means that the parabolic weight and residue of $\mathcal{E}^{\lambda}$ are respectively

$$\mathfrak{p}(\lambda, a', \alpha') \quad \text{and} \quad \mathfrak{e}(\lambda, a', \alpha').$$

By inspection, the functions $\mathfrak{p}, \mathfrak{e}$ of [**39**, §2.1.7] are the same as the parabolic weight functions and residue functions occurring in Proposition 5.3:

$$(\varpi_p, \mathrm{res}_p)(\psi(a,\alpha)) = (\mathfrak{p}(p,a,\alpha), \mathfrak{e}(p,a,\alpha)). \tag{6.8}$$

Recall that $\lambda \mapsto (\mathcal{E}^\lambda, \nabla^\lambda)$ is exactly our preferred section $\mathcal{P}(\mathcal{E})$ (lifted over the gauge group action). Let (a,α) denote the Higgs coordinates for the residue section, so

$$\mathrm{res}_{D_i}(\mathcal{P}(\mathcal{E})) = \psi(a,\alpha) \in \Gamma(\mathbb{P}^1, \mathbb{T}(-1,\log))^\sigma \cong \mathbb{R} \times \mathbb{C}.$$

The value of this section at the point λ, which is the residue of the logarithmic λ-connection ∇^λ, is given by the residue function $\mathrm{res}_\lambda(\psi(a,\alpha))$ calculated in §5.5 above. We conclude that for all $\lambda \in \mathbb{A}^1$,

$$\mathrm{res}_\lambda(\psi(a,\alpha)) = \mathfrak{e}(\lambda, a', \alpha').$$

The identity (6.8) between the functions $\mathrm{res}_\lambda(\psi(a,\alpha))$ and $\mathfrak{e}(\lambda,a,\alpha)$, writing them out according to §5.5, means that

$$\alpha - a\lambda - \overline{\alpha}\lambda^2 = \alpha' - a'\lambda - \overline{\alpha}'\lambda^2$$

for all $\lambda \in \mathbb{A}^1$. It follows that $a = a'$ and $\alpha = \alpha'$. This proves the statement of the theorem at $p = \lambda = 0$.

At a general value of $\lambda = p$, the parabolic weight and residue of the harmonic bundle are given as $\mathfrak{p}(\lambda,a,\alpha)$ and $\mathfrak{e}(\lambda,a,\alpha)$ respectively, by Corollary 7.71, [**39**, §7.3.3]. The identity (6.8) shows that these are the same as $(\varpi_p, \mathrm{res}_p)_{D_i}(\mathcal{P}(\mathcal{E}))$. Modulo the action of the gauge group (which absorbs our initial choice of extension of the bundle), this gives the statement of the theorem. □

We now have enough to do half of the isomorphism in Theorem 6.4.

COROLLARY 6.8. *The preferred-sections morphism $\mathcal{P}$ in Theorem 6.4 is injective.*

PROOF. Suppose $\mathcal{E}$ and $\mathcal{F}$ are rank one harmonic bundles, such that $\mathcal{P}(\mathcal{E}) \cong \mathcal{P}(\mathcal{F})$. The local parabolic weight and residue data of the harmonic bundles coincide, because these functions factor through $\mathcal{P}$ by Theorem 6.7. The line bundles with connection ($\lambda = 1$) associated to $\mathcal{E}$ and $\mathcal{F}$ correspond to filtered local systems of rank 1 [**50, 39, 24**]. The filtration weight of the filtered local system is obtained from the parabolic weight and residues of the line bundles with connection; see the third column of the table in [**50**, p. 720]. Therefore, the filtration weights of the filtered local systems associated to $\mathcal{E}$ and $\mathcal{F}$ are the same. The fact that $\mathcal{P}(\mathcal{E}) \cong \mathcal{P}(\mathcal{F})$ at $\lambda = 1$ restricted over $\lambda = 1$ means that the associated logarithmic connections are the same up to local meromorphic gauge transformation, hence the associated monodromy representations are the same. Now, in rank one a filtered local system is determined uniquely by its monodromy representation and its filtration weight. A filtered local system corresponds to a unique harmonic bundle. Therefore $\mathcal{E} \cong \mathcal{F}$. □

6.8. Comparison with [50]. It is interesting to comment on the particular cases $p = 0$ and $p = 1$. The transformation for going from $p = 0$ to $p = 1$ gives back the transformation between the first two columns of the table on p. 720 of [**50**], which has remained mysterious to me up until now. The parabolic weights and residues for the different points $\lambda = p$ are different coordinate systems on the same three-dimensional space $\Gamma(\mathbb{P}^1, \mathbb{T}(-1,\log))^\sigma$. Going between two different values of

p gives a change of coordinates. In [**50**] only the values $\lambda = 0$ (Higgs bundles) and $\lambda = 1$ (logarithmic connections) were considered. However there are some changes of notation: we have adopted Mochizuki's coordinates [**39**, §6.1.1] at the Higgs point (a, α) for the reader's convenience. We have also adopted the standard convention that the parabolic structure is indexed by an increasing filtration.

In [**50**], the parabolic structure was given by a decreasing filtration. So, here $a \in \mathbb{R}$ is the parabolic weight of the Higgs bundle in the increasing sense, which corresponds to $-\alpha$ in the notation of [**50**]. In [**50**] the sheaf E_α corresponded to sections whose growth was bounded by $|z|^\alpha$ whereas here E_a corresponds to sections whose growth is bounded by $|z|^{-a}$.

And here, α is the residue of the Higgs field, which was denoted by $b + ci$ on p. 720 of [**50**]. These coordinates coincide with our parabolic weight and residue at $p = 0$. The parabolic weight and residue at $p = 1$ are given by the formulae (5.6) (5.3)

$$\varpi_1(a, \alpha) = a + \alpha + \overline{\alpha}, \quad \mathrm{res}_1(a, \alpha) = \alpha - \overline{\alpha} - a.$$

In terms of the notation (α, b, c) of [**50**]—where α has a different meaning from the rest of the present paper, and where $i \in \mathbb{C}$ is chosen—we get

$$\varpi_1 = -\alpha + 2b, \quad \mathrm{res}_1 = \alpha + 2ic.$$

These are the values in the second column of the table on page 720 of [**50**], taking into account that the "jump" there is $-\varpi_1$.

The conclusion is that the three-dimensional space, and the coordinate transformation in the table of [**50**], come from the fact that the twistor bundle of residues is $\mathbb{T}(-1, \log) \cong \mathcal{O}_{\mathbb{P}^1}(2)$, in other words the local monodromy around singular divisors is in a weight-two twistor bundle.

6.9. The weight filtration. Because we look at line bundles, the moduli spaces form groups under tensor product. For M_{DH} we get a group structure relative to $\mathbb{P}^1$. Define the *weight filtration*:

$$W_1 M_{\mathrm{DH}}(X, \log D) := M_{\mathrm{DH}}(X)^o, \quad W_2 M_{\mathrm{DH}}(X, \log D) := M_{\mathrm{DH}}(X, \log D).$$

In our arguments below, it has seemed most natural to use only the connected component of the identity representation in the weight 1 piece.

Define the second graded piece as the quotient using the group structure

$$Gr_2^W(M_{\mathrm{DH}}(X, \log D)) := W_2/W_1 = \frac{M_{\mathrm{DH}}(X, \log D)}{M_{\mathrm{DH}}(X)^o}.$$

Using only the connected component $M_{\mathrm{DH}}(X)^o$ for W_1 leads to a nontrivial finite group as Gr_2^W even in the compact case $D = \emptyset$.

There is a version modulo the gauge group:

$$Gr_2^W(M_{\mathrm{DH}}(U)) = \frac{M_{\mathrm{DH}}(U)}{M_{\mathrm{DH}}(X)^o} = \frac{Gr_2^W(M_{\mathrm{DH}}(X, \log D))}{\mathcal{G}}.$$

In the second equality the quotient is taken in a stacky sense over $\lambda = 0, \infty$. The inclusion $M_{\mathrm{DH}}(X)^o \subset M_{\mathrm{DH}}(U)$ is strict, injective on each fiber over $\lambda \in \mathbb{P}^1$. There is also an induced weight filtration on the connected component given by

$$W_1 M_{\mathrm{DH}}(X, \log D)^o := M_{\mathrm{DH}}(X)^o.$$

LEMMA 6.9. *The exponential map giving an isomorphism in Theorem 6.3 is strictly compatible with the weight filtrations for connected components on both sides, in other words it sends the usual weight filtration on $H^1(U,\mathbb{C})$ to the weight filtration on $M_{\mathrm{DH}}(X,\log D)^o$.*

PROOF. The exact diagrams in §6.4 extend to exact diagrams of bundles over $\mathbb{P}^1$, with the Tate twistor structure $\mathbb{T}(-1,\log)$ inserted in place of $\mathbb{C}$ at appropriate places. The exponential map is one of the middle vertical maps in the exact squares. The weight foliation is the kernel foliation of the map R, and the weight filtration on abelian cohomology is the kernel of the corresponding map $H^1(U,\mathbb{C}) \to \mathbb{C}^k$. Exactness then implies that the exponential isomorphism is compatible with weight filtrations. □

The goal of this subsection is to identify $Gr_2^W M_{\mathrm{DH}}(X,\log D)$ and show that it has "weight 2".

Write $NS(X)$ for the Néron-Severi group of divisors modulo algebraic equivalence, which is contained in $H^2(X,\mathbb{Z})$. Let $NS(X,D) \subset NS(X)$ be the subgroup generated by divisor components D_i of D. Let $NS(X,D)^{\mathrm{sat}}$ be the saturation of this subgroup, in other words the subgroup of all elements $A \in NS(X)$ such that some multiple mA is in $NS(X,D)$. This may be seen as the kernel in the sequence

$$0 \to NS(X,D)^{\mathrm{sat}} \to NS(X) \to \frac{H^{1,1}(X,\mathbb{C})}{\mathbb{C}\cdot[D_1]+\mathbb{C}\cdot[D_k]}$$

and it includes the torsion subgroup $NS(X)^{\mathrm{tors}}$. For example if $D=\emptyset$ then $NS(X,D)^{\mathrm{sat}} = NS(X)^{\mathrm{tors}}$. We close this paragraph by noting that $NS(X,D)^{\mathrm{sat}}$ is the preimage of $NS(U)^{\mathrm{tors}}$ under the restriction map $NS(X) \to NS(U)$, and

$$\frac{NS(X,D)^{\mathrm{sat}}}{NS(X,D)} = NS(U)^{\mathrm{tors}}. \tag{6.9}$$

Putting together the residue maps at divisor components $D_1,\ldots,D_k$, we get a map

$$R : M_{\mathrm{DH}}(X,\log D) \to \mathbb{T}(-1,\log)^k.$$

The condition (6.1) on $c_1(L)$ for a point $(L,\nabla) \in M_{\mathrm{DH}}(X,\log D)$ is equivalent to saying that $c_1(L) \in NS(X,D)^{\mathrm{sat}}$. These give a map

$$(c_1,R) : M_{\mathrm{DH}}(X,\log D) \to NS(X,D)^{\mathrm{sat}} \times \mathbb{T}(-1,\log)^k.$$

Consider the map defined using the divisor components

$$\Sigma : \mathbb{C}^k \to H^{1,1}(X,\mathbb{C}), \quad (a_1,\ldots,a_k) \mapsto \sum a_i[D_i].$$

The cohomology $H^2(X,\mathbb{C})$ has a pure weight two Hodge structure, whose twistor bundle is semistable of slope 2. The twistor bundle $\xi(H^2(X,\mathbb{C}),F,\overline{F})$ has a natural subbundle corresponding to $H^{1,1}$, and since that space is pure of Hodge type $(1,1)$ we have a natural isomorphism

$$\xi(H^{1,1}(X,\mathbb{C}),F,\overline{F}) \cong H^{1,1}(X,\mathbb{C}) \otimes \mathbb{T}(-1,\log).$$

With respect to these constructions, the map Σ extends over $\mathbb{P}^1$ to give a map

$$\Sigma_{\mathrm{DH}} : \mathbb{T}(-1,\log)^k \to H^{1,1}(X,\mathbb{C}) \otimes \mathbb{T}(-1,\log).$$

There is a natural morphism of groups over $\mathbb{P}^1$,

$$\Lambda_{\mathrm{DH}} : NS(X) \times \mathbb{P}^1 \to H^{1,1}(X,\mathbb{C}) \otimes \mathbb{T}(-1,\log) \cong \xi(H^{1,1}(X,\mathbb{C}),F,\overline{F})$$

which sends an element of $NS(X)$ to a section which has a simple pole at 0 and another simple pole at ∞. In terms of the usual trivialization of $\mathbb{T}(-1,\log)$ over $\mathbb{A}^1$, this corresponds to multiplying by λ the usual map from $NS(X)$ to $H^{1,1}(X,\mathbb{C})$. Restrict it to the subgroup $NS(X,D)^{\rm sat}$. Adding to $\Sigma_{\rm DH}$ gives a morphism

$$\Lambda_{\rm DH}+\Sigma_{\rm DH}: NS(X,D)^{\rm sat}\times \mathbb{T}(-1,\log)^k \rightarrow H^{1,1}(X,\mathbb{C})\otimes \mathbb{T}(-1,\log).$$

The Chern classes and residues of logarithmic λ-connections on line bundles (L,∇) satisfy the condition (6.2), which in terms of the present notation says that the composed map $(\Lambda_{\rm DH}+\Sigma_{\rm DH})\circ(c_1,R)$ is zero.

The following exact sequence is an analogue of the basic exact sequence (6.4) and the diagram following it in §6.4.

PROPOSITION 6.10. *The weight filtrations for $M_{\rm DH}(X,\log D)$ and $M_{\rm DH}(U)$, with group structure given by tensor product, fit into a strict exact sequence of analytic groups over $\mathbb{P}^1$*

$$(6.10)\qquad 1\rightarrow M_{\rm DH}(X)^o\rightarrow M_{\rm DH}(X,\log D)\stackrel{(c_1,R)}{\longrightarrow} NS(X,D)^{\rm sat}\times\mathbb{T}(-1,\log)^k\cdots$$

$$\cdots\stackrel{\Lambda_{\rm DH}+\Sigma_{\rm DH}}{\longrightarrow} H^{1,1}(X,\mathbb{C})\otimes\mathbb{T}(-1,\log).$$

PROOF. It suffices to prove this in the fiber over a fixed λ, and we may assume $\lambda\in\mathbb{A}^1$. Injectivity on the left is easy and was mentioned previously: given (L,∇) and (L',∇') on X, an isomorphism of logarithmic λ-connections on (X,D) between them is also an isomorphism of λ-connections on X.

For exactness at $M_{\rm DH}(X,\log D)$, suppose (L,∇) is a logarithmic λ-connection such that $c_1(L)=0$ and $R(\nabla)=0$, that is, ${\rm res}_{D_i}(\nabla)=0$ for each component D_i. Then ∇ is a connection over X and even if $\lambda=0$, the condition $c_1=0$ ensures inclusion in $M_{\rm DH}(X)$. The fact that $c_1=0$ in the Néron-Severi group means that L is algebraically equivalent to 0. From the structure of the Picard group this implies that L is in the connected component of the trivial bundle and $M_{\rm DH}(X)_\lambda\rightarrow {\rm Pic}(X)$ is smooth, so (L,∇) is in the connected component $M_{\rm DH}(X)^o_\lambda$.

We prove exactness at $\mathbb{T}(-1,\log)^k\times NS(X,D)^{\rm sat}$. Use the standard frame for $\mathbb{T}(-1,\log)$. A point in $\ker(\Lambda_{\rm DH}+\Sigma_{\rm DH})$ is $\zeta\in NS(X,D)^{\rm sat}$ together with a k-uple $(a_1,\ldots,a_k)\in\mathbb{C}^k$ such that

$$\lambda\zeta+\sum a_i[D_i]=0\quad\text{in}\quad H^{1,1}(X,\mathbb{C}).$$

Choose a line bundle L such that $c_1(L)=\zeta$. The elements of $NS(X,D)^{\rm sat}$ restrict to torsion elements on U by (6.9). Line bundles whose Chern class are torsion have flat regular singular connections, which in the rank 1 case are automatically logarithmic. Thus we can choose an initial λ-connection ∇' on L logarithmic with respect to (X,D). Let a_i' denote the residues of ∇' along D_i. Then

$$\sum(a_i-a_i')[D_i]=0\quad\text{in}\quad H^{1,1}(X,\mathbb{C}).$$

Hence there is a logarithmic one-form β on (X,D) having residues a_i-a_i' along D_i. Now $\nabla=\nabla'+\beta$ is a logarithmic λ-connection with $(c_1,R)(L,\nabla)=(\zeta,(a_1,\ldots,a_k))$. □

COROLLARY 6.11. *The exact sequence* (6.10) *identifies the graded piece of the weight filtration as*

$$Gr^W_2 M_{\rm DH}(X,\log D)=$$

$$\ker\left(NS(X,D)^{\text{sat}} \times \mathbb{T}(-1,\log)^k \overset{\Lambda_{\text{DH}}+\Sigma_{\text{DH}}}{\longrightarrow} H^{1,1}(X,\mathbb{C}) \otimes \mathbb{T}(-1,\log)\right).$$

We can now describe the weight two phenomenon in the title of the paper:

LEMMA 6.12. *There is b such that*

$$\ker\left(\mathbb{T}(-1,\log)^k \to H^{1,1}(X,\mathbb{C}) \otimes \mathbb{T}(-1,\log)\right) \cong \mathbb{T}(-1,\log)^b.$$

There is an exact sequence

$$0 \to \mathbb{T}(-1,\log)^b \to Gr_2^W M_{\text{DH}}(X,\log D) \to NS(X,D)^{\text{sat}} \to 0.$$

On the connected component of the identity representation

$$\mathbb{T}(-1,\log)^b \cong Gr_2^W M_{\text{DH}}(X,\log D)^o.$$

Modulo the gauge group we have

$$0 \to \mathbb{G}_m(-1)^b \to Gr_2^W M_{\text{DH}}(U) \to NS(U)^{\text{tors}} \to 0.$$

PROOF. A map between pure twistor structures of weight 2 has a kernel which is again a pure twistor structure of weight 2. Hence, there is b as in the first claim. The first exact sequence comes from Proposition 6.10 and the fact that every element of $NS(X,D)^{\text{sat}}$ goes into $H^{1,1}(X,\mathbb{C})$ to something which comes from $\mathbb{C}^k$. The third statement follows from this exact sequence because

$$Gr_2^W\left(M_{\text{DH}}(X,\log D)^o\right) = \left(Gr_2^W M_{\text{DH}}(X,\log D)\right)^o$$

since $W_1 M_{\text{DH}}(X,\log D)^o = W_1 M_{\text{DH}}(X,\log D) = M_{\text{DH}}(X)^o$.

For the last exact sequence, divide out by the gauge group $\mathcal{G} = \mathbb{Z}^k$, which means dividing the first exact sequence by the exact sequence

$$0 \to \mathbb{Z}^b \to \mathbb{Z}^k \to NS(X,D) \to 0.$$

Equation (6.9) identifies the quotient of $NS(X,D)^{\text{sat}}$ by $NS(X,D)$ with $NS(U)^{\text{tors}}$. $\square$

The weight equivalence relation induces an equivalence relation on sections: two sections are equivalent if and only if their values are equivalent over each $\lambda \in \mathbb{P}^1$. The weight filtrations of $M_{\text{DH}}(X,\log D)$ and $M_{\text{DH}}(U)$ induce filtrations on their spaces of sections and we can form the graded pieces.

LEMMA 6.13. *There is a finite abelian group K and an exact sequence*

$$0 \to \Gamma(\mathbb{P}^1, \mathbb{T}(-1,\log)^b)^\sigma \to Gr_2^W \Gamma(\mathbb{P}^1, M_{\text{DH}}(X,\log D))^\sigma \to K \to 0.$$

For any $p \in \mathbb{A}^1$ the parabolic weight and residue give an exact sequence

$$0 \to \left(\frac{\mathbb{R}\times\mathbb{C}}{(1,-p)\mathbb{Z}}\right)^b \to Gr_2^W \Gamma(\mathbb{P}^1, M_{\text{DH}}(U))^\sigma \to K \to 0.$$

PROOF. The second exact sequence comes from the first via Proposition 5.3. The map

$$Gr_2^W \Gamma(\mathbb{P}^1, M_{\text{DH}}(X,\log D))^\sigma \to \Gamma(\mathbb{P}^1, Gr_2^W M_{\text{DH}}(X,\log D))^\sigma \tag{6.11}$$

is injective. Suppose we have a σ-invariant section of $Gr_2^W M_{\text{DH}}(X,\log D)$. The obstruction to lifting it to a section in $Gr_2^W \Gamma(\mathbb{P}^1, M_{\text{DH}}(X,\log D))^\sigma$ lies in the analytic cohomology group $H^1(\mathbb{P}^1, M_{\text{DH}}(X)^o)$. In view of the exact sequence used in Lemma 3.3, we have $H^1(\mathbb{P}^1, M_{\text{DH}}(X)^o) = H^2(\mathbb{P}^1, A)$, which is discrete. Therefore, on the connected component of the space of sections, the map (6.11) is surjective. There is a finite subgroup $K \subset NS(U)^{\rm tors}$ representing the components in the image of (6.11). In fact $A = H^1(X, \mathbb{Z})$ and there is an exact sequence of length four

$$0 \to \Gamma(\mathbb{P}^1, \mathbb{T}(-1, \log)^b)^\sigma \to Gr_2^W \Gamma(\mathbb{P}^1, M_{\rm DH}(X, \log D))^\sigma \to$$
$$\to NS(U)^{\rm tors} \to H^2(\mathbb{P}^1, H^1(X, \mathbb{Z})).$$

I don't know whether there are any examples where the last connecting map is nonzero. □

6.10. Proof of Theorem 6.4. Injectivity is proven in Corollary 6.8.

Suppose we are given a σ-invariant section in the target of the map $\mathcal{P}$. Lift it over the quotient of the action of the gauge group $\mathcal{G}$, to get a section

$$\epsilon \in \Gamma(\mathbb{P}^1, M_{\rm DH}(X, \log D))^\sigma.$$

We would like to construct a harmonic bundle mapping to ϵ. For this, we will use the correspondence [**50, 39, 40**] between harmonic bundles and parabolic logarthmic λ-connections for some fixed $\lambda \in \mathbb{A}^1$. It would be sufficient to use the Higgs case $\lambda = 0$ or the de Rham case $\lambda = 1$ but it is interesting to treat a general λ.

The value $\epsilon(\lambda)$ corresponds to a logarithmic λ-connection (E, ∇), with residue ${\rm res}_{\lambda, D_i}(\epsilon)$ along each D_i. On the other hand, consider the parabolic weight parameter

$$b_i := \varpi_{\lambda, D_i}(\epsilon) \in \mathbb{R}.$$

Put a parabolic structure onto (E, ∇) using these weights. This gives a parabolic logarithmic λ-connection $(E(\sum b_i D_i), \nabla)$.

We claim that $c_1(E(\sum b_i D_i)) = 0$. To see this, look at the exact sequence of Proposition 6.10. Take spaces of σ-invariant sections, and use the identification of Proposition 5.3 at our fixed λ. In these terms, ϵ maps to an element of

$$\ker \left(NS(X, D)^{\rm sat} \times (\mathbb{R} \times \mathbb{C})^k \to H^{1,1}(X, \mathbb{R}) \otimes_{\mathbb{R}} (\mathbb{R} \times \mathbb{C}) \right).$$

The coefficient in $NS(X, D)^{\rm sat}$ is $\zeta = c_1(E)$, whereas the coefficient in $\mathbb{R}^k$ is $(b_1, \ldots, b_k)$. The image in the first factor $H^{1,1}(X, \mathbb{R}) \otimes_{\mathbb{R}} \mathbb{R}$ is

$$\varpi_\lambda(\Lambda_{\rm DH}(\zeta) + \Sigma_{\rm DH}(b_1, \ldots, b_k)).$$

Notice that $\varpi_\lambda \circ \Lambda_{\rm DH}$ is equal to the usual map $NS(X, D)^{\rm sat} \to H^{1,1}(X, \mathbb{R})$. This is because of the normalization condition that $\varpi_\lambda(1, 0) = 1$ used in §5.8. Similarly,

$$\varpi_\lambda \Sigma_{\rm DH}(b_1, \ldots, b_k) = b_1[D_1] + \ldots + b_k[D_k] \quad \in H^{1,1}(X, \mathbb{R}).$$

We conclude that $c_1(E(\sum b_i D_i)) = c_1(E) + b_1[D_1] + \cdots + b_k[D_k] = 0$ as claimed.

Then the harmonic theory for parabolic logarithmic connections [**18, 11, 17, 50, 40, 8**] provides a rank 1 tame harmonic bundle $\mathcal{E}$ over U, whose associated parabolic λ-connection is $(E(\sum b_i D_i), \nabla)$. By Theorem 6.7, the parabolic weight of $\mathcal{E}$ is the same as the parabolic weight of the harmonic bundle, that is,

$$\varpi_{1, D_i}(\mathcal{P}(\mathcal{E})) = b_i.$$

This coincides with the parabolic weight of ϵ. Furthermore, by construction the values of $\mathcal{P}(\mathcal{E})$ and ϵ at $p = 1$ are the same, both equal to (E, ∇). We conclude that $\mathcal{P}(\mathcal{E}) = \varepsilon$, which concludes the proof of Theorem 6.4, by the following lemma.

LEMMA 6.14. *Suppose $p \in \mathbb{P}^1$, and suppose ξ, ϵ are two σ-invariant sections of $M_{\rm DH}(X, \log D)$. Suppose that for each D_i, the parabolic weights agree: $\varpi_{p, D_i}(\xi) = \varpi_{p, D_i}(\epsilon)$. Suppose furthermore that $\xi(p) = \epsilon(p)$. Then $\xi = \epsilon$.*

PROOF. The weight filtration exact sequence (6.10) gives an exact sequence on spaces of σ-invariant sections. Then, identify the space of sections of $\mathbb{T}(-1, \log)^k$ with $(\mathbb{R} \times \mathbb{C})^k$ using $(\varpi_p, \mathrm{res}_p)$ as in Proposition 5.3. If $\xi(p) = \epsilon(p)$ then their residues at p agree. By hypothesis the parabolic weight coordinates agree. Therefore, ξ and ϵ go into the same section of $\mathbb{T}(-1, \log)^k$.

They go to the same section of the discrete group $NS(X, D)^{\mathrm{sat}}$, because the values at p are the same by hypothesis. By Lemma 6.12, ξ and ϵ go to the same section of $Gr_2^W M_{\mathrm{DH}}(X, \log D)$.

Therefore the difference $\xi \otimes \epsilon^{-1}$ comes from a σ-invariant section of $M_{\mathrm{DH}}(X)$. As was noted in Lemma 3.3, the weight 1 property of $M_{\mathrm{DH}}(X)$ says that the space of σ-invariant sections here maps isomorphically to any fiber. The condition $\xi(p) = \epsilon(p)$ thus implies that $\xi \otimes \epsilon^{-1}$ is trivial. □

6.11. Alternate proof. The referee has suggested an alternative proof of Theorem 6.4 which we sketch here. Given $\epsilon : \mathbb{P}^1 \to M_{\mathrm{DH}}(X, \log D)$ a σ-invariant section, we can lift it to a section of the corresponding moduli stack. Indeed the moduli stack is a $\mathbb{G}_m$-gerb over the moduli space so the obstruction lies in $H^2(\mathbb{P}^1, (\mathcal{O}^{\mathrm{an}})^*)$, which vanishes. We obtain a bundle $\mathcal{V}$ on $U \times \mathbb{P}^1$ with appropriate operators [**54**] making it into a variation of pure twistor structure. Purity is due to the fact that the restriction to any $\{x\} \times \mathbb{P}^1$ is a holomorphic line bundle, whose degree is independent of x—and tensoring with a line bundle from $\mathbb{P}^1$ we can assume it is pure of weight 0. The σ-invariance allows us to choose a pairing

$$S : \mathcal{V} \otimes \sigma_{\mathbb{P}^1}^* \mathcal{V} \to \mathbb{T}(0).$$

Notice that the inclusion of the duality operation D in the definition of σ in §3.3 means that $\epsilon = \sigma(\epsilon)$ translates into a pairing as above. After an initial choice, one can modify S so as to satisfy the required symmetry property, and then notice that the sign is the same everywhere, so again after multiplying by a constant we get a polarization. This polarized variation of pure twistor structure corresponds [**54, 47, 39**] to a harmonic bundle $\mathcal{E}$ on U. By construction, for any $\lambda \in \mathbb{A}^1$, the associated λ-connection on U is the same as the one corresponding to $\epsilon(\lambda)$. At $\lambda = 0$ this allows us to check tameness: θ has logarithmic poles. Uniqueness of the Deligne canonical extension at any $\lambda \neq 0$ implies that $\mathcal{P}(\mathcal{E})(\lambda)$ coincides with $\epsilon(\lambda)$ up to the action of the meromorphic gauge group. This proves surjectivity. Injectivity may be proved using an argument such as in §6.7 but with [**39**, Lemma 7.110] and [**40**].

7. Strictness consequences

One of the most useful things about weights in Hodge theory is that they lead to a notion of strictness. Here we formulate a conjecture which would be the corresponding strictness property coming from the weight two piece of the nonabelian H^1. Since it is just a conjecture, we consider representations of any rank.

Suppose (X, D) and (Y, E) are smooth projective varieties with simple normal crossings divisors, such that D has k components and E has m components. Suppose $\mathcal{F}$ is some natural construction from local systems on $U := X - D$ to local systems on $V := Y - E$. This could include any combination of pullbacks, higher direct images, tensor products, duals, etc. For the present purposes, denote by $M_{\mathrm{B}}(U)$ and $M_{\mathrm{B}}(V)$ the full unions of spaces of representations of all ranks. There will be a stratification of $M_{\mathrm{B}}(U)$ into locally closed subsets such that $\mathcal{F}$ is algebraic on each stratum. Assume that this stratification is maximal, that is, $M_{\mathrm{B}}(U)_\alpha$ is the full subset of representations ρ on U of a given rank, such that the image $\mathcal{F}(\rho)$ has a given rank on V.

In the higher rank case, the eigenvalues of the local monodromy transformations may be considered all at once, with their multiplicities, as divisors on $\mathbb{G}_m^{\perp}$. The group of such divisors is denoted $\mathrm{Div}(\mathbb{G}_m^{\perp})$, and for a divisor D decomposing into k irreducible components, the full collection of residual data is a point in $\mathrm{Div}(\mathbb{G}_m^{\perp})^k$.

CONJECTURE 7.1. *Let $M_{\mathrm{B}}(U)_\alpha$ be a stratum on which $\mathcal{F} = \mathcal{F}_\alpha$ is defined as an algebraic map into $M_{\mathrm{B}}(V)$. There should be a diagram expressing the effect of the construction $\mathcal{F}_\alpha$ on residues:*

$$\begin{array}{ccc} M_{\mathrm{B}}(U)_\alpha & \rightarrow & \mathrm{Div}(\mathbb{G}_m^{\perp})^k \\ \downarrow & & \downarrow \\ M_{\mathrm{B}}(V)_\alpha & \rightarrow & \mathrm{Div}(\mathbb{G}_m^{\perp})^m . \end{array}$$

Strictness: *if $\rho_1, \rho_2 \in M_{\mathrm{B}}(U)_\alpha$ are two semisimple representations such that $\mathcal{F}_\alpha(\rho_1)$ and $\mathcal{F}_\alpha(\rho_2)$ have the same residues in* $\mathrm{Div}(\mathbb{G}_m^{\perp})^m$, *then there should exist a semisimple representation $\rho_3 \in M_{\mathrm{B}}(U)_\alpha$ such that ρ_3 has the same residues as ρ_1 in* $\mathrm{Div}(\mathbb{G}_m^{\perp})^k$, *but $\mathcal{F}(\rho_3) \cong \mathcal{F}(\rho_2)$.*

To phrase it differently, this conjecture says that any variation of the image representation $\mathcal{F}(\rho)$, within a locus of representations on V all having the same residues, obtained by possibly varying the residues of ρ, can equally well be obtained while keeping the residues of ρ fixed.

It would be the analogue of the same statement in abelian Hodge theory for the diagram

$$\begin{array}{ccc} H^1(U) & \rightarrow & Gr_2^W(H^1(U)) \\ \downarrow & & \downarrow \\ H^1(V) & \rightarrow & Gr_2^W(H^1(V)) \end{array} .$$

In the abelian case, pretty much the only possibility for the construction $\mathcal{F}$ is pullback for a map $V \to U$. The strictness statement says that if a_1, a_2 are classes in $H^1(U)$ whose pullbacks to V have the same residues along D, then there is a class a_3 with the same residues as a_2, whose pullback coincides with the pullback of a_2. Our observation of the weight two phenomenon in the case of rank one local systems should provide a proof of this conjecture for the rank one case. We don't discuss that here: it would go beyond the scope of the paper.

One can also expect an infinitesimal formulation of the strictness property, which might be easier to prove. It would be the same statement, in the case where ρ_1 and ρ_2 are infinitesimally close, and we would look for ρ_3 also infinitesimally close. This should be a consequence of having a mixed Hodge structure on the local deformation theory [**7, 21, 45, 46**], plus a compatibility of the construction $\mathcal{F}$ with this mixed Hodge structure. Again, this goes out of the scope of the present discussion.

One should also be able to formulate a similar conjecture for harmonic bundles with the parabolic residual data characterized by points in $\mathrm{Div}(\frac{\mathbb{R}}{\mathbb{Z}} \times \mathbb{C})^k$.

We have been vague about what happens in the case of non-semisimple residues: is there a way to take into account the unipotent piece of the residue in the strictness statement? It doesn't seem completely clear what is the right thing to say.

References

[1] D. Arapura. *Geometry of cohomology support loci for local systems. I.* J. Algebraic Geom. **6** (1997), 563-597.

[2] V. Balaji, I. Biswas, D. Nagaraj. *Principal bundles over projective manifolds with parabolic structure over a divisor.* Tohoku Math. J. **53** (2001), 337-367.

[3] Biswas, I. *Parabolic bundles as orbifold bundles*, Duke Math. J., **88** (1997), 305-325.

[4] O. Biquard. *Sur les fibrés paraboliques sur une surface complexe.* J. London Math. Soc. **53** (1996), 302-316.

[5] P. Boalch. *From Klein to Painlevé via Fourier, Laplace and Jimbo.* Proc. London Math. Soc. **90** (2005), 167-208.

[6] H. Boden, K. Yokogawa. *Moduli spaces of parabolic Higgs bundles and parabolic $K(D)$ pairs over smooth curves, I.* Internat. J. Math. **7** (1996), 573-598.

[7] J. L. Brylinski, P. Foth. *Moduli of flat bundles on open Kähler manifolds.* J. Algebraic Geom. **8** (1999), 147-168.

[8] N. Budur. *Unitary local systems, multiplier ideals, and polynomial periodicity of Hodge numbers.* Preprint `arXiv:math/0610382`.

[9] S. Cecotti, C. Vafa. *Topological–anti-topological fusion.* Nuclear Phys. B **367** (1991), 359-461.

[10] A. Connes. *Cohomologie cyclique et foncteur Ext^n.* Comptes Rendues Acad. Sci. Paris Sér. A-B, 296 (1983), 953-958.

[11] K. Corlette. *Flat G-bundles with canonical metrics.* J. Diff. Geom. **28** (1988), 361-382.

[12] P. Deligne. Letter to the author (March 20, 1989).

[13] C. Deninger. *On the Γ-factors attached to motives.* Invent. Math. **104** (1991), 245-263.

[14] A. Dimca. *Characteristic varieties and constructible sheaves.* Atti Accad. Lincei Mat. Appl. **18** (2007), 365-389.

[15] A. Dimca, P. Maisonobe, M. Saito. *Spectrum and multiplier ideals of arbitrary subvarieties.* Preprint `arXiv:0705.4197v1`.

[16] A. Dimca, S. Papadima, A. Suciu. *Alexander polynomials: Essential variables and multiplicities.* Preprint `arXiv:0706.2499v1`.

[17] S. Donaldson. *Twisted harmonic maps and the self-duality equations.* Proc. London Math. Soc. **55** (1987), 127-131.

[18] J. Eells, J. Sampson.*Harmonic mappings of Riemannian manifolds.* Amer. J. Math. **86** (1964), 109-160.

[19] H. Esnault. *Characteristic classes of flat bundles* Topology **27** (1988), 323-352.

[20] H. Esnault, E. Viehweg. *Deligne-Beilinson cohomology.* Beilinson's conjectures on special values of L-functions, Perspect. Math., 4, Academic Press, Boston (1988), 43-91.

[21] P. Foth. *Deformations of representations of fundamental groups of open Kähler manifolds.* J. Reine Angew. Math. **513** (1999), 17-32.

[22] P. Gajer. *Geometry of Deligne cohomology.* Invent. Math. **127** (1997), 155-207.

[23] W. Goldman, E. Xia. *Rank One Higgs Bundles and Representations of Fundamental Groups of Riemann Surfaces.* Preprint `math/0402429`.

[24] S. Gukov, E. Witten. *Gauge Theory, Ramification, And The Geometric Langlands Program.* Preprint `arXiv:hep-th/0612073v1`.

[25] C. Hertling. *tt* geometry, Frobenius manifolds, their connections, and the construction for singularities.* J. Reine Angew. Math. 555 (2003), 77-161.

[26] N. Hitchin. *The self-duality equations on a Riemann surface.* Proc. London Math. Soc. (3) **55** (1987), 59-126.

[27] N. Hitchin, A. Karlhede, U. Lindström, M. Roček. *Hyperkähler metrics and supersymmetry.* Comm. Math. Phys. **108** (1987), 535-559.

[28] M. Inaba, K. Iwasaki, Masa-Hiko Saito. *Moduli of stable parabolic connections, Riemann-Hilbert correspondence and geometry of Painlevé equation of type VI, Part I.* Publ. Res. Inst. Math. Sci. **42** (2006), 987-1089.

[29] M. Inaba, K. Iwasaki, Masa-Hiko Saito. *Moduli of stable parabolic connections, Riemann-Hilbert correspondence and geometry of Painlevé equation of type VI, Part II.* Moduli spaces and arithmetic geometry. Adv. Stud. Pure Math. **45**, Math. Soc. Japan (2006), 387–432.

[30] J. Iyer, C. Simpson. *The Chern character of a parabolic bundle, and a parabolic corollary of Reznikov's theorem.* To appear, Progr. in Math. **265**, Birkäuser (2007), 437-483.

[31] J. Jost, K. Zuo.*Harmonic maps and $SL(r,\mathbb{C})$ representations of fundamental groups of quasi-projective manifolds.* J. Algebraic Geom. **5** (1996), 77-106.

[32] D. Kaledin. *Non-commutative Hodge-to-de Rham degeneration via the method of Deligne-Illusie.* Preprint `arXiv:math/0611623v3`.

[33] H. Konno. *Construction of the moduli space of stable parabolic Higgs bundles on a Riemann surface.* J. Math. Soc. Japan **45** (1993), 253-276.

[34] Jiayu Li. *Hermitian-Einstein metrics and Chern number inequalities on parabolic stable bundles over Kähler manifolds.* Comm. Anal. Geom. **8** (2000), 445-475.

[35] J. Li, M.S. Narasimhan. *Hermitian-Einstein metrics on parabolic stable bundles.* Acta Math. Sin. **15** (1999), 93-114.

[36] A. Libgober. *Problems in topology of the complements to plane singular curves.* Singularities in geometry and topology, World Sci. Publ. (2007), 370-387.

[37] A. Lubotzky, A. Magid. *Varieties of representations of finitely generated groups.* Mem. Amer. Math. Soc. **58** (1985).

[38] M. Maruyama, K. Yokogawa. *Moduli of parabolic stable sheaves*, Math. Ann. **293** (1992), 77-99.

[39] T. Mochizuki. *Asymptotic behaviour of tame harmonic bundles and an application to pure twistor D-modules, parts I and II.* Memoirs of the AMS **869-870** (2007).

[40] T. Mochizuki. *Kobayashi-Hitchin correspondence for tame harmonic bundles and an application.* Astérisque **309** (2006).

[41] H. Nakajima. *Hyper-Kähler structures on moduli spaces of parabolic Higgs bundles on Riemann surfaces.* Moduli of vector bundles (Sanda, Kyoto, 1994), Lect. Notes Pure Appl. Math. **179** (1996), 199-208.

[42] J. Neisendorfer, L. Taylor. *Dolbeault homotopy theory.* Trans. Amer. Math. Soc. 245 (1978), 183-210.

[43] N. Nitsure. *Moduli of semistable logarithmic connections.* J. Amer. Math. Soc. **6** (1993), 597-609.

[44] D. Panov. *Polyhedral Kähler manifolds.* Thesis, Ecole Polytechnique (2005).

[45] J. Pridham. *The deformation theory of representations of the fundamental group of a smooth variety.* Preprint `arXiv:math/0401344v1`.

[46] J. Pridham. *Deforming l-adic representations of the fundamental group of a smooth variety.* J. Algebraic Geom. **15** (2006), 415-442.

[47] C. Sabbah. *Polarizable twistor $\mathcal{D}$-modules.* Astérisque **300**, (2005).

[48] L. Schäfer. *tt∗-bundles in para-complex geometry, special para-Kähler manifolds and para-pluriharmonic maps* Differential Geometry and its Applications **24** (2006), 60-89.

[49] C. Seshadri. *Moduli of π-vector bundles over an algebraic curve.* Questions on Algebraic Varieties (C.I.M.E., III Ciclo, Varenna, 1969), Edizioni Cremonese, Rome (1970) 139-260.

[50] C. Simpson.*Harmonic bundles on noncompact curves.* J. Amer. Math. Soc. **3** (1990), 713-770.

[51] C. Simpson. *Nonabelian Hodge theory.* Proceedings of the International Congress of Mathematicians, Vol. I, II (Kyoto, 1990), Math. Soc. Japan, Tokyo, (1991), 747-756.

[52] C. Simpson.*Higgs bundles and local systems.* Publ. Math. I.H.E.S. **75** (1992), 5-95.

[53] C. Simpson. *The Hodge filtration on nonabelian cohomology.* Algebraic geometry—Santa Cruz 1995, , Proc. Sympos. Pure Math., **62**, Part 2, A.M.S. (1997), 217-281.

[54] C. Simpson.*Mixed twistor structures.* Preprint `arXiv:alg-geom/9705006v1`.

[55] B. Steer, A. Wren.*The Donaldson-Hitchin-Kobayashi correspondence for parabolic bundles over orbifold surfaces.* Canad. J. Math. **53** (2001), 1309-1339.

[56] M. Thaddeus.*Variation of moduli of parabolic Higgs bundles.* J. Reine Angew. Math. **547** (2002), 1-14.

[57] A. Voros. *Problème spectral de Sturm-Liouville: le cas de l'oscillateur quartique.* Séminaire Bourbaki Vol. 1982/83, Astérisque, **105-106**, S.M.F., Paris, (1983), 95–104,.

[58] K. Yokogawa. *Compactification of moduli of parabolic sheaves and moduli of parabolic Higgs sheaves.* J. Math. Kyoto Univ. **33** (1993), 451-504.

CNRS, Laboratoire J. A. Dieudonné, UMR 6621, Université de Nice-Sophia Antipolis, 06108 Nice, Cedex 2, France

E-mail address: `carlos@unice.fr`

URL: `http://math.unice.fr/~carlos/`

Proceedings of Symposia in Pure Mathematics
Volume **78**, 2008

Associativity for the Neumann system

L.K. Hoevenaars

ABSTRACT. We derive a residue formula for the derivative of the period map associated with algebraically integrable systems and use this formula to prove that the base of the Neumann system carries the structure of an almost Frobenius manifold.

Introduction

We study families of Jacobians of curves $\pi : Jac(\Sigma) \to B$ which support an algebraically integrable system: the total space $|Jac(\Sigma)|$ is the phase space and π is Lagrangian. In section 1 we briefly review some different views on algebraic integrability (affine special geometry, VHS) and their relations. In this situation the derivative of the period map simplifies and we give a certain residue formula for it in section 2, using algebraic de Rham cohomology and differentials of the second kind. Similar formulas were found earlier in singularity theory [**Sai83**], Hitchin systems [**DP92, Bal06**], Whitham hierarchies and matrix models [**Kri94, EO07**], and Hurwitz spaces [**Dub96**].

Our aim is to use this residue formula to prove that the base B of the Neumann system can naturally be endowed with the structure of an almost Frobenius manifold. The most important constructions of (almost) Frobenius manifolds come from singularity theory, Gromov-Witten theory and generalized deformations of complex structures (see e.g. [**Man00**] for a review). For any Frobenius manifold, there exists a dual almost Frobenius manifold. Applying this result to the case of simple singularities, one obtains a structure on the base of the open Toda chain which is very similar to ours [**MMM00, Dub04**]. Our construction is therefore closest in spirit to singularity theory.

Theorem 4.4 is the main result of this paper. Here we describe this result in condensed form: consider the family of spectral curves of the Neumann system, i.e. hyperelliptic genus g curves given by the affine equation

$$y^2 - R(x)Q(x) = y^2 - \prod_{k=1}^{g}(x - b_k)\prod_{\alpha=1}^{g+1}(x - q_\alpha) = 0 \tag{0.1}$$

2000 *Mathematics Subject Classification.* Primary: 14H70, 14D07, 53D45; Secondary: 32G20, 37J35.

Research supported by NWO VENI grant.

where the q_α are treated as fixed and the b_k are the parameters of the family. Locally the base B is a vector space with coordinates b_k, and expanding vector fields on B in terms of these coordinates as $X = X_k\partial_{b_k}, Y = Y_k\partial_{b_k}$ we define a metric

$$(X,Y) = \underset{R=0}{Res}\left[\frac{X(R)Y(R)dx}{RQ\partial_x R}\right] = \sum_k \frac{X_k Y_k}{Q(b_k)}$$

For generic values of the parameters, this is a flat nondegenerate metric whose Levi-Civita connection we denote by ∇^{LC}. We define an associative commutative multiplication $*$ on the holomorphic tangent space $\mathcal{T}_b B$ implicitly via

$$(X * Y, Z) = \underset{R=0}{Res}\left[\frac{X(R)Y(R)Z(R)dx}{RQ\partial_x R}\right]$$

This expression is the derivative of the period map $dp(X,Y,Z)$ for the family of spectral curves (0.1) and therefore symmetric in X, Y, Z. The potentiality condition

$$\left(\nabla^{LC} dp\right)(W, X, Y, Z) = \quad \text{symmetric}$$

is satisfied and therefore B is an almost Frobenius manifold as in definition 3.2.

1. Integrability, special geometry and VHS

From the study of $N = 2$ supersymmetric Yang-Mills theory it has become clear that an alternative way to formulate algebraic integrability locally is through affine special geometry [**DW96, Fre99**]. In turn, affine special geometry is equivalent to certain variations of Hodge structures of weight one [**Her03**]. We briefly review these correspondences.

1.1. Algebraic integrability. Even though we describe integrable systems of classical mechanics, we work over the complex numbers.

DEFINITION 1.1. An integrable system is a holomorphic proper Lagrangian map $\pi : A \to B$ from a $2n$-dimensional complex manifold A carrying a symplectic form $\eta \in \Omega^{2,0}(A, \mathbf{C})$ to an n-dimensional complex manifold B. The (connected components of the) fibers $A_b = \pi^{-1}(b)$ are tori by the Arnol'd-Liouville theorem. For an *algebraically* integrable system the fibers are moreover required to be abelian varieties, i.e. A_b carries a polarization $\rho_b \in \Omega^{1,1}(A_b, \mathbf{Z})$.

Donagi and Markman [**DM96a**] found the following result.

THEOREM 1.2. *A proper map $\pi : A \to B$ with $\dim(A) = 2\dim(B)$ whose fibers are abelian varieties is an algebraically integrable system if and only if the derivative of the period map*

$$dp : \mathcal{T}B \to Sym^2(\mathcal{T}^*B \otimes \mathcal{O}_B) \tag{1.1}$$

comes from a cubic

$$dp \in H^0\left(B, Sym^3\mathcal{T}^*B \otimes \mathcal{O}_B\right) \tag{1.2}$$

Here we give only a local characterization of dp and postpone a more detailed discussion to section 2.1.1. There exist holomorphic action variables z_i on B and the relative period matrix $\Pi_{jk}(z)$ in terms of which the condition is that

$$\frac{\partial \Pi_{jk}}{\partial z_i} = \frac{\partial \Pi_{ik}}{\partial z_j} \tag{1.3}$$

Locally there exists a holomorphic function $\mathcal{F}(z)$ on B such that

$$dp(\partial_{z_i}, \partial_{z_j}, \partial_{z_k}) = \frac{\partial^3 \mathcal{F}(z)}{\partial z_i \partial z_j \partial z_k} \tag{1.4}$$

The prepotential $\mathcal{F}(z)$ depends on choices and is not uniquely defined, its third order derivatives however form the global $(3,0)$-tensor dp.

1.2. Special geometry. We follow Freed [**Fre99**].

DEFINITION 1.3. An affine special Kähler manifold (B, ω, J, ∇) is a Kähler manifold (B, ω, J) with Kähler form $\omega \in H^{1,1}(B, \mathbf{C})$ together with a flat torsion-free connection ∇ on the tangent bundle such that

- $\nabla \omega = 0$
- $d_\nabla J = 0$

Here J is the complex structure on B and d_∇ is the de Rham differential twisted by ∇.

Note that $\nabla J = 0$ would imply that ∇ is the Levi-Civita connection for the Kähler metric. In that sense Kähler manifolds endowed with their Levi-Civita connection are trivial examples of special Kähler manifolds. There is a natural $(3,0)$-tensor on B, given as follows.

DEFINITION 1.4. Let $\pi^h \in \Omega^{1,0}(B, TB)$ be projection onto the holomorphic tangent bundle $\mathcal{T}B$. Define $\Xi \in H^0(B, \mathcal{T}^*B^{\otimes 3} \otimes \mathcal{O}_B)$ by

$$\Xi(X, Y, Z) = -\omega(\pi^h(X), \left(\nabla \pi^h\right)(Y, Z)) \tag{1.5}$$

LEMMA 1.5.

$$\Xi \in H^0(B, Sym^3 \mathcal{T}^*B \otimes \mathcal{O}_B) \tag{1.6}$$

PROOF. To show symmetry in Y and Z one uses the fact that ∇ is torsion-free and ω is of type $(1,1)$

$$\omega(X, \nabla_Y Z - \pi^h(\nabla_Y Z)) = \omega(X, \nabla_Y Z) = \omega(X, [Y, Z]) + \omega(X, \nabla_Z Y) = \omega(X, \nabla_Z Y)$$

Symmetry in X and Z uses $\nabla \omega = 0$ and the fact that ω is antisymmetric and of type $(1,1)$

$$\omega(X, \nabla_Y Z) = Y\omega(X, Z) - \omega(\nabla_Y X, Z) = \omega(Z, \nabla_Y X)$$

□

Again there exist special holomorphic coordinates z_i and a function $\mathcal{F}$ on B such that Ξ can be integrated

$$\Xi(\partial_{z_i}, \partial_{z_j}, \partial_{z_k}) = \frac{\partial^3 \mathcal{F}(z)}{\partial z_i \partial z_j \partial z_k} \tag{1.7}$$

The main topic of this section is the equivalence between affine special geometry, integrable systems and variations of Hodge structure of weight 1. There also exists a projective version of special geometry, which is equivalent to a certain variation of Hodge structure of weight 3 [**BG83, Fre99**] and at the same time is the natural geometry on the base of pure Hitchin systems (without marked points). Although pure Hitchin systems are not the main theme of this paper, we give here this notion of projective special geometry.

Recall that a triple (M, ω, L) is called a Hodge manifold if (M, ω) is a Kähler manifold with its Kähler form and $\pi : L \to M$ is a holomorphic hermitean line bundle with curvature $-2\pi i\omega$.

DEFINITION 1.6. An n-dimensional Hodge manifold (M, ω, L) carries projective special geometry if there exists a triple (H, ∇, S) where

- $H \to M$ is a rank $n+1$ holomorphic vector bundle with a given holomorphic inclusion $L \hookrightarrow H$.
- ∇ is a flat holomorphic connection on H such that $\nabla(L) \subset H$ and the section
$$\begin{aligned} M &\to \mathbf{P}\left[(H_{\mathbf{R}})_{\mathbf{C}}\right] \\ m &\to L_m \end{aligned}$$
is an immersion with respect to ∇.
- S is a ∇-flat nondegenerate skew form on $H_{\mathbf{R}}$ of type $(1,1)$. Furthermore, we assume that $S|_{L \times \overline{L}}$ is $\frac{i}{2\pi}$ times the hermitean metric on L.

1.3. Variations of Hodge structure. We consider pure polarized variations of Hodge structure of weight w on B.

DEFINITION 1.7. A pure polarized variation of Hodge structure $(H, H^{\mathbf{R}}, F^\bullet, \nabla, S)$ of weight w on B is a holomorphic vector bundle $H \to B$ together with the following structure:

i) A flat holomorphic connection $\nabla : H \to \Omega^1_B \otimes H$.
ii) A flat real subbundle $H^{\mathbf{R}}$ with decomposition of fibers
$$H_b = H_b^{\mathbf{R}} \oplus iH_b^{\mathbf{R}}$$
iii) A decreasing filtration $F^\bullet$ of H by holomorphic subbundles
$$0 = F^{w+1} \subset \cdots \subset F^1 \subset F^0 = H$$
such that
$$H = F^p \oplus \overline{F^{w+1-p}}$$
and (Griffiths transversality)
$$\nabla : F^p \to \Omega^1_B \otimes F^{p-1}$$
One can deduce from this the usual Hodge decomposition $H = \bigoplus_p H^{p,w-p} = \bigoplus_p F^p \cap \overline{F^{w-p}}$.
iv) A ∇-flat $(-1)^w$-symmetric $\mathcal{O}_B$-bilinear pairing S (the polarization) on H which takes real values on $H^{\mathbf{R}}$ and satisfies
$$\begin{aligned} S(F^p, F^q) &= 0 \qquad \text{for } p+q > w \\ i^{2p-w} S(v, \overline{v}) &> 0 \qquad \text{for } v \in H^{p,w-p} - \{0\} \end{aligned}$$
and the sesquilinear form $\mathcal{I} : H \times H \to \mathbf{C}$ given by $\mathcal{I}(a,b) = \frac{1}{2\pi} S(a, \overline{b})$ is Hermitean, positive definite and real on $H^{\mathbf{R}}$.

1.4. Equivalences.

THEOREM 1.8 ([**Fre99**]). *There is a local 1-1 correspondence between:*

1. *Algebraically integrable systems* $\pi : A \to B$
2. *Affine special Kähler manifolds* (B, ω, ∇) *together with a family of* ∇*-flat Lagrangian lattices* $\Lambda \subset T^*B$.

Under this correspondence dp *and* Ξ *are identified.*

We briefly sketch how to go from *(1)* to *(2)*. The idea is that the symplectic form η gives an isomorphism between T_b^*B and the invariant vector fields on the abelian variety A_b. The integer homology of A_b is flat with respect to the Gauss-Manin connection (see also section 2.1.1) and defines a Lagrangian lattice $\Lambda_b \subset T_b^*B$ with respect to the canonical symplectic form. Locally this extends to a symplectomorphism $A \to T^*B/\Lambda$ which allows to replace the extrinsic formulation in terms of abelian varieties by an intrinsic structure on B and its (co)tangent bundle[1]. The lattice bundle gives a flat connection ∇ on TB which coincides with the special Kähler connection; the Kähler form ω is given by the holomorphic symplectic form and the polarization via

$$\omega = \int_\pi \eta \wedge \overline{\eta} \wedge \frac{\rho^{n-1}}{(n-1)!} \tag{1.8}$$

where the integration is over the fibers of π and ρ is the family of polarizations on the fibers.

THEOREM 1.9 ([**Her03**]). *There is a 1-1 correspondence between:*

1. *Affine special Kähler manifolds* (B, J, ω, ∇)
2. *Pure polarized variations of Hodge structure* $(T^{\mathbf{C}}B, T^{\mathbf{R}}B, F^\bullet, \nabla, S)$ *of weight 1 on* B.

To go from *(1)* to *(2)* one notes that viewing B as a complex manifold naturally endows its tangent bundle with a holomorphic subbundle $\mathcal{T}B$ and a filtration

$$0 = F^2 \subset F^1 = \mathcal{T}B \subset F^0 = TB \tag{1.9}$$

The requirement that F^1 is also holomorphic with respect to the holomorphic structure defined by ∇ precisely amounts to $d_\nabla J = 0$. The polarization is given by $S = 2\pi\omega$. Griffiths transversality is trivial for weight 1. The trivial case $\nabla J = 0$ corresponds to the classical Hodge decomposition of $H^1(B, \mathbf{C})$.

2. Integrable systems from families of curves

We restrict our attention to algebraically integrable systems which come from a family of smooth curves $\Sigma \to B$ via a fiberwise Abel-Jacobi map. We call such families of curves *special* since they give rise to special Kähler geometry. This setup excludes many integrable systems, for example those associated Prym varieties rather than Jacobians. We believe however that our methods can be adapted to those cases as well.

2.1. Families of curves and the period map.

[1] Global obstructions for this identification were studied for example in [**Dui80**]

2.1.1. *The derivative of the period map.* We review the definitions of the Gauss-Manin connection, Kodaira-Spencer map and period map for general families of curves, see e.g. [**Gri68**]. Consider a holomorphic family of smooth projective curves of genus g over a contractible g-dimensional base B

$$f : \Sigma \to B$$

We use $|\Sigma|$ to denote the total space of the family, Σ for a fixed fiber and Σ_b to denote the fiber at b. Since the base is contractible, the vector bundle $H = R^1 f_* \mathbf{C} \to B$ with fiber $H^1(\Sigma_b, \mathbf{C})$ over b is topologically trivial. The integral cohomology provides a trivialization and hence a flat connection ∇^{GM} on H called the *Gauss-Manin connection.* Using

$$R^1 f_* \mathbf{C} \cong R^1 f_* \mathbf{Z} \otimes \mathcal{O}_B$$

the Gauss-Manin connection can be expressed as

$$\nabla^{GM} = 1 \otimes d_B : R^1 f_* \mathbf{Z} \otimes \mathcal{O}_B \to R^1 f_* \mathbf{Z} \otimes \Omega^1_B \tag{2.1}$$

One defines the *Kodaira-Spencer map*

$$\xi : H^0(\Sigma_b, f^* \mathcal{T} B|_{\Sigma_b}) \cong \mathcal{T}_b B \to H^1(\Sigma_b, \mathcal{T}_{\Sigma_b})$$

as follows: there is a short exact sequence of sheaves over $|\Sigma|$

$$0 \to \mathcal{T}_\Sigma \to \mathcal{T}_{|\Sigma|} \to f^* \mathcal{T} B \to 0$$

Here $\mathcal{T}_\Sigma$ is the holomorphic tangent sheaf of the fiber. The restriction of the connecting homomorphism of the long exact sequence to Σ_b is the Kodaira-Spencer map.

Although the fibration is topologically trivial, the Hodge decomposition of the cohomology does vary with the base. In fact we have a geometric realization of the abstract VHS of weight 1 as in definition 1.7, taking as filtration $F^\bullet$ the Hodge filtration, as S the intersection form and as ∇ the Gauss-Manin connection. There is a natural decomposition

$$\nabla^{GM} = \nabla^H + \overline{\nabla} \tag{2.2}$$

where ∇^H is an endomorphism of $H^{1,0}(\Sigma)$ and $\overline{\nabla} : H^{1,0}(\Sigma) \to H^{0,1}(\Sigma)$ denotes the graded part, i.e. the part which actually shifts in the Hodge filtration.

The *period map* $p : B \to \mathrm{Grass}(g, H^1(\Sigma, \mathbb{C}))$ associates to a point $b \in B$ the space $H^{1,0}(\Sigma_b, \mathbb{C})$, considered as a subspace of a generic fiber $H^1(\Sigma, \mathbb{C})$ using parallel transport with the Gauss-Manin connection. The period map, Gauss-Manin connection and Kodaira-Spencer map are related as follows.

THEOREM 2.1. [**Gri68**] *Using Serre duality, the derivative of the period map at $b \in B$ can be viewed as*

$$dp : \mathcal{T}_b B \otimes H^0(\Sigma_b, K_{\Sigma_b}) \to H^0(\Sigma_b, K_{\Sigma_b})^*$$

given by the Kodaira-Spencer map composed with interior product

$$dp(X \otimes \alpha) = \xi(X) \lrcorner \alpha \quad \in H^1(\Sigma_b, \mathcal{O}_{\Sigma_b}) \cong H^0(\Sigma_b, K_{\Sigma_b})^*$$

Moreover

$$dp(X \otimes \alpha)(\beta) = S(\beta, \nabla^{GM}_X \alpha) = \int_f \beta \wedge \nabla^{GM}_X \alpha \tag{2.3}$$

where S denotes the intersection product on cohomology and we integrate over the fiber $f^{-1}(b) = \Sigma_b$.

Note that $S(\beta,.)$ projects $\nabla_X\alpha$ onto its antiholomorphic part $\overline{\nabla}_X\alpha = \xi(X)\lrcorner\,\alpha$.

2.1.2. *Algebraic de Rham cohomology.* If the family of curves is given by an algebraic equation, a more algebraic description of the derivative of the period map is appropriate. This description is available in terms of algebraic de Rham theory [**Gro66, Gri69**]. We review this theory for arbitrary smooth projective hypersurfaces with an eye towards algebraic curves. Given an n-dimensional degree d smooth projective hypersurface

$$M = \{P = 0\} \subset \mathbf{P}^{n+1}$$

we consider the following algebraic de Rham complex of rational differentials on $\mathbf{P}^{n+1}$ with poles along M. Let $A_k^{n+1}(M)$ denote the space of meromorphic $(n+1)$-forms with poles of order at most k along M

$$\frac{\phi\Omega}{P^k}$$

where

$$\Omega = \sum_i (-1)^i x_i dx_1 \wedge \cdots \wedge \widehat{dx_i} \wedge \cdots \wedge dx_{n+2}$$

and ϕ is a homogeneous polynomial of degree $kd-n-2$. The de Rham differential can be used to define the algebraic de Rham cohomology groups

$$\mathcal{H}_k^{n+1}(M) = A_k^{n+1}(M)/dA_{k-1}^n(M)$$

whose elements are equivalence classes of meromorphic $(n+1)$-forms with poles of order up to k. One obtains a filtration according to pole order

$$\mathcal{H}_1^{n+1}(M) \subset \mathcal{H}_2^{n+1}(M) \subset \cdots \subset \mathcal{H}_k^{n+1}(M) \subset \cdots$$

which stabilizes at $k = n+1$.

Denoting by $\mathcal{S} = \mathbf{C}^h[x_1, \ldots, x_{n+2}]$ the ring of homogeneous polynomials and by $\mathcal{S}^{(r)}$ its part in degree r, there are the obvious maps

$$\phi \in \mathcal{S}^{(kd-n-2)} \to A_k^{n+1}(\mathbf{P}^{n+1}\backslash M) \ni \frac{\phi\Omega}{P^k} \tag{2.4}$$

Let us denote by $\mathcal{J}(P)$ the Jacobi ideal spanned by the first order derivatives of P and by $\mathcal{R}_P = \mathcal{S}/\mathcal{J}(P)$ the Jacobi ring.

THEOREM 2.2. [**Gri69, CG80**] *The map (2.4) gives an isomorphism*

$$\mathcal{R}_P^{(kd-n-2)} \cong \mathcal{H}_k^{n+1}(M)$$

Moreover, there is a Poincaré residue map to the primitive[2] *cohomology of* M

$$PRes : \mathcal{H}_k^{n+1}(M) \to H_0^n(M, \mathbf{C})$$

which is an isomorphism of cohomologies for $k = n+1$ *and sends the pole filtration to the Hodge filtration. The subring* $R_P^{((n+2)(d-2))}$ *is one-dimensional and the pairing*

$$\mathcal{R}_P^{(kd-n-2)} \times \mathcal{R}_P^{((n+2-k)d-n-2)} \to \mathcal{R}_P^{((n+2)(d-2))} \tag{2.5}$$

given by multiplication in $\mathcal{R}_P$ *is identified with the intersection pairing.*

[2]We will assume from now on that n is odd, so that the primitive cohomology is the whole cohomology.

Another explicit formula for the intersection pairing is given via the Grothendieck residue pairing:

$$\mathcal{R}_P^{(kd-n-2)} \times \mathcal{R}_P^{((n+2-k)d-n-2)} \ni (\phi, \psi) \to \int_\Gamma \frac{\phi\psi\Omega}{\partial_{x_1} P \cdot \dots \cdot \partial_{x_{n+2}} P}$$

where

$$\Gamma = \bigcup_{k=1}^{n+2} \left\{ |\partial_{x_1} P| = \epsilon_1, \dots, \widehat{|\partial_{x_k} P| = \epsilon_k}, \dots, |\partial_{x_{n+2}} P| = \epsilon_{n+2} \right\}$$

is the boundary of a small tubular neighbourhood of the common zeroes of the $\partial_{x_i} P$. Note that P itself is in the Jacobi ideal and that the assumption of smoothness of M implies that there are no points where all partial derivatives of P vanish. A suitable orientation of Γ is assumed.

For a *family* of projective varieties

$$f : M \to B$$

the algebraic de Rham theory is very suitable as well. The family can locally be given by a polynomial equation

$$P(x_1, \dots, x_{n+2}, b_1, \dots, b_N) = 0$$

Here the b_k are coordinates on B and the coefficients of P, considered as a polynomial in the x_i, are sections of $\mathcal{O}_B$, i.e. holomorphic functions of the b_k. A central fiber of the family is given by fixing the values of these b_k. In terms of algebraic de Rham theory, the Gauss-Manin connection is given by representing a cohomology class by a rational differential form and taking

$$\nabla_X^{GM} \left(\frac{\phi\Omega}{P^k} \right) = X \left(\frac{\phi\Omega}{P^k} \right) = \frac{X(\phi)\Omega}{P^k} - k \frac{\phi X(P)\Omega}{P^{k+1}} \tag{2.6}$$

where $X(\phi)$ and $X(P)$ are obtained by differentiating the sections of $\mathcal{O}_B$ that make up the coefficients of ϕ and P. Note that the Gauss-Manin connection shifts a differential form in the pole filtration maximally by one, and therefore satisfies the Griffiths transversality condition. This gives a VHS of weight n.

2.1.3. *Differentials of the second kind.* In the remainder of this section we restrict our attention to families of curves in $\mathbf{P}^2$, i.e. $n = 1$. Consider then a homogeneous algebraic equation

$$P(x, y, z) = 0$$

where we use $[x : y : z]$ to denote a point in $\mathbf{P}^2$. Any smooth algebraic curve Σ has such a model, i.e. there exists a (possibly singular) curve in $\mathbf{P}^2$ whose regularization is isomorphic to Σ. We will be interested in the singular curves as well, which is why we need a local formula for the intersection pairing rather than the global expression (2.5). This can be done using so-called differentials of the second kind, which we shall now discuss.

In the short exact sequence

$$0 \to \mathbf{C} \to \mathfrak{M}_\Sigma \to \mathfrak{M}_\Sigma^{1,0} \to 0$$

$\mathfrak{M}_\Sigma$ is the sheaf of meromorphic functions on Σ and $\mathfrak{M}_\Sigma^{1,0}$ the sheaf of meromorphic differentials of the *second kind*, i.e. having no residues at their poles. The maps are given by the de Rham differential. From the long exact sequence one obtains (see e.g. [**Gun67**])

$$H^1(\Sigma, \mathbf{C}) \cong \frac{H^0(\Sigma, \mathfrak{M}^{1,0}_\Sigma)}{dH^0(\Sigma, \mathfrak{M}_\Sigma)} \tag{2.7}$$

In a neighbourhood U_p of any point $p \in \Sigma$ a differential form of the second kind can be expressed as the derivative of a meromorphic function. Given two such differentials α, β, consider a covering $\{U_p\}_p$ of Σ such that each set U_p contains no more than one pole of α, β and in each U_p write $\alpha = df$ and $\beta = dg$. The *residue pairing* $S : H^1(\Sigma, \mathbf{C})^2 \to \mathbf{C}$ is given by the sum over residues

$$S(\alpha, \beta) = \sum_p Res\,[fdg] = -\sum_p Res\,[gdf]$$

This is the equivalent of the intersection pairing when the cohomology $H^1(\Sigma)$ is interpreted in terms of differentials of the second kind. If the embedding of Σ in $\mathbf{P}^2$ has an isolated singular point p, its contribution to the intersection pairing can be calculated using the method of Puiseux expansions (see e.g. [**Kir92**] and section 4.1).

2.1.4. *A modified Poincaré residue.* The derivative of the period map uses the Gauss-Manin connection (best described in algebraic de Rham theory) and the intersection pairing (best described in terms of differentials of the second kind). We need a modified Poincaré residue map to relate these two. The usual Poincaré residue maps algebraic de Rham theory to ordinary de Rham theory on Σ: locally around a point $p \in \Sigma$ one can always write

$$\frac{\phi\Omega}{P} = \frac{\phi(-xdy \wedge dz + ydx \wedge dz - zdx \wedge dy)}{P} = \frac{\chi_\phi \wedge dP}{P} + \psi_\phi$$

where χ_ϕ, ψ_ϕ are holomorphic. The Poincaré residue is by definition

$$PRes\left(\frac{\phi\Omega}{P}\right) = \chi_\phi|_\Sigma$$

The differential form χ_ϕ is not unique, but the restriction of its cohomology class to Σ is. More explicitly, since the curve is smooth we know $\forall p \in \Sigma$ that $(\partial_x P(p), \partial_y P(p), \partial_z P(p)) \neq (0,0,0)$. Assume for example that $\partial_y P(p) \neq 0$ and note that if this is the case, then in a small open neighbourhood of p one finds

$$\frac{\phi\Omega}{P} = \frac{\chi_\phi \wedge dP}{P} + d_P \cdot \frac{\phi dx \wedge dz}{\partial_y P}$$

where d_P is the degree of P and

$$\chi_\phi = \frac{\phi(xdz - zdx)}{\partial_y P}$$

If also $\partial_x P(p) \neq 0$, one analogously finds the cohomologous expression

$$\tilde{\chi}_\phi = \frac{\phi(zdy - ydz)}{\partial_x P}$$

The local expressions we found patch together globally to a holomorphic differential on Σ.

The usual Poincaré residue is less explicit when it comes to differentials with second order poles. But if one allows further poles on $\mathbf{P}^2$ along the zeroes of

$\partial_x P, \partial_y P, \partial_z P$ it is possible to bring down the pole order of P and use the Poincaré residue for differentials with first order poles as above. One finds for example

$$\frac{\phi\Omega}{P^2} = d\left[\frac{-\phi(zdx - xdz)}{P\partial_y P}\right] - \frac{\tilde{\phi}\Omega}{P} = d[\ldots] - \partial_y\left(\frac{\phi}{\partial_y P}\right)\frac{\Omega}{P}$$

This allows the residue to be defined as above.

DEFINITION 2.3. We define a Poincaré residue map

$$PRes : \mathcal{H}_2^2(\Sigma) \to \frac{H^0(\Sigma, \mathfrak{M}_\Sigma^{1,0})}{dH^0(\Sigma, \mathfrak{M}_\Sigma)}$$

via the local representatives

$$\begin{aligned} PRes\left(\frac{\phi\Omega}{P}\right) &= \chi_\phi|_\Sigma = \left.\frac{\phi(xdz - zdx)}{\partial_y P}\right|_\Sigma \\ PRes\left(\frac{\phi\Omega}{P^2}\right) &= \chi_{\tilde{\phi}}\Big|_\Sigma = -\partial_y\left(\frac{\phi}{\partial_y P}\right)\left.\frac{zdx - xdz}{\partial_y P}\right|_\Sigma \end{aligned} \tag{2.8}$$

The last expression indeed has second or higher order poles in the points where $\partial_y P = 0$ (unless ϕ is divisible by $\partial_y P$) and the modified Poincaré residue lands in the cohomology (2.7). In terms of the residue pairing, the derivative of the period map is given by

$$dp(X \otimes \alpha)(\beta) = \sum_{\substack{\text{poles of} \\ \nabla_X^{GM}\alpha}} Res\,[\beta g_{X,\alpha}] \tag{2.9}$$

where $dg_{X,\alpha} = \nabla_X^{GM}\alpha$ locally. We will apply this formula, in combination with the description of the Gauss-Manin connection using algebraic de Rham theory and the Poincaré residue, to the situation of algebraically integrable systems.

2.2. Special families of curves. Due to theorem 1.2, not every family of curves gives rise to an integrable system.

DEFINITION 2.4. A differential λ on the total space $|\Sigma|$ of a family of curves $\Sigma \to B$ is called a Seiberg-Witten differential if its derivatives with respect to the Gauss-Manin connection are all holomorphic and the map

$$\begin{aligned} \mathcal{T}_b B &\to H^0(\Sigma_b, K_{\Sigma_b}) \\ X &\to \nabla_X^{GM}\lambda|_{\Sigma_b} \end{aligned} \tag{2.10}$$

is an isomorphism.

Existence of such a differential puts restrictions on the family of curves. In particular, the dimension of B must equal the genus of the fibers.

LEMMA 2.5. *A family of Jacobians $Jac(\Sigma) \to B$ corresponding to a family of smooth curves $\Sigma \to B$ gives rise to an algebraically integrable system if and only if it can be equipped with a Seiberg-Witten differential.*

PROOF. First we assume that $Jac(\Sigma) \to B$ is an algebraically integrable system. The holomorphic 2-form $\eta = d\psi$ is exact over an open contractible part of the base. One pulls back ψ using the Abel-Jacobi map to a differential λ. Holomorphicity and nondegeneracy of η and the fact that we are dealing with a Lagrangian fibration then ensure that

$$X \to \left(\tilde{X} \lrcorner\, d\lambda\right)\Big|_{\Sigma_b} = \nabla_X^{GM}\lambda|_{\Sigma_b}$$

gives the required isomorphism, where $\tilde{X}$ is any lift of X.

Conversely, $d\lambda$ (considered as a differential on $|Jac(\Sigma)|$) can be used to define η which satisfies all the necessary properties. □

DEFINITION 2.6. If a family of curves can be equipped with a Seiberg-Witten differential (and therefore gives rise to algebraic integrability and affine special geometry), we call it a *special* family.

Using the isomorphism (2.10), one can pull back ∇^{GM} to a nonholomorphic connection ∇.

LEMMA 2.7. *The pull-back ∇ is the special geometry connection. In other words, ∇^{GM} and ∇ are related by*

$$\nabla_X^{GM}\nabla_Y^{GM}\lambda = \nabla_{\nabla_X Y}^{GM}\lambda \tag{2.11}$$

where λ is the Seiberg-Witten differential.

PROOF. Recall from theorem 1.8 that the (integral) homology of the fibers of $Jac(\Sigma) \to B$ defines a flat lattice which in turn defines the special Kähler connection via η. Let $\Gamma \in H_1(Jac(\Sigma), \mathbf{Z})$ and $X, Y \in H^0(B, \mathcal{T}_B)$. Then

$$X\left(\int_\Gamma Y \lrcorner\, \eta\right) = \int_\Gamma \nabla_X^{GM}(Y \lrcorner\, \eta) =: \int_\Gamma (\nabla_X Y) \lrcorner\, \eta = \int_\Gamma \nabla_{\nabla_X Y}^{GM}\eta$$

On the other hand, by the definition of the Gauss-Manin connection

$$\int_\Gamma \nabla_X^{GM}(Y \lrcorner\, \eta) = \int_\Gamma \nabla_X^{GM}(Y \lrcorner\, d\lambda) = \int_\Gamma \nabla_X^{GM}\nabla_Y^{GM}\lambda$$

□

2.2.1. *The derivative of the period map.* We consider a special family of curves $f : \Sigma \to B$ carrying a Seiberg-Witten differential λ. Due to the isomorphism (2.10) one can view the derivative of the period map (2.3) as a cubic on $\mathcal{T}B$

$$\begin{aligned} dp(X,Y,Z) = S(\nabla_Z^{GM}\lambda, \nabla_X^{GM}\nabla_Y^{GM}\lambda) &= \int_f \nabla_Z^{GM}\lambda \wedge \nabla_X^{GM}\nabla_Y^{GM}\lambda \\ &= \sum_{\substack{\text{poles of} \\ \nabla_X^{GM}\nabla_Y^{GM}\lambda}} Res\left[\nabla_Z^{GM}\lambda g_{XY}\right] \end{aligned} \tag{2.12}$$

where locally $dg_{XY} = \nabla_X^{GM}\nabla_Y^{GM}\lambda$. The notation $\int_f$ means that in the point $b \in B$, the derivative of the period map is given by integrating over the fiber $f^{-1}(b)$. The above expression is indeed symmetric in X, Y, Z as it should be according to theorem 1.2: using part *iv)* of definition 1.7 one finds that for any $X \in \mathcal{T}_b B$ and $\alpha, \beta \in H^{1,0}(\Sigma_b, \mathbf{C})$

$$\begin{aligned} 0 &= \left(\nabla^{GM} S\right)(X, \beta, \alpha) = XS(\beta, \alpha) - S(\nabla_X^{GM}\beta, \alpha) - S(\beta, \nabla_X^{GM}\alpha) \\ &= 0 + S(\alpha, \nabla_X^{GM}\beta) - S(\beta, \nabla_X^{GM}\alpha) \end{aligned}$$

which proves symmetry in Y, Z. The fact that ∇^{GM} is torsion-free together with $S(H^{1,0}, H^{1,0}) = 0$ gives symmetry in X, Y:

$$S(\nabla_Z^{GM}\lambda, \nabla_X^{GM}\nabla_Y^{GM}\lambda) - S(\nabla_Z^{GM}\lambda, \nabla_Y^{GM}\nabla_X^{GM}\lambda) = S(\nabla_Z^{GM}\lambda, \nabla_{[X,Y]}^{GM}\lambda) = 0$$

2.3. Calabi-Yau condition. In this section we consider a situation which is very interesting, but which we want to avoid for the purposes of this paper.

DEFINITION 2.8. If a special family of curves equipped with a Seiberg-Witten differential λ has the property that λ is a holomorphic differential, we say that it satisfies the *Calabi-Yau condition*.

PROPOSITION 2.9. *If a special family of curves satisfies the Calabi-Yau condition, the corresponding VHS of weight* 1 *(with filtration $F^1 \subset F^0 = TB$) can be refined to a VHS of weight* 3 *in a natural way. The only difference between the two VHS is the definition of the filtration, which becomes*

$$\begin{aligned}\tilde{F}^3 &:= \lambda \cdot \mathbf{C}\\ \tilde{F}^2 &:= \nabla\tilde{F}^3 = F^1\\ \tilde{F}^1 &:= \left(\tilde{F}^3\right)^\circ\\ \tilde{F}^0 &:= H = F^0\end{aligned}$$

where $\tilde{F}^1$ is the orthocomplement to $\tilde{F}^3$ with respect to S. Moreover, there is a natural $\mathbf{C}^*$-action on B and the VHS of weight 3 gives a projective special geometry on the quotient $\mathbf{P}(B)$ as in definition 1.6. From definition 1.7 one sees that a projective special geometry indeed gives a particular type of VHS of weight 3, *namely of the type introduced by Bryant and Griffiths* [**BG83**].

PROOF. We need to check Griffiths transversality $\nabla\tilde{F}^2 \subset \tilde{F}^3$. Because $\nabla S = 0$ one finds that for any $\alpha \in \tilde{F}^2$ and any $X \in \mathcal{T}B$

$$S(\lambda, \nabla_X\alpha) = XS(\lambda, \alpha) + S(\nabla_X\lambda, \alpha) = 0$$

The terms vanish separately because $\lambda, \nabla_X\lambda, \alpha \in F^1$ and $S(F^1, F^1) = 0$. Due to the isomorphism $\mathcal{T}_B \cong F^1$ given by

$$X \to \nabla_X\lambda$$

there is a vector field $\mathcal{E}$ such that $\nabla_{\mathcal{E}}\lambda = \lambda$. This vector field generates a $\mathbf{C}^*$-action on B whose quotient is denoted by $\mathbf{P}(B)$. The VHS of weight 3 on it satisfies all the axioms of a projective special Kähler structure. □

REMARK 2.10. If the Calabi-Yau condition is satisfied for a special family $f : \Sigma \to B$, one can write the derivative of the period map in a very suggestive way: using $\tilde{F}^1 = \left(\tilde{F}^3\right)^\circ$ and the flatness of S one finds

$$\begin{aligned}dp(X, Y, Z) = S(\nabla_Z\lambda, \nabla_X\nabla_Y\lambda) &= ZS(\lambda, \nabla_X\nabla_Y\lambda) - S(\lambda, \nabla_Z\nabla_X\nabla_Y\lambda)\\ &= 0 + S(\nabla_X\nabla_Y\nabla_Z\lambda, \lambda) = \int_f (\nabla_X\nabla_Y\nabla_Z\lambda) \wedge \lambda\end{aligned}$$

which is reminiscent of the Yukawa coupling on the complex structure moduli space of a Calabi-Yau threefold, with λ playing the role of the holomorphic 3-form.

The resulting VHS of weight 3 satisfies some of the conditions of Hertling and Manin [**HM04**]. Their general unfolding result can therefore (partially) be applied in this case to give many structures of Frobenius manifold on $B \times B$. This situation occurs for pure Hitchin systems (without marked points), which we will pursue elsewhere [**HHP**].

In this paper we will not be concerned with this application and the Frobenius manifold resulting from it, because it puts no further conditions on the projective special Kähler manifold. In particular, one might expect that in terms of local coordinates the associativity of the product imposes WDVV-type conditions on the prepotential $\mathcal{F}$ but these turn out to be trivial.

By contrast, we would like to construct a Frobenius manifold type structure on the base B itself (not $B \times B$) which does give conditions on the prepotential. This is inspired by discoveries in Seiberg-Witten theory [**MMM96, MH03, BMMM07**] where an interesting structure was found on the base of a number of specific integrable systems such as the periodic Toda chain, the relativistic Toda chain and some spin chains. We take the point of view that these systems can be thought of as Hitchin systems with marked points.

3. Associativity

We review the definition of (almost) Frobenius manifolds. We then define a metric and multiplication on the base of any algebraically integrable system.

3.1. (Almost) Frobenius manifolds. Roughly speaking, a Frobenius manifold carries a unital associative commutative multiplication on each tangent space which is compatible with a given flat metric. More precisely, the unit vector field must be covariantly constant with respect to the Levi-Civita connection and there must be a so-called Euler vector field. Such manifolds were considered by Dubrovin (see e.g. the survey paper [**Dub96**]) as coordinate-free characterizations of families of 2-dimensional topological conformal field theories. Typically the theory of Frobenius manifolds is local, i.e. one usually considers germs of such manifolds. Our goal is to endow the base B of any algebraically integrable system with this type of structure.

Endowing a manifold with a metric and a commutative associative multiplication on each tangent space, one can impose suitable interrelations. A Frobenius algebra captures such a structure on a fixed tangent space.

Definition 3.1. A Frobenius algebra $(A, *, e, \langle \cdot, \cdot \rangle)$ is an associative commutative unital algebra A with multiplication $*$ and identity element e, together with a symmetric nondegenerate bilinear form $\langle \cdot, \cdot \rangle$ which is compatible with the multiplication:

$$\langle X * Y, Z \rangle = \langle X, Y * Z \rangle \tag{3.1}$$

In other words, the 3-form $c(X, Y, Z) = \langle X * Y, Z \rangle$ is symmetric.

Definition 3.2. A Frobenius manifold $(B, *, e, E, \langle \cdot, \cdot \rangle)$ is an n-dimensional manifold with commutative associative multiplication on each tangent space, varying smoothly with the point, and a symmetric nondegenerate complex bilinear form $\langle \cdot, \cdot \rangle$ (henceforth called the metric) which makes each $\mathcal{T}_b B$ into a Frobenius algebra. Denoting the Levi-Civita connection for the metric by ∇, one has moreover the following requirements

i) The metric $\langle\cdot,\cdot\rangle$ is flat
ii) The identity vector field e is flat: $\nabla e = 0$
iii) $(\nabla c)(W, X, Y, Z)$ is symmetric in W, X, Y, Z
iv) There is a vector field E and a number d such that
$\mathcal{L}_E(*) = *$
$\mathcal{L}_E\langle\cdot,\cdot\rangle = d\langle\cdot,\cdot\rangle$

We will call a manifold without Euler vector field and whose identity vector field is not constant, but which otherwise satisfies all the requirements an *almost Frobenius manifold.*

REMARK 3.3. In the literature on weaker versions of Frobenius manifolds, an Euler vector field is sometimes included in the standard definitions and sometimes not. Our definition of almost Frobenius manifold differs for example from the one given by Dubrovin, who does require an Euler vector field. It is closer to the definitions given by Hertling and Manin [**HM99**].

Regardless of flatness of e, property *iii)* together with the Poincaré lemma implies that there exists locally a single holomorphic function $\mathcal{F}$ such that

$$c(X, Y, Z) = (XYZ)\mathcal{F} \tag{3.2}$$

for any triple X, Y, Z of flat vector fields. Such a function $\mathcal{F}$ is called a prepotential, characterizes the Frobenius manifold but is not unique: for example, any polynomial of degree 2 can be added to it. Property *ii)* implies that one can choose the flat coordinates $\{t_k\}$ in such a way that $e = \frac{\partial}{\partial t_1}$ and therefore

$$c(\partial_{t_i}\partial_{t_j}\partial_{t_k}) = \frac{\partial^3\mathcal{F}(t_1,\ldots,t_n)}{\partial t_i \partial t_j \partial t_k} \tag{3.3}$$

$$\langle\partial_{t_i}, \partial_{t_j}\rangle = \langle e * \partial_{t_i}, \partial_{t_j}\rangle = c(e, \partial_{t_i}, \partial_{t_j}) = \partial_{t_1}\partial_{t_i}\partial_{t_j}\mathcal{F}. \tag{3.4}$$

Naturally, not just any function can occur as a prepotential

LEMMA 3.4 ([**Dub96**]). *A holomorphic function $\mathcal{F}(t_1, ..., t_n)$ locally characterizes a Frobenius manifold if and only if it is quasihomogeneous and satisfies the following properties:*

- *The matrix $[\mathcal{F}_1]$ is constant and nondegenerate, where $[\mathcal{F}_i]$ denotes the Hesse matrix of the first order derivative $\partial_{t_i}\mathcal{F}$:*

$$[\mathcal{F}_i]_{jk} = \frac{\partial^3\mathcal{F}(t)}{\partial_{t_i}\partial_{t_j}\partial_{t_k}}$$

- *$\mathcal{F}$ satisfies the Witten-Dijkgraaf-Verlinde-Verlinde differential equations (also called the associativity equations)*

$$[\mathcal{F}_i][\mathcal{F}_1]^{-1}[\mathcal{F}_j] = [\mathcal{F}_j][\mathcal{F}_1]^{-1}[\mathcal{F}_i] \qquad \forall i, j = 1, ..., n \tag{3.5}$$

In terms of the coordinates t_k, the matrix elements of the metric are defined to be $\langle\partial_i, \partial_j\rangle = [\mathcal{F}_1]_{ij}$ and the multiplication is defined implicitly through

$$\langle\partial_{t_i} * \partial_{t_j}, \partial_{t_k}\rangle = \frac{\partial^3\mathcal{F}}{\partial_{t_i}\partial_{t_j}\partial_{t_k}}$$

The quasihomogeneity $\mathcal{F}(c^{d_1}t_1, \ldots, c^{d_n}t_n) = c^d\mathcal{F}(t)$ ensures that the vector field

$$E = \sum_{k=1}^{n} d_k t_k \partial_{t_k}$$

satsfies all the requirements of an Euler vector field. The first property in the lemma ensures flatness and nondegeneracy of the metric and existence of a constant identity vector field $e = \partial_{t_1}$ for the multiplication. The second property in the lemma is equivalent to associativity of the multiplication.

3.2. Metric and multiplication on the base of an integrable system. In this section we assume that the space B is the base of an algebraically integrable system, or equivalently an affine special Kähler manifold, or equivalently a manifold with a natural weight 1 VHS on its tangent bundle TB.

If one considers the 3-form c and the vector field e of a Frobenius manifold as given, one can define the multiplication and metric from them. We will mimic this construction with the derivative of the period map playing the role of c, but in general there is no natural choice for e. In the definition of Frobenius manifold, one can replace the identity e by any other unit e' (rescaling the multiplication) but neither e' nor the resulting metric is guaranteed to be flat. We will start with this weaker structure.

DEFINITION 3.5. For any affine special Kähler manifold B and a section $V \in H^0(B, \mathcal{T}B)$ we define a holomorphic symmetric 2-tensor $(\cdot,\cdot)_V$ given by

$$(X, Y)_V = dp(V, X, Y) \tag{3.6}$$

We assume from now on that there is a choice of V such that the metric $(\cdot,\cdot)_V$ is nondegenerate[3]. It is in general not flat.

DEFINITION 3.6. Given $V \in H^0(B, \mathcal{T}B)$ we define a commutative unital multiplication $*_V$ on each holomorphic tangent space $\mathcal{T}_bB$, varying smoothly with the point $b \in B$, by the following formula

$$(X, Y *_V Z)_V = dp(X, Y, Z) \tag{3.7}$$

where the right hand side is independent of V. Nondegeneracy of $(\cdot,\cdot)_V$ ensures that $*_V$ is well-defined, unital with identity element V and compatible with $(.,.)_V$.

Whether or not the algebra is associative is independent of the choice of V, but flatness of $(\cdot,\cdot)_V$ depends on it. We are interested in associative multiplications, or equivalently algebraically integrable systems whose prepotentials satisfy the WDVV equations. Any function whose holomorphic Hesse matrix $\frac{\partial^2\mathcal{F}}{\partial z_i \partial z_j}$ is in the Siegel upper half space defines an integrable system locally. The WDVV equations are a further highly nontrivial set of conditions, certainly not met by all integrable systems.

4. Hitchin systems with marked points

We review the theory of Hitchin systems with marked points following [**Hit87, Mar94, Bot95, Mar00**]. Given a curve $\mathcal{C}$ of genus $g_{\mathcal{C}} \geq 2$ and a complex reductive group G, a Higgs bundle is a pair (P, Φ) consisting of a principal G-bundle together with a Higgs field $\Phi \in H^0(\mathcal{C}, ad(P) \otimes K_{\mathcal{C}})$. The moduli space M of stable principal G-bundles is a smooth quasiprojective variety whose cotangent bundle T^*M by deformation theory is an open dense subspace of the moduli space of Higgs bundles. The ring of G-invariant polynomials on the Lie algebra $\mathfrak{g}$ is finitely generated by a

[3]This fails to hold for example for pure Hitchin systems because of the Calabi-Yau condition, which ensures that $dp(X, Y, \mathcal{E}) = 0$ where $\mathcal{E}$ is the generating vector field of the $\mathbf{C}^*$-action on B.

set $p_1, \ldots, p_n$ where n is the rank of $\mathfrak{g}$ and p_k is homogeneous of a certain degree d_k. Applying these invariant polynomials fibrewise one obtains the Hitchin map

$$\begin{aligned} h : T^*M \to B &:= \bigoplus_{k=1}^{n} H^0(\mathcal{C}, K_{\mathcal{C}}^{d_k}) \\ (P, \Phi) &\to \bigoplus_{k} p_k(\Phi) \end{aligned} \tag{4.1}$$

Hitchin showed [**Hit87**] that this map is proper and Lagrangian with respect to the canonical symplectic form and therefore defines an algebraically integrable system.

Markman and Bottacin generalized this setup to the moduli space $M(D)$ of stable principal bundles with level structure for which $\Phi \in H^0(\mathcal{C}, ad(P) \otimes K_{\mathcal{C}}(D))$ and D is an effective divisor. Let $\tilde{G}_D$ denote the group of maps from D to G. The group $G_D := \tilde{G}_D/Z(G)$ acts freely and symplectically on the symplectic space $T^*M(D)$ and the symplectic quotient, which carries a Poisson structure, is a dense open subset of the moduli space of stable meromorphic Higgs pairs (P, Φ). Given a coadjoint orbit $\mathcal{O} \subset \mathfrak{g}_D^*$, the Hitchin map restricted to a symplectic leaf $\mu^{-1}(\mathcal{O})//G_D$ is again an algebraically integrable system. Denoting the moment map of the G_D-action by μ, one has the following commutative diagram (see e.g. [**DM96b**])

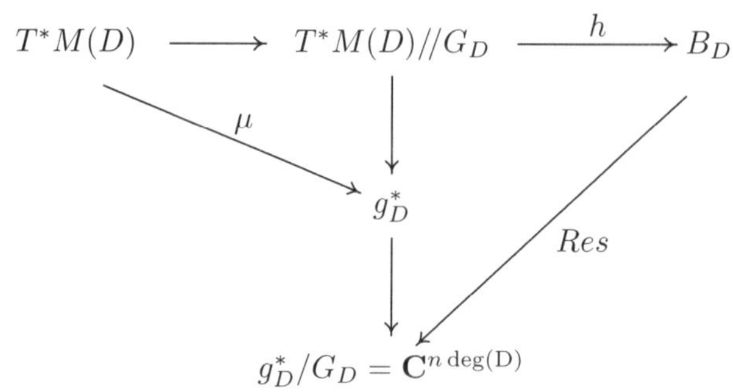

where

$$Res : B_D := \bigoplus_{k=1}^{n} H^0\left(\mathcal{C}, K_{\mathcal{C}}(D)^{\otimes d_k}\right) \to \mathbf{C}^{n \deg(D)} \tag{4.2}$$

is given in terms of a local coordinate ζ on $\mathcal{C}$ as follows: in degree d_k, one defines for a meromorphic function $f(\zeta)$

$$\underset{\zeta=0}{Res}^{d_k}\left(f(\zeta)\frac{(d\zeta)^{\otimes d_k}}{\zeta^{d_k}}\right) := f(0)$$

In particular, the quadratic residue is given by

$$\underset{\zeta=0}{Res}^{2}\left(f(\zeta)\frac{(d\zeta)^{2}}{\zeta^{2}}\right) := f(0)$$

According to [**Mar94**], Main Theorem, in the case of GL_n and SL_n the fibers of the Hitchin map can be identified with the Jacobians of spectral curves. A complication arises if D contains points p with multiplicity $m_p > 1$ since the dimension of the image of Res does not have the expected dimension $n \deg(D)$. In this situation one should take the residues

$$\underset{\zeta=0}{Res}^{d_k}\left(\zeta^{(m_p-1)d_k-j}p_k(\Phi)\right) \qquad j = 0, \ldots, m_p - 1$$

For pure Hitchin systems one is forced to take $g_{\mathcal{C}} \geq 2$ for stability reasons, but with marked points this condition can be dropped if $\deg(D)$ is large enough. In particular, if $\deg(D) \geq 3$ it is possible to consider $\mathcal{C} = \mathbf{P}^1$ and make contact with the formalism of Lax pairs with a spectral parameter.

4.1. The Neumann system. We now consider the Neumann system as an example of a Hitchin system with marked points, see also [**Bea90, DM96a, Bot95, Mum84**]. As input data for the system we take

$$(\mathcal{C}, G, D) = (\mathbf{P}^1, SL_2, \sum_{\alpha=1}^{g+1} q_\alpha + 2 \cdot \infty)$$

We denote a point $p \in \mathbf{P}^1$ by $p = [x : z]$ and the Higgs field by

$$\Phi(x,z) = \begin{bmatrix} V(x,z) & U(x,z) \\ W(x,z) & -V(x,z) \end{bmatrix} \frac{zdx - xdz}{z^2 \prod_{\alpha=1}^{g+1}(x - q_\alpha z)}$$

with U, V, W homogeneous polynomials of degree $g+1$. We have identified $q_\alpha = [q_\alpha : 1]$ and $\infty = [1 : 0]$. Consider the Hitchin map

$$\Phi \to \operatorname{tr}\left[\frac{1}{2}\Phi^2\right] = \left(V^2 + UW\right) \frac{(zdx - xdz)^2}{z^4 \prod_{\alpha=1}^{g+1}(x - q_\alpha z)^2}$$

and the values of the residue map

$$\begin{aligned}
\underset{q_\alpha}{Res^2}\left(\operatorname{tr}\left[\frac{1}{2}\Phi^2\right]\right) &= \frac{V(q_\alpha)^2 + U(q_\alpha)W(q_\alpha)}{\prod_{\beta \neq \alpha}(q_\alpha - q_\beta)^2} = 0 \\
\underset{\infty}{Res^2}\left(z^2 \operatorname{tr}\left[\frac{1}{2}\Phi^2\right]\right) &= V(\infty)^2 + U(\infty)W(\infty) = 0 \\
\underset{\infty}{Res^2}\left(z \operatorname{tr}\left[\frac{1}{2}\Phi^2\right]\right) &= \left[\frac{V(1,z)^2 + U(1,z)W(1,z)}{z}\right]_+ (0) = 1
\end{aligned}$$

where $[\cdot]_+$ denotes the polynomial part. The first residue at ∞ requires $V^2 + UW$ to be divisible by z, while the second requires the coefficient of zx^{2g+1} to be equal to 1. The fiber $Res^{-1}(0, \ldots, 0, 1)$ is therefore given by the polynomials $V^2 + UW$ of the form

$$V(x,z)^2 + U(x,z)W(x,z) = z \prod_{k=1}^{g}(x - b_k z) \prod_{\alpha=1}^{g+1}(x - q_\alpha z)$$

where the b_k are g free parameters. Equipped with the appropriate Hamiltonian the corresponding symplectic leaf gives rise to the Neumann system. We will not give this Hamiltonian here since it plays no role whatsoever in our constructions, which only depend on the Lagrangian fibration.

We define the spectral curve by

$$\Sigma = \{\det\left(\Phi - id \otimes \tilde{y}\right) = 0\} \subset |K_{\mathbf{P}^1}(D)|$$

with

$$\tilde{y} = \frac{z^g y(zdx - xdz)}{z^2 \prod_{\alpha=1}^{g+1}(x - q_\alpha z)} \in H^0(\mathbf{P}^1, K_{\mathbf{P}^1}(D)).$$

The equation for the spectral curve contains a factor of z, and its maximal irreducible component is given by the homogeneous polynomial equation

(4.3)

$$\begin{aligned} P(x,y,z) &= y^2 z^{2g-1} - \frac{1}{z}(V^2 + UW) = y^2 z^{2g-1} - \prod_{k=1}^{g}(x - b_k z)\prod_{\alpha=1}^{g+1}(x - q_\alpha z) \\ &= y^2 z^{2g-1} - R(x,z)Q(x,z) = 0 \end{aligned}$$

The dimension of the base of the integrable system is $\dim(B) = \dim B_D - \deg(D) = 2g+3-(g+3) = g$, which indeed equals the genus of the spectral curve. The parameters b_k serve as good local coordinates on B away from the discriminant locus where the curve develops singularities. We are now in the situation sketched in section 2.2 and we can define a Seiberg-Witten differential.

DEFINITION 4.1. The Seiberg-Witten differential for the Neumann system is given by

$$\lambda = \tilde{y} = \frac{R(zdx - xdz)}{z^{g+1}y} = PRes\left(\frac{2z^{g-2}R\Omega}{P}\right)$$

Its covariant derivatives are given by (recall the notations from section 2.1.2)

$$\begin{aligned} \nabla_X^{GM}\lambda &= PRes\left(\frac{2z^{g-2}X(R)\Omega}{P} + \frac{2z^{g-2}QRX(R)\Omega}{P^2}\right) \\ &= PRes\left(\frac{2z^{g-2}X(R)\Omega}{P} + d\left[\frac{yz^{g-2}X(R)(zdx - xdz)}{P}\right] - \frac{z^{g-2}X(R)\Omega}{P}\right) \\ &= PRes\left(\frac{z^{g-2}X(R)\Omega}{P}\right) = \frac{X(R)(zdx - xdz)}{2z^{g+1}y} \end{aligned}$$

and

$$\begin{aligned} \nabla_X^{GM}\nabla_Y^{GM}\lambda &= PRes\left(\frac{z^{g-2}X(Y(R))\Omega}{P} + \frac{z^{g-2}QX(R)Y(R)\Omega}{P^2}\right) \\ &= PRes\left(\frac{z^{g-2}X(Y(R))\Omega}{P} + d\left[\frac{yz^{g-2}X(R)Y(R)(zdx - xdz)}{2RP}\right]\right. \\ &\qquad \left. - \frac{z^{g-2}X(R)Y(R)\Omega}{2RP}\right) \\ &= \left(X(Y(R)) - \frac{X(R)Y(R)}{2R}\right)\frac{(zdx - xdz)}{2z^{g+1}y} \end{aligned}$$

It is well-known that for a hyperelliptic curve given by an equation

$$y^2 z^{2g-1} = F(x,z)$$

the forms

$$\frac{z^{g-k}x^k(zdx - xdz)}{z^{g+1}y} \qquad k = 0,\dots,g-1$$

form a basis of the holomorphic differentials. The derivatives of λ are therefore holomorphic, and by the Lagrange interpolation formula they are independent since the $X(R)$ have pairwise distinct zeroes. So they indeed form a basis of the holomorphic differentials, as required in definition 2.4.

PROPOSITION 4.2. *For the Neumann system, the derivative of the period map is given by*

$$dp(X,Y,Z) = \operatorname*{Res^2}_{\lambda=0}\left[\frac{\nabla_X^{GM}\lambda\otimes\nabla_Y^{GM}\lambda\otimes\nabla_Z^{GM}\lambda}{\lambda}\right] = \operatorname*{Res^2}_{R=0}\left[\frac{X(R)Y(R)Z(R)dx\otimes dx}{4R^2Q}\right]$$

where the first expression is a sum over residues on the spectral curve Σ_b and the second is a sum over residues on the base curve $\mathcal{C} = \mathbf{P}^1$.

PROOF. The spectral curves have one singular point $[x:y:z] = [0:1:0]$. We will first show that the derivatives of λ don't have a pole in this point. To find out what happens at the singularity we normalize the curve using the method of Puiseux expansions, see e.g. [**Kir92**]. In the coordinate patch $y \neq 0$ the curve is given by the affine equation

$$z^{2g-1} - \prod_{k=1}^{g}(x - b_k z)\prod_{\alpha=1}^{g+1}(x - q_\alpha z) = 0$$

and around the origin we can use a local coordinate t in terms of which

$$x = t^{2g-1}f(t)$$
$$z = t^{2g+1}g(t)$$

where f, g are holomorphic functions in a neighbourhood of $t = 0$ with nonzero constant term. Using this coordinate, one finds that

$$\lambda = \frac{\prod_{k=1}^{g}(x-b_k z)(zdx - xdz)}{z^{g+1}} = t^{-2}h(t)dt$$

$$\nabla_X^{GM}\lambda = \frac{X(R)(zdx - xdz)}{2z^{g+1}} = h_X(t)dt$$

$$\nabla_X^{GM}\nabla_Y^{GM}\lambda = \left(X(Y(R)) - \frac{X(R)Y(R)}{2R}\right)\frac{(zdx - xdz)}{2z^{g+1}} = t^2 h_{XY}(t)dt$$

where h, h_X, h_{XY} are holomorphic functions in a neighbourhood of $t = 0$. This shows that λ is not holomorphic since it has a pole in $[0:1:0]$, which is consistent with the description of the holomorphic differentials. Moreover the derivatives of λ don't have a pole in $[0:1:0]$, and therefore one can apply formula (2.12) directly.

All the poles of $\nabla_X^{GM}\nabla_Y^{GM}\lambda$ lie in the coordinate patch given by $z = 1$ at the branch points $R = 0$ of the spectral cover, viewed as a cover of the base curve $\mathbf{P}^1$. A good local coordinate is given by y and in terms of that coordinate one finds that locally

$$\begin{aligned}\nabla_X^{GM}\nabla_Y^{GM}\lambda &= \frac{X(R)Y(R)dy}{2R\partial_x P} + \text{holomorphic terms}\\ &= \frac{X(R)Y(R)Qdy}{2y^2\partial_x P} + \text{hol.} = d\left[-\frac{X(R)Y(R)Q}{2y\partial_x P}\right] + \text{hol.}\end{aligned}$$

The derivative of the period map is therefore given by

$$\begin{aligned}dp(X,Y,Z) &= -\operatorname*{Res}_{R=0}\left[\frac{X(R)Y(R)Q}{2y\partial_x P}\frac{Z(R)dy}{\partial_x P}\right] = \operatorname*{Res}_{R=0}\left[\frac{X(R)Y(R)Z(R)dx}{4R(Q\partial_x R + R\partial_x Q)}\right]\\ &= \operatorname*{Res}_{R=0}\left[\frac{X(R)Y(R)Z(R)dx}{4RQ\partial_x R}\right] = \operatorname*{Res^2}_{R=0}\left[\frac{X(R)Y(R)Z(R)dx}{4R^2Q}\right]\end{aligned}$$

□

Remark 4.3. This type of residue formula is known for example from singularity theory [**Sai83**], Hitchin systems [**DP92, Bal06**], Whitham hierarchies and matrix models [**Kri94, EO07**], and Hurwitz spaces [**Dub96**]. We conjecture that the formula

$$dp(X,Y,Z) = \operatorname*{Res^2}_{\lambda=0}\left[\frac{\nabla_X^{GM}\lambda\otimes\nabla_Y^{GM}\lambda\otimes\nabla_Z^{GM}\lambda}{\lambda}\right]$$

holds for all Hitchin systems with marked points.

4.2. The base of the Neumann system is an almost Frobenius manifold. Throughout the rest of the section, we work in the coordinate patch $z=1$. Consider the vector field e such that $e(R)=4$. In terms of the coordinates b_k, it is given by

$$e=\sum_k\left(\prod_{m\neq k}\frac{4}{b_k-b_m}\right)\partial_{b_k}$$

Using proposition 4.2, we find that the metric associated to e in section 3.2 is given by

$$(X,Y)_e = dp(X,Y,e) = \operatorname*{Res^2}_{R=0}\left[\frac{X(R)Y(R)}{R^2Q}dx^2\right] = \sum_k\frac{X_kY_k}{Q(b_k)} \tag{4.4}$$

where X_k, Y_k are the coefficients of expansions of X, Y in terms of coordinate vector fields ∂_{b_k}. For generic values of b_k, q_α the metric is well-defined, and nondegenerate since

$$\det\left[(\partial_{b_i},\partial_{b_j})_e\right] = \prod_{k=1}^{g}\frac{1}{Q(b_k)} = \prod_{k=1}^{g}\prod_{\alpha=1}^{g+1}\frac{1}{b_k-q_\alpha}$$

Diagonal metrics whose matrix elements (∂_k,∂_k) depend only on the coordinate b_k are well-known to be of Darboux-Egoroff type and therefore flat [**Dub96**].

The multiplication as defined in section 3.2 is characterized by

$$(X*_eY,Z)_e = dp(X,Y,Z) = \operatorname*{Res^2}_{R=0}\left[\frac{X(R)Y(R)Z(R)}{R^2Q}dx^2\right]$$

One finds

$$(X*_eY,Z)_e=\sum_k\left(X_kY_k\prod_{m\neq k}(b_k-b_m)\right)\frac{Z_k}{Q(b_k)}$$

and therefore

$$X*_eY=\sum_k X_kY_k\prod_{m\neq k}(b_k-b_m)\partial_{b_k} = \sum_k\frac{X_kY_k}{e_k}\partial_{b_k}$$

In fact, this is just the multiplication of the quotient ring $\mathbf{C}[x]/\langle R\rangle$ pulled back via the isomorphism

$$H^0(B,\mathcal{T}B)\cong\mathbf{C}[x]/\langle R\rangle$$
$$X\to X(R)$$

and indeed the pull-back of (four times) the identity $1\in\mathbf{C}[x]/\langle R\rangle$ is the identity vector field e. So the metric is nondegenerate and flat, the multiplication is commutative & associative and by definition the two are compatible.

The Levi-Civita connection ∇^{LC} for the metric $(\cdot,\cdot)_e$ is given by the standard formula

$$\begin{aligned}(\nabla^{LC}_X Y, Z)_e &= \frac{1}{2}\left[X(Y,Z)_e + Y(X,Z)_e - Z(X,Y)_e\right.\\ &\qquad \left.-(Y,[X,Z])_e - (X,[Y,Z])_e + (Z,[X,Y])_e\right]\\ &= \sum_k \left(X(Y)_k - \frac{1}{2}X_k Y_k \frac{\partial_{b_k} Q(b_k)}{Q(b_k)}\right)\frac{Z_k}{Q(b_k)}\end{aligned}$$

Let us write

$$\nabla^{LC}_X Y = X(Y) - X \circ Y$$

Since $X \circ Y$ is symmetric in X and Y, the connection ∇^{LC} is indeed torsion-free. Introducing the Euclidean connection $\nabla^E_X Y = X(Y)$, flatness is equivalent with symmetry in X, Y, Z of

$$(\nabla^E \circ)(X,Y,Z) = X(Y\circ Z) - X(Y)\circ Z - Y \circ X(Z) = \frac{1}{2}\sum_k X_k Y_k Z_k \partial_k \left(\frac{\partial_{b_k} Q(b_k)}{Q(b_k)}\right)$$

together with associativity of $\circ$. This is indeed an associative multiplication, in fact it's just a rescaling of $*_e$.

We arrive at the following main theorem.

THEOREM 4.4. *The base of the Neumann system is an almost Frobenius manifold.*

PROOF. We have a flat metric $(\cdot,\cdot)_e$ and a compatible unital commutative associative algebra with multiplication $*_e$. We consider the only remaining requirement *iii)* of an almost Frobenius manifold, where $c = dp$. An explicit calculation shows that

$$\begin{aligned}\left(\nabla^{LC} dp\right)(W,X,Y,Z) &= W dp(X,Y,Z) - dp(W(X) - W\circ X, Y, Z)\\ &\quad - dp(X, W(Y) - W\circ Y, Z) - dp(X,Y,W(Z) - W\circ Z)\\ &= -\sum_k X_k Y_k Z_k W_k \frac{\partial_{b_k} Q(b_k)}{Q^2(b_k)} + dp(W\circ X, Y, Z)\\ &\quad + dp(X, W\circ Y, Z) + dp(X,Y,W\circ Z)\\ &= 2dp(W\circ X, Y, Z) = \sum_k X_k Y_k Z_k W_k \frac{\partial_{b_k} Q(b_k)}{Q^2(b_k)}\end{aligned}$$

which is indeed symmetric in W, X, Y, Z. □

If the identity vector field e were constant with respect to ∇^{LC}, this would imply that

$$X(e,e)_e = (\nabla^{LC}_X e, e)_e + (e, \nabla^{LC}_X e)_e = 0 \qquad \forall X$$

which is not the case since

$$X(e,e)_e = X\left(\sum_k \frac{1}{Q(b_k)\prod_{m\neq k}(b_k - b_m)^2}\right) \neq 0$$

and therefore (at least with our choice of vector field e) we are not dealing with a Frobenius manifold. This is similar to the case of the open Toda chain [**Dub04**], whose base can be equipped with the structure of an almost Frobenius manifold. This is an example of Dubrovin's almost duality which takes a Frobenius manifold (of a simple singularity in this case) to its almost dual, which has the general property that the identity vector field is not flat.

It would be interesting to see whether the almost Frobenius structure on the base of the Neumann system is the almost dual of a genuine Frobenius manifold.

5. Conclusion

We hope that our result, an almost Frobenius manifold structure on the base of the Neumann system, holds true for any Hitchin system with marked points. For any such system, one must specify a choice of vector field e. Once this is done, using the definitions in section 3.2 the metric $(\cdot,\cdot)_e$ and multiplication $*_e$ are fixed and we need to prove flatness of $(\cdot,\cdot)_e$, associativity of $*_e$ and potentiality of $(\cdot *_e \cdot,\cdot)_e$. We start by discussing associativity.

For the Neumann system, we find that

$$Spec\,(\mathbf{C}[x]/\langle R(x,1)\rangle) = \bigcup_k \{b_k\}$$

can be identified with the zero divisor of λ. This gives a coordinate-free interpretation for the identity vector field e in terms of the identity of a ring whose spectrum gives the (second order) zeroes of λ.

For a more general Hitchin system with marked points we assume that we can work with a spectral curve, which is naturally embedded in the total space $|K_C(D)|$ and covers the base curve C. According to Markman [**Mar94**] this happens for instance if $G = SL_n$ or GL_n. The genus of the spectral curve equals the dimension of B and for a generic point in B the spectral cover has second order branch points. The zero divisor of λ consists of two[4] types of points: the branch points of the spectral cover (these occur with multiplicity two) and the intersections of the spectral curve with the zero section of $K_C(D)$ (these occur with multiplicity one). We split the zero divisor accordingly $D_\lambda = 2D_{br} + D_{int}$. One can now define the vector field e to be the identity element in the local ring $\mathcal{O}_{D_{br}}$. This automatically makes the multiplication associative and commutative, and by definition compatible with the metric.

In this paper we have not yet exploited the fact that potentiality is a direct consequence of the existence of special coordinates z_k in terms of which $dp(\partial_i,\partial_j,\partial_k) = \partial_i\partial_j\partial_k\mathcal{F}(z)$. Potentiality is therefore not an issue.

Flatness of the metric is another matter. The position of the second order zeroes of λ serve as local coordinates on B and in terms of them the metric may perhaps be diagonal, depending only on the divisor D of the marked points as in the case of the Neumann system where it depends on D through Q. This would then prove flatness. The best way to approach this question is to try to identify the metric in terms of the intersection pairing and the Gauss-Manin connection, thereby proving flatness in a coordinate-free setting.

Acknowledgement. It is a pleasure to thank Harry Braden, Andrei Marshakov and Hessel Posthuma for stimulating discussions. This research was supported by the VENI-program of NWO.

[4]The last type of points is absent for the Neumann system.

References

[Bal06] D. Balduzzi. Donagi-Markman cubic for Hitchin systems. *Math. Res. Lett.*, (13):923–933, 2006.

[Bea90] Arnaud Beauville. Jacobiennes des courbes spectrales et systèmes hamiltoniens complètement intégrables. *Acta Math.*, 164(3-4):211–235, 1990.

[BG83] Robert L. Bryant and Phillip A. Griffiths. Some observations on the infinitesimal period relations for regular threefolds with trivial canonical bundle. In *Arithmetic and geometry, Vol. II*, volume 36 of *Progr. Math.*, pages 77–102. Birkhäuser Boston, Boston, MA, 1983.

[BMMM07] H. W. Braden, A. Marshakov, A. Mironov, and A. Morozov. WDVV equations for 6D Seiberg-Witten theory and bi-elliptic curves. *Acta Appl. Math.*, 99(3):223–244, 2007.

[Bot95] Francesco Bottacin. Symplectic geometry on moduli spaces of stable pairs. *Ann. Sci. École Norm. Sup. (4)*, 28(4):391–433, 1995.

[CG80] James A. Carlson and Phillip A. Griffiths. Infinitesimal variations of Hodge structure and the global Torelli problem. In *Journées de Géometrie Algébrique d'Angers, Juillet 1979/Algebraic Geometry, Angers, 1979*, pages 51–76. Sijthoff & Noordhoff, Alphen aan den Rijn, 1980.

[DM96a] Ron Donagi and Eyal Markman. Cubics, integrable systems, and Calabi-Yau threefolds. In *Proceedings of the Hirzebruch 65 Conference on Algebraic Geometry (Ramat Gan, 1993)*, volume 9 of *Israel Math. Conf. Proc.*, pages 199–221, Ramat Gan, 1996. Bar-Ilan Univ.

[DM96b] Ron Donagi and Eyal Markman. Spectral covers, algebraically completely integrable, Hamiltonian systems, and moduli of bundles In *Integrable systems and quantum groups (Montecatini Terme, 1993)*, volume 1620 of *Lecture Notes in Math.*, pages 1–119. Springer, Berlin, 1996.

[DP92] Ron Donagi and Tony Pantev. Unpublished, 1992

[Dub96] Boris Dubrovin. Geometry of 2D topological field theories. In *Integrable systems and quantum groups (Montecatini Terme, 1993)*, pages 120–348. Springer, Berlin, 1996.

[Dub04] Boris Dubrovin. On almost duality for Frobenius manifolds. In *Geometry, topology, and mathematical physics*, volume 212 of *Amer. Math. Soc. Transl. Ser. 2*, pages 75–132. Amer. Math. Soc., Providence, RI, 2004.

[Dui80] J. J. Duistermaat. On global action-angle coordinates. *Comm. Pure Appl. Math.*, 33(6):687–706, 1980.

[DW96] Ron Donagi and Edward Witten. Supersymmetric Yang-Mills theory and integrable systems. *Nuclear Phys. B*, 460(2):299–334, 1996.

[EO07] B. Eynard and N. Orantin. Invariants of algebraic curves and topological expansion. *Commun. Number Theory Phys.*, 1(2):347–452, 2007.

[Fre99] Daniel S. Freed. Special Kähler manifolds. *Comm. Math. Phys.*, 203(1):31–52, 1999.

[Gri68] Phillip A. Griffiths. Periods of integrals on algebraic manifolds. I,II. *Amer. J. Math.*, 90, 1968.

[Gri69] Phillip A. Griffiths. On the periods of certain rational integrals. I, II. *Ann. of Math. (2) 90 (1969), 460-495; ibid. (2)*, 90:496–541, 1969.

[Gro66] A. Grothendieck. On the de Rham cohomology of algebraic varieties. *Inst. Hautes Études Sci. Publ. Math.*, (29):95–103, 1966.

[Gun67] R. C. Gunning. *Lectures on vector bundles over Riemann surfaces.* University of Tokyo Press, Tokyo, 1967.

[Her03] Claus Hertling. tt^* geometry, Frobenius manifolds, their connections, and the construction for singularities. *J. Reine Angew. Math.*, 555:77–161, 2003.

[HHP] Claus Hertling, Luuk Hoevenaars, and Hessel Posthuma, to appear.

[Hit87] Nigel Hitchin. Stable bundles and integrable systems. *Duke Math. J.*, 54(1):91–114, 1987.

[HM99] C. Hertling and Yu. Manin. Weak Frobenius manifolds. *Internat. Math. Res. Notices*, (6):277–286, 1999.

[HM04] Claus Hertling and Yuri Manin. Unfoldings of meromorphic connections and a construction of Frobenius manifolds. In *Frobenius manifolds*, Aspects Math., E36, pages 113–144. Vieweg, Wiesbaden, 2004.

[Kir92] Frances Kirwan. *Complex algebraic curves.* Cambridge University Press, Cambridge, 1992.

[Kri94] Igor Moiseevich Krichever. The τ-function of the universal Whitham hierarchy, matrix models and topological field theories. *Comm. Pure Appl. Math.*, 47(4):437–475, 1994.

[Man00] Yu. I. Manin. Three constructions of Frobenius manifolds: a comparative study [MR1701927 (2001b:14087)]. In *Surveys in differential geometry*, Surv. Differ. Geom., VII, pages 497–554. Int. Press, Somerville, MA, 2000.

[Mar94] Eyal Markman. Spectral curves and integrable systems. *Compositio Math.*, 93(3):255–290, 1994.

[Mar00] Eyal Markman. Algebraic geometry, integrable systems, and Seiberg-Witten theory. In *Integrability: the Seiberg-Witten and Whitham equations (Edinburgh, 1998)*, pages 23–41. Gordon and Breach, Amsterdam, 2000.

[MH03] R. Martini and L. K. Hoevenaars. Trigonometric solutions of the WDVV equations from root systems. *Lett. Math. Phys.*, 65(1):15–18, 2003.

[MMM96] A. Marshakov, A. Mironov, and A. Morozov. WDVV-like equations in N = 2 SUSY Yang-Mills theory. *Phys. Lett.*, B389:43–52, 1996.

[MMM00] A. Marshakov, A. Mironov, and A. Morozov. More evidence for the WDVV equations in N = 2 SUSY Yang- Mills theories. *Int. J. Mod. Phys.*, A15:1157, 2000.

[Mum84] David Mumford. *Tata lectures on theta. II*, volume 43 of *Progress in Mathematics*. Birkhäuser Boston Inc., Boston, MA, 1984. Jacobian theta functions and differential equations, With the collaboration of C. Musili, M. Nori, E. Previato, M. Stillman and H. Umemura.

[Sai83] Kyoji Saito. The higher residue pairings $K_F^{(k)}$ for a family of hypersurface singular points. In *Singularities, Part 2 (Arcata, Calif., 1981)*, volume 40 of *Proc. Sympos. Pure Math.*, pages 441–463. Amer. Math. Soc., Providence, RI, 1983.

Department of Mathematics, Universiteit Utrecht, Budapestlaan 6, 3584 CD Utrecht, The Netherlands

Proceedings of Symposia in Pure Mathematics
Volume **78**, 2008

Two-dimensional Gauge Theories and Quantum Integrable Systems

Anton A. Gerasimov[1,2,3] and Samson L. Shatashvili[2,3,4]

Abstract. In this paper the relation between 2d topological gauge theories and Bethe Ansatz equations is reviewed.[1] In addition we present some new results and clarifications. We hope the relations discussed here are particular examples of more general relations between quantum topological field theories in dimensions $d \leq 4$ and quantum integrable systems.

1. Introduction

In [**1**] a relation between a certain type of two-dimensional Yang-Mills theory and the Bethe Ansatz equations for the quantum theory of the Nonlinear Schrödinger equation was uncovered. The topological Yang-Mills-Higgs theory considered in [**1**] captures the hyperkähler geometry of the moduli space of Higgs bundles introduced in [**2**] by Hitchin. A further step in the understanding of the role of Bethe Ansatz equations in two-dimensional gauge theories was made in [**3**] where the wave functions in gauge theory were identified with eigenfunctions of the quantum Hamiltonian in the N-particle sector for the quantum theory of the Nonlinear Schrödinger equation, constructed in the framework of the coordinate Bethe Ansatz. This implies the equivalence between two seemingly different quantum field theories. Moreover, in [**3**] it was argued that this phenomenon is not isolated and it was shown how to generalize the considerations of [**1**] from the Yang-Mills-Higgs theory to the G/G gauged WZW model with Higgs fields. The corresponding partition function is expressed in terms of solutions of the Bethe Ansatz equations describing a particular limit of XXZ spin chains. One expects that generic two-dimensional topological quiver gauge theories with additional matter fields in various representations should reproduce the Bethe Ansatz equations of other known quantum systems with a finite number of degrees of freedom.

2000 *Mathematics Subject Classification.* Primary 81T45, Secondary 53D20, 81R12, 81T13, 81T70.

The research is supported by a Science Foundation Ireland grant.

[1]These are notes of the talk delivered by second author at *"From tQFT to tt* and integrability"*, Augsburg, May 2007.

Let us remark that topological quantum field theories on manifolds fibered over the circle S^1 can be effectively reduced to quantum mechanical systems with a finite number of degrees of freedom. Upon being perturbed by topological observables, the higher-dimensional topological theories lead to multi-parameter evolution in the effective, one-dimensional, quantum theories. One can expect that their integrability also is a property of three- and four-dimensional gauge theories, among which the topological theories calculating Donaldson invariants look especially interesting.

2. On symplectic and hyperkähler reductions

To motivate the introduction of the topological Yang-Mills-Higgs theory in the next section we start by recalling the representation of integrals over symplectic and hyperkähler quotients of a G-manifold X in terms of integrals over X. This is basically covered by [**1**]; here we give integrals over X, which are expected to be invariant under $SO(3)$-rotations of symplectic structures.

Consider a symplectic manifold X of dimension $d = 2n$ with a Hamiltonian action of a compact Lie group G. Define a volume form on X as the maximal non-zero exterior product $\mathrm{vol}_X = \omega_X \wedge \cdots \wedge \omega_X$ of the symplectic two-form ω_X. Given a Hamiltonian action of a Lie group G one has a corresponding moment map:

$$\mu : X \to \mathfrak{g}^*, \qquad \mathfrak{g} = \mathrm{Lie}(G), \tag{2.1}$$

and the symplectic quotient over the co-adjoint orbit $\mathcal{O}_u$ of an element $u \in \mathfrak{g}^*$ is given by

$$Y = \mu^{-1}(\mathcal{O}_u)/G = \mu^{-1}(u)/G_u. \tag{2.2}$$

Here $G_u \subset G$ is the stabilizer of u. Let $\pi : \mu^{-1}(u) \to Y = \mu^{-1}(u)/G_u$ be the projection on Y and $\mathrm{vol}_{\mu^{-1}(u)} = \mathrm{vol}_X / \wedge^{\dim \mathfrak{g}} d\mu$ be the volume form on $\mu^{-1}(u)$ induced by the volume form vol_X on X. Then the volume form vol_Y on the quotient Y is given by

$$\mathrm{vol}_Y(y) = \frac{1}{\mathrm{Vol}(G_u)} \pi_*(\mathrm{vol}_{\mu^{-1}(u)}) = \frac{1}{\mathrm{Vol}(G_u)} \int_{\pi^{-1}(y)} \mathrm{vol}_{\mu^{-1}(u)}. \tag{2.3}$$

More generally the integral of any function f over Y can be expressed as an integral over X as follows:

$$\langle f \rangle = \int_Y f\, \mathrm{vol}_Y = \frac{1}{\mathrm{Vol}(G_u)} \int_{X \times \mathfrak{g}^*} F\, e^{i\langle \varphi, (\mu(x) - u)\rangle} \mathrm{vol}_X \wedge d\varphi, \tag{2.4}$$

where $\langle , \rangle$ is the natural pairing between $\mathfrak{g}^*$ and $\mathfrak{g}$, F is an arbitrary smooth extension of $\pi^*(f)$ to a function on X and $d\varphi$ is a flat measure on $\mathfrak{g}^*$. The integrand is explicitly invariant with respect to the action of G_u and thus instead of integration over G_u one can use the "Faddeev-Popov" gauge fixing procedure:

$$\begin{aligned} \langle f \rangle \;\; = \;\; & \int_{X \times \mathfrak{g}^* \times \Pi(\mathfrak{g}_{\mathbb{C}})} F\, e^{i\langle \varphi, (\mu(x) - u)\rangle + i\langle \chi, \nu(x)\rangle + \sum_{a,b=1}^{\dim(G_u)} \overline{\mathcal{C}}^a i_{V^b} d\nu_a(y) \mathcal{C}^b} \times \\ & \times\, \mathrm{vol}_X \wedge d\varphi \wedge d\chi \wedge d\overline{\mathcal{C}}^a \wedge d\mathcal{C}^a, \end{aligned} \tag{2.5}$$

where $\Pi(\mathfrak{g}_{\mathbb{C}})$ is the Lie algebra considered as an odd vector space, $V^b \in \mathrm{Vect}_X$, $a = 1, \ldots, \dim(G_u)$ is a basis of vector fields generating the action of $\mathfrak{g}_u = \mathrm{Lie}(G_u)$, $\nu_a(x)$, $a = 1, \ldots, \dim(G_u)$ is a generic set of gauge fixing functions, $(\overline{\mathcal{C}}, \mathcal{C})$ is a pair of odd $\mathfrak{g}_u$-valued ghosts and χ is a $\mathfrak{g}_u$-valued even ghost.

Now consider a generalization of the representation (2.5) to the case of hyperkähler reduction. Given a manifold X with three compatible symplectic structures ω_α, $\alpha = 1, 2, 3$ and an action of a compact Lie group G which is Hamiltonian with respect to any of the symplectic structures. We have a family of symplectic structures with an obvious action of the group $SO(3)$ through a three-dimensional representation. It will be useful to consider a pair of real and complex symplectic forms $\omega_{\mathbb{R}} = \omega_3$, $\omega_{\mathbb{C}} = \omega_1 + i\omega_2$. The hyperkähler reduction of X with respect to G is defined as follows:

$$Y = (\mu_{\mathbb{C}}^{-1}(u_c) \cap \overline{\mu_{\mathbb{C}}^{-1}(u_c)} \cap \mu_{\mathbb{R}}^{-1}(u_r))/G_{(u_c,u_r)}, \tag{2.6}$$

where $\mu_{\mathbb{C}}$ and $\mu_{\mathbb{R}}$ are moment maps for $\omega_{\mathbb{C}}$ and $\omega_{\mathbb{R}}$ and $G_{(u_c,u_r)} \subset G$ is the stabilizer of the triple $u_r \in \mathfrak{g}^*$, $u_c \in \mathfrak{g}^* \otimes \mathbb{C}$. The same space (up to some subtleties with compactification of the space of stable orbits) can be obtained as a holomorphic symplectic manifold by a holomorphic version of symplectic reduction:

$$Y = \mu_{\mathbb{C}}^{-1}(u_c)/\widetilde{G}_u^c, \tag{2.7}$$

where $\widetilde{G}_u^c \subset G^c$ is the stabilizer of $u_c \in \mathfrak{g}^* \otimes \mathbb{C}$ with respect to the action of the complexification G^c of G. Note that $\widetilde{G}_u^c$ is not necessary a complexification of $G_{(u_c,u_r)}$. However in the following we will consider only the case when $\widetilde{G}_u^c$ is indeed a complexification of $G_{(u_c,u_r)}$.

It would be natural to define the integral of a function f over the quotient Y by an analog of (2.4):

$$\int_Y f \sim \frac{1}{\mathrm{Vol}(\widetilde{G}_u^c)} \int_{\pi^{-1}(y)} \pi^*(f)\mathrm{vol}_{\mu_{\mathbb{C}}^{-1}(u_c)}, \tag{2.8}$$

where $\pi : \mu_{\mathbb{C}}^{-1}(u_c) \to Y$, $\mathrm{vol}_X = (\omega_{\mathbb{C}} \wedge \overline{\omega_{\mathbb{C}}}) \wedge \cdots \wedge (\omega_{\mathbb{C}} \wedge \overline{\omega_{\mathbb{C}}})$ is a volume form on X. However this expression is not well defined because the volume of $\widetilde{G}_u^c$ is not finite and the fibres of the projection π are not compact.

To represent the integral over the hyperkähler quotient one should consider an analog of (2.5). It is useful to impose gauge-fixing conditions only on a troublesome non-compact part of G_u^c so that the compact subgroup $G_{(u_c,u_r)} \subset G_u^c$ remains unfixed. The natural choice of the gauge-fixing conditions is a set of real moments $\mu_{\mathbb{R}}$. Thus we arrive at the following representation:

$$\begin{aligned} \langle f \rangle &= \int_Y f \\ &= \frac{1}{\mathrm{Vol}(G_{(u_c,u_r)})} \int_{X \times (\mathfrak{g}^* \otimes \mathbb{R}^3)} F \times \\ &\quad \times e^{i \sum_{\alpha=1}^3 \langle \varphi_\alpha, (\mu^\alpha(x) - u^\alpha) \rangle + \sum_{a,b=1}^{\dim(G_{(u_c,u_r)})} \overline{\mathcal{C}}^a i_{V^b} d\mu_{\mathbb{R}}^a(y) \mathcal{C}^b} \times \\ &\quad \times \mathrm{vol}_X \wedge d\varphi_\alpha \wedge d\overline{\mathcal{C}} \wedge d\mathcal{C} \end{aligned} \tag{2.9}$$

This integral allows an explicitly $SU(2)$-invariant reformulation:

$$\begin{aligned} \langle f\rangle &= \frac{1}{\mathrm{Vol}(G_{(u_c,u_r)})}\int_{X\times(\mathfrak{g}^*\otimes\mathbb{R}^3)\times\Pi(\mathfrak{g}_{\mathbb{C}})} F\times \\ &\quad \times e^{i\mathcal{F}(x,\phi,\mathcal{C},\overline{\mathcal{C}})}\,\mathrm{vol}_X\wedge d\varphi_\alpha\wedge d\overline{\mathcal{C}}^a\wedge d\mathcal{C}^a, \\ \mathcal{F}(x,\phi,\mathcal{C},\overline{\mathcal{C}}) &= \sum_{\alpha=1}^{3}\langle\varphi_\alpha,(\mu^\alpha(x)-u^\alpha)\rangle \\ &\quad + \sum_{\alpha,\beta,\gamma=1}^{3}\ \sum_{a,b=1}^{\dim(G_{(u_c,u_r)})} \overline{\mathcal{C}}^a\,\{\mu^a_\alpha,\mu^b_\beta\}_\gamma\mathcal{C}^b\,\epsilon^{\alpha\beta\gamma} \end{aligned} \tag{2.10}$$

where μ_α, $\alpha=1,2,3$ is a triple of moment maps corresponding to the symplectic forms ω_α, $\{\,,\,\}_\alpha$ are Poisson brackets given by the inverse of the symplectic forms ω_α and F is the extension to X of a lift of f to $\mu_{\mathbb{C}}^{-1}(u_c)\cap\overline{\mu_{\mathbb{C}}^{-1}(u_c)}\cap\mu_{\mathbb{R}}^{-1}(u_r)$.

In the next section we consider the example of an infinite-dimensional hyperkähler quotient given by Hitchin's equation.

3. Topological Yang-Mills-Higgs theory

In [**1**] a two-dimensional gauge theory was proposed such that the space of classical solutions of the theory on a Riemann surface Σ_h of genus h is closely related to the cotangent space to the space of solutions of the dimensionally reduced (to 2d) four-dimensional self-dual Yang-Mills equations studied by Hitchin [**2**]. Consider a principal G-bundle P_G over a Riemann surface Σ endowed with a complex structure. Then we have an associated vector bundle $\mathrm{ad}_{\mathfrak{g}}=(P_G\times\mathfrak{g})/G$ with the fiber $\mathfrak{g}=\mathrm{Lie}(G)$ endowed with the coadjoint action of G. Consider the pairs (A,Φ) where A is a connection on P_G and Φ is a one-form taking values in $\mathrm{ad}_{\mathfrak{g}}$. Then Hitchin's equations are given by

$$F(A)-\Phi\wedge\Phi=0,\quad \nabla_A^{(1,0)}\Phi^{(0,1)}=0,\quad \nabla_A^{(0,1)}\Phi^{(1,0)}=0. \tag{3.1}$$

The space of Hitchin's solutions has a natural hyperkähler structure and admits a compatible $U(1)$ action. The correlation functions in the theory introduced in [**1**] can be described by the integrals of the products of $U(1)\times G$-equivariant cohomology classes over the moduli space of solutions of (3.1). Note that the $U(1)$-equivariance makes the path integral well-defined.

The field content of the theory introduced in [**1**] can be described as follows. In addition to the triplet (A,ψ_A,φ_0) of the topological Yang-Mills theory one has:

$$(\Phi,\psi_\Phi):\quad \Phi\in\mathcal{A}^1(\Sigma_h,\mathrm{ad}_{\mathfrak{g}}),\quad \psi_\Phi\in\mathcal{A}^1(\Sigma_h,\mathrm{ad}_{\mathfrak{g}}) \tag{3.2}$$
$$(\varphi_\pm,\chi_\pm):\quad \varphi_\pm\in\mathcal{A}^0(\Sigma_h,\mathrm{ad}_{\mathfrak{g}}),\quad \chi_\pm\in\mathcal{A}^0(\Sigma_h,\mathrm{ad}_{\mathfrak{g}}) \tag{3.3}$$

where Φ, $\varphi_\pm$ are even and ψ_Φ, $\chi_\pm$ are odd fields. We will use also another notation $\varphi_\pm=\varphi_1\pm i\varphi_2$.

The theory is described by the following path integral:

(3.4)
$$Z_{\mathrm{YMH}}(\Sigma_h)=\frac{1}{\mathrm{Vol}(\mathcal{G}_{\Sigma_h})}\int D\varphi_0\,D\varphi_\pm\,DA\,D\Phi\,D\psi_A\,D\psi_\Phi D\chi_\pm\;e^{S(\varphi_0,\varphi_\pm,A,\Phi,\psi_A,\psi_\Phi,\chi_\pm)},$$

where $S = S_0 + S_1$ with

$$
\begin{aligned}
S_0(\varphi_0, \varphi_\pm, A, \Phi, \psi_A, \psi_\Phi, \chi_\pm) &= \frac{1}{2\pi}\int_{\Sigma_h} d^2z\, \mathrm{Tr}(i\varphi_0\,(F(A) - \Phi\wedge\Phi) - c\Phi\wedge *\Phi) \\
&\quad + \varphi_+ \nabla_A^{(1,0)}\, \Phi^{(0,1)} + \varphi_- \nabla_A^{(0,1)}\, \Phi^{(1,0)}
\end{aligned} \tag{3.5}
$$

and

$$
S_1(\varphi_0, \varphi_\pm, A, \Phi, \psi_A, \psi_\Phi, \chi_\pm) = \frac{1}{2\pi}\int_{\Sigma_h} d^2z\, \mathrm{Tr}\left(\frac{1}{2}\psi_A\wedge\psi_A + \frac{1}{2}\psi_\Phi\wedge\psi_\Phi \right. \tag{3.6}
$$
$$
\left. + \chi_+[\psi_A^{(1,0)}, \Phi^{(0,1)}] + \chi_-[\psi_A^{(0,1)}, \Phi^{(1,0)}] + \chi_+\nabla_A^{(1,0)}\psi_\Phi^{(0,1)} + \chi_-\nabla_A^{(0,1)}\psi_\Phi^{(1,0)}\right)
$$

where the decompositions $\Phi = \Phi^{(1,0)} + \Phi^{(0,1)}$ and $\psi_\Phi = \psi_\Phi^{(1,0)} + \psi_\Phi^{(0,1)}$ correspond to the decomposition of the space of one-forms $\mathcal{A}^1(\Sigma_h) = \mathcal{A}^{(1,0)}(\Sigma_h) \oplus \mathcal{A}^{(0,1)}(\Sigma_h)$ defined in terms of a fixed complex structure on Σ_h.

The theory is invariant under the action of the following even vector field:

$$
\begin{aligned}
\mathcal{L}_v\Phi^{(1,0)} &= +\Phi^{(1,0)}, \quad \mathcal{L}_v\Phi^{(0,1)} = -\Phi^{(1,0)}, \quad \mathcal{L}_v\psi_\Phi^{(1,0)} = +\psi_\Phi^{(1,0)}, \\
\mathcal{L}_v\psi_\Phi^{(0,1)} &= -\psi_\Phi^{(0,1)} \quad \mathcal{L}_v\varphi_\pm = \mp\varphi_\pm, \quad \mathcal{L}_v\chi_\pm = \pm\chi_\pm,
\end{aligned} \tag{3.7}
$$

$$
\begin{cases} \mathcal{L}_{\varphi_0} A = -\nabla_A\varphi_0, & \mathcal{L}_{\varphi_0}\psi_A = -[\varphi_0, \psi_A], \\ \mathcal{L}_{\varphi_0}\Phi = -[\varphi_0, \Phi], & \mathcal{L}_{\varphi_0}\psi_\Phi = -[\varphi_0, \psi_\Phi], \end{cases} \tag{3.8}
$$
$$
\mathcal{L}_{\varphi_0}\varphi_0 = 0, \quad \mathcal{L}_{\varphi_0}\varphi_\pm = -[\varphi_0, \varphi_\pm], \quad \mathcal{L}_{\varphi_0}\chi_\pm = -[\varphi_0, \chi_\pm],
$$

and an odd vector field generated by the BRST operator:

$$
\begin{aligned}
QA &= i\psi_A, \quad Q\psi_A = -\nabla_A\varphi_0, \quad Q\varphi_0 = 0, \\
&\begin{cases} Q\Phi = i\psi_\Phi, & Q\psi_\Phi^{(1,0)} = -[\varphi_0, \Phi^{(1,0)}] + c\Phi^{(1,0)}, \\ & Q\psi_\Phi^{(0,1)} = -[\varphi_0, \Phi^{(0,1)}] - c\Phi^{(0,1)}, \end{cases} \\
Q\chi_\pm &= i\varphi_\pm, \quad Q\varphi_\pm = -[\varphi_0, \chi_\pm] \pm c\chi_\pm.
\end{aligned} \tag{3.9}
$$

We have $Q^2 = i\mathcal{L}_{\varphi_0} + c\mathcal{L}_v$ and Q can be considered as a BRST operator on the space of $\mathcal{L}_{\varphi_0}$- and $\mathcal{L}_v$-invariant functionals. The action functional of the topological Yang-Mills-Higgs theory can be represented as a sum of the action functional of the topological pure Yang-Mills theory (written in terms of the fields ϕ_0, A, ψ_A) and an additional part which can be represented as a Q-anti-commutator:

(3.10)
$$
S_{\mathrm{YMH}} = S_{\mathrm{YM}} + \left[Q, \int_{\Sigma_h} d^2z\, \mathrm{Tr}\left(\frac{1}{2}\Phi\wedge\psi_\Phi + \varphi_+\nabla_A^{(1,0)}\Phi^{(0,1)} + \varphi_-\nabla_A^{(0,1)}\Phi^{(1,0)}\right)\right].
$$

The theory given by (3.4) is a quantum field theory whose correlation functions are given by the intersection pairings of the equivariant cohomology classes on the moduli space of Higgs bundles.

To simplify the calculations it is useful to consider a more general action given by

$$
\begin{aligned}
S_{\mathrm{YMH}} &= S_{\mathrm{YM}} + \left[Q, \int_{\Sigma_h} d^2z\, \mathrm{Tr}\left(\frac{1}{2}\Phi\wedge\psi_\Phi \right.\right. \\
&\quad \left.\left. + \tau_1\,(\varphi_+\nabla_A^{(1,0)}\Phi^{(0,1)} + \varphi_-\nabla_A^{(0,1)}\Phi^{(1,0)}) + [\tau_2(\chi_+\varphi_- + \chi_-\varphi_+)\,\mathrm{vol}_{\Sigma_h}]\right)\right]
\end{aligned} \tag{3.11}
$$

Cohomological localization of the functional integral takes the simplest form for $\tau_1 = 0,\ \tau_2 \neq 0$. Note that it is not obvious that the theory for $\tau_1 = 0,\ \tau_2 \neq 0$ is equivalent to that for $\tau_1 \neq 0,\ \tau_2 = 0$. However, taking into account that the action functionals in these two cases differ by an equivariantly exact form and for $c \neq 0$ the space of fields is essentially compact one can expect that the theories are equivalent.

For $\tau_1 = 0$ the path integrals over Φ, $\varphi_\pm$ and $\chi_\pm$ are quadratic. Thus we have for the partition function the following formal representation:

$$\begin{aligned} Z_{\rm YMH}(\Sigma_h) &= \frac{1}{\mathrm{Vol}(\mathcal{G}_{\Sigma_h})} \int DA\, D\varphi_0\, D\psi_A \times \\ &\quad\times e^{\frac{1}{2\pi}\int_{\Sigma_h} d^2z\, \mathrm{Tr}\,(i\varphi_0 F(A) + \frac{1}{2}\psi_A \wedge \psi_A)}\ \mathrm{Sdet}_V(\mathrm{ad}_{\varphi_0} + ic), \end{aligned} \tag{3.12}$$

where the super-determinant is taken over the superspace:

$$V = V_{even} \oplus V_{odd} = \mathcal{A}^0(\Sigma_h, \mathrm{ad}_{\mathfrak{g}}) \oplus \mathcal{A}^{(1,0)}(\Sigma_h, \mathrm{ad}_{\mathfrak{g}}). \tag{3.13}$$

and should be properly understood using a regularization compatible with Q-symmetry of the path integral (e.g. $\tau_1 \neq 0$). Thus the Yang-Mills-Higgs theory can be considered as a pure Yang-Mills theory deformed by a non-local gauge invariant observable. Note that for $c \to \infty$ the action functional (3.5) reduces to that of 2d topological Yang-Mills theory, with corresponding consequences. At the same time for $c = 0$ it is again the 2d topological Yang-Mills action but now for the complexified gauge group; thus the theory interpolates for non-zero c between topological Yang-Mills theory for compact and complex gauge groups, which also becomes apparent from the final results.

The partition function (3.4) of the Yang-Mills-Higgs theory on a compact Riemann surface can be calculated using the standard methods of cohomological localization. As in the case of Yang-Mills theory we consider the deformation of the action of the theory:

$$\Delta S_{\rm YMH} = -\sum_{k=1}^{\infty} t_k \int_{\Sigma_h} d^2z\ \mathrm{Tr}\,\varphi_0^k\ \mathrm{vol}_{\Sigma_h}. \tag{3.14}$$

where we impose the condition that $t_k \neq 0$ for a finite subset of indices $k \in \mathbb{Z}$.

The path integral with the action (3.11) at $\tau_1 = 0$ and $\tau_2 = 1$ is easily reduced to the integral over abelian gauge fields. The contribution of the additional nonlocal observable in (3.12) can be calculated as follows. The purely bosonic part of the nonlocal observable after reduction to abelian fields can be easily evaluated using any suitable regularization (i.e. zeta function regularization) and the result is the change of the bosonic part of the abelian action $\int d^2z (\varphi_0)_i F^i(A)$ by

$$\begin{aligned} \Delta S &= \int_{\Sigma_h} d^2z \sum_{i,j=1}^{N} \log\left(\frac{(\varphi_0)_i - (\varphi_0)_j + ic}{(\varphi_0)_i - (\varphi_0)_j - ic}\right) F(A)^i \\ &\quad + \frac{1}{2}\int_{\Sigma_h} d^2z \sum_{i,j=1}^{N} \log((\varphi_0)_i - (\varphi_0)_j + ic) R^{(2)}\sqrt{g}, \end{aligned} \tag{3.15}$$

where $F(A)^i$ is the i-th component of the curvature of the abelian connection A and $R^{(2)}(g)$ is the curvature on Σ_h for the 2d metric g used to regularize the nonlocal observable. We will use the notation $(\varphi_0)_i = \lambda_i$ in the remaining part of the paper. This leads to the unique Q-closed completion. The completion of the term in

(3.15) containing the curvature of the gauge field is given by the two-observable $\mathcal{O}_f^{(2)}$ corresponding to the descendant of the following function on the Cartan subalgebra isomorphic to $\mathbb{R}^N$:

$$f(\mathrm{diag}(\lambda_1,\ldots,\lambda_N)) = \sum_{k,j=1}^{N} \int_0^{\lambda_j-\lambda_k} \mathrm{arctg}\,(\lambda/c)\,d\lambda. \tag{3.16}$$

Thus, the abelianized action is defined by the two-observable descending from

$$I(\lambda) = \sum_{j=1}^{N}\left(\frac{1}{2}\lambda_i^2 - 2\pi n_j\lambda_j\right) + \sum_{k,j=1}^{N}\int_0^{\lambda_j-\lambda_k} \mathrm{arctg}\ (\lambda/c)\,d\lambda. \tag{3.17}$$

On the other hand the term containing the metric curvature R in (3.15) is Q-closed and thus does not need any completion. It can be considered as an integral of the zero observable

$$\mathcal{O}^{(0)} = \sum_{i,j=1}^{N} \log(\lambda_i - \lambda_j + ic) \tag{3.18}$$

over Σ_h weighted by half of the metric curvature. Note that the function $I(\lambda)$ plays an important role in Nonlinear Schrödinger theory, which we explain in the next section.

After integrating out the fermionic partners of the abelian connection A the standard localization procedure leads to the following final finite-dimensional integral representation for the partition function [**1**]:

$$\begin{aligned} Z_{\mathrm{YMH}}(\Sigma_h) &= \frac{e^{(1-h)a(c)}}{|W|}\int_{\mathbb{R}^N} d^N\lambda\ \mu(\lambda)^h \sum_{(n_1,\ldots,n_N)\in\mathbb{Z}^N} e^{2\pi i\sum_{m=1}^{N}\lambda_m n_m}\ \times \\ &\quad \times\prod_{k\neq j}(\lambda_k-\lambda_j)^{n_k-n_j+1-h}\ \times \\ &\quad \prod_{k,j}(\lambda_k-\lambda_j-ic)^{n_k-n_j+1-h}\, e^{-\sum_{k=1}^{\infty} t_k\, p_k(\lambda)}, \end{aligned} \tag{3.19}$$

where

$$\mu(\lambda) = \det\left\|\frac{\partial^2 I(\lambda)}{\partial\lambda_i\,\partial\lambda_j}\right\|, \tag{3.20}$$

and $p_k(\lambda)$ are S_N-invariant polynomial functions of degree k on $\mathbb{R}^N$ and $a(c)$ is an h-independent constant defined by the appropriate choice of the regularization of the functional integral. One can write the n_i-dependent parts of the products in (3.19) as the exponent of the sum:

$$\begin{aligned} Z_{\mathrm{YMH}}(\Sigma_h) &= \frac{e^{(1-h)a(c)}}{|W|}\int_{\mathbb{R}^N} d^N\lambda\ \mu(\lambda)^h \sum_{(n_1,\ldots,n_N)\in\mathbb{Z}^N} e^{2\pi i\sum_j n_j\alpha_j(\lambda)} \\ &\quad \times\prod_{k\neq j}(\lambda_k-\lambda_j)^{1-h}\ \prod_{k,j}(\lambda_k-\lambda_j-ic)^{1-h}\, e^{-\sum_{k=1}^{\infty} t_k\, p_k(\lambda)}, \end{aligned} \tag{3.21}$$

with the notation

$$e^{2\pi i\alpha_j(\lambda)} = \mathcal{F}_j(\lambda) \equiv e^{2\pi i\lambda_j}\prod_{k\neq j}\frac{\lambda_k-\lambda_j-ic}{\lambda_k-\lambda_j+ic} \tag{3.22}$$

After taking the sum over $(n_1, \ldots, n_N) \in \mathbb{Z}^N$ using

$$\begin{aligned}\mu(\lambda) \sum_{(n_1,\ldots,n_N)\in\mathbb{Z}^N} e^{2\pi i \sum_j n_j \alpha_i(\lambda)} &= \mu(\lambda) \sum_{(m_1,\ldots,m_N)\in\mathbb{Z}^N} \prod_j \delta(\alpha_j(\lambda) - m_j) \\ &= \sum_{(\lambda_1^*,\ldots,\lambda_N^*)\in\mathcal{R}_N} \prod_j \delta(\lambda_j - \lambda_j^*) \qquad (3.23)\end{aligned}$$

(see the definition of $\mathcal{R}_N$ below) and the integral over $(\lambda_1, \ldots, \lambda_N) \in \mathbb{R}^N$ we see that only $\alpha_j(\lambda) \in \mathbb{Z}$, or equivalently $\mathcal{F}_j(\lambda) = 1$, contribute to the partition function, which now can be written as

$$Z_{\mathrm{YMH}}(\Sigma_h) = e^{(1-h)a(c)} \sum_{\lambda\in\mathcal{R}_N} D_\lambda^{2-2h} e^{-\sum_{k=1}^\infty t_k\, p_k(\lambda)}, \qquad (3.24)$$

where

$$D_\lambda = \mu(\lambda)^{-1/2} \prod_{i<j} (\lambda_i - \lambda_j)(c^2 + (\lambda_i - \lambda_j)^2)^{1/2}, \qquad (3.25)$$

and the $\mathcal{R}_N$ in (3.23) and (3.24) denotes a set of the solutions of the Bethe Ansatz equations $\mathcal{F}_j(\lambda) = 1$:

$$e^{2\pi i \lambda_j} \prod_{k\neq j} \frac{\lambda_k - \lambda_j + ic}{\lambda_k - \lambda_j - ic} = 1, \qquad k = 1, \ldots, N, \qquad (3.26)$$

for the N-particle sector of the quantum theory of the Nonlinear Schrödinger equation. Note that the sum in (3.24) is taken over the classes of solutions up to the action of the symmetric group on λ_i. This set can be enumerated by the multiplets of integers $(p_1, \ldots, p_N) \in \mathbb{Z}^N$ such that $p_1 \geq p_2 \geq \cdots \geq p_N$, $p_i \in \mathbb{Z}$. Thus, the sum in (3.24) is the sum over the same set of partitions as in 2d Yang-Mills theory.

4. N-particle wave functions in Nonlinear Schrödinger theory

The appearance of a particular form of Bethe Ansatz equations (3.26) strongly suggests the relevance of quantum integrable theories in the description of wave-functions in topological Yang-Mills-Higgs theory. Precisely this form of Bethe Ansatz equations (3.26) arises in the description of the N-particle wave functions for the quantum Nonlinear Schrödinger theory with the coupling constant $c \neq 0$ [**4, 5, 6, 7**]. In this section we recall the standard facts about the construction of the these wave functions using the coordinate Bethe Ansatz. We also discuss the relation with the representation theory of degenerate (double) affine Hecke algebras and the representation theory of Lie groups over the complex and p-adic numbers. For the application of the quantum inverse scattering method to Nonlinear Schrödinger theory see [**8, 9**]. One can also recommend [**10**] as a quite readable introduction to the Bethe Ansatz machinery.

The Hamiltonian of Nonlinear Schrödinger theory with a coupling constant c is given by:

$$\mathcal{H}_2 = \int dx\, (\frac{1}{2} \frac{\partial \phi^*(x)}{\partial x} \frac{\partial \phi(x)}{\partial x} + c(\phi^*(x)\phi(x))^2), \qquad (4.1)$$

with the following Poisson structure for bosonic fields:

$$\{\phi^*(x), \phi(x')\} = \delta(x - x'). \qquad (4.2)$$

The operator of the number of particles:

$$\mathcal{H}_0 = \int dx\, \phi^*(x)\phi(x), \tag{4.3}$$

commutes with the Hamiltonian $\mathcal{H}_2$ and thus one can solve the eigenfunction problem in the sub-sector for a given number of particles $\mathcal{H}_0 = N$. We will consider both the theory on the infinite interval ($x \in \mathbb{R}$) and its periodic version $x \in S^1$. The equation for eigenfunctions in the N-particle sector (Yang system) has the following form:
(4.4)

$$(-\frac{1}{2}\sum_{i=1}^{N}\frac{\partial^2}{\partial x_i^2} + c\sum_{1\leq i<j\leq N}\delta(x_i - x_j))\Phi_\lambda(x) = 2\pi^2(\sum_{i=1}^{N}\lambda_i^2)\Phi_\lambda(x) \qquad i = 1,\dots,N.$$

This equation is obviously symmetric with respect to the action of the symmetric group S_N on the coordinates x_i. Thus the solutions are classified according to the representations of S_N. Quantum integrability of the Nonlinear Schrödinger theory implies the existence of a complete set of commuting Hamiltonian operators. The corresponding eigenvalues are given by the symmetric polynomials $p_k(\lambda)$.

Finite-particle subsectors of the Nonlinear Schrödinger theory can be described in terms of the representation theory of a particular kind of Hecke algebra [**11, 12, 13**]. Let $R = \{\alpha_1,\dots,\alpha_l\}$ be a root system, W the corresponding Weyl group and P a weight lattice. The degenerate affine Hecke algebra $\mathcal{H}_{R,c}$ associated to R is defined as the algebra with basis S_w, $w \in W$ and $\{D_\lambda, \lambda \in P\}$ such that S_w, $w \in W$ generate a subalgebra isomorphic to the group algebra $\mathbb{C}[W]$ and the elements D_λ, $\lambda \in P$ generate the group algebra $\mathbb{C}[P]$ of the weight lattice P. In addition one has the relations

$$S_{s_i} D_\lambda - D_{s_i(\lambda)}\, S_{s_i} = c\,\frac{2(\lambda,\alpha_i)}{(\alpha_i,\alpha_i)}, \qquad i = 1,\dots,n. \tag{4.5}$$

Here s_i are the generators of the Weyl algebra corresponding to the reflection with respect to the simple roots α_i. The center of $\mathcal{H}_{R,c}$ is isomorphic to the algebra of W-invariant polynomial functions on $R\otimes\mathbb{C}$. The degenerate affine Hecke algebras were introduced by Drinfeld [**14**] and independently by Lusztig [**15**].

Below we consider only the case of the $\mathfrak{gl}_N$ root system and thus we have $W = S_N$. Let us introduce the following differential operators (Dunkl operators [**16**]):

$$\mathcal{D}_i = -i\frac{\partial}{\partial x_i} + i\frac{c}{2}\sum_{j=i+1}^{N}(\epsilon(x_i - x_j) + 1)s_{ij}. \tag{4.6}$$

Here $\epsilon(x)$ is a sign-function and $s_{ij} \in S_N$ is the transposition (ij). These operators together with the action of the symmetric group (4.6) provide a representation of the degenerate affine algebra $\mathcal{H}_{N,c}$ for $\mathfrak{g} = \mathfrak{gl}(N)$:

$$S_{s_i} \to s_i, \quad D_i \to \mathcal{D}_i, \quad i = 1,\dots,N. \tag{4.7}$$

The image of the quadratic element of the center is given by:

$$\frac{1}{2}\sum_{i=1}^{N}\mathcal{D}_i^2 = -\frac{1}{2}\sum_{i=1}^{N}\frac{\partial^2}{\partial x_i^2} + c\sum_{1\leq i<j\leq N}\delta(x_i - x_j) \tag{4.8}$$

and thus coincides with the restriction of the quantum Hamiltonian to the N-particle sector of Nonlinear Schrödinger theory on the infinite interval.

We are interested in S_N-invariant solutions of (4.4). They play the role of spherical vectors (with respect to the subalgebra $\mathbb{C}[W] \in \mathcal{H}_{N,c}$) in the representation theory of degenerate affine Hecke algebras.

The eigenvalue problem (4.4) allows an equivalent reformulation as an eigenvalue problem in the domain $x_1 \leq x_2 \leq \cdots \leq x_N$ for the differential operator

$$(-\frac{1}{2}\sum_{i=1}^{N}\frac{\partial^2}{\partial x_i^2})\Phi_\lambda(x) = 2\pi^2(\sum_{i=1}^{N}\lambda_i^2)\Phi_\lambda(x) \quad i = 1,\ldots,N, \tag{4.9}$$

with the boundary conditions:

$$(\partial_{x_{i+1}}\Phi_\lambda(x) - \partial_{x_i}\Phi_\lambda(x))_{x_{i+1}-x_i=+0} = 4\pi c\Phi_\lambda(x)_{x_{i+1}-x_i=0}. \tag{4.10}$$

The solution is given by

$$\Phi_\lambda^{(0)}(x) = \sum_{w\in W}\prod_{1\leq i<j\leq N}\left(\frac{\lambda_{w(i)}-\lambda_{w(j)}+ic}{\lambda_{w(i)}-\lambda_{w(j)}}\right)\exp(2\pi i\sum_k \lambda_{w(k)}x_k), \tag{4.11}$$

or equivalently, by

$$\Phi_\lambda^{(0)}(x) = \frac{1}{\Delta_{\mathfrak{g}}(\lambda)}\sum_{w\in W}(-1)^{l(w)}\prod_{1\leq i<j\leq N}(\lambda_{w(i)}-\lambda_{w(j)}+ic)\exp(2\pi i\sum_k\lambda_{w(k)}x_k). \tag{4.12}$$

where $\Delta_{Fg}(\lambda) = \prod_{1\leq i<j\leq N}(\lambda_i - \lambda_j)$. Note that the wave-function is explicitly symmetric under the action of the symmetric group S_N on $\lambda = (\lambda_1,\ldots,\lambda_N)$. This solution can also be constructed using the representation theory of the degenerate affine Hecke algebra $\mathcal{H}_{N,c}$ (see [**11, 12, 13**]).

Given a solution of the equations (4.9) with boundary conditions (4.10), S_N-symmetric solutions of (4.4) on $\mathbb{R}^N$ can be represented in the following form:

$$\Phi_\lambda^{(0)}(x) = \sum_{w\in W}\left(\prod_{i<j}\left(\frac{\lambda_{w(i)}-\lambda_{w(j)}+ic\epsilon(x_i-x_j)}{\lambda_{w(i)}-\lambda_{w(j)}}\right)\exp(2\pi i\sum_k\lambda_{w(k)}x_k)\right), \tag{4.13}$$

where $\epsilon(x)$ is a sign-function. This gives the full set of solutions of (4.4) for $(\lambda_1 \leq \cdots \leq \lambda_N) \in \mathbb{R}^N$ satisfying the orthogonality condition with respect to the natural pairing:

$$\langle\Phi_\lambda,\Phi_\mu\rangle = \frac{1}{N!}\int dx_1\cdots dx_N\,\overline{\Phi}_\lambda(x)\Phi_\mu(x) = G(\lambda)\prod_{i=1}^{N}\delta(\lambda_i-\mu_i), \tag{4.14}$$

where

$$G(\lambda) = \prod_{1\leq i<j\leq N}\frac{(\lambda_i-\lambda_j)^2+c^2}{(\lambda_i-\lambda_j)^2}. \tag{4.15}$$

Therefore the normalized wave functions are given by

$$\Phi_\lambda(x) = \sum_{w\in W}(-1)^{l(w)}\prod_{i<j}\left(\frac{\lambda_{w(i)}-\lambda_{w(j)}+ic\epsilon(x_i-x_j)}{\lambda_{w(i)}-\lambda_{w(j)}-ic\epsilon(x_i-x_j)}\right)^{\frac{1}{2}}\exp(2\pi i\sum_k\lambda_{w(k)}x_k). \tag{4.16}$$

The eigenvalue problem for the periodic N-particle Hamiltonian of Nonlinear Schrödinger theory can be reformulated in the following way. Consider the eigenfunction problem for the differential operator

(4.17)
$$(-\frac{1}{2}\sum_{i=1}^{N}\frac{\partial^2}{\partial x_i^2}+c\sum_{n\in\mathbb{Z}}\sum_{1\leq i<j\leq N}\delta(x_i-x_j+n))\Phi_\lambda(x)=2\pi^2(\sum_{i=1}^{N}\lambda_i^2)\Phi_\lambda(x)\qquad i=1,\ldots,N.$$

The wave functions of the periodic Nonlinear Schrödinger equation are the eigenfunctions of (4.17) satisfying the following invariance conditions:

$$\begin{aligned}\Phi_\lambda(x_1,\ldots,x_j+1,\ldots,x_N) &= \Phi_\lambda(x_1,\ldots,x_N),\quad j=1,\ldots,N,\\ \Phi_\lambda(x_{w(1)},\ldots,x_{w(N)}) &= \Phi_\lambda(x_1,\ldots,x_N),\quad w\in S_N.\end{aligned}\tag{4.18}$$

These are the conditions of invariance under the action of the affine Weyl group on the space of wave functions.

The solutions can be obtained by imposing additional periodicity conditions on the wave functions (4.16). This leads to the following set of Bethe Ansatz equations for $(\lambda_1,\ldots,\lambda_N)$:

$$\mathcal{F}_j(\lambda)\equiv e^{2\pi i\lambda_j}\prod_{k\neq j}\frac{\lambda_k-\lambda_j-ic}{\lambda_k-\lambda_j+ic}=1,\qquad j=1,\ldots,N.\tag{4.19}$$

The set of solutions of these equations can be enumerated by sets of integers $(p_1\geq\cdots\geq p_N)$: for each ordered set of these integers there is exactly one solution to the Bethe Ansatz equations [**7**]. Let us remark that there is the following equivalent representation for the periodic wave functions:

(4.20)
$$\widetilde{\Phi}_\lambda(x)=\sum_{w\in W}(-1)^{l(w)}\prod_{i<j}\left(\frac{\lambda_{w(i)}-\lambda_{w(j)}+ic}{\lambda_{w(i)}-\lambda_{w(j)}-ic}\right)^{\frac{1}{2}+[x_i-x_j]}\exp(2\pi i\sum_k\lambda_{w(k)}x_k),$$

where $[x]$ is the greatest integer less or equal to x. It easy to see that these wave functions are periodic and descend to the wave functions (4.16) if $\lambda=(\lambda_1,\ldots,\lambda_N)$ satisfy (4.19). The normalized wave functions in the periodic case are given by

$$\Phi_\lambda^{norm}(x)=\left(\det\left\|\frac{\partial\log\mathcal{F}_j(\lambda)}{\partial\lambda_k}\right\|\right)^{-1/2}\Phi_\lambda(x)=\mu(\lambda)^{-1/2}\Phi_\lambda(x).\tag{4.21}$$

Note that the normalization factor is closely related to the factor (3.20) arising in the representation of the partition function of Yang-Mills-Higgs theory. Indeed, the function $I(\lambda)$ introduced in (3.20) is known in the theory of Nonlinear Schrödinger equations as the Yang function [**7**]; critical points of the Yang function are in one to one correspondence with the solutions of the Bethe Ansatz equations:

$$\alpha_j(\lambda)=\log\mathcal{F}_j(\lambda)=\frac{\partial I(\lambda)}{\partial\lambda_j}=n_j.\tag{4.22}$$

Below we will see that this is not accidental. Finally note that the periodic Nonlinear Schrödinger theory has an interpretation in terms of the representation theory of degenerate double affine Hecke algebras introduced by Cherednik [**17**]. For the details in this regard see [**13**].

5. Wave function in topological Yang-Mills-Higgs theory

In this section we provide the identification of a natural basis of wave functions of topological Yang-Mills-Higgs theory for $G = U(N)$ (given by a path integral on a disk with the insertion of observables in the center) with the eigenfunctions of the N-particle Hamiltonian operator of Nonlinear Schrödinger theory. First, by counting the observables of the theory we show that the phase space of Yang-Mills-Higgs theory can be considered as a deformation of the phase space of Yang-Mills theory. We compute the cylinder path integral (Green function) and torus partition function in Nonlinear Schrödinger theory (with all higher Hamiltonians) and show that the latter coincides with the torus partition function in Yang-Mills-Higgs theory (with arbitrary observables turned on). Then, using the explicit representation of the partition function on the two-dimensional torus we derive the transformation properties of the wave functions under large gauge transformations. They are in agreement with the known explicit transformation properties of wave functions in Nonlinear Schrödinger theory. Finally, we demonstrate that eigenfunctions of the N-particle Hamiltonian operator of Nonlinear Schrödinger theory indeed coincide with the basis of Yang-Mills-Higgs wave functions in an appropriate polarization.

5.1. Local Q-cohomology. We start with a description of the Hilbert space of Yang-Mills-Higgs theory using the operator-state correspondence. In the simplest form the operator-state correspondence is as follows. Each operator, by acting on the vacuum state, defines a sate in the Hilbert space. In turn, for each state there is an operator, creating the state from the vacuum state. Moreover, for the maximal commutative subalgebra of operators this correspondence should be one to one. For example, the space of local gauge-invariant Q-cohomology classes in topological $U(N)$ Yang-Mills theory is spanned, linearly, by the operators

$$\mathcal{O}_k^{(0)} = \frac{1}{(2\pi i)^k} \operatorname{Tr} \varphi^k, \tag{5.1}$$

and thus this space coincides with the space of Ad_G-invariant regular functions on the Lie algebra $\mathfrak{u}_N$. This is in accordance with the description of the Hilbert space of the theory given in Section 2.

We would like to apply the same reasoning to topological Yang-Mills-Higgs theory. To get an economical description of the Hilbert space of the theory one should find a maximal (Poisson) commutative subalgebra of local Q-cohomology classes (where Q is given by (3.9) and acts on the space of functions invariant under the symmetries generated by (3.7), (3.8)). Obviously the operators (5.1) provide non-trivial cohomology classes. One can show that these operators provide a maximal commutative subalgebra for $c \neq 0$ and therefore the reduced phase space in the Yang-Mills-Higgs system can be identified with the phase space of pure Yang-Mills theory. Thus the Hilbert space of Yang-Mills-Higgs theory ($c \neq 0$) can be naturally identified with the Hilbert space of Yang-Mills theory (considered as a Yang-Mills-Higgs theory in the limit $c \to \infty$). The fact that the Hilbert space of Yang-Mills-Higgs theory is the same for all $c \neq 0$ implies that the basis of wave functions for $c \neq 0$ should be a deformation of the basis for $c = \infty$.

One should emphasise that this reasoning is not applicable to the case $c = 0$. The local cohomology for $c = 0$ contains additional operators. For example, the following operators provide non-trivial cohomology classes for arbitrary $t \in \mathbb{C}$ and

$c = 0$:

$$\mathcal{O}_k^{(0)}(t) = \frac{1}{(2\pi i)^k}\mathrm{Tr}(\varphi_0 + t\varphi_+ - \frac{t}{2}\chi_+^2)^k. \tag{5.2}$$

This is a manifestation of the fact that the $c = 0$ theory is a Yang-Mills theory for the complexified group G^c and thus its phase space is given by $\mathcal{M}_{\mathbb{C}} = T^*H^c/W$.

5.2. Gauge transformations of wave function. The discrete character of the spectrum of the N-particle Hamiltonian operator in periodic Nonlinear Schrödinger theory is due to a periodicity condition on wave functions. Thus the eigenfunctions in the periodic case are given by a subset of the eigenfunctions on $\mathbb{R}^N$ descending to S_N-invariant functions on $(S^1)^N$. The eigenfunctions (4.16) of the Hamiltonian operator on $\mathbb{R}^N$ are represented as the sum over elements of the symmetric group of simple wave functions. For generic eigenvalues each term of the sum is multiplied by some function under the shift $x_i \to x_i + n_i$, $n_i \in \mathbb{Z}$ of the coordinates. Below we will show how these multiplicative factors arising in Nonlinear Schrödinger theory can be derived in Yang-Mills-Higgs gauge theory.

We start with the simple case of Yang-Mills theory. The partition function of $U(N)$ Yang-Mills theory on the torus Σ_1 is given by

$$\begin{aligned} Z_{\mathrm{YM}}(\Sigma_1) &= \int_{\mathbb{R}^N/S_N} d^N\lambda \sum_{(n_1,\dots,n_N)\in\mathbb{Z}^N} e^{2\pi i\sum_{m=1}^N \lambda_m n_m}\, e^{-\sum_{k=1}^\infty t_k\, p_k(\lambda)} \\ &= \sum_{(n_1,\dots,n_N)\in P_{++}} e^{-\sum_{k=1}^\infty t_k\, p_k(m+\rho)} \end{aligned} \tag{5.3}$$

where P_{++} is the set of dominant weights of $U(N)$. The sum over $(n_1, \dots, n_N) \in \mathbb{Z}^N$ has the meaning of the sum over topological classes of $U(1)^N$-principal bundles on the torus Σ_1. It results in the replacement of the integration over λ by a sum over a discrete subset. This should be compared with the partition function of dimensionally reduced $U(N)$ Yang-Mills theory on S^1:

$$Z_{\mathrm{QM}}(S^1) = \int_{\mathbb{R}^N/S_N} d^N\lambda\; e^{-\sum_{k=1}^\infty t_k\, p_k(\lambda)}. \tag{5.4}$$

Contrary to the two-dimensional Yang-Mills theory in the last case we do not have any additional restriction on the spectrum $(\lambda_1, \dots, \lambda_N) \in \mathbb{R}^N/S_N$. The appearance of the additional sum can be traced back to the difference between the Hilbert spaces of the dimensionally reduced and non-reduced theories.

The mechanism of spectrum restriction via the sum over the topological sectors can be explained in terms of the structure of the Hilbert space of the theory as follows. In the Hamiltonian formalism the partition function on a torus is given by the trace of the evolution operator over the Hilbert space of the theory. Let us first consider the dimensionally reduced $U(N)$ Yang-Mills theory. The phase space of the theory is $T^*\mathbb{R}^N/S_N$, where we divide by the Weyl group $W = S_N$. To construct the Hilbert space we quantize the phase space using the following polarization. Consider the Lagrangian projection $\pi : T^*\mathbb{R}^N \to \mathbb{R}^N$ supplied with a section. We choose the coordinates on the base as position variables and the coordinates on the fibers as the corresponding momenta. Thus the Hilbert space in this polarization is realized as the space of S_N (skew)-invariant functions on the base $\mathbb{R}^N$ of the projection.

Now consider two-dimensional Yang-Mills theory. For the phase space we have T^*H/S_N, where H is the Cartan subgroup. We use the polarization associated with the projection $\pi : T^*H \to H$. Thus the wave functions are S_N-invariant functions on a torus H or equivalently the functions on $\mathbb{R}^N$ invariant under the action of the semidirect product of the lattice $P_0 = \pi_1(H)$ and Weyl group $W = S_N$ (i.e. under the action of the affine Weyl group W^{aff}). The lattice P_0 can be interpreted as the lattice of $\mathbb{R}^N$-valued constant connections on S^1 which are gauge equivalent to the zero connection. The corresponding gauge transformations act on the wave functions by the shifts $x_j \to x_j + n_j$, $n_j \in \mathbb{Z}$ of the argument of the wave functions in the chosen polarization and the wave functions in two-dimensional Yang-Mills theory can be obtained by averaging over these gauge transformations and global gauge transformations by the nontrivial elements of the normalizer of the Cartan torus $W = N(H)/H$.

It is possible to relate the averaging over the topologically non-trivial transformations with the sum over topological classes of an H-bundle on the torus. Note that the maps of S^1 to the gauge group H are topologically classified by $\pi_1(H) = \mathbb{Z}^N$. Consider a connection $A = (A_1, \dots, A_N)$ on an H-bundle over a cylinder L, $\partial L = S^1_+ \cup S^1_-$ such that the holonomies along the boundaries S^1_+ and S^1_- are in the different topological classes $[(m_1, \dots, m_N)] \in \pi_1(H)$ and $[(m_1 + n_1, \dots, m_N + n_N)] \in \pi_1(H)$. Gluing boundaries of the cylinder L we obtain a torus supplied with a connection ∇_A such that the first Chern classes of the bundles corresponding to each $U(1)$-factor are given by $c_1(\nabla_{A_i}) = \frac{1}{2\pi i}\int_L F(A_j) = n_j$, $j = 1, \dots, N$. Thus we see that the sum over the topologically non-trivial gauge transformations on S^1 can be translated into the sum over topological classes of H-bundles on the torus.

Let us rederive the partition function of Yang-Mills theory on the torus (5.3) using the averaging procedure. We start with the dimensionally reduced theory. Let us choose a basis in the Hilbert space of the dimensionally reduced Yang-Mills theory given by the S_N skew-invariant eigenfunctions of the quadratic operator $H_2^{(0)} = \operatorname{Tr}\varphi^2$. In the polarization discussed above we have

$$H_2^{(0)}\,\psi_\lambda(\theta) = -\frac{1}{2}\left(\sum_{j=1}^{N}\frac{\partial^2}{\partial\theta_j^2}\right)\psi_\lambda(\theta) = 2\pi^2\sum_{j=1}^{N}\lambda_i^2\,\psi_\lambda(\theta), \tag{5.5}$$

where $(\theta_1, \dots, \theta_N) \in \mathbb{R}^N$. The set of normalized skew-invariant eigenfunctions is given by

$$\psi_\lambda(\theta) \;=\; \sum_{w\in S_N}(-1)^{l(w)}\exp(2\pi i\sum_{j=1}^{N}\lambda_{w(j)}\theta_j), \quad (\lambda_1, \dots, \lambda_N) \in \mathbb{R}^N/W. \tag{5.6}$$

$$\begin{aligned}\frac{1}{N!}\int_{\mathbb{R}^N} d^N\theta\,\overline{\psi_\lambda}(\theta)\,\psi_{\lambda'}(\theta) &= (2\pi)^N\sum_{w\in S_N}(-1)^{l(w)}\prod_{j=1}^{N}\delta(\lambda_{w(j)} - \lambda'_j)\\ &= \delta^{(S_N)}(\lambda - \lambda'). \end{aligned}\tag{5.7}$$

The integral kernel of the identity operator acting on the skew-symmetric functions can be represented (due to translation invariance it is a function of the difference

$\theta-\theta'$) as

$$K_0(\theta,\theta')=K_0(\theta-\theta')=\delta^{(S_N)}(\theta-\theta')=\int_{\mathbb{R}^N/S_N}d^N\lambda\,\overline{\psi_\lambda}(\theta)\,\psi_\lambda(\theta'). \tag{5.8}$$

The partition function of the dimensionally reduced Yang-Mills theory on S^1 is given by the trace of an evolution operator and can be written explicitly as

$$\begin{aligned} Z_{\mathrm{QM}}(S^1) &= \mathrm{Tr}\,e^{-t_2H_2(\hat{p},\hat{q})}=\int_{(\mathbb{R}^N\times\mathbb{R}^N)/S_N}d^N\theta\,d^N\lambda\;\overline{\psi_\lambda}(\theta)\,e^{-t_2H_2^{(0)}(i\partial_\theta,\theta)}\,\psi_\lambda(\theta)\\ &= \int_{(\mathbb{R}^N\times\mathbb{R}^N)/S_N}d^N\theta\,d^N\lambda\;e^{-t_2p_2(\lambda)}. \end{aligned} \tag{5.9}$$

The Green function of the theory is

$$\begin{aligned} G_0(\theta,\theta') &= \int_{\mathbb{R}^N/S_N}d^N\lambda\;\overline{\psi_\lambda}(\theta')\,e^{-t_2H_2^{(0)}(i\partial_\theta,\theta)}\,\psi_\lambda(\theta)\\ &= \int_{\mathbb{R}^N/S_N}d^N\lambda\;\overline{\psi_\lambda}(\theta')e^{-t_2p_2(\lambda)}\psi_\lambda(\theta). \end{aligned} \tag{5.10}$$

Up to the infinite factor given by the integral over $\theta=(\theta_1,\dots,\theta_N)\in\mathbb{R}^N$ the integral in (5.9) coincides with the expression (5.4) for the partition function for $t_i=0$, $i\neq 2$.

Now consider two-dimensional Yang-Mills theory. In this case we have the periodic eigenvalue problem for (5.5). Then for the normalized eigenfunctions of H_2 we have

$$\psi_n(\theta)=\sum_{w\in S_N}(-1)^{l(w)}\,\exp(2\pi i\sum_{j=1}^N(n_{w(j)}+\rho_{w(j)})\theta_j),\ (n_1,\dots,n_N)\in P_{++}, \tag{5.11}$$

$$\frac{1}{N!}\int_{(S^1)^N}d^N\theta\,\overline{\psi_n}(\theta)\,\psi_{n'}(\theta)=\sum_{w\in S_N}(-1)^{l(w)}\prod_{j=1}^N\delta_{n_{w(j)},n'_j}=\delta^{S_N}_{n,n'}. \tag{5.12}$$

Here $\rho=(\rho_1,\dots,\rho_N)$ is the half-sum of the positive roots of $\mathfrak{u}_N$. The integral kernel of the identity operator can be represented as

$$K(\theta,\theta')=K(\theta-\theta')=\delta_{(S_N)}(\theta-\theta')=\sum_{n\in P_{++}}\overline{\psi_n}(\theta)\,\psi_n(\theta'). \tag{5.13}$$

The partition function of Yang-Mills theory on a torus Σ_1 is given by the trace of an evolution operator and

$$Z_{\mathrm{YM}}(\Sigma_1)=\mathrm{Tr}\,e^{-t_2H_2(\hat{p},\hat{q})}=\sum_{n\in P_{++}}\int_{(S^1)^N}d^Nx\,\overline{\psi_n}(\theta)\,e^{-t_2H_2^{(0)}(i\partial_\theta,\theta)}\,\psi_n(\theta). \tag{5.14}$$

The kernel for the periodic case can be obviously represented as a matrix element of the projection operator as follows:

$$K(\theta,\theta')=\int_{\mathbb{R}^N/S_N}d^N\lambda\,\overline{\psi_\lambda}(\theta)\;P(\lambda)\,\psi_\lambda(\theta'), \tag{5.15}$$

where the wave functions $\psi_\lambda(\theta)$ are given by (5.6) and

$$P(\lambda)=\sum_{m\in\mathbb{Z}^N}\prod_{j=1}^N\delta(\lambda_j-m_j)=\sum_{k\in\mathbb{Z}^N}e^{2\pi i\sum_{j=1}^N\lambda_jk_j}. \tag{5.16}$$

Equivalently we have

$$K(\theta,\theta') = \sum_{k\in\mathbb{Z}^N}\int_{\mathbb{R}^N/S_N} d^N\lambda\,\overline{\psi_\lambda}(\theta)\; e^{2\pi i\sum_{j=1}^N \lambda_j k_j}\;\psi_\lambda(\theta') = \sum_{k\in\mathbb{Z}^N}\int_{\mathbb{R}^N/S_N} d^N\lambda\,\overline{\psi_\lambda}(\theta)\;\psi_\lambda(\theta'+k). \tag{5.17}$$

We conclude that the Green function $G(\theta,\theta')$ (the path integral on the cylinder with the insertion of $\exp(-t_2 H_2^{(0)})$) is represented as

$$G_{\rm YM}(\theta,\theta') = \sum_{n\in\mathbb{Z}^N}\int_{\lambda\in\mathbb{R}^N/S_N} d^N\lambda\,\overline{\psi_\lambda}(\theta)\; e^{2\pi i\sum_{j=1}^N \lambda_j n_j} e^{-t_2 p_2(\lambda)}\;\psi_\lambda(\theta') \tag{5.18}$$

or equivalently as

$$G_{\rm YM}(\theta,\theta') = \sum_{k\in\mathbb{Z}^N}\int_{\lambda\in\mathbb{R}^N/S_N} d^N\lambda\,\overline{\psi_\lambda}(\theta) e^{-t_2 p_2(\lambda)}\;\psi_\lambda(\theta'+k). \tag{5.19}$$

Let us note that the identities in (5.17), (5.19) are based on the following transformation property of the complete set of skew-symmetric normalized wave functions on $\mathbb{R}^N$:

$$\psi_\lambda(\theta+k) = \sum_{w\in S_N}(-1)^{l(w)}\, e^{2\pi i\sum_{j=1}^N \lambda_{w(j)}k_j} e^{2\pi i\sum_{j=1}^N \lambda_{w(j)}\theta_j}. \tag{5.20}$$

Thus each elementary term in the averaging over S_N is multiplied by the simple exponential factor entering the description of the projector (5.16). Let us also note that the shift transformations in (5.20) can be interpreted as large gauge transformations in Yang-Mills theory discussed above.

The representation (5.19) can be written in the following form:

$$G_{\rm YM}(\theta,\theta') = \sum_{k\in\mathbb{Z}^N} G_0(\theta,\theta'+k). \tag{5.21}$$

If we set the coupling t_2 to zero, $t_2=0$, we recover the formula (5.17) for $K(\theta,\theta')$:

$$K(\theta,\theta') = \sum_{k\in\mathbb{Z}^N} K_0(\theta,\theta'+k). \tag{5.22}$$

For the partition function of Yang-Mills theory on a torus we get (after setting $\theta=\theta'$ above and integrating over x):

$$Z_{\rm YM}(\Sigma_1) = \sum_{n\in\mathbb{Z}^N}\int_{\lambda\in\mathbb{R}^N/S_N} d^N\lambda\int_{(S^1)^N} d^N\theta\,\overline{\psi_\lambda}(\theta)\; e^{2\pi i\sum_{j=1}^N \lambda_j n_j} e^{-t_2 p_2(\lambda)}\;\psi_\lambda(\theta) = \sum_{m\in P_{++}} e^{-t_2 p_2(m+\rho)}.$$

and this coincides with the representation (5.3). Note the obvious relations between (5.17), (5.21), (5.22) and the averaging over the topologically non-trivial gauge transformations discussed above.

Now we are finally ready to consider the case of Yang-Mills-Higgs theory. As was conjectured above, one can choose as a basis of the wave functions the basis of eigenfunctions of the set of Hamiltonian operators in N-particle subsector of Nonlinear Schrödinger theory. Below we construct the Green function and partition function in Nonlinear Schrödinger theory and demonstrate that identifying the

Hamiltonian operator with the quadratic observable $\mathcal{O}_2^{(0)} = \frac{1}{(2\pi)^2} \mathrm{Tr}\, \varphi_0^2$ in Yang-Mills-Higgs theory we reproduce the partition function of Yang-Mills-Higgs theory on a torus.

Let us start with the construction of the kernel of the identity operator in the basis of the N-particle eigenfunctions of the Nonlinear Schrödinger theory. The representation for the kernel (5.17) can be straightforwardly generalized to this case:

(5.23)

$$\widetilde{K}(x,x') = \sum_{(\lambda_1,\dots,\lambda_N)\in\mathcal{R}_N} \overline{\Phi_\lambda^{norm}}(x)\Phi_\lambda^{norm}(x') = \int_{\mathbb{R}^N/S_N} d^N\lambda \;\; \overline{\Phi_\lambda}(x)\, P(\lambda)\, \Phi_\lambda(x'),$$

where $\Phi_\lambda(x)$ are normalized skew-invariant eigenfunctions on $\mathbb{R}^N$ given by (4.16), $\Phi_\lambda^{norm}(x)$ are normalized periodic eigenfunctions given by (4.21) and the sum goes over the set $\mathcal{R}_N$ of solutions of the Bethe Ansatz equations. The projector here is given by

$$P(\lambda) = \mu(\lambda) \sum_{m\in\mathbb{Z}^N} \prod_{j=1}^{N} \delta(\alpha_j(\lambda) - m_j) = \sum_{(\lambda_1^*,\dots,\lambda_N^*)\in\mathcal{R}_N} \prod_j \delta(\lambda_j - \lambda_j^*) \tag{5.24}$$

where $\alpha_j(\lambda)$ are defined as follows (compare with (3.22)):

$$\alpha_j(\lambda) = \lambda_j + \frac{1}{2\pi i}\sum_{k\neq j} \log\left(\frac{\lambda_k - \lambda_j - ic}{\lambda_k - \lambda_j + ic}\right). \tag{5.25}$$

Then we have:

$$\begin{aligned}\widetilde{K}(x,x') &= \sum_{n\in\mathbb{Z}^N} \int_{\mathbb{R}^N/S_N} d^N\lambda \; \mu(\lambda) \;\times \\ &\quad \times \overline{\Phi}_\lambda(x) \; e^{2\pi i \sum_{m=1}^N \lambda_m n_m} \prod_{l\neq j} \left(\frac{\lambda_l - \lambda_j - ic}{\lambda_l - \lambda_j + ic}\right)^{n_j} \quad \Phi_\lambda(x') \\ &= \sum_{n\in\mathbb{Z}^N} \int_{\mathbb{R}^N/S_N} d^N\lambda \; \overline{\Phi}_\lambda(x) \;\; \Phi_\lambda(x'+n).\end{aligned} \tag{5.26}$$

The last equality is a consequence of the following property of the eigenfunctions (4.20) of the N-particle Hamiltonian in Nonlinear Schrödinger theory:

$$\begin{aligned}\Phi_\lambda(x+n) &= \sum_{w\in W} (-1)^{l(w)} \prod_{i<j} \left(\frac{\lambda_{w(i)} - \lambda_{w(j)} + ic}{\lambda_{w(i)} - \lambda_{w(j)} - ic}\right)^{n_i} \exp(2\pi i \sum_m \lambda_{w(m)} n_m) \;\times \\ &\quad \times \prod_{i<j} \left(\frac{\lambda_{w(i)} - \lambda_{w(j)} + ic}{\lambda_{w(i)} - \lambda_{w(j)} - ic}\right)^{\frac{1}{2} + [x_i - x_j]} \exp(2\pi i \sum_k \lambda_{w(k)} x_k),\end{aligned} \tag{5.27}$$

These wave functions are periodic and descend to the wave functions (4.16) if $\lambda = (\lambda_1, \dots, \lambda_N)$ satisfy (4.19).

The representation for the kernel (5.26) leads to the following representation for the Green function (cylinder path integral) and torus partition function for

$t_{i\neq 2}=0$:

$$(5.28)\quad G_{\mathrm{NLS}}(x,x') = \sum_{n\in\mathbb{Z}^N}\int_{\mathbb{R}^N/S_N} d^N\lambda\;\mu(\lambda)\;\overline{\Phi_\lambda(x)}\,e^{2\pi i\sum_{m=1}^N \lambda_m n_m}\times$$
$$\times\prod_{l\neq j}\left(\frac{\lambda_l-\lambda_j-ic}{\lambda_l-\lambda_j+ic}\right)^{n_j} e^{-t_2p_2(\lambda)}\Phi_\lambda(x'),$$

or what is the same:

$$G_{\mathrm{NLS}}(x,x') = \sum_{k\in\mathbb{Z}^N} G^0_{\mathrm{NLS}}(x,x'+k) = \sum_{k\in\mathbb{Z}^N}\int_{\mathbb{R}^N/S_N} d^N\lambda\;\overline{\Phi}_\lambda(x)e^{-t_2p_2(\lambda)}\Phi_\lambda(x'+k).$$

Similarly for the kernel: $\widetilde{K}(x,x') = \sum_{k\in\mathbb{Z}^N}\widetilde{K}_0(x,x'+k)$ since the kernel is a Green function at $t_2=0$. Integrating over x after setting $x=x'$ we obtain the representation for the partition function on the torus:

$$(5.29)\quad Z_{\mathrm{NLS}}(\Sigma_1) = \int_{\mathbb{R}^N/S_N} d^N\lambda\;\mu(\lambda)\sum_{(n_1,\cdots,n_N)\in\mathbb{Z}^N} e^{2\pi i\sum_{m=1}^N\lambda_m n_m}\times$$
$$\times\prod_{l\neq j}\left(\frac{\lambda_l-\lambda_j-ic}{\lambda_l-\lambda_j+ic}\right)^{n_j} e^{-t_2p_2(\lambda)}.$$

This is in complete agreement with the representation for the partition function of the $U(N)$ Yang-Mills-Higgs theory on a torus discussed in Section 3, formula (3.24). Note that one can repeat the same arguments for all observables and higher differential operators of Nonlinear Schrödinger theory, traces of higher powers of the Dunkl operator from Section 4, by simply turning on all other couplings t_k. The identification of the representation of the partition function of the Nonlinear Schrödinger operator and Yang-Mills-Higgs theory on the torus strongly suggests the full equivalence of the theories.

5.3. More precise relation between wave functions. Now we establish a precise relation between wave functions in Yang-Mills-Higgs $U(N)$ gauge theory and the N-particle sector in Nonlinear Schrödinger equation theory. Let us stress that the localization procedure in topological theories is most straightforward in the case of compact manifolds. Thus for example in the case of the torus in Yang-Mills-Higgs theory we recover unambiguously the spectral invariants of the operators $\mathcal{O}_k^{(0)}$ acting on the Hilbert space of the theory. On the classical level this means that the results do not depend on the total derivative terms in the Lagrangian density. It was shown above that the spectrum of the operators $\mathcal{O}_k^{(0)} = \frac{1}{(2\pi i)^k}\operatorname{Tr}\varphi_0^k$ coincides with the spectrum of the quantum Hamiltonians of Nonlinear Schrödinger theory restricted to the N-particle subsector. The identification of the basis of eigenfunctions implies a fixing of additional data – the choice of polarization for the symplectic form in the Hamiltonian quantization. Below we provide a choice of polarization in Yang-Mills-Higgs theory leading to the identification of wave functions with eigenfunctions in Nonlinear Schrödinger theory. To simplify the presentation below we omit symmetrization of wave functions.

Let us consider Yang-Mills-Higgs theory on Σ_h. The result of abelianization can be formulated in the form of an effective gauge theory with the gauge group

$U(1)^N$ and the bosonic part of the action (3.15):

$$\Delta S = \int_{\Sigma_h} d^2z \sum_{i,j=1}^{N} \log\left(\frac{(\varphi_0)_i - (\varphi_0)_j + ic}{(\varphi_0)_i - (\varphi_0)_j - ic}\right) F(A)^i \tag{5.30}$$
$$+\frac{1}{2}\int_{\Sigma_h} d^2z \sum_{i,j=1}^{N} \log((\varphi_0)_i - (\varphi_0)_j + ic)R^{(2)}\sqrt{g};$$

We consider a family of deformations (3.14) of the theory:

$$\Delta S_{\rm YMH} = -\sum_{k=1}^{\infty} t_k \int_{\Sigma_h} d^2z \; {\rm Tr}\,\varphi_0^k \; {\rm vol}_{\Sigma_h} = -\sum_{k=1}^{\infty}\sum_{j=1}^{N} t_k \int (\varphi_0)_j^k {\rm vol}_{\Sigma_h} \tag{5.31}$$

where we impose the condition that $t_k \neq 0$ for a finite subset of indices $k \in \mathbb{Z}$.

In the case of $\Sigma_h = T^2$ the Feynman path integral in the two-dimensional quantum field theory with the action (5.30) is equivalent to the path integral in the one-dimensional theory on S^1 with the phase space T^*H supplied with the following symplectic form:

$$\omega = d(\sum_{j=1}^{N}(\lambda_j + \sum_{k\neq j}\log\,(\frac{\lambda_j - \lambda_k + ic}{\lambda_j - \lambda_k - ic}))d\theta_j), \qquad \theta_j \sim \theta_j + 2\pi i n_j, \;\; n_j \in \mathbb{Z}, \tag{5.32}$$

which is a c-deformation of the standard symplectic form $\omega = \sum_{j=1}^N d\lambda_j \wedge d\theta_j$ on T^*H where $\{\theta_j\}$ are coordinates on H and $\{\lambda_j\}$ are coordinates in the fiber of the projection $T^*H \to H$. Note that λ_j can be identified with $(z,\bar{z})$-independent $(\varphi_0)_i$.

This symplectic form (5.32) can be easily transformed into a canonical one:

$$\omega = d(\sum_{j=1}^{N}\alpha_j(\lambda)d\theta_j) = \sum_{j=1}^{N} d\alpha_j(\lambda)\wedge d\theta_j, \tag{5.33}$$

where

$$\alpha_j(\lambda) = \lambda_j + \sum_{k=1,k\neq j}^{N} \log\,(\frac{\lambda_j - \lambda_k + ic}{\lambda_j - \lambda_k - ic}). \tag{5.34}$$

Quantization in the θ-polarization is given by the following realization:

$$\widehat{\alpha}_j = -i\frac{\partial}{\partial\theta_j}, \qquad \widehat{\theta}_j = \theta_j. \tag{5.35}$$

A basis of wave functions can then be constructed using a complete set of common eigenfunctions of the operators $\widehat{\alpha}_j$:

$$\Psi_\lambda(\theta) = e^{i\sum_{j=1}^N \alpha_j(\lambda)\theta_j} = \prod_{j,k=1}^{N}(\frac{\lambda_j - \lambda_k + ic}{\lambda_j - \lambda_k - ic})^{\theta_j}\, e^{i\sum_{j=1}^N \lambda_j\theta_j}, \tag{5.36}$$

$$\widehat{\alpha}_j\,\Psi_\lambda(\theta) = \alpha_j(\lambda)\,\Psi_\lambda(\theta). \tag{5.37}$$

The deformation (5.31) leads to a consideration of the following set of Hamiltonians in the effective one-dimensional theory:

$$\mathcal{H}_k = \sum_{j=1}^{N}\widehat{\lambda}_j^k, \tag{5.38}$$

where $\hat{\lambda}$ is related to the derivative in θ through

$$\widehat{\alpha}_j = \widehat{\lambda}_j + \sum_{k\neq j} \log\left(\frac{\widehat{\lambda}_j - \widehat{\lambda}_k + ic}{\widehat{\lambda}_j - \widehat{\lambda}_k - ic}\right) = -i\frac{\partial}{\partial\theta_j}. \tag{5.39}$$

Computations from Section 3 show that the eigenvalues of the Hamiltonians are given by $E_k = \sum_{j=1}^N \lambda_j^k$ with the λ's subject to (for periodicity conditions $\theta_j \sim \theta_j + 2\pi m_j$, $m_j \in \mathbb{Z}$):

$$\alpha_j(\lambda) = \lambda_j + \sum_{k\neq j} \log\left(\frac{\lambda_j - \lambda_k + ic}{\lambda_j - \lambda_k - ic}\right) = n_j, \qquad n_j \in \mathbb{Z}. \tag{5.40}$$

Let us remark that in the θ-representation the Hamiltonians (5.38) act on wave functions by complicated integral operators due to the non-trivial relations (5.34) between $\widehat{\alpha}_j$ and $\widehat{\lambda}_k$.

Compare now the basis (5.36) of common eigenfunctions of $\mathcal{H}_k$, $k \in \mathbb{Z}_+$ with the basis of common eigenfunctions for the Yang N-particle system:

$$\Psi_\lambda^Y(x) = \prod_{j,k=1}^N \left(\frac{\lambda_j - \lambda_k + ic}{\lambda_j - \lambda_k - ic}\right)^{[x_j]} e^{i\sum_{j=1}^N \lambda_j x_j}, \tag{5.41}$$

where $[x]$ is the integer part of x. Here we also impose periodicity conditions $x_j \sim x_j + 2\pi m_j$, $m_j \in \mathbb{Z}$ and thus

$$\lambda_j + \sum_{k\neq j} \log\left(\frac{\lambda_j - \lambda_k + ic}{\lambda_j - \lambda_k - ic}\right) = n_j, \qquad n_j \in \mathbb{Z}. \tag{5.42}$$

The Yang system (4.4) can be considered as a quantization of T^*H with the symplectic structure

$$\omega = \sum_{j=1}^N dp_j \wedge dx_j, \qquad x_j \sim x_j + 2\pi n_j, \quad n_j \in \mathbb{Z}. \tag{5.43}$$

In the x-polarization we have

$$\widehat{p}_j = -i\frac{\partial}{\partial x_j}, \qquad \widehat{x}_j = x_j. \tag{5.44}$$

The Hamiltonians $\mathcal{H}_k$ of the Yang system can be expressed as symmetric functions of the Dunkl operators $\mathcal{D}_j$ (see Section 4), and for $k = 2$ we have

$$\mathcal{H}_2 = \frac{1}{2}\sum_{i=1}^N \mathcal{D}_j^2 = \frac{1}{2}\sum_{i=1}^N \widehat{p}_j^2 + c \sum_{1\leq i<j\leq N} \delta(x_i - x_j), \tag{5.45}$$

The eigenvalues E_k of the Hamiltonians H_k are given by

$$E_k = \sum_{j=1}^N \lambda_j^k, \qquad \lambda_j + \sum_{k\neq j} \log\left(\frac{\lambda_j - \lambda_k + ic}{\lambda_j - \lambda_k - ic}\right) = n_j, \tag{5.46}$$

and coincide with the spectrum in Yang-Mills-Higgs theory.

The two bases of wave functions (5.36) and (5.41) are related by a unitary operator with the integral kernel given by

$$K(\theta, x) = \sum_{\mathbf{n}\in\mathbb{Z}^N} \Psi_{\lambda_\mathbf{n}}(\theta)\overline{\Psi^Y_{\lambda_\mathbf{n}}}(x), \tag{5.47}$$

where the sum is over solutions $\lambda_{\mathbf{n}} = (\lambda_1(\mathbf{n}), \ldots, \lambda_N(\mathbf{n}))$ of the equations (5.42) for all $\mathbf{n} = (n_1, \ldots, n_N) \in \mathbb{Z}^N$. The only invariant that does not depend on the choice of the basis of wave functions is the common spectrum of the operators $\mathcal{H}_k$, which is captured by the partition function. The coincidence of the partition functions for Yang-Mills-Higgs theory and the Yang system was demonstrated in the previous subsection.

Let us stress that in the θ-representation there is no simple way to write the eigenfunction problem for wave functions in terms of the differential operators while in the x-representation the eigenfunction problem for Hamiltonians $\mathcal{H}_k$ acting on $\Psi_\lambda^Y(x)$ is given by the Yang equations and its higher-derivative analogs. This explains the relevance of the x-polarization in Yang-Mills-Higgs theory deformed by the local operators $\mathcal{O}_k^{(0)}$.

6. Generalization of G/G gauged WZW model

In this section we describe following [**3**] another instance of the relation between two-dimensional gauge theories and quantum integrability in many-body integrable systems. We consider a generalization of GWZW theory and demonstrate that the partition function can be represented as a sum over solutions of a certain generalization of the Bethe Ansatz equation.

We start with the definition of the set of fields and the action of the odd and even symmetries in the spirit of (3.7), (3.8), (3.9). Let us note that the gauged Wess-Zumino-Witten model can be obtained from the topological Yang-Mills theory by using the group-valued field g instead of the algebra-valued field φ. Introduce the set of fields $(A, \psi_A, \Phi, \psi_\Phi, \chi_\pm, \varphi_\pm, g)$ and $t \in \mathbb{R}^*$ with the following action of the odd and even symmetries:

$$\begin{cases} \mathcal{L}_{(g,t)}\, A^{(1,0)} = (A^g)^{(1,0)} - A^{(1,0)}, \\ \mathcal{L}_{(g,t)}\, A_A^{(0,1)} = -(A^{g^{-1}})^{(0,1)} + A^{(0,1)}, \end{cases} \tag{6.1}$$

$$\begin{aligned}
\mathcal{L}_{(g,t)}\, \psi_A^{(1,0)} &= -g\psi_A^{(1,0)}g^{-1} + \psi_A^{(1,0)}, \quad \mathcal{L}_{(g,t)}\, \psi_A^{(0,1)} = g^{-1}\psi_A^{(0,1)}g - \psi_A^{(0,1)}, \\
\mathcal{L}_{(g,t)}\, g &= 0, \\
\mathcal{L}_{(g,t)}\, \Phi^{(1,0)} &= tg\Phi^{(1,0)}g^{-1} - \Phi^{(1,0)}, \quad \mathcal{L}_{(g,t)}\, \Phi^{(0,1)} = -t^{-1}g^{-1}\Phi^{(0,1)}g + \Phi^{(0,1)}, \\
\mathcal{L}_{(g,t)}\, \psi_\Phi^{(1,0)} &= tg\psi_\Phi^{(1,0)}g^{-1} - \psi_\Phi^{(1,0)}, \quad \mathcal{L}_{(g,t)}\, \psi_\Phi^{(0,1)} = -t^{-1}g^{-1}\psi_\Phi^{(0,1)}g + \psi_\Phi^{(0,1)}, \\
\mathcal{L}_{(g,t)}\, \chi_+ &= tg\chi_+g^{-1} - \chi_+, \quad \mathcal{L}_{(g,t)}\chi_- = -t^{-1}g^{-1}\chi_-g + \chi_- \\
\mathcal{L}_{(g,t)}\, \varphi_+ &= t^{-1}g\varphi_+g^{-1} - \varphi_+, \quad \mathcal{L}_{(g,t)}\varphi_+ = -tg^{-1}\varphi_+g + \varphi_+
\end{aligned}$$

$$\begin{cases} QA = i\psi_A, \quad Q\,\psi_A^{(1,0)} = i(A^g)^{(1,0)} - iA^{(1,0)}, \\ \qquad\qquad\quad\; Q\,\psi_A^{(0,1)} = -i(A^{g^{-1}})^{(0,1)} + iA^{(0,1)}, \end{cases} \tag{6.2}$$

$$Qg = 0,$$

$$\begin{cases} Q\Phi = i\psi_\Phi, \quad Q\psi_\Phi^{(1,0)} = tg\Phi^{(1,0)}g^{-1} - \Phi^{(1,0)}, \\ \qquad\qquad\quad\; Q\psi_\Phi^{(0,1)} = -t^{-1}g^{-1}\Phi^{(0,1)}g - \Phi^{(0,1)}, \end{cases} \tag{6.3}$$

$$Q\chi_\pm = i\varphi_\pm, \quad Q\varphi_+ = tg\chi_+g^{-1} - \chi_+, \quad Q\varphi_- = -t^{-1}g^{-1}\chi_-g + \chi_-.$$

We have $Q^2 = \mathcal{L}_{(g,t)}$ and Q can be considered as a BRST operator on the space of $\mathcal{L}_{(g,t)}$-invariant functionals.

We define the action of the theory in analogy with the construction of the action for Yang-Mills-Higgs theory as follows:

$$\begin{aligned} S &= S_{GWZW} + [Q, \int_{\Sigma_h} d^2 z \, \mathrm{Tr}\,(\frac{1}{2}\Phi \wedge \psi_\Phi \\ &\quad + \tau_1 \, (\varphi_+ \nabla_A^{(1,0)} \Phi^{(0,1)} + \varphi_- \nabla_A^{(0,1)} \Phi^{(1,0)}) + \tau_2 (\chi_+ \varphi_- + \chi_- \varphi_+) \mathrm{vol}_{\Sigma_h})]_+ . \end{aligned} \tag{6.4}$$

Taking $\tau_1 = 0$, $\tau_2 = 1$ and applying the standard localization technique to this theory we obtain for the partition function:

$$\begin{aligned} Z_{GWZWH}(\Sigma_h) &= \frac{e^{(1-h)a(t)}}{|W|} \int_H d^N\lambda \; \mu_q(\lambda)^h \sum_{(n_1,\ldots,n_N)\in\mathbb{Z}^N} e^{2\pi i \sum_{m=1}^N \lambda_m n_m (k+c_v)} \times \\ &\quad \times \prod_{j\neq k} (e^{2\pi i(\lambda_j-\lambda_k)} - 1)^{n_j-n_k+1-h} \\ &\quad \times \prod_{j,k} (te^{2\pi i(\lambda_j-\lambda_k)} - 1)^{n_j-n_k+1-h}, \end{aligned} \tag{6.5}$$

where $a(t)$ is an h-independent constant, the integral goes over the Cartan torus $H = (S^1)^N$ and

$$\mu_q(\lambda) = \det \| \frac{\partial \beta_j(\lambda)}{\partial \lambda_k} \|, \tag{6.6}$$

with

$$e^{2\pi i \beta_j(\lambda)} = e^{2\pi i \lambda_j (k+c_v)} \prod_{k\neq j} \frac{te^{2\pi i(\lambda_j-\lambda_k)} - 1}{te^{2\pi i(\lambda_k-\lambda_j)} - 1}. \tag{6.7}$$

We can rewrite this formula in a form similar to (3.21):

$$\begin{aligned} Z_{GWZWH}(\Sigma_h) &= \frac{e^{(1-h)a(t)}}{|W|} \int_H d^N\lambda \; \mu_q(\lambda)^h \sum_{(n_1,\cdots,n_N)\in\mathbb{Z}^N} e^{2\pi i \sum_{m=1}^N \beta_m(\lambda) n_m} \times \\ &\quad \times \prod_{j<k} (e^{i\pi(\lambda_j-\lambda_k)} - e^{i\pi(\lambda_k-\lambda_j)})^{2-2h} \times \\ &\quad \times \prod_{j<k} |te^{i\pi(\lambda_j-\lambda_k)} - e^{i\pi(\lambda_k-\lambda_j)}|^{2-2h}. \end{aligned} \tag{6.8}$$

Summation over the integers in (6.5) leads to the following restriction on the integration parameters:

$$e^{2\pi i \lambda_j (k+c_v)} \prod_{k\neq j} \frac{te^{2\pi i(\lambda_j-\lambda_k)} - 1}{te^{2\pi i(\lambda_k-\lambda_j)} - 1} = 1 \;, \qquad i = 1, \ldots, N, \tag{6.9}$$

It is useful to rewrite the equations (6.9) in the standard form of the Bethe Ansatz equations:

$$e^{2\pi i \lambda_j (k+c_v)} \prod_{k\neq j} \frac{\sin(i\pi(\lambda_j - \lambda_k + ic))}{\sin(i\pi(\lambda_j - \lambda_k - ic))} = 1 \;, \qquad i = 1, \ldots, N, \tag{6.10}$$

This clearly shows that we are dealing with a kind of XXZ quantum integrable chain. The particular form (6.10) can be obtained by the taking the limit $s \to -i\infty$ in the following Bethe equations:

(6.11)
$$\left(\frac{\sin(i\pi(\lambda_j - isc))}{\sin(i\pi(\lambda_j + isc))}\right)^{(k+c_v)} \prod_{k\neq j} \frac{\sin(i\pi(\lambda_j - \lambda_k + ic))}{\sin(i\pi(\lambda_j - \lambda_k - ic))} = 1 \quad , \qquad i = 1, \cdots, N,$$

corresponding to the formal limit of the infinite spin s of the XXZ chain.

The partition function is the generalization of Yang-Mills-Higgs theory, discussed above, and can be written in the following form:

$$Z_{GWZWH}(\Sigma_h) = \sum_{\lambda_i \in \mathcal{R}_q} (D^q_\lambda)^{2-2h}, \tag{6.12}$$

where $\mathcal{R}_q$ is the set of solutions of (6.9) and

$$D^q_\lambda = \mu_q(\lambda)^{-1/2} \prod_{i<j} (q^{\frac{1}{2}(\lambda_i - \lambda_j)} - q^{\frac{1}{2}(\lambda_j - \lambda_i)}) \prod_{i<j} |tq^{\frac{1}{2}(\lambda_i - \lambda_j)} - q^{\frac{1}{2}(\lambda_j - \lambda_i)}|, \tag{6.13}$$

where we use the standard parametrization $q = \exp(2\pi i/(k + c_v))$. Note that in the limit $t \to \infty$ equation (6.9) and the expression for the partition function (6.13) up to an overall scaling factor become the corresponding expressions for a gauged Wess-Zumino-Witten model. Finally note that the form of (6.9) and the explicit expressions for the q-Casimir operators, playing the role of the Hamiltonians, strongly imply the description of the wave functions of the theory in terms of the wave functions in a particular XXZ finite spin chain. This proposition will be discussed in detail elsewhere.

Acknowledgements. We are grateful to J. Bernstein, M. Kontsevich, W. Nahm, N. Nekrasov, F. Smirnov and L. Takhtajan for discussions. The research is supported by a Science Foundation Ireland grant.

References

[1] G. Moore, N. Nekrasov and S. Shatashvili, *Integrating over Higgs Branches,* Commun. Math. Phys. **209** (2000) 97, [arXiv:hep-th/9712241].

[2] N. J. Hitchin, *The self-duality equations on a Riemann surface*, Proc. London Math. Soc. 55 (1987), 59-126.

[3] A. Gerasimov and S. Shatashvili, *Higgs bundles, gauge theories and quantum groups,* [arXiv:hep-th/0609024].

[4] E. H. Lieb, W. Liniger, *Exact analysis of an interacting Bose gas I. The general solution and the ground state*, Phys. Rev. (2), **130** (1963), 1605.

[5] F. A. Berezin, G. P. Pohil, V. M. Finkelberg, Vestnik MGU **1**, 1 (1964) 21.

[6] C. N. Yang, *Some exact results for the many-body problems in one dimension with repulsive delta-function interaction*, Phys.Rev.Lett. **19** (1967), 1312.

[7] C. N. Yang, C. P. Yang, *Thermodynamics of a one-dimensional system of bosons with repulsive delta-function interaction*, J. Math. Phys. **10** (1969), 1115.

[8] P. P. Kulish, S. V. Manakov, L. D. Faddeev, *Comparison of the exact quantum and quasi-classical results for the nonlinear Schrödinger equation*, Theor. Math. Phys. **28** (1976) 615.

[9] L. D. Faddeev, L. A. Takhtajan, *Hamiltonian Methods in the Theory of Solitons*, Springer-Verlag, (1980).

[10] L. D. Faddeev, *How the algebraic Bethe Ansatz works for integrable models*, "Relativistic gravitation and gravitational radiation", Proceedings of Les Houches School of Physics (1995) p. 149, [arXiv:hep-th/9605187].

[11] K. Hikami, *Notes on the structure of the δ-function interacting gas. Intertwining operator in the degenerate affine Hecke algebra*, J. Phys. A: Math. Gen **31** (1998) No. 4.

[12] G. J. Heckman, E. M. Opdam, *Yang's system of particles and Hecke algebras*, Annals of Math. **145** (1997), 139.

[13] E. Emsiz, E. M. Opdam, J. V. Stokman, *Periodic integrable systems with delta-potentials*, Comm. Math. Phys. **264** (2006), 191, [arXiv:math.RT/0503034].
[14] V. G. Drinfeld, *Degenerate affine Hecke algebras and Yangians*, Funct. Anal. Appl. **20** (1986) 58.
[15] G. Lusztig, *Affine Hecke algebras and their graded version*, J. AMS **2**, (1989) 3, 599.
[16] T. C. Dunkl, *Differential-difference operators associated to reflection groups*, Trans. Amer. Math. Soc. **311**, no.1 (1989), 167.
[17] I. Cherednik, *A unification of Knizhnik-Zamolodchikov and Dunkl operators via affine Hecke algebras*, Invent. Math. J. **106** (1991), 411.

(1) Institute for Theoretical and Experimental Physics, Moscow, 117259, Russia; (2) School of Mathematics, Trinity College, Dublin 2, Ireland; (3) The Hamilton Mathematics Institute TCD, Dublin 2, Ireland; (4) IHES, 35 route de Chartres, Bures-sur-Yvette, FRANCE

E-mail address: anton@maths.tcd.ie, samson@maths.tcd.iee

Proceedings of Symposia in Pure Mathematics
Volume **78**, 2008

Hurwitz numbers, matrix models and enumerative geometry

Vincent Bouchard and Marcos Mariño

ABSTRACT. We propose a new, conjectural recursion solution for Hurwitz numbers at all genera. This conjecture is based on recent progress in solving type B topological string theory on the mirrors of toric Calabi–Yau manifolds, which we briefly review to provide some background for our conjecture. We show in particular how this B-model solution, combined with mirror symmetry for the one-leg, framed topological vertex, leads to a recursion relation for Hodge integrals with three Hodge class insertions. Our conjecture in Hurwitz theory follows from this recursion for the framed vertex in the limit of infinite framing.

1. Introduction

This article has two parts that can be read independently, although they are logically related to each other. The first part presents a new conjecture in Hurwitz theory. Hurwitz numbers can be regarded as the simplest enumerative quantities in algebraic geometry, since they count covers of Riemann surfaces by Riemann surfaces. Recently, fascinating connections have emerged between this classical problem and the most modern incarnations of enumerative geometry. Hurwitz numbers turn out to be closely related to topological gravity in two dimensions [**GJV, OP**], to Hodge integrals [**ELSV, FP, GV**], and to the Gromov–Witten theory of $\mathbb{P}^1$ [**OP1**].

An interesting problem in Hurwitz theory has been to find a complete set of recursion relations solving the theory. Such relations are known to exist, and they can take many forms; see [**V, GJ, GJV, GJVa, FP, P**] for significant examples. In this paper, we provide a new conjectural recursion solution of Hurwitz theory, different from the ones that have been presented in the literature, which determines completely the simple Hurwitz numbers, counting covers of $\mathbb{P}^1$ with arbitrary genus and with arbitrary ramification at one point. Our result should be regarded as a conjecture, since we do not provide a proof. We try however to state it in a precise way and we give what we consider to be convincing evidence.

2000 *Mathematics Subject Classification.* Primary 14N35; Secondary 14N10, 81T30.
Key words and phrases. String theory, Hurwitz theory, matrix models.
We would like to thank R. Pandharipande and R. Vakil for useful discussions, and A. Klemm and S. Pasquetti for collaboration in a related work. Preprint number: CERN-PH-TH/2007-152.

This conjecture in Hurwitz theory is in fact a spinoff of some recent progress in topological string theory/Gromov–Witten theory on toric manifolds. In [**M, BKMP**], a new formalism for the type B topological string on mirrors of toric Calabi–Yau threefolds was proposed, inspired by the results of [**EO**]. The recursion relations for Hurwitz numbers that we obtain are in fact modeled on the recursion procedure discovered in [**EO**] in the context of matrix models. This procedure has already found applications in two-dimensional topological gravity, and it has been shown in [**EO1, E**] that Mirzakhani's recursion relations [**Mir**] can be regarded as particular cases of the general formalism of [**EO**].

The second part of the paper is devoted to a brief survey of this progress in matrix model theory and topological string theory, intended for a mathematical audience. In particular, we review the proposal of [**M, BKMP**] that the one-leg, framed topological vertex obeys a similar set of recursion relations by using the mirror geometry first found in [**AKV**]. This result, together with the conjectural formula in [**MV**] (later proved in [**LLZ, OP3**]) leads to a recursion relation for Hodge integrals with three Hodge class insertions. It is known that Hurwitz numbers can be obtained as a limit of the one-leg framed vertex in the case of infinite framing, and this finally leads to the conjectural recursion relation for Hurwitz numbers.

The organization of this paper is as follows. In section 2 we present our conjectural recursion relation in Hurwitz theory. In section 3 we explain how this recursion arises in the context of mirror symmetry and the reformulation of the B-model proposed in [**M, BKMP**].

2. A new conjecture in Hurwitz theory

2.1. Hurwitz numbers. In this paper we will be concerned with Hurwitz numbers counting the number of covers of $\mathbb{P}^1$ by a Riemann surface of genus g and with arbitrary ramification at one point. The ramification type of the covering map at the special point is labeled by a partition μ of length $h := \ell(\mu)$:

$$\mu = (\mu_1 \geq \cdots \geq \mu_{\ell(\mu)} > 0). \tag{2.1}$$

We will use the standard notation

$$|\mu| = \sum_{i=1}^{\ell(\mu)} \mu_i, \qquad z_\mu = |\mathrm{Aut}(\mu)| \prod_{i=1}^{\ell(\mu)} \mu_i. \tag{2.2}$$

A partition μ can be regarded as a conjugacy class of the group of permutations of $|\mu|$ elements, $S_{|\mu|}$. Given a representation R of $S_{|\mu|}$, we define,

$$f_R(\mu) = \frac{|\mu|!}{z_\mu} \frac{\chi_R(\mu)}{\dim R}, \tag{2.3}$$

where $\dim R$ is the dimension of R and $\chi_R(\mu)$ is the character of the conjugacy class μ in the representation R. Representations R of the symmetric group will be identified by Young tableaux, labeled by the number of boxes l_i in each row. The total number of boxes of R will be denoted by

$$|R| = \sum_i l_i. \tag{2.4}$$

We will set $f_R(\mu)=0$ if $|R|\neq|\mu|$. We also define the following quantity associated to a tableau R,

$$\kappa_R=\sum_i l_i(l_i-2i+1). \tag{2.5}$$

A classical result expresses the Hurwitz numbers $H^\bullet_{g,\mu}$ counting disconnected covers in terms of the representation theory of the symmetric group $S_{|\mu|}$,

$$H^\bullet_{g,\mu}=\sum_R\left(\frac{\dim R}{|\mu|!}\right)^2 f_R(\mu)(\kappa_R/2)^{2g-2+h+|\mu|}, \tag{2.6}$$

where only representations with $|R|=|\mu|$ contribute to the sum. We also introduce a generating functional for Hurwitz numbers,

$$Z(v)=1+\sum_\mu\sum_{g\geq 0} g_s^{2g-2+\ell(\mu)}\frac{H^\bullet_{g,\mu}}{(2g-2+\ell(\mu)+|\mu|)!}p_\mu(v), \tag{2.7}$$

where $v=\{v_i\}_{i\geq 1}$ is a set of infinite formal variables and $p_\mu(v)$ are the symmetric polynomials

$$p_\mu(v)=p_{\mu_1}(v)\cdots p_{\mu_h}(v), \qquad p_r=\sum_i v_i^r, \tag{2.8}$$

i.e. p_r is the rth power sum (see [**Mac**]). We will call $Z(v)$ the *partition function* of Hurwitz theory. It can be also written as

$$Z(v)=\sum_R g_s^{-|R|}\left(\frac{\dim R}{|R|!}\right)\mathrm{e}^{-g_s\kappa_R/2}s_R(v), \tag{2.9}$$

where $s_R(v)$ is the Schur polynomial. One can extract the formal logarithm of this, to obtain the *free energy* of Hurwitz theory,

$$F(v)=\sum_\mu\sum_{g\geq 0} g_s^{2g-2+\ell(\mu)}\frac{H_{g,\mu}}{(2g-2+\ell(\mu)+|\mu|)!}p_\mu(v), \tag{2.10}$$

which defines the Hurwitz numbers $H_{g,\mu}$ counting connected covers.

We will find it very convenient to collect the Hurwitz numbers at fixed g and $h=\ell(\mu)$ in the following functionals

$$H_g(x_1,\ldots,x_h)=\sum_{\mu|\ell(\mu)=h}\frac{z_\mu}{(2g-2+\ell(\mu)+|\mu|)!}H_{g,\mu}m_\mu(x), \tag{2.11}$$

where $m_\mu(x)$ are the monomial symmetric functions in the x_i, $i=1,\cdots,h$, divided by an overall factor $x_1\cdots x_h$,

$$m_\mu(x)=\frac{1}{|\mathrm{Aut}(\mu)|}\sum_{\sigma\in S_h}x_{\sigma(i)}^{\mu_i-1}. \tag{2.12}$$

We have that

$$H_g(x_1,\cdots,x_h)=\frac{h!}{(2g-2+2h)!}H^{\mathbb{P}^1}_{g,\mathbf{1^h}}+\mathcal{O}(x_1,\cdots,x_h). \tag{2.13}$$

We can use the ELSV formula to write the generating functionals (2.11) in a nice form (see [**GJV**] for a similar rearrangement). Let $\overline{M}_{g,h}$ be the Deligne-Mumford moduli space of genus g curves with h marked points, $\mathbb{E}$ the Hodge bundle, and $\lambda_i = c_i(\mathbb{E})$ its Chern classes. Define

$$\Lambda_g^\vee(t) = \sum_{i=0}^{g} (-1)^i \lambda_i t^{g-i}. \tag{2.14}$$

The ELSV formula states that

$$H_{g,\mu} = \frac{(2g-2+\ell(\mu)+|\mu|)!}{z_\mu} \prod_{i=1}^{\ell(\mu)} \frac{\mu_i^{\mu_i+1}}{\mu_i!} \int_{\overline{M}_{g,\ell(\mu)}} \frac{\Lambda_g^\vee(1)}{\prod_{i=1}^{\ell(\mu)}(1-\mu_i\psi_i)}. \tag{2.15}$$

The ψ_i classes of two-dimensional topological gravity are defined as follows. Let $\mathcal{L}_i$ be the line bundle over $\overline{M}_{g,h}$ whose fiber at a point $(C; x_1, \ldots, x_h) \in \overline{M}_{g,h}$ is the cotangent space to C at x_i. Then $\psi_i := c_1(\mathcal{L}_i)$.

We now introduce the variable y through

$$x = y\mathrm{e}^{-y}, \tag{2.16}$$

which defines the so-called tree function $T(x)$

$$y(x) = T(x) = -W(-x) \tag{2.17}$$

where $W(x)$ is the Lambert W function (see [**CGHJK**] for a very useful summary of its properties). Near $(x, y) = (0, 0)$ one has the series expansion

$$y(x) = \sum_{\mu=1}^{\infty} \frac{\mu^{\mu-1}}{\mu!} x^\mu. \tag{2.18}$$

We now define the one-forms

$$\zeta_n(y) = \mathrm{d}y \frac{1-y}{y} \left(\frac{y}{1-y} \frac{\mathrm{d}}{\mathrm{d}y} \right)^{n+2} y, \qquad n \geq 0. \tag{2.19}$$

For example,

$$\zeta_0(y) = \frac{\mathrm{d}y}{(1-y)^2}, \quad \zeta_1(y) = \frac{1+2y}{(1-y)^4} \mathrm{d}y. \tag{2.20}$$

Since

$$\frac{y}{1-y} \frac{\mathrm{d}}{\mathrm{d}y} = x \frac{\mathrm{d}}{\mathrm{d}x}, \quad \frac{\mathrm{d}x}{x} = \frac{1-y}{y} \mathrm{d}y, \tag{2.21}$$

we have the following expansion in powers of x

$$\zeta_n(x) = \frac{\mathrm{d}x}{x} \left(x \frac{\mathrm{d}}{\mathrm{d}x} \right)^{n+2} y(x) = \sum_{\mu=1}^{\infty} \frac{\mu^{\mu+1+n}}{\mu!} x^{\mu-1} \mathrm{d}x. \tag{2.22}$$

It then follows that

$$\begin{aligned} &H_g(x_1, \ldots, x_h)\mathrm{d}x_1 \cdots \mathrm{d}x_h \\ &= \sum_{n_i \geq 0} \sum_{\mu_i \geq 1} \left(\prod_{i=1}^{h} \frac{\mu_i^{\mu_i+1+n_i}}{\mu_i!} x_i^{\mu_i-1} \mathrm{d}x_i \right) \int_{\overline{M}_{g,h}} \Lambda_g^\vee(1) \prod_{i=1}^{h} \psi_i^{n_i} \\ &= \sum_{n_i=0}^{3g-3+h} \langle \tau_{n_1} \cdots \tau_{n_h} \Lambda_g^\vee(1) \rangle \prod_{i=1}^{h} \zeta_{n_i}(y_i), \end{aligned} \tag{2.23}$$

where we introduced the Hodge integrals

$$\langle \tau_{n_1} \cdots \tau_{n_h} \Lambda_g^\vee(1) \rangle = \int_{\overline{M}_{g,h}} \Lambda_g^\vee(1) \prod_{i=1}^h \psi_i^{n_i}. \tag{2.24}$$

2.2. A recursion relation for Hurwitz theory. Our conjecture will describe a recursion solution for the generating functionals of Hurwitz numbers (2.11). This recursion relation was first proposed by Eynard and Orantin in a different context, namely for correlation functions of matrix models [**EO**]. Then, in [**M, BKMP**] it was conjectured that a suitable generalization of this relation also describes B-model topological string theory on the mirrors of toric Calabi-Yau threefolds. In the present paper we conjecture that this relation also generates recursively the generating functionals of Hurwitz numbers.

We will review these developments in section 3, and motivate our conjecture for Hurwitz numbers in section 3.4 by using the relation between Gromov-Witten theory and Hurwitz theory. We will see that the recursion relation in Hurwitz theory is in fact a particular case of a similar recursion relation describing Gromov-Witten potentials (or topological string amplitudes) on toric Calabi–Yau threefolds. But, for the moment, let us explain the recursion in the context of Hurwitz theory and state our conjecture.

Generically, the recursion relations of Eynard and Orantin provide a formalism to generate an infinite sequence of meromorphic differentials and symplectic invariants associated to a curve. Our application to Hurwitz theory consists in a particular case of this construction, where we consider the curve introduced earlier,

$$C : \{x = y\mathrm{e}^{-y}\}, \tag{2.25}$$

which defines the tree function $T(x)$

$$y(x) = T(x). \tag{2.26}$$

To formulate the recursion relations we need a few geometric ingredients associated to the curve C. First, the x-projection of C has one ramification point, which we will denote by ν, at

$$(x(\nu), y(\nu)) = (\mathrm{e}^{-1}, 1). \tag{2.27}$$

Near the ramification point ν there are two points $q, \bar{q} \in C$ with the same x-projection, that is, $x(q) = x(\bar{q})$. Write

$$y(q) = 1 + z, \qquad y(\bar{q}) = 1 + S(z), \tag{2.28}$$

with

$$S(z) = -z + \mathcal{O}(z^2). \tag{2.29}$$

Note that in these variables, the ramification point is at $z = 0$. By definition of q and $\bar{q}$, we have

$$(1 + z)\mathrm{e}^{-z} = (1 + S(z))\mathrm{e}^{-S(z)}, \tag{2.30}$$

which we can solve to get $S(z)$ as a power series in z:

$$S(z) = -z + \frac{2\,z^2}{3} - \frac{4\,z^3}{9} + \frac{44\,z^4}{135} - \frac{104\,z^5}{405} + \frac{40\,z^6}{189} - \frac{7648\,z^7}{42525} + \mathcal{O}(z^8). \tag{2.31}$$

This series arises in the study of random graphs; see for example [**JKLP**].

The first ingredient is the Bergman kernel of C, which is the unique meromorphic differential $B(x_1, x_2)$ on C with a double pole at $x_1 = x_2$. Since C has genus 0, the Bergman kernel is given, in terms of the local coordinate y, by

$$B(y_1, y_2) = \frac{\mathrm{d}y_1 \mathrm{d}y_2}{(y_1 - y_2)^2}, \tag{2.32}$$

where y_i is implicitly related to x_i by the tree function, that is, $y_i = y(x_i) = T(x_i)$. The next ingredient is built up from this, and it is the one-form

$$\mathrm{d}E_q(x) = \frac{1}{2} \int_q^{\bar{q}} B(\xi, x), \tag{2.33}$$

which is well-defined locally near the ramification point ν. For C, we have, in the coordinates z and y,

$$\mathrm{d}E_z(y) = \frac{1}{2} \mathrm{d}y \Big[\frac{1}{y - 1 - z} - \frac{1}{y - 1 - S(z)} \Big]. \tag{2.34}$$

Finally, we need the one-form

$$\omega(q) = (\log y(q) - \log y(\bar{q})) \frac{\mathrm{d}x(q)}{x(q)}, \tag{2.35}$$

which becomes, in terms of z,

$$\begin{aligned} \omega(z) &= \Big(\log(1 + S(z)) - \log(1 + z) \Big) \frac{z \mathrm{d}z}{1 + z} \\ &= (S(z) - z) \frac{z \mathrm{d}z}{1 + z}. \end{aligned} \tag{2.36}$$

With these ingredients, we define recursively a set of meromorphic differentials $W_g(y_1, \ldots, y_h)$, with $g \geq 0$, $h \geq 1$, using residue calculus at the ramification point $z = 0$ of the x-projection of the curve C. We first define the starting point of the recursion as

$$W_0(y) = 0, \qquad W_0(y_1, y_2) = B(y_1, y_2). \tag{2.37}$$

The meromorphic differentials are then defined by

$$\begin{aligned} W_g(y, y_1 \ldots, y_h) = \mathrm{Res}_{z=0} \frac{\mathrm{d}E_z(y)}{\omega(z)} \Big[& W_{g-1}(1 + z, 1 + S(z), y_1, \ldots, y_h) \\ & + \sum_{l=0}^{g} \sum_{J \subset H} W_{g-l}(1 + z, y_J) W_l(1 + S(z), y_{H \backslash J}) \Big], \end{aligned} \tag{2.38}$$

where $H = 1, \ldots, h$, and given any subset $J = \{i_1, \ldots, i_j\} \subset H$ we set

$$y_J = \{y_{i_1}, \ldots, y_{i_j}\}. \tag{2.39}$$

The recursion relation (2.38) can be represented graphically as in Fig. 1.

We are now ready to state our main conjecture:

Conjecture 2.1. The generating functionals of Hurwitz theory in the differential form (2.23) are explicitly given as follows (we use the notation above). For $g = 0, h = 1$ and $g = 0, h = 2$ one has

$$H_0(x) \mathrm{d}x = y(x) \frac{\mathrm{d}x}{x}, \qquad H_0(x_1, x_2) \mathrm{d}x_1 \mathrm{d}x_2 = B(y_1, y_2) - \frac{\mathrm{d}x_1 \mathrm{d}x_2}{(x_1 - x_2)^2}, \tag{2.40}$$

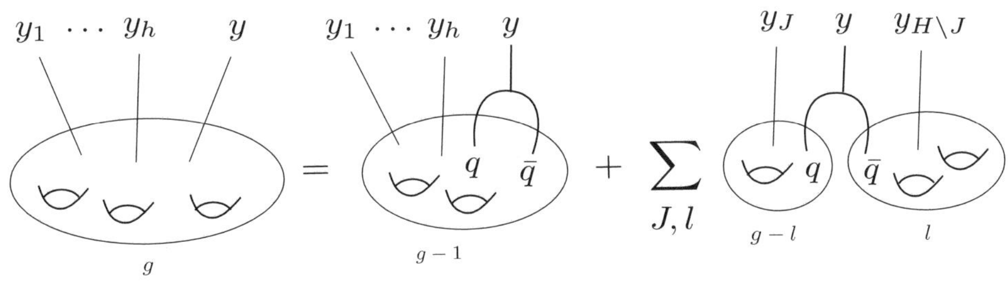

FIGURE 1. A graphic representation of the recursion relation (2.38).

where $y(x) = T(x)$ and the Bergman kernel $B(y_1, y_2)$ was defined in (2.32). For the remaining cases, one has that

$$H_g(x_1, \ldots, x_h)\mathrm{d}x_1 \cdots \mathrm{d}x_h = W_g(y_1, \ldots, y_h) \tag{2.41}$$

where $W_g(y_1, \ldots, y_h)$ are defined by the recursion relation (2.38).

This conjecture will be further motivated in section 3.4, using the relations between Hurwitz theory, Gromov–Witten theory and B-model topological string theory. Let us now present some examples of the generating functionals.

We find, for example, that

$$\begin{aligned} W_1(y) &= -\frac{1}{2}\mathrm{Res}_{z=0}\left[\frac{1+z}{z}\Big(\frac{1}{y-1-z} - \frac{1}{y-1-S(z)}\Big)\frac{S'(z)\mathrm{d}z}{(z-S(z))^3}\right]\mathrm{d}y \\ &= \frac{4y-y^2}{24(1-y)^2}\mathrm{d}y. \end{aligned} \tag{2.42}$$

When expanded in terms of x, we recover known results on Hurwitz numbers. Notice that it is more convenient to work with the y variable, rather than the x variable, and this is in turn very useful in order to extract the Hodge integrals in (2.15) from the results, since one only has to write them in terms of the one-forms $\zeta_n(y)$ and compare them to (2.23). For instance,

$$W_1(y) = \frac{1}{24}\big(-\zeta_0(y) + \zeta_1(y)\big). \tag{2.43}$$

Similarly, one can find by using the residue calculus,

$$\begin{aligned} W_0(y_1, y_2, y_3) &= \prod_{i=1}^{3}\zeta_0(y_i), \\ W_0(y_1, \ldots, y_4) &= \sum_{i=1}^{4}\zeta_1(y_i)\prod_{j\neq i}\zeta_0(y_j), \end{aligned} \tag{2.44}$$

for $g = 0$. For $g = 1, h = 2$ the result of the recursion gives

$$\begin{aligned} W_1(y_1, y_2) = \frac{1}{24}\Big(&-\zeta_0(y_1)\zeta_1(y_2) + \zeta_0(y_1)\zeta_2(y_2) + (y_1 \leftrightarrow y_2) \\ &+ \zeta_1(y_1)\zeta_1(y_2)\Big). \end{aligned} \tag{2.45}$$

For $g = 2$ and $h = 1, 2$, one obtains,

$$W_2(y) = \frac{1}{5760}\Big(7\zeta_2(y) - 12\zeta_3(y) + 5\zeta_4(y)\Big),$$

$$\begin{aligned} W_2(y_1, y_2) = \frac{1}{5760}\Big(&7\zeta_0(y_1)\zeta_3(y_2) - 12\zeta_0(y_1)\zeta_4(y_2) + 5\zeta_0(y_1)\zeta_5(y_2) \\ &+ 21\zeta_1(y_1)\zeta_2(y_2) - 36\zeta_1(y_1)\zeta_3(y_2) + 15\zeta_1(y_1)\zeta_4(y_2) \\ &+ 29\zeta_2(y_1)\zeta_3(y_2) + (y_1 \leftrightarrow y_2) - 50\zeta_2(y_1)\zeta_2(y_2)\Big), \end{aligned} \tag{2.46}$$

and finally for $g = 3$, $h = 1$ one has

$$W_3(y) = \frac{1}{2903040}\Big(- 93\zeta_4(y) + 205\zeta_5(y) - 147\zeta_6(y) + 35\zeta_7(y)\Big). \tag{2.47}$$

By comparing these results with the expected form (2.23), it is easy to check that the recursion relation reproduces known results for the Hodge integrals involved in (2.15). This implies in particular that the above conjecture provides a recursion relation for the Hodge integrals appearing in the ELSV formula (which include all the correlation functions of 2d gravity).

Some remarks concerning these results are in order:

(1) In view of the results of [**EO1, E**], our conjectural recursion formula is probably the analogue in Hurwitz theory of Mirzakhani's recursion for the Weil–Petersson volume of moduli space [**Mir**]. It is known that Mirzakhani's recursion is essentially equivalent to the KdV/Virasoro structure of 2d topological gravity [**MS, Mir2, LZ**]. In the same way, we expect our recursion relation to be equivalent to the Toda structure governing Hurwitz theory [**P, O**].

(2) It may be worth mentioning that the conjectural recursion that we propose is different from the cut-and-join equation for simple Hurwitz numbers (see for example Lemma 3.1 in [**GJVa**]). The latter is more easily formulated in the "representation basis" (consisting of the symmetric polynomials $p_\mu(v)$), while our recursion is naturally formulated in the "winding number basis" (consisting of the monomial symmetric functions $m_\mu(x)$); moreover, the initial conditions of the recursions are different. In fact, the cut-and-join equation should be regarded as a differential equation governing the generating functional of Hurwitz numbers, while our recursive conjecture should be regarded as an explicit solution of this equation. Nevertheless, it should be possible to derive one from the other. In particular, it would be very interesting to prove our recursion using the cut-and-join equation as a starting point. However, the relation between the two equations seems to involve highly non-trivial combinatorics, which appear to be nicely encoded in the function $S(z)$ and the residue calculus.

3. Toric geometry, matrix models, and mirror symmetry

To provide further motivation for the recursion conjecture 2.1 in Hurwitz theory, we need to introduce a new framework for B-model topological string theory on mirrors of toric Calabi–Yau threefolds, inspired by matrix models. Let us start by reviewing known results in the matrix model realm.

3.1. Matrix models. The recursion relation (2.38) which we use with small modifications in Hurwitz theory was first found in the context of matrix models. A matrix model for a Hermitian $N \times N$ matrix M is defined by a partition function

$$Z = \int \mathrm{d}M \, \mathrm{e}^{-N \mathrm{Tr}\, V(M)}, \tag{3.1}$$

where $V(x)$ is a polynomial, or more generally a power series in x. The main goal of matrix model theory is to compute (3.1) as well as the connected correlation functions

$$W(z_1, \ldots, z_h) = \left\langle \mathrm{Tr}\, \frac{1}{z_1 - M} \cdots \mathrm{Tr}\, \frac{1}{z_h - M} \right\rangle^{(c)}. \tag{3.2}$$

It turns out that these quantities have an asympotic expansion in inverse powers of N (the rank of the matrix), with the structure

$$\begin{aligned} \log Z &= \sum_{g=0}^{\infty} F_g N^{2-2g}, \\ W(z_1, \ldots, z_h) &= \sum_{g=0}^{\infty} N^{2-2g+h} W_g(z_1, \ldots, z_h). \end{aligned} \tag{3.3}$$

We will call F_g the genus g closed amplitude, and $W_g(z_1, \ldots, z_h)$ the genus g open amplitude with h holes. The reason for these names is their diagrammatic interpretation in terms of double-line diagrams or fatgraphs (see for example [**BIZ**] for an exposition of the diagrammatics of matrix models).

The $1/N$ expansion of matrix models has been studied for thirty years now, starting with the seminal paper [**BIPZ**]. The main result that emerges is that the genus g amplitudes can be obtained from a single object, an algebraic curve

$$\{P(x, y) = 0\} \subset \mathbb{C}^2 \tag{3.4}$$

called the *classical spectral curve* of the matrix model. This curve is closely related to the generating function of disk amplitudes (also called the resolvent):

$$W_0(x) = \frac{1}{2}(V'(x) + y(x)). \tag{3.5}$$

It turns out that the genus g open amplitudes of the matrix model can be constructed recursively as follows (see [**EO**] for a detailed exposition and a list of references).[1]

We first need some necessary ingredients, which were introduced in section 2.2 for the curve relevant for Hurwitz theory. To start with, we notice that, generically, the x-projection of the curve C will have a set of *ramification points* $q_i \in C$ defined by the condition

$$\mathrm{d}x(q_i) = 0. \tag{3.6}$$

Near a ramification point[2] there are two points, q and $\bar{q}$, with the same x-coordinate $x(q) = x(\bar{q})$. The first ingredient is the *Bergman kernel* of C, which is the unique

[1]In fact, the genus g amplitudes constructed in this manner on any algebraic curve in $\mathbb{C}^2$ provide symplectic invariants of the curve, whether the curve is the spectral curve of a matrix model or not. This was the main insight of [**EO**].

[2]We consider only simple ramification points.

meromorphic differential on C with a double pole at $p = q$, and normalized such that

$$\oint_{A^I} B(p, q) = 0, \tag{3.7}$$

where (A^I, B_I) is a canonical basis of cycles on the Riemann surface. If C has genus 0 (which is the only case we will consider in some detail in this paper), it is given explicitly by, in terms of the local coordinate y,

$$B(p, q) = \frac{\mathrm{d}y(p)\mathrm{d}y(q)}{(y(p) - y(q))^2}. \tag{3.8}$$

The next ingredient is the one-form

$$\mathrm{d}E_q(p) = \frac{1}{2}\int_q^{\bar{q}} B(\xi, p), \tag{3.9}$$

which is well-defined locally near a ramification point q_i. In genus 0, we have

$$\mathrm{d}E_q(p) = \frac{1}{2}\mathrm{d}y(p)\Big[\frac{1}{y(p) - y(q)} - \frac{1}{y(p) - y(\bar{q})}\Big]. \tag{3.10}$$

Finally, we need the one-form

$$\omega(q) = (y(q) - y(\bar{q}))\mathrm{d}x(q). \tag{3.11}$$

With these ingredients, [**EO**] define recursively a set of meromorphic differentials $W_g(p_1, \ldots, p_h)$, with $g \geq 0$, $h \geq 1$, as follows. We first define the starting point of the recursion as

$$W_0(p) = 0, \qquad W_0(p_1, p_2) = B(p_1, p_2). \tag{3.12}$$

The meromorphic differentials are then defined by

$$\begin{aligned} W_g(p, p_1 \ldots, p_h) = \sum_{q_i} \mathrm{Res}_{q=q_i} \frac{\mathrm{d}E_q(p)}{\omega(q)} \Big[& W_{g-1}(q, \bar{q}, p_1, \ldots, p_h) \\ & + \sum_{l=0}^{g} \sum_{J \subset H} W_{g-l}(q, p_J) W_l(\bar{q}, p_{H \backslash J}) \Big], \end{aligned} \tag{3.13}$$

where the sum is over all ramification points defined by (3.6), and the notation is as in (2.38).

Note that it is also possible to calculate the closed amplitudes F_g by the following formula,

$$F_g = \sum_i \mathrm{Res}_{p=q_i} \Phi(q) W_g(q), \tag{3.14}$$

where $\Phi(q)$ is any primitive of $y(q)$.

3.2. Topological strings on toric Calabi–Yau threefolds. Topological string theory on a Calabi–Yau manifold X comes in two varieties, called the A- and the B-models. In the A-model, the genus g closed amplitudes F_g are generating functionals for Gromov–Witten invariants $N_{g,Q}$ "counting" holomorphic maps from genus g Riemann surfaces Σ_g to X

$$F_g(t) = \sum_{Q \in H_2(X)} N_{g,Q} \mathrm{e}^{-Q \cdot t} \tag{3.15}$$

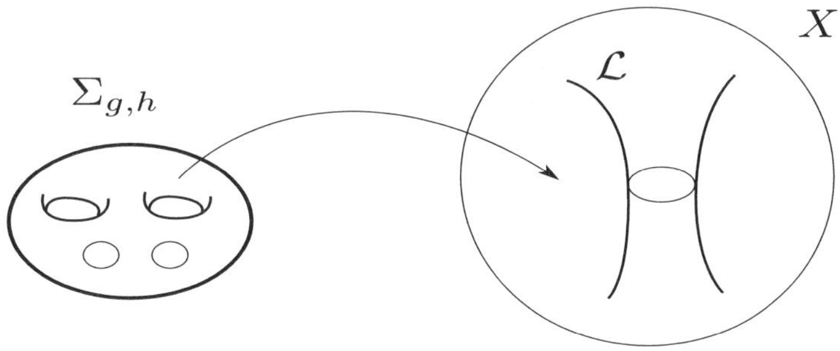

FIGURE 2. In the presence of topological D-branes, we can consider maps from Riemann surfaces with boundaries.

where t are complexified Kähler parameters. In the B-model we study variation of complex structures on a mirror Calabi–Yau manifold $\widetilde{X}$. The basic buiding blocks are the period integrals of the holomorphic 3-form Ω on $\widetilde{X}$

$$t^I = \oint_{A^I} \Omega, \qquad \frac{\partial F_0}{\partial t^I} = \oint_{B_I} \Omega \tag{3.16}$$

where A^I, B_I, $I = 1, \ldots, b_3$ is a symplectic basis of $H_3(\widetilde{X})$. This defines the genus zero closed amplitude $F_0(t)$, where t now parameterizes complex structures. The genus one amplitude $F_1(t)$ can be defined in terms of the Ray–Singer torsion of $\widetilde{X}$. The higher genus $F_g(t)$ are not well-defined mathematically, although they can be partially constructed from the holomorphic anomaly equations of [**BCOV**].

We can introduce an open sector in topological string theory by considering topological D-branes. Again, the description is easier to make in the A-model, where the topological D-branes must wrap a Lagrangian submanifold $\mathcal{L} \subset X$. In this way we can also consider maps from Riemann surfaces with genus g and h holes $\Sigma_{g,h}$ to X with boundary conditions set by $\mathcal{L}$, as shown in Fig. 2.

Let us now assume that $b_1(\mathcal{L}) = 1$. The counting of maps is encoded in a topological open string amplitude $F_{g,h}(x_i)$,

$$F_{g,h}(x_1, \ldots, x_h) = \sum_{w_i} F_{g,w}(t) x_1^{w_1} \cdots x_h^{w_h}, \tag{3.17}$$

where the w_i are the winding numbers of the holes of $\Sigma_{g,h}$ around the nontrivial cycle in $\mathcal{L}$, and

$$F_{g,w}(t) = \sum_{Q \in H_2(X)} N_{g,w,Q} \mathrm{e}^{-Q \cdot t} \tag{3.18}$$

counts holomorphic maps with boundary conditions set by $\mathcal{L}$ in the topological sector specified by Q, w, g. The quantities $N_{g,w,Q}$ are called open Gromov–Witten invariants; in general, they are not well–defined mathematically. In the toric case, however, they can be fully formalized in terms of relative Gromov–Witten theory [**LLLZ**]. The variables x_i in (3.17) are called *open moduli*.

By mirror symmetry, there should be similar amplitudes in the B-model on $\widetilde{X}$, for appropriate objects which are mirror to Lagrangian submanifolds (these objects are, in general, coherent sheaves). In general situations it is difficult to define these amplitudes — although in the case of rigid Lagrangian submanifolds in compact

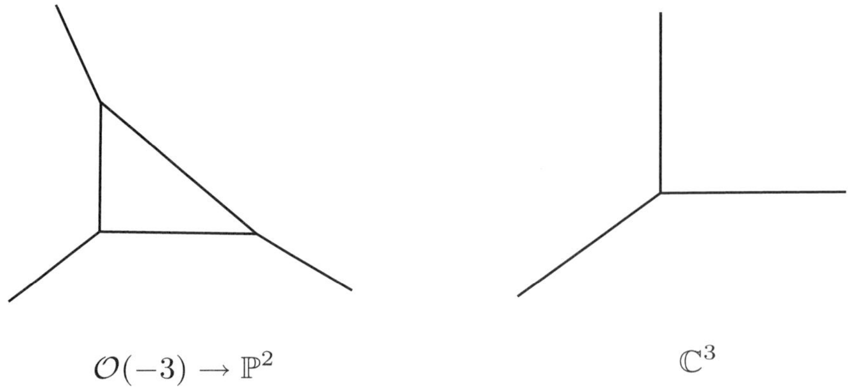

FIGURE 3. Two simple examples of toric manifolds.

Calabi–Yau manifolds one can write down generalizations of the BCOV equations [**W**]. As we will see in a moment, in the toric case these open amplitudes can be defined by a recursion similar to (3.13).

Let us then focus on the case in which X is toric. A toric Calabi–Yau is necessarily noncompact, and can be represented by a toric diagram which encodes its fibration structure (see for example [**M1**] for details). In Fig. 3 we show some simple examples of toric diagrams: the total space of the bundle $\mathcal{O}(-3) \to \mathbb{P}^2$, also known as local $\mathbb{P}^2$, and the even simpler case of $\mathbb{C}^3$.

In toric geometries there is an important class of D-branes wrapping Lagrangian submanifolds with the topology $\mathbb{C} \times S^1$, which we will call for short *toric branes.* These Lagrangian submanifolds are generalizations to the toric realm of the Lagrangian submanifolds of $\mathbb{C}^3$ considered by Harvey and Lawson in [**HL**]. In the physics literature they were first studied in [**AV**] and further explored in [**AKV**]. In the toric diagram these Lagrangian submanifolds simply project to points in the edges, and one can study the open string amplitudes with the Lagrangian boundary conditions that they define.

The A-model open and closed topological string amplitudes on toric Calabi–Yau threefolds can be computed with the so-called topological vertex [**AKMV**]. In this formalism, the total closed partition function

$$Z(t, g_s) = \exp\Big[\sum_{g=0}^{\infty} F_g(t) g_s^{2g-2}\Big] \tag{3.19}$$

is obtained as a sum over partitions or Young tableaux,

$$Z(t, g_s) = \sum_{R_i} C(R_i, g_s) \mathrm{e}^{-\sum_i \ell(R_i)t}. \tag{3.20}$$

In this expression, i runs over the edges of the toric diagram, $\ell(R)$ is the length of the partition R, and $C(R_i, g_s)$ is a quantity that depends on the representations, the variable g_s, and the shape of the toric diagram, and which can be computed in a very precise way by using the theory of the topological vertex. The expression for the partition function in (3.20) is exact in g_s (therefore it contains the all-genus information of the topological string), but it is perturbative in e^{-t}. The expansion

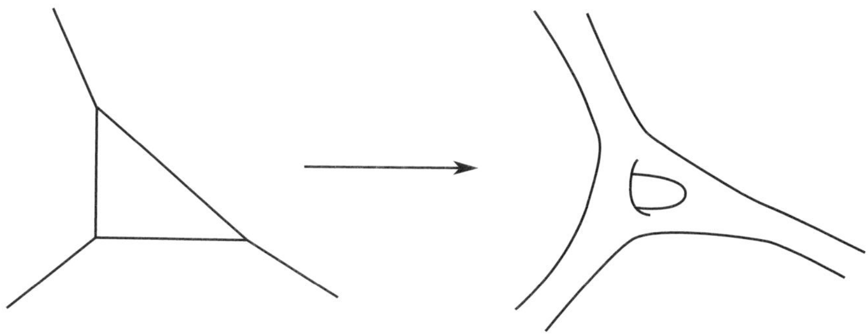

FIGURE 4. Local $\mathbb{P}^2$ and its mirror.

is around the large radius limit $t \to \infty$ of the Kähler moduli space. For the open amplitudes, the result is perturbative in both e^{-t} and the open moduli z_i.

Therefore, the vertex formalism is only appropriate for some particular point in the open/closed moduli space (i.e. large radius for t, and the point $z_i = 0$). It is not well-suited for studying other points in the moduli space like the conifold and orbifold points. This is where the B-model becomes useful, since it is nonperturbative in the Kähler parameter. Let us now review what is known about the B-model on the mirrors of toric manifolds.

The Calabi–Yau mirrors of toric geometries have been known for some time. Generically, they are given by conic fibrations over $\mathbb{C}^* \times \mathbb{C}^*$, where the conic fiber degenerates to two lines over an algebraic curve (see for example [**KKV, CKYZ**], and [**HV**] for a physics derivation). The geometry is essentially captured by the algebraic curve in $\mathbb{C}^* \times \mathbb{C}^*$, which is usually called the *mirror curve*, and is given by a polynomial

$$\Sigma : \{H(x, y) = 0\} \subset \mathbb{C}^* \times \mathbb{C}^*. \tag{3.21}$$

It is crucial to note that, in contrast with the spectral curve (3.4) of matrix models, the mirror curve is a Riemann surface embedded in $\mathbb{C}^* \times \mathbb{C}^*$ rather than $\mathbb{C}^2$.

The mirror curve of a toric Calabi–Yau threefold X can be visualized as a thickening of the toric diagram of X. The example of local $\mathbb{P}^2$ and its mirror is shown in figure Fig. 4; in this case, the mirror curve is a genus 1 Riemann surface with three punctures.

The closed sector of the B-model on these geometries can be analyzed by using techniques similar to those developed for the compact case. In particular, the prepotential is determined by the equations

$$t^I = \oint_{A^I} \log y \frac{\mathrm{d}x}{x}, \qquad \frac{\partial F_0}{\partial t^I} = \oint_{B_I} \log y \frac{\mathrm{d}x}{x} \tag{3.22}$$

where A^I, B_I is a symplectic basis of the mirror curve. Higher genus amplitudes F_g can be studied with a toric version of the holomorphic anomaly equations; see for example [**KZ**].

What about open string amplitudes? It turns out that the toric branes become, under mirror symmetry, just points on the mirror curve [**AV**], and the open string moduli appearing in (3.17) are local coordinates for the curve. The choice of local coordinate — which is equivalent to a choice of projection $\Sigma \to \mathbb{C}^*$ — is related to the edge on which the brane is located in the toric diagram of the A-model.

More precisely, as a curve in $\mathbb{C}^* \times \mathbb{C}^*$, the mirror curve Σ has reparameterization group $SL(2,\mathbb{Z})$ acting as

$$(x,y) \to (\tilde{x},\tilde{y}) = (x^a y^b, x^c y^d), \qquad \begin{pmatrix} a & b \\ c & d \end{pmatrix} \in SL(2,\mathbb{Z}). \tag{3.23}$$

Changing the parameterization of the curve corresponds to choosing different local coordinates (or open string moduli), hence moving the position of the toric brane on the toric diagram on the mirror side [**AV, AKV**]. In fact, not all reparameterizations lead to different geometric branes; fixing the position of the brane only fixes the reparameterization of the curve up to a one-parameter subgroup of $SL(2,\mathbb{Z})$. This extra integer corresponds to the *framing* of the brane, which is an ambiguity in the computation of open string amplitudes. The framing transformations correspond to reparameterizations of the form

$$(x,y) \mapsto (\tilde{x},\tilde{y}) = (xy^f, y), \qquad f \in \mathbb{Z}. \tag{3.24}$$

In [**AV**] it was further shown that the disk amplitude of a toric brane can be computed in terms of the defining equation of the mirror curve

$$F_0(x)\mathrm{d}x = \log y \frac{\mathrm{d}x}{x}, \tag{3.25}$$

which is nothing but the one-form appearing in (3.16).

EXAMPLE 3.1. A relevant example in the following will be $\mathbb{C}^3$ with a brane in one leg. The mirror curve to $\mathbb{C}^3$ reads

$$1 - y - x = 0, \tag{3.26}$$

which is $\mathbb{P}^1$ with three punctures. The open modulus can be taken to be simply x, the local coordinate. The D-brane configuration and its mirror are shown in Fig. 5.

The framed brane in one leg of $\mathbb{C}^3$, which we will call the *framed vertex*, can be obtained by applying the framing transformation (3.24) to the mirror curve (3.26). We get

$$-\tilde{y}^{f+1} + \tilde{y}^f - \tilde{x} = 0. \tag{3.27}$$

The open modulus is now $\tilde{x}$, and the expression (3.25) is still valid in the new parameterization of the curve given by $(\tilde{x},\tilde{y})$.

A natural question is how to compute $F_{g,h}(x_i)$ in the B-model, for arbitrary g, h. The main claim of [**M**] (backed by various nontrivial examples), which was extensively developed and clarified in [**BKMP**], is that these amplitudes can be computed by using a slight modification of the recursion relation (3.13) first found in the context of matrix models. The classical spectral curve is now taken to be the mirror curve. The only difference in the recursion relation comes from the fact that the mirror curve is embedded in $\mathbb{C}^* \times \mathbb{C}^*$ rather than $\mathbb{C}^2$; consequently, the one-form $\omega(q)$ becomes

$$\omega(q) = \big(\log y(q) - \log y(\bar{q})\big)\frac{\mathrm{d}x(q)}{x(q)}, \tag{3.28}$$

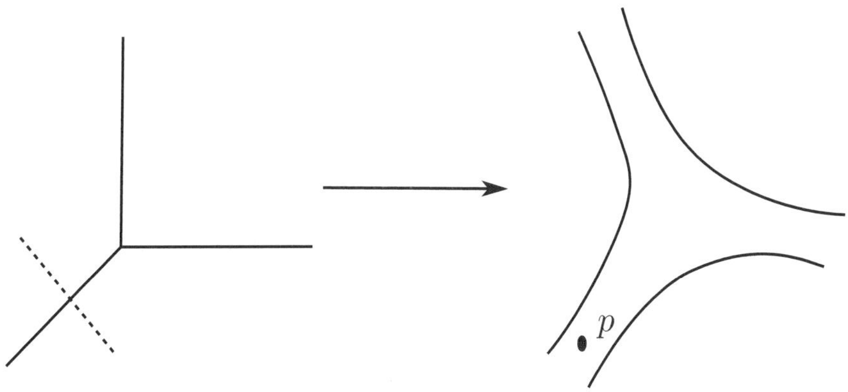

FIGURE 5. $\mathbb{C}^3$ with a brane in a leg, and its mirror.

rather than $\omega(q) = (y(q) - y(\bar{q}))\mathrm{d}x(q)$, which reflects the fact that the symplectic form on $\mathbb{C}^* \times \mathbb{C}^*$ is

$$\frac{\mathrm{d}x}{x} \wedge \frac{\mathrm{d}y}{y}. \tag{3.29}$$

The physical reason for the claim in [**M**] is that the theory of matrix models is just an example of a more general construction, namely a chiral boson on a "quantum" Riemann surface, as was suggested in [**ADKMV**]. In that paper it was also argued that the B-model on the mirror of a toric Calabi–Yau threefold is another example of this theory, where the chiral boson now lives on the mirror curve. The fact that the invariants defined in [**EO**] depended only on the Riemann surface indicated that they must be the amplitudes for the chiral boson theory, which led to the claim of [**M**]. Note that after the appearance of [**M, BKMP**], this line of thought was further refined in [**DV**], where a physical reinterpretation of the recursion relation directly in terms of the chiral boson theory was proposed.

From a more mathematical point of view, one can take the modified recursion relation (3.13) with the one-form (3.28) to be the definition of the B-model on the mirrors of toric geometries, and the claim of [**M, BKMP**] is then a conjecture that these quantities correctly reproduce the open and closed Gromov–Witten invariants of toric Calabi–Yau manifolds. From this point of view, the recursive formalism can be understood as a gluing procedure for open and closed topological string amplitudes, where the basic building blocks are the disk and the annulus amplitudes. A very particular case of this conjecture, involving the disk amplitude of a certain class of branes in local $\mathbb{P}^2$, was already proved in [**GZ**].

We will now consider in some more detail the simplest possible case of this formalism, namely the framed vertex introduced in example 3.1. As a spinoff we will derive our conjecture for the recursion relation of Hurwitz numbers.

3.3. Framed vertex. The A-model topological string on $\mathbb{C}^3$ with a framed brane on one leg of the toric diagram has been studied extensively in the physics literature. For instance, it was first conjectured in [**MV**], and then proved in [**LLZ, OP3**], that the open amplitudes can be written explicitly in terms of Hodge integrals, paralleling the ELSV formula for Hurwitz numbers.

First, recall that we defined in the previous section the open string amplitude with h holes[3]

$$F_{g,h}(x_1,\dots,x_h;f)=\sum_{w_i}F_{g,w}(f)x_1^{w_1}\cdots x_h^{w_h}, \tag{3.30}$$

where we wrote explicitly the dependence on the framing $f\in\mathbb{Z}$ of the brane. These amplitudes can be rewritten as integrals of differential forms,

$$F_{g,h}(x_1,\dots,x_h;f)=\int W_g(x_1,\dots,x_h;f)\mathrm{d}x_1\cdots\mathrm{d}x_h, \tag{3.31}$$

where we introduced, in terms of partitions μ (in the following we use the notation introduced in section 2)

$$W_g(x_1,\dots,x_h;f)=\sum_{\mu|\ell(\mu)=h}z_\mu W_{g,\mu}(f)m_\mu(x). \tag{3.32}$$

The amplitudes $W_{g,\mu}(f)$ can be written in terms of Hodge integrals as follows [**MV, LLZ**]:

$$W_{g,\mu}(f)=\frac{(-1)^{g+\ell(\mu)}}{|\mathrm{Aut}(\mu)|}(f(f+1))^{\ell(\mu)-1}\prod_{i=1}^{\ell(\mu)}\left(\frac{\mu_i\prod_{j=1}^{\mu_i-1}(\mu_i f+j)}{\mu_i!}\right)\times\int_{\overline{M}_{g,h}}\frac{\Lambda_g^\vee(1)\Lambda_g^\vee(-f-1)\Lambda_g^\vee(f)}{\prod_{i=1}^{\ell(\mu)}(1-\mu_i\psi_i)}. \tag{3.33}$$

What we would like to do now is to use the formalism proposed in [**M, BKMP**] to construct recursively the open amplitudes, hence providing an infinite set of relations between the above Hodge integrals. We will then explain how these relations motivate our conjecture 2.1 for Hurwitz numbers.

Recall that the framed mirror curve (3.27) reads

$$-y^{f+1}+y^f-x=0. \tag{3.34}$$

The open string modulus is given by x, hence y is a good local coordinate. In order to apply the recursion relation (3.13), we will need to find $y=y(x)$. We can invert (3.34) near $(x,y)=(0,1)$ to get

$$y(x)=1-\sum_{n=1}^{\infty}\frac{x^n\prod_{j=0}^{n-2}(nf+j)}{n!}. \tag{3.35}$$

We now define the one-forms

$$\zeta_n(y,f)=\mathrm{d}y\frac{(1+f)y-f}{y(y-1)}\left(\frac{y(y-1)}{(1+f)y-f}\frac{\mathrm{d}}{\mathrm{d}y}\right)^{n+1}\frac{1}{(1+f)((1+f)y-f)},\qquad n\geq 0. \tag{3.36}$$

For example,

$$\zeta_0(y,f)=-\frac{\mathrm{d}y}{(y+f(y-1))^2},\quad \zeta_1(y,f)=\frac{-f+y(y-2)+fy^2}{(y+f(y-1))^4}\mathrm{d}y. \tag{3.37}$$

It is straightforward to check that we have the following expansion in powers of x:

$$\zeta_n(x,f)=\sum_{\mu=1}^{\infty}\frac{\mu^{n+2}\prod_{j=1}^{\mu-1}(\mu f+j)}{\mu!}x^{\mu-1}\mathrm{d}x. \tag{3.38}$$

---

[3] Note that there is no dependence on t anymore since $\mathbb{C}^3$ has no Kähler parameter.

It then follows that

$$(3.39)\quad W_g(x_1,\ldots,x_h)\mathrm{d}x_1\cdots\mathrm{d}x_h = (-1)^{g+h}(f(f+1))^{h-1} \times \sum_{n_i=0}^{3g-3+h} \langle \tau_{n_1}\cdots\tau_{n_h}\Lambda_g^\vee(1)\Lambda_g^\vee(-f-1)\Lambda_g^\vee(f)\rangle \prod_{i=1}^{h}\zeta_{n_i}(y_i,f).$$

Now, it was proposed in [**M, BKMP**] that these differentials can be computed as follows. Take the framed curve (3.34), which gives the expression (3.35) for $y(x)$. Then,

$$(3.40)\qquad W_0(x)\mathrm{d}x = \log y(x)\frac{\mathrm{d}x}{x}, \qquad W_0(x_1,x_2) = B(y_1,y_2) - \frac{\mathrm{d}x_1\mathrm{d}x_2}{(x_1-x_2)^2},$$

where the y_i are defined implicitly in terms of x_i through the series $y(x_i)$ given by (3.35). The remaining differentials correspond precisely to the differentials generated through the modified recursion relation (3.13) with the one-form (3.28).

Let us now be a little more explicit. To perform the recursive calculations, we first need to determine the ramification points of the framed curve (3.34). There is only one ramification point of the x-projection, which we denote by ν. It is given by

$$(3.41)\qquad y(\nu) = \frac{f}{f+1}.$$

We then find the two points q and $\bar{q}$ on the curve such that $x(q) = x(\bar{q})$ near the ramification point. Write first

$$(3.42)\qquad y(q) = \frac{f}{f+1} + z, \qquad y(\bar{q}) = \frac{f}{f+1} + P(z),$$

with

$$(3.43)\qquad P(z) = -z + \mathcal{O}(z^2).$$

Then $\bar{q}$ must satisfy

$$(3.44)\qquad -y(q)^{f+1} + y(q)^f = -y(\bar{q})^{f+1} + y(\bar{q})^f,$$

which we can solve to extract $P(z)$ as a power series in z. We obtain:

$$(3.45)\quad P(z) = -z - \frac{2\left(-1+f^2\right)}{3\,f}z^2 - \frac{4\left(-1+f^2\right)^2}{9\,f^2}z^3 - \frac{2\left(1+f\right)^3\left(-22+57\,f-57\,f^2+22\,f^3\right)}{135\,f^3}z^4 + \cdots$$

Using these ingredients, the one-form $\omega(q)$ defined in (3.28) becomes, in terms of the coordinate z:

$$(3.46)\quad \omega(z) = \left(\log\left(\frac{f}{f+1}+z\right) - \log\left(\frac{f}{f+1}+P(z)\right)\right)\frac{(f+1)^3 z\mathrm{d}z}{(f+(f+1)z)(-1+(f+1)z)}.$$

The other one-form that we need, $\mathrm{d}E_q(p)$, was defined, for genus 0 curves, in (2.34). Using the results above we obtain, in terms of z and y:

$$(3.47)\qquad \mathrm{d}E_z(y) = \frac{1}{2}\mathrm{d}y\left(\frac{1}{y-\frac{f}{f+1}-z} - \frac{1}{y-\frac{f}{f+1}-P(z)}\right).$$

We can now compute explicitly the differentials generated through the recursion (3.13), which becomes

(3.48)

$$W_g(y, y_1 \dots, y_h) = \mathrm{Res}_{z=0} \frac{\mathrm{d}E_z(y)}{\omega(z)} \Big(W_{g-1} \Big(\frac{f}{f+1} + z, \frac{f}{f+1} + P(z), y_1, \dots, y_h \Big) + \sum_{l=0}^{g} \sum_{J \subset H} W_{g-l} \Big(\frac{f}{f+1} + z, y_J \Big) W_l \Big(\frac{f}{f+1} + P(z), y_{H \setminus J} \Big) \Big).$$

We perform the residue calculation, and express our results in terms of the one-forms $\zeta_n(y,f)\mathrm{d}y$ introduced earlier. This makes it straightforward to extract the results for the Hodge integrals (3.33), by using the expression (3.39) for the differentials W_g.

Here are some of the results we obtained.

(3.49)

$$W_0(y_1, y_2, y_3) = -(f(f+1))^2 \prod_{i=1}^{3} \zeta_0(y_i),$$

$$W_0(y_1, \dots, y_4) = (f(f+1))^3 \sum_{i=1}^{4} \zeta_1(y_i) \prod_{j \neq i} \zeta_0(y_j),$$

$$W_1(y) = \frac{1}{24}\Big((1+f+f^2)\zeta_0(y,f) - f(1+f)\zeta_1(y,f) \Big),$$

$$W_1(y_1, y_2) = \frac{1}{24} f(f+1) \Big(-(1+f+f^2)\zeta_0(y_1)\zeta_1(y_2) + f(1+f)\zeta_0(y_1)\zeta_2(y_2) + (y_1 \leftrightarrow y_2) + f(1+f)\zeta_1(y_1)\zeta_1(y_2) \Big),$$

$$W_2(y) = \frac{1}{5760}\Big(2f(f+1)\zeta_1(y,f) - 7(1+f+f^2)^2\zeta_2(y,f) + 12f(1+2f+2f^2+f^3)\zeta_3(y,f) - 5f^2(1+f)^2\zeta_4(y,f) \Big).$$

3.4. Back to Hurwitz theory. We are now ready to motivate our Conjecture 2.1 in Hurwitz theory. It is known that Hurwitz theory can be seen as a particular limit of topological string theory on the framed vertex studied in the previous section, where we send the framing f to infinity. This can be seen in many ways. A simple way is just to look at the explicit expression for the topological vertex and take the limit directly [**CGMPS**]. Alternatively, one can look at the Hodge integral formula for the framed vertex conjectured in [**MV**] and proved in [**LLZ, OP3**]. The (appropriately scaled) limit of infinite framing involves only Hodge integrals with one Hodge class insertion, and using the ELSV formula one relates it to Hurwitz numbers.

Here we use this last approach and look at the Hodge integrals expression for the differentials that we computed. Recall that for the framed vertex, the objects we studied and generated recursively were the differentials (3.39), which we reproduce here for clarity:

$$W_g(x_1,\dots,x_h)\mathrm{d}x_1\cdots\mathrm{d}x_h = (-1)^{g+h}(f(f+1))^{h-1} \times \sum_{n_i=0}^{3g-3+h} \langle \tau_{n_1}\cdots\tau_{n_h}\Lambda_g^\vee(1)\Lambda_g^\vee(-f-1)\Lambda_g^\vee(f)\rangle \prod_{i=1}^h \zeta_{n_i}(x_i,f). \tag{3.50}$$

On the Hurwitz theory side, we computed similar differentials, which were expressed in terms of Hodge integrals in (2.23) using the ELSV formula:

$$H_g(x_1,\dots,x_h)\mathrm{d}x_1\cdots\mathrm{d}x_h = \sum_{n_i=0}^{3g-3+h} \langle \tau_{n_1}\cdots\tau_{n_h}\Lambda_g^\vee(1)\rangle \prod_{i=1}^h \zeta_{n_i}(x_i), \tag{3.51}$$

First, using the Mumford relation

$$\Lambda_g^\vee(t)\Lambda_g^\vee(-t) = (-1)^g t^{2g}, \tag{3.52}$$

we obtain the following relation between the Hodge integrals

$$\lim_{f\to\infty}\frac{1}{f^{2g}}\langle \tau_{n_1}\cdots\tau_{n_h}\Lambda_g^\vee(1)\Lambda_g^\vee(-f-1)\Lambda_g^\vee(f)\rangle = (-1)^g\langle \tau_{n_1}\cdots\tau_{n_h}\Lambda_g^\vee(1)\rangle. \tag{3.53}$$

Moreover, using the expansions (2.22) and (3.38) for the one-forms $\zeta_n(x)$ and $\zeta_n(x,f)$, it is easy to see that

$$\lim_{f\to\infty}\zeta_n\left(\frac{x}{f},f\right) = \zeta_n(x). \tag{3.54}$$

Notice that we renormalized the variable x by $x\mapsto x/f$ before taking the limit $f\to\infty$.

As a result, we obtain that

$$\lim_{f\to\infty}\left(\frac{(-1)^h}{f^{2g+2h-2}}W_g\left(\frac{x_1}{f},\dots,\frac{x_h}{f}\right)\right) = H_g(x_1,\dots,x_h), \tag{3.55}$$

which relates the Gromov-Witten potentials to the Hurwitz generating functions. This relation can also be seen directly by computing the limits of the expressions (3.49) obtained through the recursion relation.

Finally, as we have seen in the previous subsection, following the proposal of [**M, BKMP**] the differentials $W_g(x_1,\dots,x_h)\mathrm{d}x_1\cdots\mathrm{d}x_h$ can be computed in the recursive formalism discussed previously. This implies that the differentials $H_g(x_1,\dots,x_h)\mathrm{d}x_1\cdots\mathrm{d}x_h$ in Hurwitz theory can also be computed through this recursive formalism, after taking the limit $f\to\infty$ appropriately. We can work out this limit explicitly for all the objects entering in the recursive relation. For example, the curve (3.34), after setting $x\to x/f$ and $y=1-T/f$, reads,

$$x = T\Big(1-\frac{T}{f}\Big)^f, \tag{3.56}$$

which as $f\to\infty$ becomes precisely the definition of the tree function $x=T(x)\mathrm{e}^{-T(x)}$. By working out the infinite framing limit of the rest of the ingredients, we obtain the statement of Conjecture 2.1.

References

[ADKMV] M. Aganagic, R. Dijkgraaf, A. Klemm, M. Mariño and C. Vafa, "Topological strings and integrable hierarchies," Commun. Math. Phys. **261**, 451 (2006) [arXiv:hep-th/0312085].

[AKMV] M. Aganagic, A. Klemm, M. Mariño and C. Vafa, "The topological vertex," Commun. Math. Phys. **254**, 425 (2005) [arXiv:hep-th/0305132].

[AKV] M. Aganagic, A. Klemm and C. Vafa, "Disk instantons, mirror symmetry and the duality web," Z. Naturforsch. A **57**, 1 (2002) [arXiv:hep-th/0105045].

[AV] M. Aganagic and C. Vafa, "Mirror symmetry, D-branes and counting holomorphic discs," arXiv:hep-th/0012041.

[BCOV] M. Bershadsky, S. Cecotti, H. Ooguri and C. Vafa, "Kodaira-Spencer theory of gravity and exact results for quantum string amplitudes," Commun. Math. Phys. **165**, 311 (1994) [arXiv:hep-th/9309140].

[BIZ] D. Bessis, C. Itzykson and J. B. Zuber, "Quantum field theory techniques in graphical enumeration," Adv. Appl. Math. **1**, 109 (1980).

[BIPZ] E. Brézin, C. Itzykson, G. Parisi and J. B. Zuber, "Planar Diagrams," Commun. Math. Phys. **59**, 35 (1978).

[BKMP] V. Bouchard, A. Klemm, M. Mariño and S. Pasquetti, "Remodeling the B-model," arXiv:0709.1453 [hep-th].

[CGMPS] N. Caporaso, L. Griguolo, M. Mariño, S. Pasquetti and D. Seminara, "Phase transitions, double-scaling limit, and topological strings," Phys. Rev. D **75**, 046004 (2007) [arXiv:hep-th/0606120].

[CKYZ] T. M. Chiang, A. Klemm, S. T. Yau and E. Zaslow, "Local mirror symmetry: Calculations and interpretations," Adv. Theor. Math. Phys. **3**, 495 (1999) [arXiv:hep-th/9903053].

[CGHJK] R.M. Corless, G.H. Gonnet, D.E.G. Hare, D.J. Jeffrey and D.E. Knuth, "On the Lambert W function," Advances in Computational Mathematics **5**, 329 (1996).

[DV] R. Dijkgraaf and C. Vafa, "Two Dimensional Kodaira-Spencer Theory and Three Dimensional Chern-Simons Gravity," arXiv:0711.1932 [hep-th].

[ELSV] T. Ekedahl, S. Lando, M. Shapiro and A. Vainshtein, "Hurwitz numbers and intersections on moduli spaces of curves," Invent. Math. **146**, 297 (2001) [arXiv:math.AG/0004096].

[E] B. Eynard, "Recursion between volumes of moduli spaces," arXiv:0706.4403 [math-ph].

[EMO] B. Eynard, M. Mariño and N. Orantin, "Holomorphic anomaly and matrix models," JHEP **0706**, 058 (2007) [arXiv:hep-th/0702110].

[EO] B. Eynard and N. Orantin, "Invariants of algebraic curves and topological expansion," Commun. Number Theory Phys. **1**, 347 (2007) [arXiv:math-ph/0702045].

[EO1] B. Eynard and N. Orantin, "Weil-Petersson volume of moduli spaces, Mirzakhani's recursion and matrix models," arXiv:0705.3600 [math-ph].

[FP] B. Fantechi and R. Pandharipande, "Stable maps and branch divisors," Compositio Math. **130**, 345 (2002) [arXiv:math.AG/9905104].

[GJ] I. P. Goulden and D.M. Jackson, "The number of ramified coverings of the sphere by the double torus, and a general form for higher genera," J. Combin. Theory Ser. A, **88**, 259 (1999) [arXiv:math.AG/9902011].

[GJVa] I. P. Goulden, D. M. Jackson and A. Vainshtein, The number of ramified coverings of the sphere by the torus and surfaces of higher genera, Annals of Comb. **4**, 27 (2000) [arXiv:math.AG/9902125].

[GJV] I. P. Goulden, D.M. Jackson and R. Vakil, "The Gromov-Witten potential of a point, Hurwitz numbers, and Hodge integrals," Proc. Lond. Math. Soc., III. Ser. 83, No.3, 563-581 (2001) [arXiv:math.AG/9910004].

[GV] T. Graber and R. Vakil, "Hodge integrals and Hurwitz numbers via virtual localization," Compositio Math. **135**, 25 (2003) [arXiv:math.AG/0003028].

[GZ] T. Graber and E. Zaslow, "Open string Gromov-Witten invariants: Calculations and a mirrors 'theorem'," Orbifolds in mathematics and physics (Madison, WI, 2001), 107, Contemp. Math. **310**, Amer. Math. Soc., Providence, RI (2002) [arXiv:hep-th/0109075].

[HL] R. Harvey and H. B. Lawson, "Calibrated Geometries," Acta Mathematica **148**, 47.

[HV] K. Hori and C. Vafa, "Mirror symmetry," arXiv:hep-th/0002222.

[JKLP] S. Janson, D.E. Knuth, T. Luczak and B. Pitte, "The birth of the giant component," Random Structures and Algorithms **4**, 233 (1993) [arXiv:math.PR/9310236].

[KKV] S. H. Katz, A. Klemm and C. Vafa, "Geometric engineering of quantum field theories," Nucl. Phys. B **497**, 173 (1997) [arXiv:hep-th/9609239].

[KZ] A. Klemm and E. Zaslow, "Local mirror symmetry at higher genus," Winter School on Mirror Symmetry, Vector Bundles and Lagrangian Submanifolds (Cambridge, MA, 1999), 183, AMS/IP Stud. Adv. Math. **23**, Amer. Math. Soc., Providence, RI (2001). [arXiv:hep-th/9906046].

[LLLZ] J. Li, C.-C. Liu, K. Liu and J. Zhou, "A mathematical theory of the topological vertex," [arXiv:math.AG/0408426].

[LLZ] C.-C. Liu, K. Liu and J. Zhou, "A proof of a conjecture of Mariño-Vafa on Hodge integrals", J. Differential Geom. **65**, 289 (2003); "Mariño-Vafa formula and Hodge integral identities," J. Algebraic Geom. **15**, 379 (2006) [arXiv:math.AG/0308015].

[LZ] K. Liu and H. Xu, "A simple proof of Mirzakhani's recursion formula of Weil–Petersson volumes," arXiv:0705.2086 [math.AG].

[Mac] I. Macdonald, *Symmetric functions and Hall polynomials*, Oxford University Press, Oxford.

[M] M. Mariño, "Open string amplitudes and large order behavior in topological string theory," arXiv:hep-th/0612127.

[M1] M. Mariño, "Chern-Simons theory and topological strings," Rev. Mod. Phys. **77** (2005) 675 [arXiv:hep-th/0406005]. *Chern–Simons theory, matrix models, and topological strings*, Oxford University Press, 2005.

[MV] M. Mariño and C. Vafa, "Framed knots at large N," Orbifolds in mathematics and physics (Madison, WI, 2001), Contemp. Math. **310**, 185, Amer. Math. Soc., Providence, RI (2002) [arXiv:hep-th/0108064].

[Mir] M. Mirzakhani, "Simple geodesics and Weil–Petersson volumes of moduli spaces of bordered Riemann surfaces," Invent. Math. **167**, 179 (2007).

[Mir2] M. Mirzakhani, "Weil–Petersson volumes and intersection theory on the moduli space of curves," J. Amer. Math. Soc. **20**, 1 (2007).

[MS] M. Mulase and B. Safnuk, "Mirzakhani's recursion relations, Virasoro constraints and the KdV hierarchy," arXiv:math.QA/0601194.

[O] A. Okounkov, "Toda equations for Hurwitz numbers," Math. Res. Lett. **7**, 447 (2000) [arXiv:math.AG/0004128].

[OP] A. Okounkov and R. Pandharipande, "Gromov-Witten theory, Hurwitz numbers, and matrix models, I," arXiv:math.AG/0101147.

[OP1] A. Okounkov and R. Pandharipande, "Gromov–Witten theory, Hurwitz theory, and completed cycles," Ann. of Math. **163**, 517 (2006) [arXiv:math.AG/0204305].

[OP3] A. Okounkov and R. Pandharipande, "Hodge integrals and invariants of the unkot," Geometry and Topology **8**, 675 (2004) [arXiv:math.AG/0307209].

[OV] H. Ooguri and C. Vafa, "Knot invariants and topological strings," Nucl. Phys. **B 577**, 419 (2000) [arXiv:hep-th/9912123].

[P] R. Pandharipande, "The Toda equation and the Gromov–Witten theory of the Riemann sphere," Lett. Math. Phys. **53**, 59 (2000) [arXiv:math.AG/9912166].

[V] R. Vakil, "Recursion formulas, and graph-theoretic interpretations of ramified coverings of the sphere by surfaces of genus 0 and 1," arXiv:math.CO/9812105.

[W] J. Walcher, "Extended Holomorphic Anomaly and Loop Amplitudes in Open Topological String," arXiv:0705.4098 [hep-th].

Jefferson Physical Laboratory, Harvard University, 17 Oxford St., Cambridge, MA 02138, USA

E-mail address: `bouchard@physics.harvard.edu`

Department of Physics, Theory Division, CERN, Geneva 23, CH-1211 Switzerland

E-mail address: `marcos@mail.cern.ch`

Proceedings of Symposia in Pure Mathematics
Volume **78**, 2008

Background Independence and the Open Topological String Wavefunction

Andrew Neitzke and Johannes Walcher

ABSTRACT. The open topological string partition function in the background of a D-brane on a Calabi-Yau threefold specifies a state in the Hilbert space associated with the quantization of the underlying special geometry. This statement is a consequence of the extended holomorphic anomaly equation after an appropriate shift of the closed string variables, and can be viewed as the expression of background independence for the open-closed topological string. We also clarify various other aspects of the structure of the extended holomorphic anomaly equation. We conjecture that the collection of all D-branes furnishes a basis of the Hilbert space, and revisit the BPS interpretation of the open topological string wavefunction in this light.

1. Introduction

In this paper, we study topological strings on Calabi-Yau threefolds with background D-branes. Our most basic motivation is to understand the properties of the open topological string amplitudes $\mathcal{F}^{(g,h)}$ that are implied by the extended holomorphic anomaly equations recently found in [**1**]. Two of the applications we have in mind are open-closed duality in the topological string, and the relation of the topological string to BPS state counting.

1.1. Background. Without attempting a complete history of the subject, we note that the holomorphic anomaly equations were originally obtained by BCOV [**2, 3**] as a kind of generalization of the tt^*-equations of Cecotti and Vafa [**4**] to higher-genus amplitudes. These equations arise as constraints on the amplitudes of certain topologically twisted $\mathcal{N} = 2$ superconformal field theories, and the coupling of the latter to topological gravity. The original tt^*-equations, which apply more generally also to topological twists of massive theories, have an appealing geometrical interpretation in the context of Frobenius manifolds [**5**], and have attracted much interest over the years. The mathematical scope of the holomorphic anomaly equations is somewhat less well understood, although of course when combined with mirror symmetry, they are a powerful tool to access the enumerative geometry of Calabi-Yau manifolds; see [**6, 7**] for the state of the art. Physically, the holomorphic anomaly equations have an elegant interpretation as the realization

2000 *Mathematics Subject Classification.* 81T45, 53C25, 81S10, 81T30, 81T50.

of "quantum background independence" of the topological string [**8**], also known as the "wavefunction interpretation". This slightly mysterious notion has recently played a central role in relating the topological string to BPS state counting [**9**].

One of the recurring themes in the literature on the tt^*-equations, the topological string, and the holomorphic anomaly equations is the relation to classical integrable systems and their quantization. Recent examples include the solution of the topological string on certain local Calabi-Yau manifolds in terms of matrix models [**10**], especially when viewed in the broader context of the duality between open and closed topological strings [**11**].

Our present work is concerned with some extensions of these structures to the situation with D-branes in the background. We believe that our results provide further hints of how the various topics listed above might be related at a deeper level. In particular, our results reinforce the point of view that D-branes and open-closed duality are central to a complete understanding of the wavefunction interpretation of the topological string, background independence, as well as the underlying integrable structure. Although there are notable differences from previous works, we attribute most of them to the distinction between local and compact setups. For example, the recent works [**12, 13**] have shown that topological string amplitudes that can be obtained from matrix models satisfy a set of equations essentially equivalent to the holomorphic anomaly equations.[1] In this situation, the Calabi-Yau geometry reduces to the spectral curve of the matrix model, and the resulting simplification in the Hodge structure should be the origin of the remaining discrepancies. Via open-closed duality for matrix models, we also anticipate a connection to the larger framework of [**11**], even though concrete attempts in this direction have so far been largely unsuccessful.

As another example, we mention that the wavefunction properties of the open topological string amplitudes on local Calabi-Yau manifolds have been previously discussed in [**11, 16**]; see also [**17**]. In that context, the main discrepancies from our work seem to originate from the generic decoupling of open string moduli, or else the inaccessibility of certain D-brane charge sectors, on compact Calabi-Yau manifolds.

We hope that the clarification of the relation of our results with those and other works will help to isolate the central structures, and possibly even shed some light on the much more important problem of background independence in physical string theory, maybe along the lines of [**18, 19, 20**].

1.2. Results. In [**21**], it was noted that the general solution of the extended holomorphic anomaly equation of [**1**] can be mapped to a solution of the ordinary (BCOV) holomorphic anomaly equation by a certain shift of the closed string variables. This observation was then used to give a proof of the Feynman rule computation of the open topological string amplitudes given in [**1**], following [**3**]. Other recent work on the holomorphic anomaly for open strings includes [**22, 23, 24**], and see [**25**] for some earlier work.

---

[1] More precisely, the approach of [**12, 13**], which is based on new techniques in [**14**], should work quite generally for local Calabi-Yau manifolds which are either local curves or conic bundles over $\mathbb{C}^2$ or $(\mathbb{C}^*)^2$ which degenerate over a complex curve in the base, whether or not there is an underlying matrix model. See also [**15**] for recent progress in this direction.

Could such a simple relation between the open and closed topological strings also hold for the *physical solutions* of the anomaly equations, namely, for the topological amplitudes themselves? The holomorphic anomaly equations do not constrain the holomorphic dependence of the topological amplitudes, so there is clearly something non-trivial to check. On the other hand, the holomorphic part of the amplitudes is to a large extent constrained by a second set of equations, which express the statement that adding closed string insertions in worldsheet diagrams is equivalent to taking holomorphic (covariant) derivatives with respect to the moduli. In this way, one is actually reduced to the problem of determining, at each order in perturbation theory, a finite number of constants specifying the holomorphic part of the vacuum amplitudes. Fixing this "holomorphic ambiguity" normally requires additional information not contained in the holomorphic anomaly equation itself.

It turns out that when the open string amplitudes for non-trivial D-branes are shifted as in [**21**], they in fact *do not* obey the second, holomorphic, set of constraints. We hasten to emphasize that this does not affect the proof of the Feynman rules given in [**21**], because that proof only depends on the validity of the antiholomorphic constraints, namely the holomorphic anomaly equations. We can further mitigate the disappointment by revealing that there is a *different shift* of the open string amplitudes that leads to a solution of both the holomorphic and the antiholomorphic constraints.

We can give a brief summary of this result in formulas. First we recall that, by a transformation described in [**26**], the BCOV "master holomorphic anomaly equations" of [**3**] can be simplified into a heat equation and a holomorphy condition:

$$
\begin{aligned}
\left[\frac{\partial}{\partial X^I}-\frac{1}{2}C_{IJK}\frac{\partial^2}{\partial y_J\partial y_K}\right]\Psi_{\text{closed}} &= 0\,,\\
\frac{\partial}{\partial \bar{X}^I}\Psi_{\text{closed}} &=0\,.
\end{aligned}
\tag{1}
$$

Here $\Psi_{\text{closed}} := \Psi_{\text{closed}}(X^I, y_I)$, as a function of the y_I, is a generating function for the (closed) topological string amplitudes $\mathcal{F}^{(g)}_{i_1,\ldots,i_n} := \mathcal{F}^{(g,0)}_{i_1,\ldots,i_n}$, themselves functions of homogeneous coordinates X^I on the closed string moduli space.[2] In (1), C_{IJK} is the three-point function on the sphere (Yukawa coupling), which is the basic data of the closed topological string at tree level.

As we will show, the extended holomorphic anomaly equations of [**1**] can be similarly transformed into a heat equation extended by a "convection term,"

$$
\begin{aligned}
\left[\frac{\partial}{\partial X^I}-\frac{1}{2}C_{IJK}\frac{\partial^2}{\partial y_J\partial y_K}-i\mu\nu_{IJ}\frac{\partial}{\partial y_J}\right]\Psi_{\text{open}} &= 0\,,\\
\frac{\partial}{\partial \bar{X}^I}\Psi_{\text{open}} &= 0\,,
\end{aligned}
\tag{2}
$$

for a generating function Ψ_{open} of the topological amplitudes $\mathcal{F}^{(g,h)}_{i_1,\ldots,i_n}$. Here ν_{IJ} is (part of) the disk amplitude with two bulk insertions, which is the basic holomorphic data specifying the D-brane background [**1**]. The tensor ν_{IJ} can be integrated to the

[2]To minimize confusion we emphasize that Ψ_{closed} is not exactly the generating function introduced in [**3**]; the relation between the two is described in [**26**]. In particular, the coordinates y_I are related to those appearing in [**3**] by $y_I = (\operatorname{Im}\tau)_{IJ}\bar{z}^J$.

superpotential or domain wall tension, $\nu_{IJ} = \partial_I \nu_J = \partial_I \partial_J \mathcal{T}$. We also introduced a formal real parameter μ counting the number of worldsheet boundaries.

It is fairly obvious that (2) can be transformed into (1) by a simple shift of the closed string variables, $y_I \to y_I - i\mu\nu_I$, as anticipated in [**21**].[3] In other words, given an open string background (specified by ν_{IJ}), we define a *shifted* open string partition function by $\Psi^\nu(X^I, y_I) = \Psi_{\text{open}}(X^I, y_I - i\mu\nu_I)$ (in this notation, $\Psi_{\text{closed}} = \Psi^0$). This Ψ^ν then satisfies the ordinary heat equation (1), independent of ν_{IJ}.

It is now meaningful to ask whether the shifted open string partition function is equal to the closed string partition function, in other words, whether $\Psi^\nu \stackrel{?}{=} \Psi^0$ is in fact independent of ν. As it turns out, the answer is in the negative,[4] but it shines in a positive light when viewed instead as an answer to a long-standing question raised by Witten's interpretation of the holomorphic anomaly equation in the context of background independence, which we now recall.

1.3. A positive attitude. For fixed background X^I, Witten proposed [**8**] to view the topological string partition function $\Psi(X^I, y_I)$, as a function of the y_I, as a "wavefunction" specifying a quantum mechanical state in a particular presentation of a certain Hilbert space. Witten's Hilbert space, which we denote by $\mathcal{H}_W$, arises from the quantization of the symplectic vector space of topological ground states of the underlying worldsheet theory. The choice of background is equivalent to specifying a complex polarization of this vector space, and hence a particular presentation of the wavefunctions. The wavefunction depends on the background, but in a way that is completely fixed, according to the heat equation (1), by the variation of the polarization. The abstract quantum mechanical state itself, $|\Psi\rangle \in \mathcal{H}_W$, is independent of the background.

Our punchline might be clear already. A priori, the physical significance of Witten's auxiliary Hilbert space $\mathcal{H}_W$ is obscure if the closed topological string specifies only one particular state in it. But if for any D-brane configuration whose topological amplitudes satisfy the extended holomorphic anomaly equation, the shifted partition function Ψ^ν satisfies the ordinary heat equation, this means precisely that any such D-brane specifies a state $|\Psi^\nu\rangle \in \mathcal{H}_W$. In fact, in section 4 we will describe evidence that, as ν varies over all possible D-brane configurations, the set of $|\Psi^\nu\rangle$ furnishes a basis of $\mathcal{H}_W$, thus filling Witten's entire Hilbert space with life.

It remains to be understood what physical principle selects the basis of states $|\Psi^\nu\rangle$, and in particular the closed string ground state $|\Psi^0\rangle$. To give some hints at the nature of this question, we note that when our topological string is the B-model on some Calabi-Yau threefold Y, then the basis $|\Psi^\nu\rangle$ is, at least partially, indexed by the set of all possible holomorphic curves in Y (see section 4 for details). In other words, understanding the quantum Hilbert space of the topological B-model on Y involves knowledge of all holomorphic curves in Y. Of course, the topological string knows a great deal about holomorphic curves on Calabi-Yau threefolds. Remarkably though, this knowledge arises from studying the A-model on Y, whereas we would here be trying to answer a B-model question (on the same manifold, not its mirror). This is suggestive of an intimate relation between

---

[3]More precisely, the shift proposed in [**21**] reads in the present notation as $y_I \to y_I - i\mu(\nu_I - \bar{\nu}_I)$. We will discuss the difference between the two shifts extensively in section 3.1.

[4]Interestingly, the discrepancy between Ψ^ν and Ψ^0 arises *before* taking into account the holomorphic ambiguity.

the topological A- and B-models on the same Calabi-Yau manifold, once D-branes are appropriately taken into account. This basic point has been emphasized by many people, beginning with [**27**], and is an ingredient in the topological S-duality proposal of [**28, 29**]. We also note that speculations along the above lines first appeared in the work of Donagi and Markman [**30**].

Let us now close this introduction and start with the derivation of the above-mentioned results. We will return to their interpretation in section 4, where we'll also include some more concrete speculations on the relation to BPS state counting.

2. Derivations

We are interested in the topological string obtained by twisting an $\mathcal{N} = 2$ superconformal field theory of central charge $\hat{c} = 3$ and all-integral $U(1)$ charges. As shown in [**4, 3**] in great generality, the space of chiral deformations of such a superconformal field theory carries the structure of a special Kähler manifold, which we will denote by $\mathcal{M}$, and which forms the basic holomorphic arena for the topological string. We will use local coordinates t^i, $i = 1, 2, \ldots, n = \dim_{\mathbb{C}}(\mathcal{M})$.

We note that our conventions in this section differ for convenience from those in [**1, 26**], which will result in various factors of i appearing differently.

2.1. Special geometry with D-branes. The central data of special geometry of $\mathcal{M}$ are the Hodge line bundle $\mathcal{L}$, with a Hermitian metric whose curvature is the special Kähler form of $\mathcal{M}$, and the Yukawa cubic C, which is a holomorphic section of $\mathcal{L}^{-2} \otimes \mathrm{Sym}^3 T^*\mathcal{M}$. The metric on $\mathcal{L}$ is denoted as e^{-K} w.r.t. some local trivialization, providing a Kähler potential for the special Kähler metric on $\mathcal{M}$, $G_{i\bar{j}} = \partial_i \partial_{\bar{j}} K$. We will write D generically for the metric-compatible connections on products of powers of $\mathcal{L}$ and T.

A crucial object is the "bundle of ground states" which we define as

$$V_{\mathbb{C}} := \mathcal{L} \oplus \mathcal{L} \otimes T\mathcal{M} \oplus \bar{\mathcal{L}} \otimes \bar{T}\mathcal{M} \oplus \bar{\mathcal{L}}\,, \tag{3}$$

where $T\mathcal{M}$ is the holomorphic tangent bundle of $\mathcal{M}$. $V_{\mathbb{C}}$ has an obvious conjugation operator $\bar{\cdot}$ which defines a real subbundle $V_{\mathbb{R}}$, a Hermitian metric $\langle \cdot | \cdot \rangle$ induced from those on $T\mathcal{M}$ and $\mathcal{L}$, and an antisymmetric bilinear form $\langle \cdot, \cdot \rangle$ defined by

$$\langle a, b \rangle := \langle \bar{a} | \sigma b \rangle \quad \text{where} \quad \sigma := \begin{cases} +i \text{ on } \mathcal{L} \oplus \bar{\mathcal{L}} \otimes \bar{T}\mathcal{M}\,, \\ -i \text{ on } \bar{\mathcal{L}} \oplus \mathcal{L} \otimes T\mathcal{M}\,. \end{cases} \tag{4}$$

We introduce an operator $\hat{C} : T\mathcal{M} \otimes V_{\mathbb{C}} \to V_{\mathbb{C}}$; for $v \in T\mathcal{M}$, $\hat{C}(v) \in \mathrm{End}(V_{\mathbb{C}})$ maps each space in (3) to its successor,

$$\hat{C}(v)|_{\mathcal{L}} = \cdot \otimes v\,, \quad \hat{C}(v)|_{\mathcal{L} \otimes T\mathcal{M}} = C(v) e^K G^{-1}\,, \quad \hat{C}(v)|_{\bar{\mathcal{L}} \otimes \bar{T}\mathcal{M}} = G(v, \cdot)\,, \quad \hat{C}(v)|_{\bar{\mathcal{L}}} = 0\,. \tag{5}$$

Similarly $\bar{\hat{C}}(v)$ maps each space in (3) to its predecessor. Using $\hat{C}$ we can define the "Gauss-Manin connection" on $V_{\mathbb{C}}$ in terms of its holomorphic and antiholomorphic parts,

$$\nabla = D - i\hat{C}\,, \quad \bar{\nabla} = \bar{D} + i\bar{\hat{C}}\,. \tag{6}$$

The connection preserves $V_{\mathbb{R}}$. Moreover, it is flat, as one verifies using the special geometry formula for the curvature of G. Hence it makes $V_{\mathbb{C}}$ into a holomorphic vector bundle, and there is a natural filtration by holomorphic subbundles

$$0 \subset F^3 V_{\mathbb{C}} \subset F^2 V_{\mathbb{C}} \subset F^1 V_{\mathbb{C}} \subset F^0 V_{\mathbb{C}} = V_{\mathbb{C}}\,, \tag{7}$$

where $F^k V_{\mathbb{C}}$ is the sum[5] of the first $4-k$ summands in (3).

When the special Kähler manifold arises from twisting of an $\mathcal{N}=2$ field theory, we have more data than what was mentioned above. The first extra datum is a twisted chiral ring, which in particular allows us to define the Euler characteristic,

$$\chi := 2n - 2\dim\big((a,c)\text{-ring}\big)\,. \tag{8}$$

The second extra datum is a lattice $V_{\mathbb{Z}}^* \subset V_{\mathbb{R}}^*$ preserved by the Gauss-Manin connection. We assume that $V_{\mathbb{Z}}^*$ is self-dual for the skew-symmetric pairing. Then we can choose an integral basis $\{\alpha^I, \beta_I\}$ for $V_{\mathbb{Z}}^*$, obeying

$$\langle \alpha^I, \alpha^J\rangle = 0\,, \quad \langle \alpha^I, \beta_J\rangle = \delta^I_J\,, \quad \langle \beta_I, \beta_J\rangle = 0\,. \tag{9}$$

Having fixed such a basis and a local section Ω of $\mathcal{L}$ one gets $2n+2$ functions on $\mathcal{M}$, the "periods"

$$X^I = \alpha^I(\Omega)\,, \quad F_I = \beta_I(\Omega)\,, \tag{10}$$

with $I = 0, \ldots, n$. The Kähler potential on $\mathcal{M}$ can then be written as

$$e^{-K} = i\big(X^I \bar{F}_I - \bar{X}^I F_I\big)\,. \tag{11}$$

Now we consider the open string sector. As argued in [**31, 1**], certain data of topological D-brane configurations on Calabi-Yau threefolds can be encoded holomorphically in the Hodge theoretic concept of *normal functions.* By definition, a normal function ν is a holomorphic section of the intermediate Jacobian fibration

$$J^3 := V_{\mathbb{C}}/(V_{\mathbb{Z}} + F^2 V_{\mathbb{C}}) \simeq (F^2 V_{\mathbb{C}})^*/V_{\mathbb{Z}}^* \tag{12}$$

satisfying Griffiths transversality,

$$\nabla \tilde{\nu} \in F^1 V_{\mathbb{C}}\,, \tag{13}$$

where $\tilde{\nu}$ is any lift of ν to $V_{\mathbb{C}}$, as well as certain growth conditions at infinities in $\mathcal{M}$ that will not be of central importance for our considerations. We shall not review here the discussion of the relation of D-branes with normal functions. We emphasize, however, that one of the fundamental ingredients in this identification is the decoupling of any continuous open string moduli from the topological string amplitudes.

The arguments [**1**] for this decoupling convolve facts from the obstruction theory for D-branes on threefolds (namely, $\dim \mathrm{Ext}^1(B,B) = \dim \mathrm{Ext}^2(B,B)$ for any brane B on a compact Calabi-Yau) with the identification of the obstruction map $F^2V_{\mathbb{C}}/F^3V_{\mathbb{C}} \to \mathrm{Ext}^2(B,B)$ as the dual of the infinitesimal Abel-Jacobi map $\mathrm{Ext}^1(B,B) \to F^1V_{\mathbb{C}}/F^2V_{\mathbb{C}}$. The precise statement is therefore that for generic values of the closed string moduli, the normal function corresponding to some given D-brane configuration does not depend on any continuous open string moduli. It is clear that this decoupling (and the extrapolation to loop amplitudes) might fail in special situations. In particular, the dependence of the topological amplitudes on open string moduli is essential in the non-compact settings of [**32, 33, 34, 35**]. It is also clear that the normal function need not capture the full information contained in any given D-brane. To simplify our presentation, we assume decoupling of open string moduli, and use ν as a shorthand for the D-brane configuration.

[5]This is a direct sum of complex vector bundles; we emphasize however that its holomorphic structure induced from the Gauss-Manin connection is not a direct sum, because $\bar{\bar{C}}$ mixes the summands.

The central point of [**1**] was the identification of the open string counterpart of the Yukawa cubic, which as reviewed above is the defining holomorphic data of the closed topological string at tree level. Recall that the Yukawa coupling is computed as the three-point function on the sphere. The open analogue is the two-point function on the disk, given geometrically by a particular non-holomorphic lift of the Griffiths infinitesimal invariant of the normal function ν. A pedagogical reference on normal functions and their infinitesimal invariants is [**36**]. This lift is a section of $\mathcal{L}^{-1} \otimes \mathrm{Sym}^2 T^*\mathcal{M}$, and can be written as[6]

$$\Delta_{ij} = \langle \Omega, \nabla_i \nabla_j \tilde{\nu} \rangle = D_i D_j \mathcal{T} + i C_{ijk} e^K G^{k\bar{k}} D_{\bar{k}} \bar{\mathcal{T}} \,, \tag{14}$$

where $\mathcal{T}$ is a section of $\mathcal{L}^{-1}$ given by

$$\mathcal{T} = \langle \Omega, \nu \rangle \tag{15}$$

and we have chosen the unique *real* lift $\tilde{\nu} \in V_{\mathbb{R}}$.

Crucially, Δ_{ij} is not a holomorphic section. Instead, it satisfies a holomorphic anomaly equation,

$$\partial_{\bar{k}} \Delta_{ij} = i C_{ijk} \bar{\Delta}^k_{\bar{k}} \,, \tag{16}$$

where indices are raised with the metric, $\bar{\Delta}^k_{\bar{k}} = e^K G^{k\bar{j}} \bar{\Delta}_{\bar{k}\bar{j}}$.

A fundamental example of the structure described above is provided by Type II string theory on a Calabi-Yau threefold Y. In that case: $\mathcal{M}$ is the moduli space of complex structures on Y; $V_{\mathbb{C}} = H^3(Y, \mathbb{C})$ (similarly for $V_{\mathbb{R}}$, $V_{\mathbb{Z}}$); the decomposition (3) is the Hodge decomposition; and the Gauss-Manin connection is the standard flat structure provided by the deformation invariance of integer homology. The basic example of a normal function comes from a pair of homologically equivalent holomorphic curves C_+, C_- in Y varying over $\mathcal{M}$. Over every point in $\mathcal{M}$ we pick a three-cycle Γ in Y such that $\partial\Gamma = C_+ - C_-$. We obtain an element ν of $(F^2 V_{\mathbb{C}})^* \simeq (H^{3,0} \oplus H^{2,1})^*$ by associating to any $(3,0)$- or $(2,1)$-form ω the chain integral

$$\langle \omega, \nu \rangle = \int_\Gamma \omega \tag{17}$$

viewed as a function over moduli space. The chain integral of cohomology classes in (17) is well-defined by Dolbeault's theorem and the holomorphicity of $\partial\Gamma$. Changing the choice of Γ by a closed three-cycle, the integral changes by a period, which precisely accounts for the quotient by $V^*_{\mathbb{Z}}$ in (12).

The association (17) is known as the Abel-Jacobi map. It was first used in the context of mirror symmetry with D-branes on non-compact Calabi-Yau manifolds in [**32, 33**], and in a compact setting in [**37, 31**]. The relation (16) can be viewed as an open string analogue of special geometry [**1**]. See [**34, 35**] for a related proposal.

2.2. Holomorphic anomaly of topological string amplitudes. We now fix some topological string background with D-branes characterized at tree-level by a normal function ν living over the special Kähler moduli space $\mathcal{M}$. We are interested in the perturbative topological string amplitudes $\mathcal{F}^{(g,h)}_{i_1, i_2, \ldots, i_n}$. These amplitudes are defined by integrating over the moduli space of Riemann surfaces of

[6] This expression differs by a factor of i from the corresponding expression in [**1**]. This factor can be traced back to a different convention for the Gauss-Manin connection; see (6).

genus g, with h boundary components, the appropriate correlators of the underlying worldsheet theory. The indices $i_1, i_2, \ldots, i_n$ stand for closed string insertions. See, e.g., [**1**] for details.

As mentioned in the introduction, the $\mathcal{F}^{(g,h)}_{i_1,i_2,\ldots,i_n}$ satisfy two sets of iterative relations. The more obvious relations, which are holomorphic in nature, iteratively relate the amplitudes with insertions to those without. The amplitudes are non-zero for $6g + 3h + 2n - 6 \geq 0$, and we have in this case

$$\mathcal{F}^{(g,h)}_{i_1,i_2,\ldots,i_{n+1}} = D_{i_{n+1}} \mathcal{F}^{(g,h)}_{i_1,\ldots,i_n} \,. \tag{18}$$

The amplitudes with $6g + 3h + 2n - 6 < 0$ (all of which are tree-level amplitudes) are customarily set to zero, and we note that the vacuum amplitudes $\mathcal{F}^{(g,h)}$ for $2g + h - 2 \geq 0$, the sphere three-point function $\mathcal{F}^{(0,0)}_{ijk} = C_{ijk}$, and the disk two-point function, $\mathcal{F}^{(0,1)}_{ij} = \Delta_{ij}$ are not constrained by (18) in any way.

The second set of relations are less obvious, but equally fundamental, and relate amplitudes on different worldsheet topologies,

$$\partial_{\bar{i}} \mathcal{F}^{(g,h)}_{i_1,\ldots,i_n} = \frac{1}{2} \sum_{\substack{g_1+g_2=g \\ h_1+h_2=h}} \bar{C}^{jk}_{\bar{i}} \sum_{s,\sigma\in S_n} \frac{1}{s!(n-s)!} \mathcal{F}^{(g_1,h_1)}_{j,i_{\sigma(1)},\ldots,i_{\sigma(s)}} \mathcal{F}^{(g_2,h_2)}_{k,i_{\sigma(s+1)},\ldots,i_{\sigma(n)}} + \tag{19}$$

$$\frac{1}{2} \bar{C}^{jk}_{\bar{i}} \mathcal{F}^{(g-1,h)}_{j,k,i_1,\ldots,i_n} + i\bar{\Delta}^j_{\bar{i}} \mathcal{F}^{(g,h-1)}_{j,i_1,\ldots,i_n} - (2g+h-2+n-1) \sum_{s=1}^{n} G_{i_s\bar{i}} \mathcal{F}^{(g,h)}_{i_1,\ldots,i_{s-1},i_{s+1},\ldots,i_n} \,.$$

These equations are valid for all $2g+h+n-2 > 0$, except for the one-point functions at one-loop $(g, h, n) = (1, 0, 1)$ or $(0, 2, 1)$, for which we have an additional term on the right hand side:

$$\begin{aligned} \partial_{\bar{i}} \mathcal{F}^{(1,0)}_j &= \frac{1}{2} C_{jkl} \bar{C}^{kl}_{\bar{i}} - \Big(\frac{\chi}{24} - 1\Big) G_{j\bar{i}} \,, \\ \partial_{\bar{i}} \mathcal{F}^{(0,2)}_j &= i\Delta_{jk} \bar{\Delta}^k_{\bar{i}} - \frac{N}{2} G_{j\bar{i}} \,, \end{aligned} \tag{20}$$

where χ is the Euler characteristic (8), and N is the number of open string Ramond ground states of zero charge. The one-loop vacuum amplitudes are not constrained by (19) directly, but indirectly by (18) and (20). The holomorphicity of the sphere three-point function $\partial_{\bar{l}} C_{ijk} = 0$, as well as the holomorphic anomaly of the disk two-point function, (16), appear as special cases of (19). (Here, one has to use the vanishing of the tree-level amplitudes with few insertions.)

The two sets of equations (18) and (19) together with their exceptional modification at one-loop and tree-level can be summarized more concisely in two "master equations" by introducing a certain generating function,

$$\Psi(t^i, \bar{t}^{\bar{i}}; x^i, \lambda^{-1}) = \lambda^{\frac{\chi}{24} - 1 - \mu^2 \frac{N}{2}} \exp\Bigg[\sum_{\substack{g,h,n \\ 2g+h+n-2>0}} \frac{\lambda^{2g+h+n-2}}{n!} \mu^h \mathcal{F}^{(g,h)}_{i_1,\ldots,i_n} x^{i_1} \cdots x^{i_n} \Bigg] . \tag{21}$$

With this definition, Ψ is a section of the pullback of $\mathcal{L}^{\frac{\chi}{24} - 1 - \mu^2 \frac{N}{2}}$ to the total space of $\mathcal{L} \oplus \mathcal{L} \otimes T\mathcal{M} \to \mathcal{M}$. (We are working formally here and do not worry about the existence of this peculiar power of $\mathcal{L}$). The variables x^i are coordinates on $\mathcal{L} \otimes T\mathcal{M}$, the inverse string coupling λ^{-1} is a coordinate on $\mathcal{L}$, and μ is a real parameter that keeps track of the number of boundaries in the expansion (21). Ψ is holomorphic on each fiber but not holomorphic on the total space.

It is not hard to check that the relations (19) are equivalent to the following equation satisfied by the generating function Ψ:

$$\left[\partial_{\bar{i}} - \frac{1}{2}\bar{C}_{\bar{i}}^{jk}\frac{\partial^2}{\partial x^j \partial x^k} - G_{j\bar{i}}x^j\frac{\partial}{\partial\lambda^{-1}} - i\mu\bar{\Delta}_{\bar{i}}^{j}\frac{\partial}{\partial x^j}\right]\Psi = 0\,. \tag{22}$$

Also, the relations (18), together with the statements about the amplitudes with $2g+h+n-2\leq 0$ (which are absent from (21)), are equivalent to

(23)

$$\left[\partial_i - \Gamma_{ij}^k x^j\frac{\partial}{\partial x^k} + \partial_i K\left(\lambda^{-1}\frac{\partial}{\partial\lambda^{-1}} + x^k\frac{\partial}{\partial x^k} + \frac{\chi}{24} - 1 - \mu^2\frac{N}{2}\right)\right.$$
$$\left. - \lambda^{-1}\frac{\partial}{\partial x^i} + \mathcal{F}_i^{(1,0)} + \mu^2\mathcal{F}_i^{(0,2)} + \frac{1}{2}C_{ijk}x^jx^k + \mu\Delta_{ij}x^j\right]\Psi = 0\,.$$

In the rest of this section, we will rewrite the equations (22) and (23) in various ways. The purpose is to show that the underlying structure is fairly simple, especially when viewed from the point of view of the so-called "large phase space". Readers interested primarily in the conceptual questions might be able to skip directly to section 3 on page 12.

2.3. The large phase space. To construct the large phase space [**39**], one first replaces the moduli space $\mathcal{M}$ by the total space $\widetilde{\mathcal{M}}$ of the line bundle $\mathcal{L}$ minus the zero section. One reason for introducing $\widetilde{\mathcal{M}}$ is that it comes equipped with very convenient coordinates, namely the X^I we considered before. In other words, a choice of nowhere vanishing section of $\mathcal{L}$ defines an embedding of $\mathcal{M}$ into $\widetilde{\mathcal{M}}$ via

$$X^I = X^I(t^i)\,, \tag{24}$$

with inverse projection

$$t^i = t^i(X^I) \tag{25}$$

which is homogeneous of degree 0,

$$X^I\frac{\partial}{\partial X^I}t^i(X^I) = 0\,. \tag{26}$$

All quantities we consider are homogeneous of fixed degree under the overall rescaling of the X^I.

As complex vector bundles over $\mathcal{M}$, $\mathcal{L}\oplus\mathcal{L}\otimes T\mathcal{M}\simeq F^2V_{\mathbb{C}}$; the pullback of $F^2V_{\mathbb{C}}$ to $\widetilde{\mathcal{M}}$ is isomorphic to $T\widetilde{\mathcal{M}}$. The map from the pullback of $\mathcal{L}\oplus\mathcal{L}\otimes T\mathcal{M}$ to $T\widetilde{\mathcal{M}}$ is[7]

$$z^I = 2\,(\lambda^{-1}X^I + x^iX^I_{;i}) \tag{27}$$

where z^I is the coordinate on the fiber of $T\widetilde{\mathcal{M}}$, and we defined

$$X^I_i := \partial_i X^I\,, \qquad X^I_{;i} := X^I_i + \partial_i K X^I\,. \tag{28}$$

Altogether, we have changed from the "small phase space" coordinates (t^i, x^i, λ^{-1}) on $(\mathcal{L}\oplus\mathcal{L}\otimes T\mathcal{M})\to\mathcal{M}$ to the "large phase space" coordinates (X^I, z^I) on $T\widetilde{\mathcal{M}}$.

---

[7]For convenience, this formula differs from that of [**26**] by an eighth root of unity.

Next we describe special geometry from the large phase space point of view, beginning with the closed string. The basic holomorphic data is encoded in the prepotential

$$F := \frac{1}{2} X^I F_I \tag{29}$$

and its first few derivatives,

$$F_I = \partial_I F\,, \qquad \tau_{IJ} := \partial_I \partial_J F\,, \qquad C_{IJK} := \partial_I \partial_J \partial_K F\,. \tag{30}$$

The homogeneity of F implies a useful relation for the Yukawa coupling,

$$X^I C_{IJK} = 0\,. \tag{31}$$

Now we describe the large phase space version of the open string data, namely, normal functions and their infinitesimal invariants. The domain wall tension $\mathcal{T} = \langle \Omega, \nu \rangle$ is a period-like object, homogeneous of degree 1 in the large phase space:

$$\mathcal{T} = X^I \nu_I\,, \qquad \text{where } \nu_I := \partial_I \mathcal{T}\,. \tag{32}$$

The large phase space infinitesimal invariant is

$$\Delta_{IJ} := \langle \Omega, \partial_I \partial_J \nu \rangle = \nu_{IJ} - C_{IJ}^K \mathrm{Im} \nu_K\,, \tag{33}$$

where of course

$$\nu_{IJ} := \partial_I \partial_J \mathcal{T}\,, \qquad \text{with } X^I \nu_{IJ} = X^I \Delta_{IJ} = 0\,. \tag{34}$$

The last relation is very similar to (31) satisfied by the Yukawa coupling.

In Appendix A we give some useful relations between the large and small phase space data.

2.4. The anomaly equations in the large phase space. In this section we will rewrite the anomaly equations in the large phase space, which turns out to be the most convenient setting for understanding the relation between the open and closed anomaly equations.

We follow [**38**] and first solve the holomorphic anomaly equations for the one-loop amplitudes (20). Recall that the torus anomaly can be integrated to [**2**]

$$\mathcal{F}^{(1,0)} = -\frac{1}{2} \log \det \mathrm{Im} \tau_{IJ} - \Big(\frac{\chi}{24} - 1\Big) K + f^{(1,0)} + \bar{f}^{(1,0)}\,, \tag{35}$$

where $f^{(1,0)}$ is a holomorphic ambiguity. The holomorphic anomaly equation of the annulus (as well as higher (g,h)) can be integrated by a procedure very similar to that in [**3**], see [**1**]. To this end, let us introduce the large phase space analogue, δ^J, of the "terminator" of [**1**]. This quantity is defined by the equation

$$\bar{\Delta}_I^J = \bar{\partial}_I \delta^J\,, \tag{36}$$

which can be locally solved by

$$\delta^J = -2i \mathrm{Im} \tau^{JK} \mathrm{Im} \nu_K = \mathrm{Im} \tau^{JK} (\bar{\nu}_K - \nu_K)\,. \tag{37}$$

This choice of δ^J also satisfies

$$\partial_I \delta^J = -\Delta_I^J. \tag{38}$$

For future reference, we also record the holomorphic anomaly of the disk two-point function in the large phase space:

$$\bar{\partial}_K \Delta_{IJ} = -\frac{i}{2} C_{IJL} \bar{\partial}_K \delta^L = -\frac{i}{2} C_{IJL} \bar{\Delta}_K^L\,. \tag{39}$$

With these definitions, the holomorphic anomaly equation of the annulus, which in the large phase space takes the form

$$\bar\partial_J \partial_I \mathcal{F}^{(0,2)} = -\frac{i}{2}\Delta_{IK}\bar\Delta^K_J + \frac{N}{2}\bar\partial_J\partial_I K\,, \tag{40}$$

can be integrated to

$$\begin{aligned} \mathcal{F}^{(0,2)}_I &= -\frac{i}{2}\Delta_{IK}\delta^K + \frac{1}{8}C_{IKL}\delta^K\delta^L + \frac{N}{2}\partial_I K + \partial_I f^{(0,2)}\,, \\ \mathcal{F}^{(0,2)} &= \frac{i}{4}\delta^K \mathrm{Im}\,\tau_{KL}\delta^L + \frac{N}{2}K + f^{(0,2)} + \bar f^{(0,2)}\,, \end{aligned} \tag{41}$$

where $f^{(0,2)}$ is another holomorphic ambiguity.

We absorb $f^{(1,0)}$ and $f^{(0,2)}$ into a redefinition of the generating function Ψ:

$$\Psi \to e^{-f^{(1,0)}-\mu^2 f^{(0,2)}}\Psi\,. \tag{42}$$

This should be interpreted as $\Psi_{\rm old} = e^{-f^{(1,0)}-\mu^2 f^{(0,2)}}\Psi_{\rm new}$; we will use this notation repeatedly in the next few sections.

It is then straightforward, using the formulas given in Appendix A, to show that the equations (22) and (23) become respectively

$$\begin{aligned} &\Big[\bar\partial_I - \frac{1}{2}\bar C_I^{JK}\frac{\partial^2}{\partial z^J\partial z^K} + i\mu\bar\Delta^J_I\frac{\partial}{\partial z^J}\Big]\Psi = 0\,, \\ &\Big[\partial_I - \frac{1}{2}\partial_I \log\det \mathrm{Im}\,\tau_{IJ} + \frac{i}{2}C^K_{IJ}z^J\frac{\partial}{\partial z^K} + \frac{1}{8}C_{IJK}z^J z^K \\ &\qquad + \frac{1}{2}\mu\Delta_{IJ}z^J - \frac{i}{2}\mu^2\Delta_{IJ}\delta^J + \frac{1}{8}\mu^2 C_{IJK}\delta^J\delta^K\Big]\Psi = 0\,. \end{aligned} \tag{43}$$

The closed string version of these equations was first obtained in [**39**]. Now note that by using

$$\Delta_{IJ} = \nu_{IJ} - C^K_{IJ}\mathrm{Im}\,\nu_K = \nu_{IJ} - \frac{i}{2}C_{IJK}\delta^K \tag{44}$$

we can rewrite the last four terms in (43) as

$$\begin{aligned} &\frac{1}{8}C_{IJK}z^J z^K + \frac{1}{2}\mu\nu_{IJ}z^J - \frac{i}{4}\mu C_{IJK}z^J\delta^K - \frac{i}{2}\mu^2\nu_{IJ}\delta^J - \frac{1}{8}\mu^2 C_{IJK}\delta^J\delta^K \\ &\qquad = \frac{1}{8}C_{IJK}(z^J - i\mu\delta^J)(z^K - i\mu\delta^K) + \frac{1}{2}\mu\nu_{IJ}(z^J - i\mu\delta^J)\,. \end{aligned} \tag{45}$$

So if we now resubstitute

$$\Psi \to \sqrt{\det \mathrm{Im}\,\tau_{IJ}}\exp\Big[\frac{i}{4}(z^J - i\mu\delta^J)\mathrm{Im}\,\tau_{JK}(z^K - i\mu\delta^K)\Big]\Psi \tag{46}$$

and use the above relations for Δ_{IJ}, δ^J, etc., the equations take the form

$$\begin{aligned} &\Big[\bar\partial_I - \frac{i}{2}\bar C^K_{IJ}z^J\frac{\partial}{\partial z^K} - \frac{1}{2}\bar C_I^{JK}\frac{\partial^2}{\partial z^J\partial z^K} + i\mu\bar\nu^J_I\frac{\partial}{\partial z^J}\Big]\Psi = 0\,, \\ &\Big[\partial_I + \frac{i}{2}C^K_{IJ}z^J\frac{\partial}{\partial z^K}\Big]\Psi = 0\,. \end{aligned} \tag{47}$$

Finally, we change variables to $\bar y_I = \mathrm{Im}\,\tau_{IJ}z^J$ [**26**]. Geometrically this amounts to considering Ψ as defined on $T^*\widetilde{\mathcal{M}}$ instead of $T\widetilde{\mathcal{M}}$. We then arrive at the simplest form of the anomaly equations,

$$\begin{aligned} &\Big[\bar\partial_I - \frac{1}{2}\bar C_{IJK}\frac{\partial^2}{\partial\bar y_J\partial\bar y_K} + i\mu\bar\nu_{IJ}\frac{\partial}{\partial\bar y_J}\Big]\Psi = 0\,, \\ &\partial_I\Psi = 0\,. \end{aligned} \tag{48}$$

For completeness, we summarize the sequence of redefinitions of Ψ leading from (21) to (48):

(49)

$$\Psi_{(48)} = \frac{1}{\sqrt{\det \operatorname{Im}\tau_{IJ}}}\exp\Big[-\frac{i}{4}(z^J - i\mu\delta^J)\operatorname{Im}\tau_{JK}(z^K - i\mu\delta^K) + f^{(1,0)} + \mu^2 f^{(0,2)}\Big]\Psi_{(21)}\,.$$

After all these transformations Ψ has turned out to be purely antiholomorphic. We can take $\Psi \to \overline{\Psi}$ to get

$$\begin{aligned} &\Big[\partial_I - \frac{1}{2}C_{IJK}\frac{\partial^2}{\partial y_J\partial y_K} - i\mu\nu_{IJ}\frac{\partial}{\partial y_J}\Big]\Psi = 0\,, \\ &\bar\partial_I\Psi = 0\,. \end{aligned} \tag{50}$$

To conclude this section we briefly discuss the global properties of Ψ. Before the redefinition Ψ represented a section of the pullback of $\mathcal{L}^{\frac{\chi}{24}-1-\mu^2\frac{N}{2}}$, i.e. under a change of local section $\Omega \to e^f\Omega$ for $\mathcal{L}$ it transformed by $\Psi \to e^{(\frac{\chi}{24}-1-\mu^2\frac{N}{2})f}\Psi$. After the redefinition this transformation is canceled by the explicit transformations of $f^{(1,0)}$ and $f^{(0,2)}$ determined by (35) and (41). However, $f^{(1,0)}$, $f^{(0,2)}$, z^J, δ^J and τ appearing in (49) are all defined using a symplectic basis for $V_{\mathbb{Z}}$, and such a choice cannot be made globally on $\widetilde{\mathcal{M}}$ because $V_{\mathbb{Z}}$ can have global $Sp(2n+2,\mathbb{Z})$-valued monodromies; so the new Ψ has to be considered as a section of a bundle which transforms appropriately under $Sp(2n+2,\mathbb{Z})$, i.e. as a modular form.

3. Discussion

The rewriting of the holomorphic anomaly equations in the last section makes the redefinition of variables which removes the open string data from the equation completely transparent. Shifting[8]

$$\bar y_I \to \bar y_I + i\mu\bar\nu_I \tag{51}$$

in (48) eliminates ν from both equations, and maps them onto the ordinary heat equations satisfied by the closed topological string amplitudes [**26**].

The existence of a shift removing open string data was first noted in [**21**]. With slightly different conventions, it was shown there that a general solution of the master holomorphic anomaly equation (22) can be mapped to a solution of the master equation with $\Delta_{ij} \equiv 0$ (which is the master equation of BCOV) by shifting the closed string variables. The fate of the second equation (23) was however not analyzed in [**21**]. Instead, it was shown, using the above observation and the techniques of [**3**], that the solution of the perturbative holomorphic anomaly equations (19) can be written in a diagrammatic fashion using Feynman rules as noticed in [**1**].

In this section, we will first discuss the difference between the shift (51) and the shift of [**21**], at the level of the equation and at the level of the Feynman rules. We will then answer the question whether the simple shift maps open to closed

---

[8]Again, this shift should be interpreted as $(\bar y_I)_{\text{old}} = (\bar y_I)_{\text{new}} + i\mu\bar\nu_I$.

topological string also at the level of the topological string amplitudes, which are the physically relevant solutions of the anomaly equations.

3.1. Shifts. The shift studied in [**21**] reads in the small phase space

$$x^i \to x^i - i\mu\bar{\Delta}^i\,, \qquad \lambda^{-1} \to \lambda^{-1} + i\mu\bar{\Delta}\,, \tag{52}$$

where $\bar{\Delta}^i$, $\bar{\Delta}$ are the terminators of [**1**] in the small phase space, satisfying

$$\partial_{\bar{j}}\bar{\Delta}^i = \bar{\Delta}^i_{\bar{j}}\,, \qquad \partial_{\bar{j}}\bar{\Delta} = G_{\bar{j}i}\bar{\Delta}^i\,, \tag{53}$$

where $\bar{\Delta}^i_{\bar{j}} = e^K G^{i\bar{i}}\bar{\Delta}_{\bar{j}\bar{i}}$.

To understand the relation between (51) and (52), we recall that the Griffiths infinitesimal invariant is the sum of two terms, which in small phase space are $\Delta_{ij} = \Delta^{(1)}_{ij} + \Delta^{(2)}_{ij}$, with $\Delta^{(1)}_{ij} = D_iD_j\mathcal{T}$ and $\Delta^{(2)}_{ij} = iC^{\bar{k}}_{ij}D_{\bar{k}}\bar{\mathcal{T}}$. In large phase space, the corresponding two terms are $\Delta_{IJ} = \Delta^{(1)}_{IJ} + \Delta^{(2)}_{IJ}$, with $\Delta^{(1)}_{IJ} = \nu_{IJ} + \frac{i}{2}C^K_{IJ}\nu_K$, $\Delta^{(2)}_{IJ} = -\frac{i}{2}C^K_{IJ}\bar{\nu}_K$. From this, we see that our shift (51) is by a potential for only the first term, $\bar{\partial}_I\bar{\nu}^J = (\bar{\Delta}^{(1)})^J_I$, whereas the shift (52) is by a potential for the full infinitesimal invariant; see (53).

In other words, the large phase space analogue of (52) would be

$$z^J \to z^J + i\mu\delta^J = z^J + i\mu(\bar{\nu}^J - \nu^J)\,. \tag{54}$$

It is not hard to see that under this shift, the open string data is eliminated only from the first equation in (43), whereas the second equation remains with an additional term, which in the large phase space reads

$$\left(-i\mu\nu^J_I\frac{\partial}{\partial z^J} - \frac{1}{2}\mu\nu_{IJ}z^J\right)\Psi\,. \tag{55}$$

Obviously, one may also write this term in the small phase space. It is worthwhile pointing out that the difference between the shift (51) and (52) cannot be absorbed into a holomorphic ambiguity inherent in the definition of the terminators.

The proof of the Feynman rule expansion of open topological string amplitudes given in [**21**] relies on the same auxiliary finite-dimensional quantum system used in [**3**]. The dynamical variables are the $x = (x^i, \lambda^{-1})$. The quadratic part of the action is $Q(x,x)$, where Q is an inverse to the collection of propagators S^{ij}, $S^i := S^{i\lambda^{-1}}$, $S := S^{\lambda^{-1}\lambda^{-1}}$. The interactions of the system are given by $\log\Psi(t^i, \bar{t}^{\bar{i}}; x^i - i\mu\bar{\Delta}^i, \lambda^{-1} + i\mu\bar{\Delta})$ where Ψ is defined in (21) and the variables are shifted as in (52). This includes the vacuum amplitudes $\mathcal{F}^{(g,h)}$, and as non-trivial interaction vertices the infinite collection of topological amplitudes $\mathcal{F}^{(g,h)}_{i_1,\ldots,i_n}$ with $n \geq 1$. As pointed out in [**21**], the shift (52) generates the possibility of terminating indices on $\bar{\Delta}^i$, $\bar{\Delta} =: \bar{\Delta}^{\lambda^{-1}}$, or in other words, it introduces a background vev for (x^i, λ^{-1}). This precisely reproduces the Feynman rules noticed in [**1**].

Our result is that to completely remove the open string data from both equations, one should shift only by (51) and absorb the second term, originating from $(\bar{\Delta}^{(2)})^J_I$, into the redefinition of Ψ given in (49). This has a simple interpretation in terms of the finite-dimensional system discussed above. Namely, we give a smaller vev to (x^i, λ^{-1}) and instead introduce additional vertices $\sim D_i\mathcal{T}x^i$. The latter also behave like a tadpole, so that the overall Feynman rules are unaffected.

Referring to the perturbative holomorphic anomaly equations in (19), we notice that the second part of the infinitesimal invariant enters as if it were on equal footing with the closed string degenerations. Namely, it is of the form $(\bar\Delta^{(2)})_{\bar i}^{j} = -i\bar C_{\bar i}^{jk} D_k \mathcal{T}$, where $D_k\mathcal{T}$ could be viewed as a disk one-point function. The first part, $(\bar\Delta^{(1)})_{\bar i}^{j}$, however, cannot be treated in this way and has to be accounted for in the master equation by a shift of variables (51).

We remark that the holomorphic anomaly equations for open topological strings derived in [**13, 12**] using matrix model duals to certain local Calabi-Yau manifolds appear to contain only the first type of contribution described in the previous paragraph, namely that originating from $\bar\Delta^{(2)}_{\bar i\bar j}$. It would be interesting to understand the Hodge theoretic origin of this simplification, as well as possible implications for open/closed duality in this context.

3.2. Solutions. We can now answer the question whether the simple shift of variables that maps the extended holomorphic anomaly equations to the ordinary BCOV equations will also transform correctly the actual topological string amplitudes. By studying the perturbative expansion in the small phase space, we will see that this possibility is in fact not realized.

The small phase space analogue of the shift (51) is

$$x^i \to x^i - i\mu\epsilon^i\,, \qquad \lambda^{-1} \to \lambda^{-1} + i\mu\epsilon\,, \tag{56}$$

where ϵ^i, ϵ are potentials for the first part of the infinitesimal invariant. They are defined by the equations

$$\begin{aligned} \partial_{\bar i}\epsilon^j &= e^K G^{j\bar k} D_{\bar k} D_{\bar i}\bar{\mathcal{T}}\,,\\ \partial_{\bar i}\epsilon &= G_{\bar i j}\epsilon^j\,, \end{aligned} \tag{57}$$

which can easily be solved by choosing

$$\begin{aligned} \epsilon &= e^K\bar{\mathcal{T}}\,,\\ \epsilon^i &= G^{i\bar j} e^K D_{\bar j}\bar{\mathcal{T}}\,. \end{aligned} \tag{58}$$

After the shift (56), the small phase space holomorphic anomaly equations (22) and (23) are transformed into

$$\begin{aligned} &\Big[\partial_{\bar i} - \frac12 \bar C_{\bar i}^{jk}\frac{\partial^2}{\partial x^j\partial x^k} - G_{j\bar i}x^j\frac{\partial}{\partial\lambda^{-1}} - \mu\bar C_{\bar i}^{jk} D_k\mathcal{T}\frac{\partial}{\partial x^j}\Big]\Psi = 0\,,\\ &\Big[\partial_i - \Gamma^k_{ij}x^j\frac{\partial}{\partial x^k} + \partial_i K\Big(\lambda^{-1}\frac{\partial}{\partial\lambda^{-1}} + x^k\frac{\partial}{\partial x^k} + \frac{\chi}{24} - 1\Big) - \lambda^{-1}\frac{\partial}{\partial x^i} + \mathcal{F}_i^{(1,0)} +\\ &\mu^2\Big(\mathcal{F}_i^{(0,2)} - \frac{N}{2}\partial_i K\Big) + \frac12 C_{ijk}(x^j - i\mu\epsilon^j)(x^k - i\mu\epsilon^k) + \mu\Delta_{ij}(x^j - i\mu\epsilon^j)\Big]\Psi = 0\,. \end{aligned} \tag{59}$$

The remaining μ-dependent terms can be removed by the substitution

$$\Psi \to \exp\Big[-\mu D_k\mathcal{T}x^k - \mu\mathcal{T}\lambda^{-1} - \frac{\mu^2}{2}S^{jk}D_j\mathcal{T}D_k\mathcal{T} + \mu^2 S^k D_k\mathcal{T}\mathcal{T} - \mu^2 S\mathcal{T}^2 - \mu^2 f^{(0,2)}\Big]\Psi \tag{60}$$

where S^{ij}, S^i, and S are the BCOV propagators, defined up to holomorphic ambiguity by the equations

$$\begin{aligned} \partial_{\bar{i}} S^{jk} &= \bar{C}_{\bar{i}}^{jk} = e^{2K} G^{j\bar{j}} G^{k\bar{k}} \bar{C}_{\bar{i}\bar{j}\bar{k}}\,, \\ \partial_{\bar{i}} S^{j} &= G_{\bar{i}k} S^{jk}\,, \\ \partial_{\bar{i}} S &= G_{\bar{i}k} S^{k}\,. \end{aligned} \tag{61}$$

To check that this substitution indeed removes the open string data, one has to use (among other things) the identity

(62)

$$\partial_{\bar{l}}\Big(\frac{1}{2}C_{ijk}\epsilon^j\epsilon^k + i\Delta_{ij}\epsilon^j + \frac{1}{2}\partial_i\big(S^{jk}D_j\mathcal{T}D_k\mathcal{T}\big) - \partial_i\big(S^jD_j\mathcal{T}\mathcal{T}\big) + \partial_i\big(S\mathcal{T}^2\big)\Big) = i\Delta_{ij}\bar{\Delta}^j_{\bar{l}}\,,$$

which provides the small phase space integration of the annulus anomaly (cf., (41)).

Summarizing these transformations, the shifted open topological string partition function is

$$\begin{aligned} \Psi^\nu(t^i,\bar{t}^{\bar{i}};x^i,\lambda^{-1}) = \exp\Big[&\mu D_k\mathcal{T}x^k + \mu\mathcal{T}\lambda^{-1} + \frac{\mu^2}{2}S^{jk}D_j\mathcal{T}D_k\mathcal{T} \\ &-\mu^2 S^k D_k\mathcal{T}\mathcal{T} + \mu^2 S\mathcal{T}^2 + \mu^2 f^{(0,2)}\Big]\Psi(t^i,\bar{t}^{\bar{i}};x^i - i\mu\epsilon^i, \lambda^{-1} + i\mu\epsilon)\,, \end{aligned} \tag{63}$$

with Ψ as defined in (21),

(64)

$$\Psi(t^i,\bar{t}^{\bar{i}};x^i,\lambda^{-1}) = \lambda^{\frac{\chi}{24}-1-\mu^2\frac{N}{2}}\exp\Bigg[\sum_{\substack{g,h,n\\2g+h+n-2>0}}\frac{\lambda^{2g+h+n-2}}{n!}\mu^h\mathcal{F}^{(g,h)}_{i_1,\ldots,i_n}x^{i_1}\cdots x^{i_n}\Bigg]\,,$$

in terms of the open-closed topological string amplitudes with D-brane configuration given by the normal function, ν. The statement is that all Ψ^ν satisfy the same holomorphic anomaly equations, identical to the anomaly equations satisfied by the closed topological partition function $\Psi^{\text{closed}} := \Psi^0$.

To see that the Ψ^ν are not all identical, it suffices to look at a few of the terms in the expansion of (63). For example, the leading behavior as $\lambda \to 0$ is given by $\lambda^{\frac{\chi}{24}-1-\mu^2\frac{N}{2}}$, and thus depends on N. More generally, because the sum in (64) is restricted to $2g+h+n-2>0$, the leading terms in the λ-expansion are given by the exponential prefactor in (63), and hence depend explicitly on the brane configuration.

In terms of the diagrammatic expansion discussed in the previous subsection, the fact that the shifted Ψ^ν all satisfy the same differential equation means that the Feynman rules, viewed as recursion relations between the $\mathcal{F}^{(g,h)}$, are independent of ν. However, the different expansion around $\lambda = 0$ means that the initial conditions for the recursion relations are different. We stress that as a result the Ψ^ν are not determined by Ψ^0 alone, *even before taking into account the holomorphic ambiguity.*

4. Speculations

We have seen above that any D-brane configuration specified by a normal function ν determines a solution Ψ^ν of the ordinary holomorphic anomaly equation of BCOV. This is achieved by forming the generating function of open-closed topological string amplitudes, with a particular convention for disk and annulus amplitudes with few insertions, and then shifting the closed string variables in a certain way.

This shift is different from the one proposed in [**21**]. We have explained this statement in various ways in both the small and large phase space. In this final section, we present three applications of our results.

4.1. Topological string wavefunctions. Witten has proposed in [**8**] that one should interpret the BCOV holomorphic anomaly equations as a statement of background independence in the topological string. Although background independence does not hold order by order in perturbation theory, the all-genus partition function $\Psi(t^i, \bar{t}^{\bar{i}}; x^i, \lambda^{-1})$, as a function of x^i, λ^{-1}, can be thought of as a wavefunction defining a background independent state in a certain auxiliary "Hilbert space". Witten's Hilbert space, $\mathcal{H}_W$, arises by quantizing the symplectic vector space $V_{\mathbb{R}} = H^3(Y, \mathbb{R})$ in certain holomorphic polarizations, indexed by the choice of background $(t^i, \bar{t}^{\bar{i}})$. The holomorphic anomaly equations guarantee that as the background is varied, the wavefunction $\Psi(t^i, \bar{t}^{\bar{i}}; x^i, \lambda^{-1})$ varies precisely according to an infinitesimal Bogoliubov transformation, while the state $|\Psi\rangle \in \mathcal{H}_W$ is background independent.

In this interpretation of the holomorphic anomaly equation, the fact that the shifted open string partition functions Ψ^ν also satisfy the ordinary BCOV equation simply means that they also define states in the same Hilbert space,

$$|\Psi^\nu\rangle \in \mathcal{H}_W \qquad \text{for all } \nu\,. \tag{65}$$

The discussion in section 3 shows that $|\Psi^\nu\rangle$ indeed depends on ν. We find it natural to conjecture that as ν varies over the set of all D-branes, the $|\Psi^\nu\rangle$ will furnish a basis of $\mathcal{H}_W$.

In support of this conjecture, we note that it is probably true at the semi-classical level, i.e., at the level of the disk amplitude. Indeed, the topological disk partition function is given simply by the normal function itself. So we have to ask for the set of normal functions that can be realized by wrapping D-branes, where for concreteness we work in the context of the B-model on a Calabi-Yau Y. (Note that the usual definition of a normal function requires only Griffiths transversality, not that it be realizable algebraically.) It is by now well-accepted that the set of all B-branes on Y is equivalent to the derived category of coherent sheaves, $D^b(Y)$. We obtain a normal function from any object in $D^b(Y)$ that deforms with Y and whose second Chern class vanishes in $H^2(Y;\mathbb{Z})$ [**40**]. We gave the typical example of this at the end of subsection 2.1: Two homologous curves that deform with Y define a normal function by integration over a bounding three-chain.

What is known mathematically is that the image of the so-called Griffiths group $\mathrm{Griff}^2(Y)$ (homologically trivial algebraic cycles of codimension two, modulo algebraic equivalence) under the Abel-Jacobi map to the intermediate Jacobian $J^3(Y)$ is not finitely generated as a vector space over $\mathbb{Q}$ [**40, 41, 42**]. It appears likely[9] that the image of $\mathrm{Griff}^2(Y)$ in $J^3(Y)$ is in fact dense (in the analytic topology). Lifting the period ambiguity in the definition of a normal function, we conclude that the chain integrals (17) will also be dense in $V_{\mathbb{C}}/F^2V_{\mathbb{C}} \simeq T^*\widetilde{\mathcal{M}}$.

Now Witten's Hilbert space $\mathcal{H}_W$ is a kind of quantization of the symplectic vector space $V_{\mathbb{R}}$, in a complex polarization which identifies it with a fiber of $T^*\widetilde{\mathcal{M}}$. Although the exact relation is somewhat cumbersome (particularly because the formal inner product in the complex polarization is not positive definite), we view

---

[9]We thank D. Morrison for conversations on this issue.

the denseness of the image of the Abel-Jacobi map in $T^*\widetilde{\mathcal{M}}$ as evidence for the conjecture that the set of corresponding $|\Psi^\nu\rangle$ furnishes a ("Hilbert space") basis of $\mathcal{H}_W$.

4.2. BPS state counting. The wavefunction interpretation of the topological partition function plays a crucial role in the formulation of the OSV conjecture [**9**]. Namely, consider the Type IIB superstring compactified on Y. The resulting $d = 4$ supergravity theory includes electrically and magnetically charged BPS states, with charges $C \in V_{\mathbb{Z}}^*$. OSV conjectured that the degeneracies $\Omega(C)$ of these states are given by the Wigner function of $|\Psi\rangle$:

$$\Omega(C) = \langle\Psi|\mathcal{O}_C|\Psi\rangle\,, \tag{66}$$

where $\mathcal{O}_C$ is the Heisenberg group element associated to C. To write (66) more concretely one has to choose a polarization. For example, [**9**] uses the real polarization determined by a decomposition of $V_{\mathbb{R}}$ into Lagrangian subspaces, $V_{\mathbb{R}} = V_{electric} \oplus V_{magnetic}$. Then $|\Psi\rangle$ is represented by a function $\Psi(\chi)$, $\chi \in V_{magnetic}$, the charge decomposes as $C = Q + P$, and (66) becomes

$$\Omega(Q,P) = \int d^n\chi\, \bar{\Psi}(\chi) \exp\left(Q^I \frac{\partial}{\partial \chi^I} + iP_I\chi^I\right) \Psi(\chi)\,. \tag{67}$$

As we have seen, each D-brane configuration ν provides another state $|\Psi^\nu\rangle$ in $\mathcal{H}_W$; so one might ask what its Wigner function computes. We are confused about various aspects of this question, but one possible guess follows. Recall from [**1**] that ν corresponds to a brane configuration with zero net charge. The simplest example is a pair of homologous holomorphic curves $E_\pm$ in Y. Consider wrapping a D3-brane of Type IIB on a 2-cycle in Y; this will give a string in the spatial $\mathbb{R}^3$, extended say in the x^1 direction. This string supports an effective field theory, with two supersymmetric vacua corresponding to wrapping the brane on the two curves $E_\pm$. Now consider a "domain wall" configuration, i.e. Type IIB on Y plus a brane which is in the E_- vacuum as $x^1 \to -\infty$ and the E_+ vacuum as $x^1 \to \infty$. We regard this configuration as the "background". Possible BPS states in this background include particles in the bulk of $\mathbb{R}^3$ as well as ones bound to the domain wall on the string. So a minimally invasive modification of the OSV conjecture is to propose that the Wigner function of $|\Psi^\nu\rangle$ counts such states. (The quantization of charge in this sector is non-standard — there is a fractional part determined by ν. We do not understand how this affects the proposal.)

An open string extension of the OSV conjecture was also proposed in [**16**]. In that case, however, one considers branes in a non-compact Calabi-Yau. This leads to several salient differences from the compact case: the topological partition function depends on the continuous moduli of the brane, and the physical theory includes additional BPS states which couple to them. As a result the proposal there took a somewhat different form.

4.3. Hints from supersymmetry. We have found that the holomorphic anomaly equations of the open and closed topological strings can be transformed into one another by a simple shift of variables, similar to that given in [**21**]. This simple statement deserves to have a physical interpretation. In searching for one, however, we encounter an immediate puzzle: the variables we shifted were not the physical moduli $(t^i, \bar{t}^{\bar{i}})$ of the worldsheet theory, but rather the formal generating-function parameters (x^i, λ^{-1}). So to find a physical explanation of the shift we have to find a physical meaning for these parameters.

A possible clue comes from the fact that the equations and the shift take their simplest form when we trade the coordinates $(t^i, \bar{t}^{\bar{i}}, x^i, \lambda^{-1})$ for (X^I, y_I), i.e. we use the complex structure of $T^*\widetilde{\mathcal{M}}$, as was done in [**26**]. This complex manifold has a natural meaning from the spacetime point of view. In the superconformal approach to $\mathcal{N} = 2$ supergravity coupled to the vector multiplet moduli space $\mathcal{M}$, one begins by considering a rigid (non-gravitational) theory with vector multiplet moduli space $\widetilde{\mathcal{M}}$. Upon classical dimensional reduction of this rigid theory from $d = 4$ to $d = 3$, say along x^3, supersymmetry dictates that one obtains a theory with a hyperkähler moduli space. This space turns out to be (in one of its complex structures) exactly $T^*\widetilde{\mathcal{M}}$ [**43**]. So from this point of view the formal parameters y_I become physical: they represent the electric and magnetic Wilson lines $(A_I)_3, (A_I^D)_3$ of the $d = 4$ gauge fields along the x^3 direction. It would be very interesting to understand whether the shift relating the open and closed strings can be related to turning on these gauge fields. As support for this idea note that λ certainly is related to the graviphoton field strength [**3**].

Acknowledgements. We are grateful to J. Bernstein, M. Kontsevich, W. Nahm, N. Nekrasov, F. Smirnov and L. Takhtajan for discussions. The research is supported by a Science Foundation Ireland grant. We would like to thank Juan Maldacena, Hirosi Ooguri, Martin Roček, Cumrun Vafa, and Edward Witten for valuable discussions and communications. We are grateful to the Aspen Center for Physics and to the Simons Workshop in Mathematics and Physics for providing a stimulating atmosphere at some stages of this project. The work of A.N. is supported in part by the Martin A. and Helen Chooljian Membership at the Institute for Advanced Study, and by the NSF under grant number PHY-0503584. The work of J.W. is supported in part by the Roger Dashen Membership at IAS, and also by the NSF grant number PHY-0503584.

Appendix A. From large to small phase space

In this appendix we give a few formulas which are needed for passing between the large and small phase space.

Recall the definition (28),

$$X_i^I := \partial_i X^I, \qquad X_{;i}^I := X_i^I + \partial_i K X^I. \tag{68}$$

Tangent directions to $\mathcal{M}$ and $\widetilde{\mathcal{M}}$ are related by

$$\frac{\partial t^j}{\partial X^J} X_i^J = \delta_i^j, \quad \frac{\partial t^i}{\partial X^J} X_{;i}^I = \delta_J^I - e^K \bar{X}^L 2\mathrm{Im}\,\tau_{LJ} X^I. \tag{69}$$

The natural metric on $\widetilde{\mathcal{M}}$ is $\mathrm{Im}\,\tau$, related to the special Kähler data (G, e^K) on $\mathcal{M}$ by

(70)

$$e^{-K} = 2\bar{X}^I X^J \mathrm{Im}\,\tau_{IJ}, \qquad e^{-K} G_{i\bar{j}} = -X_{;i}^I \bar{X}_{;\bar{j}}^J\, 2\mathrm{Im}\,\tau_{IJ}, \qquad \bar{X}^M 2\mathrm{Im}\,\tau_{MI} X_{;j}^I = 0.$$

The inverse metrics are similarly related by

$$e^K G^{i\bar{j}} = -\frac{\partial t^i}{\partial X^I}\frac{\partial \bar{t}^{\bar{j}}}{\partial \bar{X}^J}\,\frac{1}{2}\mathrm{Im}\,\tau^{IJ}, \qquad X_{;i}^I \bar{X}_{;\bar{j}}^J e^K G^{i\bar{j}} = -\frac{1}{2}\mathrm{Im}\,\tau^{IJ} + e^K X^I \bar{X}^J. \tag{71}$$

The connection and Yukawa coupling on $\mathcal{M}$ are related to the data on $\widetilde{\mathcal{M}}$ by

$$\Gamma^k_{ij} - \partial_i K \delta^k_j = \frac{\partial t^k}{\partial X^I} \partial_i (X^I_{;j}) - \frac{i}{2} \frac{\partial t^k}{\partial X^I} X^M_{;i} X^K_{;j} C^I_{KM}\,, \tag{72}$$

$$C_{ijk} = X^I_i X^J_j X^K_k C_{IJK} = X^I_{;i} X^J_{;j} X^K_{;k} C_{IJK}\,. \tag{73}$$

where we have used the homogeneity (31) of the Yukawa coupling. Using the foregoing relations one can verify in particular that Γ^k_{ij} and C_{ijk} have the expected transformation properties.

Another useful fact, which may be verified using the last equation of (70), is

$$e^K 2\bar{X}^M 2\mathrm{Im}\,\tau_{MI} \partial_i (X^I_{;j}) x^j = \frac{i}{2} X^M_i C^I_{KM} e^K \bar{X}^N 2\mathrm{Im}\,\tau_{NI} z^K\,. \tag{74}$$

Finally, the infinitesimal invariants as defined in (14) and (33) are related by

$$\Delta_{ij} = X^I_i X^J_j \Delta_{IJ} = X^I_{;i} X^J_{;j} \Delta_{IJ}\,. \tag{75}$$

References

[1] J. Walcher, "Extended Holomorphic Anomaly and Loop Amplitudes in Open Topological String," arXiv:0705.4098 [hep-th].

[2] M. Bershadsky, S. Cecotti, H. Ooguri and C. Vafa, "Holomorphic anomalies in topological field theories," Nucl. Phys. B **405**, 279 (1993) [arXiv:hep-th/9302103].

[3] M. Bershadsky, S. Cecotti, H. Ooguri and C. Vafa, "Kodaira-Spencer theory of gravity and exact results for quantum string amplitudes," Commun. Math. Phys. **165**, 311 (1994) [arXiv:hep-th/9309140].

[4] S. Cecotti and C. Vafa, "Topological antitopological fusion," Nucl. Phys. B **367** (1991) 359.

[5] B. Dubrovin, "Geometry of 2D topological field theories," in Integrable systems and quantum groups (Montecatini Terme, 1993), 120–348, Lecture Notes in Math., 1620, Springer, Berlin, 1996.

[6] M. x. Huang, A. Klemm and S. Quackenbush, "Topological String Theory on Compact Calabi-Yau: Modularity and Boundary Conditions," arXiv:hep-th/0612125.

[7] S. Hosono and Y. Konishi, "Higher genus Gromov-Witten invariants of the Grassmannian, and the Pfaffian Calabi-Yau threefolds," arXiv:0704.2928 [math.AG].

[8] E. Witten, "Quantum background independence in string theory," arXiv:hep-th/9306122.

[9] H. Ooguri, A. Strominger and C. Vafa, "Black hole attractors and the topological string," Phys. Rev. D **70**, 106007 (2004) [arXiv:hep-th/0405146].

[10] R. Dijkgraaf and C. Vafa, "Matrix models, topological strings, and supersymmetric gauge theories," Nucl. Phys. B **644**, 3 (2002) [arXiv:hep-th/0206255].

[11] M. Aganagic, R. Dijkgraaf, A. Klemm, M. Mariño and C. Vafa, "Topological strings and integrable hierarchies," Commun. Math. Phys. **261**, 451 (2006) [arXiv:hep-th/0312085].

[12] M. Mariño, "Open string amplitudes and large order behavior in topological string theory," JHEP **0706**, 058 (2007) [arXiv:hep-th/0612127].

[13] B. Eynard, M. Mariño and N. Orantin, "Holomorphic anomaly and matrix models," arXiv:hep-th/0702110.

[14] B. Eynard and N. Orantin, "Invariants of algebraic curves and topological expansion," Commun. Number Theory Phys. **1**, no. 2, 347 (2007) [arXiv:math-ph/0702045].

[15] V. Bouchard, A. Klemm, M. Mariño and S. Pasquetti, "Remodeling the B-model," arXiv:0709.1453 [hep-th].

[16] M. Aganagic, A. Neitzke and C. Vafa, "BPS microstates and the open topological string wave function," Adv. Theor. Math. Phys. **10**, no. 5, 603 (2006) [arXiv:hep-th/0504054].

[17] A. K. Kashani-Poor, "The Wave Function Behavior of the Open Topological String Partition Function on the Conifold," JHEP **0704**, 004 (2007) [arXiv:hep-th/0606112].

[18] E. Witten, "On background independent open string field theory," Phys. Rev. D **46**, 5467 (1992) [arXiv:hep-th/9208027].

[19] S. L. Shatashvili, "Comment on the background independent open string theory," Phys. Lett. B **311**, 83 (1993) [arXiv:hep-th/9303143].
[20] M. Baumgartl, I. Sachs and S. L. Shatashvili, "Factorization conjecture and the open / closed string correspondence," JHEP **0505**, 040 (2005) [arXiv:hep-th/0412266].
[21] P. L. H. Cook, H. Ooguri and J. Yang, "Comments on the Holomorphic Anomaly in Open Topological String Theory," arXiv:0706.0511 [hep-th].
[22] G. Bonelli and A. Tanzini, "The holomorphic anomaly for open string moduli," arXiv:0708.2627 [hep-th].
[23] M. Alim and J. D. Laenge, "Polynomial Structure of the (Open) Topological String Partition Function," arXiv:0708.2886 [hep-th].
[24] Y. Konishi and S. Minabe, "On solutions to Walcher's extended holomorphic anomaly equation," arXiv:0708.2898 [math.AG].
[25] I. Antoniadis, K. S. Narain and T. R. Taylor, "Open string topological amplitudes and gaugino masses," Nucl. Phys. B **729**, 235 (2005) [arXiv:hep-th/0507244].
[26] M. Günaydin, A. Neitzke and B. Pioline, "Topological wave functions and heat equations," JHEP **0612**, 070 (2006) [arXiv:hep-th/0607200].
[27] H. Ooguri, Y. Oz and Z. Yin, "D-branes on Calabi-Yau spaces and their mirrors," Nucl. Phys. B **477**, 407 (1996) [arXiv:hep-th/9606112].
[28] A. Neitzke and C. Vafa, "N = 2 strings and the twistorial Calabi-Yau," arXiv:hep-th/0402128.
[29] N. Nekrasov, H. Ooguri and C. Vafa, "S-duality and topological strings," JHEP **0410**, 009 (2004) [arXiv:hep-th/0403167].
[30] R. Donagi and E. Markman, "Cubics, integrable systems, and Calabi-Yau threefolds," Proceedings of the Hirzebruch 65 Conference on Algebraic Geometry, Israel Math. Conf. Proc., **9**, 199–221 (1993) [arXiv:alg-geom/9408004].
[31] D. R. Morrison and J. Walcher, "D-branes and normal functions," arXiv:0709.4028 [hep-th].
[32] M. Aganagic and C. Vafa, "Mirror symmetry, D-branes and counting holomorphic discs," arXiv:hep-th/0012041.
[33] M. Aganagic, A. Klemm and C. Vafa, "Disk instantons, mirror symmetry and the duality web," Z. Naturforsch. A **57**, 1 (2002) [arXiv:hep-th/0105045].
[34] W. Lerche, P. Mayr and N. Warner, "Holomorphic N = 1 special geometry of open-closed type II strings," arXiv:hep-th/0207259;
[35] W. Lerche, P. Mayr and N. Warner, "N = 1 special geometry, mixed Hodge variations and toric geometry," arXiv:hep-th/0208039.
[36] M. L. Green, "Infinitesimal methods in Hodge theory," Algebraic cycles and Hodge theory (Torino, 1993), Lecture Notes in Math., vol. 1594, Springer, 1994, pp. 1–92.
[37] J. Walcher, "Opening Mirror Symmetry on the Quintic," Comm. Math. Phys. **276**, no. 3, 671 (2007) [arXiv:hep-th/0605162]
[38] E. P. Verlinde, "Attractors and the holomorphic anomaly," arXiv:hep-th/0412139.
[39] R. Dijkgraaf, E. P. Verlinde and M. Vonk, "On the partition sum of the NS five-brane," arXiv:hep-th/0205281.
[40] P. Griffiths, "On the periods of certain rational integrals I, II," Ann. of Math. **90** (1969) 460–541.
[41] H. Clemens, "Homological equivalence, modulo algebraic equivalence, is not finitely generated," Publ. Math. IHES 58 (1983) 19–38.
[42] C. Voisin, "The Griffiths group of a general Calabi-Yau threefold is not finitely generated," Duke Math. J. 102 (2000), 151–186.
[43] S. Cecotti, S. Ferrara and L. Girardello, "Geometry of Type II Superstrings and the Moduli of Superconformal Field Theories," Int. J. Mod. Phys. A **4**, 2475 (1989).

School of Natural Sciences, Institute for Advanced Study, Princeton, New Jersey, USA

E-mail address: `neitzke@ias.edu, walcher@ias.edu`